Technik

68,- €

'20

Vehicle–Track Coupled Dynamics

Wanming Zhai

Vehicle–Track Coupled Dynamics

Theory and Applications

Science Press
Beijing

Springer

Wanming Zhai
Train and Track Research Institute
State Key Laboratory of Traction Power
Southwest Jiaotong University
Chengdu, China

ISBN 978-981-32-9282-6 ISBN 978-981-32-9283-3 (eBook)
https://doi.org/10.1007/978-981-32-9283-3

Jointly published with Science Press
The print edition is not for sale in China. Customers from China please order the print book from: Science Press.

This Springer imprint is published by the registered company Springer Nature Singapore Pte Ltd.
The registered company address is: 152 Beach Road, #21-01/04 Gateway East, Singapore 189721, Singapore

Preface

Dynamic interaction between train and track is increasingly intensive with the rapid development of high-speed railways, heavy-haul railways, and urban rail transits, causing more critical and complex vibration problems. Higher train running speed would result in severer train and track interaction, bringing more prominent problems in terms of running safety and stability of the train moving on elastic railway track structures. It must ensure that the train has a good ride comfort when running at a high speed without overturn or derailment. Additionally, the greater the wheel–axle load of a vehicle, the stronger the dynamic effect of the vehicle on track structures, inducing more serious dynamic damage to railway tracks. This requires mitigation of the dynamic interaction between heavy-haul train and track. Obviously, seeking solutions to the abovementioned sophisticated dynamic interaction problems of the large-scale system just from the vehicle system or the track system itself is no longer sufficient. It is necessary to conduct dedicated and in-depth research on the dynamic interaction between rolling stock and track systems. Only with a deep and comprehensive understanding of the mechanism of vehicle–track dynamic interaction is it possible to implement reasonable approaches to minimize the dynamic wheel–rail interaction, to obtain optimal integrated designs of modern rolling stocks and track structures, and eventually to ensure safe, smooth, and efficient train operations. Owing to the fast development of computation technologies, it is realistic today to study and simulate such coupled dynamics problems by considering the vehicle system and track system as a large integrated system with interaction and interdependence. This is the original intention of the vehicle–track coupled dynamics theory discussed in this book.

The author proposed the concept of *Vehicle–Track Coupled Dynamics* for the first time in the late 1980s. In 1991, the author completed his doctoral thesis entitled *Vertical Vehicle–Track Coupled Dynamics*. In 1993, a research paper for investigating the vertical interaction between vehicle and track based on the vehicle–track coupled dynamics was published at the 13th Symposium of the International Association for Vehicle System Dynamics (IAVSD), and then was included in a supplement of the IAVSD journal *Vehicle System Dynamics (VSD)* in 1994. With the continuous funding from the National Natural Science Foundation of China

(NSFC), the National Outstanding Young Scientist Foundation of China (received by the author in 1995), the Ministry of Science and Technology of China (MOST), the China Railway (former China Ministry of Railway), railway industry companies, and others, the research group (including graduate students) led by the author carried out many follow-up research tasks, and published the first academic monograph in this research field entitled *Vehicle–Track Coupled Dynamics* (First edition, in Chinese) in 1997. Afterward, the second, third, and fourth editions of the monograph (in Chinese) were published in 2002, 2007, and 2015 respectively, which became the most fundamental reference books in the field of railway system dynamics and design of rolling stocks and track structures in China, especially for high-speed railways.

In recent years, with the great-leap-forward development of modern railway transportation, especially for high-speed railways, the vehicle–track coupled dynamics theory needs to address more demanding engineering requirements and many new emerging open problems. Supported by the NSFC Major Project (Grand No. 11790280), the NSFC Key Project (Grand No. 51735012), the Program of Introducing Talents of Discipline to Universities (111 Project) (Grant No. B16041) from the China Ministry of Education (MOE), the author led his group to extend the vehicle–track coupled dynamics theory through more elaborate theoretical analysis and more extensive investigations of field problems uncovered in practice. Meanwhile, worldwide research on this topic has also been extremely active and achieved much progress recently. The first English monograph re-edited from the author's Chinese monographs is published when the relevant field is undergoing rapid development in terms of theoretical research and engineering practices.

The writing of this book would not be possible without the support from various individuals and organizations. First, the author is most grateful for the continuous support from the NSFC, the MOST, the China Railway, the MOE, etc. during the past decades. The author also owes much gratitude to those who have participated in the amendment of this English monograph. They are Dr. Shengyang Zhu, Dr. Liang Ling, and Dr. Zaigang Chen from the author's group; Dr. Yunshi Zhao, Dr. Xiaoyun Liu, and Dr. Ilaria Grossoni from University of Huddersfield (UK), Dr. Guoying Tian from Xihua University (China). The author would like to thank the following scholars with special gratitude: Dr. Qing Wu and Dr. Tim Mcsweeney from Central Queensland University (Australia), Prof. Zili Li from Delft University of Technology (the Netherlands), Prof. Kelvin C. P. Wang from Oklahoma State University (USA), and Prof. Manicka Dhanasekar from Queensland University of Technology (Australia), for their extreme enthusiasm in proofreading this book. Some calculation examples performed by Dr. Liang Ling are also gratefully acknowledged. Finally, the author wants to thank his Ph.D. students, Mr. Yu Sun, Ms. Yu Guo, Mr. Jun Luo, Mr. Tao Zhang, and Ms. Mei Chen, for their assistance in carefully editing and supplying photographs, diagrams, and relevant information.

The author believes the publication of this English monograph on *Vehicle–Track Coupled Dynamics* will be conducive to both the investigation of railway engineering dynamics and the development of modern railway industry.

Chengdu, China
December 2018

Wanming Zhai

Contents

Chapter 1
Introduction

Abstract To better understand vehicle–track coupled dynamics which is a new theoretical system, it is necessary for readers to understand the following questions. What is the background under which the theory was proposed? What is the academic rationale of the theory? What are the research scopes and research methodologies? In this chapter, the author will give detailed explanations of these questions.

1.1 Background of Vehicle–Track Coupled Dynamics

Railways are major transportation arteries in many countries and play a very important role in social and economic development. The railway transportation system is a type of wheel–rail contact transportation system ("wheel–rail system" for short). Rolling stocks (including locomotives, passenger cars, and freight wagons, all referred to as "vehicles" in this book) and tracks are essential components of the railway system. The function of wheel–rail transportation is achieved via the interaction between wheels and rails. Wheel–rail interaction is the most significant feature that distinguishes the railway system from other types of transportation systems.

For a long time, studies on railway vehicle dynamics and track structure vibration were carried out separately. This resulted in two relatively independent disciplines, i.e., vehicle dynamics [1, 2] and track dynamics [3, 4].

In classic vehicle dynamics [1, 2], the vehicle system is the research object while the track structure is considered as a "rigid support foundation" (i.e., a rigidly fixed boundary), neglecting the dynamic influence of track vibrations on the vehicle system. Under this situation, geometric irregularities of the rail surface are treated as external disturbances of the vehicle system. In this research field, the dynamic behaviors of the vehicle, including the hunting stability, the running safety, the ride comfort, etc. are investigated with the assumption that the vehicle operates on a rigid rail surface. A basic model illustrating this is shown in Fig. 1.1.

In classic track dynamics [3, 4], the vehicle is usually simplified as external excitation loads $Pe^{i\omega t}$ for the track system (the harmonic vehicle loads P are applied

W. Zhai, *Vehicle–Track Coupled Dynamics*,
https://doi.org/10.1007/978-981-32-9283-3_1

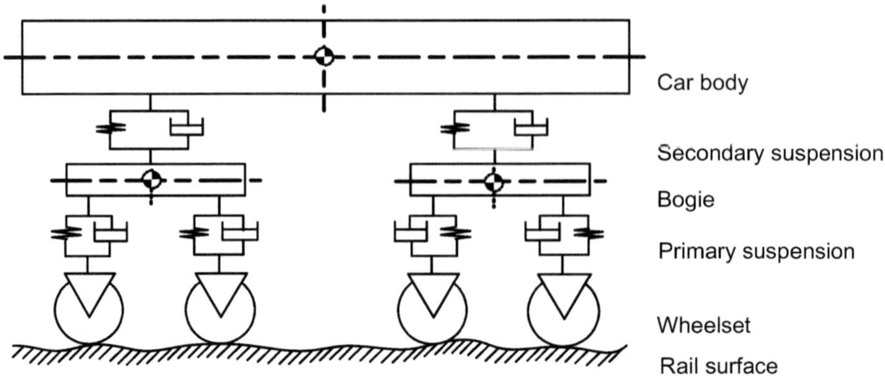

Fig. 1.1 Classic vehicle dynamics model

on fixed points of the track system, or applied as moving loads on the track at a speed of v). The characteristics of the vibration response and deformation of the track structure are analyzed correspondingly. The fundamental model of the classic track dynamics is shown in Fig. 1.2.

Thanks to the long-term studies and practices by railway scientists all over the world, the theories of vehicle dynamics and track dynamics are becoming more and more complete. Significant achievements of these systematic studies have been reported from many fields, including vehicle dynamics modeling, wheel–rail contact geometry, wheel–rail creep theory, vehicle hunting stability, curve negotiation performance, track dynamics modeling, vibration characteristics of track structure, track loading, and deformation characteristics, etc. These research outputs have laid the theoretical foundation in revealing and understanding vehicle dynamics performances and track dynamics characteristics. These outputs have also played a magnificent role in the development of the railway transportation systems.

The rapid development of modern railway transportation, especially the dramatic increases of operating speed, hauling mass, and transportation density, makes the

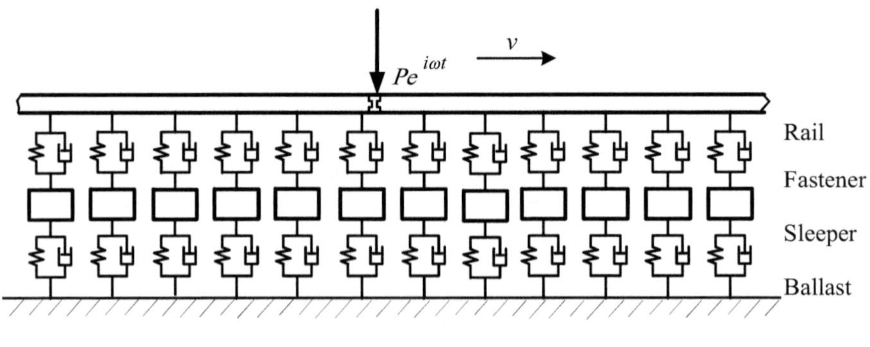

Fig. 1.2 Classic track dynamics model

dynamics problems of the railway vehicle and track systems more prominent and complicated. In general, the higher the operating speed of the train, the stronger the dynamic interaction between the vehicle and the track, the more prominent the problems of running safety and ride comfort. For one thing, it must be guaranteed that the train passes key railway sections (including horizontal curves, vertical curves, switches, turnouts, bridge approach transitions, etc.) safely at a reasonably fast speed or even at a high speed without overturn and derailment. In addition, it must be ensured that the rolling stock can operate with a good ride quality and comfort under the disturbance of track irregularities. Normally, the heavier the gross mass of the vehicle, the stronger is the dynamic interaction between wheel and rail, and the more detrimental are the dynamic effect of the vehicle on the railway infrastructure. Therefore, it is critical to significantly alleviate wheel–rail dynamic interaction by exploring efficient and economical solutions.

Here, we can take the Chinese railway as an example to illustrate this issue. Chinese railway transportation has been in a highly loaded situation for a long time. On the one hand, the railway network density is geographically relatively low, however, the total transport volume is very large. This situation results in a very high traffic density which ranks first in the world. At present, the Chinese railway is able to complete one-fourth of the transport volume of the world's railways with only 6% of the world's railway network length! On the other hand, given the need to increase operational speeds of the passenger and freight trains to satisfy the demands of rapid socioeconomic development, upgrades of existing railway lines, which were originally designed and constructed with relatively low standards have been carried out repeatedly over recent decades. From 1997 to 2007, six major speedup projects were launched and implemented, which have increased the maximum train operational speed from lower than 100 km/h to over 200 km/h. The speedups have even achieved a maximum operational speed of 250 km/h. In this way, high-speed operations were successfully achieved on those existing railway lines in China. As a result, the transportation capacity was effectively improved. However, the dynamic interaction between the rolling stock and infrastructure was seriously aggravated [5]. On the one hand, the dynamic effect of running faster trains on track structures was intensified, which directly affected the fatigue life of the infrastructure and increased the cost for maintenance and repairs. On the other hand, the track geometry deformation and the subgrade settlement of the railway lines were increased, which led to increasing detrimental effects on the dynamic behavior of running trains. In particular, the vibrations and impacts that resulted from the damaged and worn wheel–rail interface have become even more prominent, which can lead to severe safety problems of the wheel–rail system.

Therefore, it is very necessary to conduct dedicated and in-depth research on the dynamic interaction between rolling stock and track systems. Only with a deep and comprehensive understanding of the mechanism of vehicle–track dynamic interaction is it possible to achieve reasonable approaches to minimize the dynamic wheel–rail interaction, to obtain optimal integrated designs of modern rolling stocks and track structures, and eventually to ensure safety, smoothness, and efficiency of train operations. In traditional disciplines, i.e., classic vehicle dynamics and track dynamics, the vehicle–track system was divided into two relatively independent subsystems. In this circumstance, it is very difficult to use these theoretical tools to solve the dynamic interaction problem of such a complex and integrated system. Given this situation, the author proposed the new concept of "vehicle–track coupled dynamics" from the perspective of an overall integrated vehicle and track system in the late 1980s, and the theory was put into practice in the early 1990s [6–12]. In 1991, the author completed his doctoral thesis entitled "Vertical vehicle–track coupled dynamics" [6]. The basic mechanism of vehicle–track coupled dynamics was published for the first time in Chinese in 1992 [7]. In 1993, a research paper on a coupled model that was established on the basis of vehicle–track coupled dynamics for investigating the vertical interaction between vehicle and track was published at the 13th Symposium of the International Association for Vehicle System Dynamics (IAVSD). The paper was then included in a supplement of the IAVSD journal "Vehicle System Dynamics (VSD)" in 1994 [8]. In 1996, a further developed "Vertical and lateral vehicle–track coupled model" was published in VSD [9]. In 2009, the "Fundamentals of vehicle–track coupled dynamics" [11] were systematically introduced in the journal "VSD". The first academic monograph in this research field titled "Vehicle–track coupled dynamics" (First edition, in Chinese) [10] was published in 1997. The second, third and fourth editions of this monograph (in Chinese) were published in 2002, 2007, and 2015, respectively [12], which became the most fundamental reference books in the field of research on railway system dynamics and design of rolling stocks and track structures in China, especially for high-speed railways. This book is the first English monograph reedited from the author's Chinese monographs.

1.2 Academic Rationale of Vehicle–Track Coupled Dynamics

In general, the fundamental academic rationale of vehicle–track coupled dynamics is to consider the vehicle system and the track system as one interactive and integrated system coupled with the wheel–rail interaction, in which the wheel–rail interaction functions as a "link" between the two subsystems. With this approach, it is feasible to carry out comprehensive studies on the dynamic behavior of the vehicle running over an elastic and damped track structure as well as on the dynamic effect of the vehicle on the track structure, in particular on the characteristics of dynamic wheel–rail interaction.

In fact, rolling stock and track are two inseparable components of a railway transportation system. The two components constitute an integrity via the wheel–rail interaction system, as shown in Fig. 1.3. A vehicle running on the track is a complicated interactive dynamics process, involving many interactive factors from both vehicle and track aspects. For example, the geometry deformation of the track can stimulate the vibration of the vehicle system. In contrast, the propagation of the

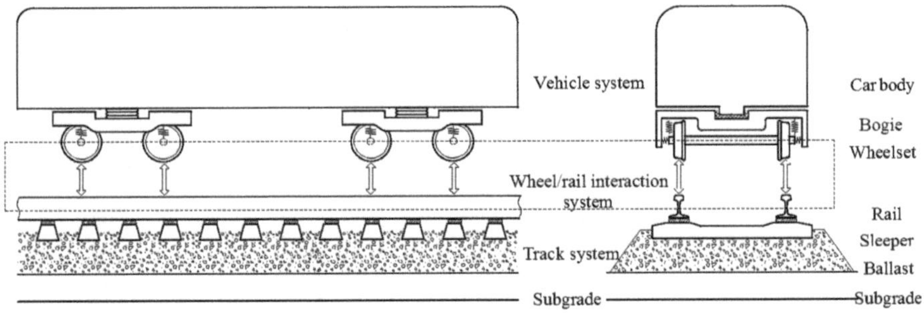

Fig. 1.3 Composition of wheel–rail system in railway transportation

vehicle vibration via the wheel–rail contact interface results in aggravated vibrations of the track structure, which in turn deteriorates the geometry condition of the track. It is evident that it is the dynamic wheel–rail contact force that significantly influences the dynamic behavior of the vehicle–track system.

Furthermore, the dynamic coupling mechanism between the vehicle system and the track system via the wheel–rail interface is illustrated in Fig. 1.4. Under the disturbance from the wheel–rail interface, the dynamic fluctuation of the wheel–rail contact force is stimulated correspondingly. The dynamic effect of the wheel–rail force can be transmitted upwards resulting in vibrations of the vehicle system. Meanwhile, the dynamic effect can also be transmitted downwards leading to vibrations of the track structure. The vibrations of wheelset and rail can directly result in dynamic variation of the wheel–rail contact geometry. Under the influence of wheelset and rail vibrations, the variation of elastic compressive deformation on the normal plane of the wheel–rail contact leads to the fluctuation of the wheel–rail normal contact force. Meanwhile, the variation of wheel–rail creepage (depending on the relative velocity between wheel and rail) in the tangential plane of the wheel–rail contact results in the fluctuation of the wheel–rail tangential creep force. The dynamic changes of the wheel–rail contact forces (wheel–rail normal force and creep force) can, in turn, affect the vibrations of the vehicle and track systems (including the vibrations of wheelset and rail). Actually, the coupled vibration of the vehicle–track system results from this interactive feedback mechanism, which ultimately determines the entire dynamic behavior of the vehicle–track system.

Obviously, the wheel–rail relationship is the essential element of the vehicle–track coupled system. The dynamic feedback between the vehicle system and track system is realized via the variations of the dynamic wheel–rail contact relationship, i.e., the dynamic wheel–rail contact deformation and contact geometry due to the vibration and deformation of wheel and rail. A typical example is given in Fig. 1.5 to clarify the influence of vehicle–track system vibrations on wheel–rail contact relationship, as well as to further demonstrate the importance of considering track system vibrations in certain problems. Figure 1.5 exhibits the dynamic variation of a wheel–rail contact point position on a wheel tread of a Chinese freight wagon negotiating a curved track with a small radius (R = 350 m). In this figure, the solid

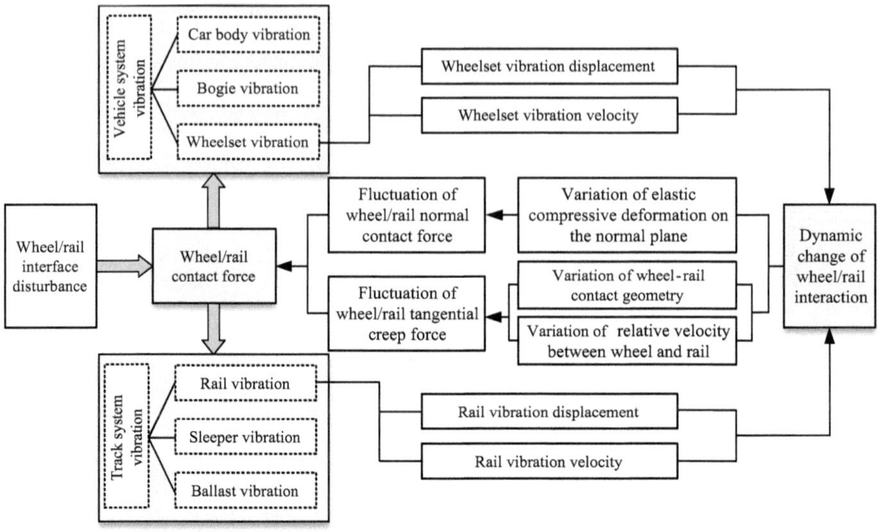

Fig. 1.4 The mechanism of dynamic vehicle–track coupled system

line is the simulation result with the consideration of track vibration effects while the dashed line is the result based on the assumption that the entire track system is remaining stationary. The results show that the contact positions are significantly different in these two cases. Noticeable changes in the position of wheel–rail contact will directly lead to dramatic variations in the magnitude and direction of the wheel–rail contact force, further affect the vibration features of the vehicle and track systems. The measured results from the Chinese railway demonstrate that the wheel–rail lateral forces can induce elastic rail displacements laterally, and then correspondingly resulting in dynamic gauge widening. For example, under a high-speed operating condition, the lateral rail displacement is about 1 mm while the track gauge is dynamically enlarged by 1–2 mm (from a high-speed test on Qinhuangdao–Shenyang passenger-dedicated line in China). Another example is from the test result on the existing Chengdu–Chongqing railway [13]. When a vehicle negotiates a small radius curve at a low speed, the lateral rail displacement and the dynamic gauge widening on a track with concrete sleepers are 1–3 mm and 2–4 mm, respectively. For the situation of timber sleepers, the corresponding rail displacement and gauge widening can even reach 6 mm and 10 mm, respectively. Obviously, for these situations where intensive vehicle–track interactions were observed, if it was assumed that the rail was absolutely stationary, the theoretical calculation result would considerably deviate from the actual situation. Therefore, the actual vibration effects of the elastic track system shall be taken into consideration in the study of the highly interactive vehicle–track system. In other words, the concept of vehicle–track coupled dynamics should be adopted.

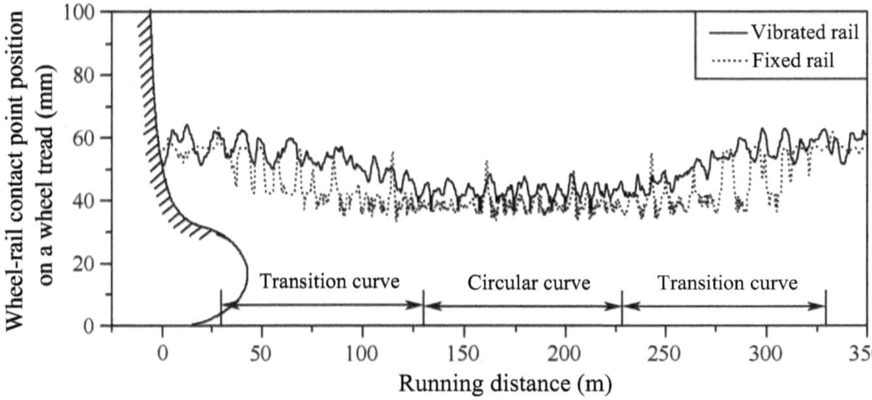

Fig. 1.5 The effect of track vibration on the dynamic wheel–rail contact geometry

Vehicle–track coupled dynamics is an interdisciplinary topic developed on the basis of classic vehicle dynamics and track dynamics. From the perspective of the discipline development, the research into vehicle–track coupled dynamics is not only necessary but also very feasible, since the theories of vehicle dynamics and track dynamics are becoming mature and the modern numerical computing techniques can also provide powerful tools for simulation analyses of such a large coupled dynamics system.

1.3 The Research Scope of Vehicle–Track Coupled Dynamics

Vehicle–track coupled dynamics involves three research aspects: vehicle dynamics, track dynamics and the wheel–rail interaction. The symbolic feature of this study is that the dynamic behaviors and interactions of the vehicle and track structure are investigated from the perspective of an overall vehicle–track system.

In general, vehicle–track coupled dynamics can be divided into three research areas: vertical, lateral, and longitudinal coupled dynamics. The vertical vehicle–track coupled dynamics is relatively independent, and can usually be studied neglecting the influence of lateral and longitudinal dynamics. However, the investigations of lateral or/and longitudinal vehicle–track coupled dynamics must be carried out considering the coupled effect of the vertical dynamics, as the vertical support system of the vehicle plays an indispensable role in the vehicle motions. In fact, the vertical, lateral, and longitudinal vehicle–track coupled dynamics are coupled interactively and are therefore inseparable from each other. They constitute a complex three-dimensional coupled dynamics system. From the perspective of the system disturbance types, vehicle–track coupled dynamics can be divided into two categories: deterministic coupled dynamics and stochastic coupled dynamics.

The vertical vehicle–track coupled dynamics is mainly related to the studies of the dynamic responses of the vehicle–track coupled system under various vertical wheel–rail disturbances and the corresponding wheel–rail vertical interaction characteristics, especially the vertical dynamic characteristics of the track affected by the vehicle. Along the vertical–longitudinal plane of the wheel–rail system, a wide variety of disturbances exist. For instance, there are local defects on the rail surface, such as squats, spalling, etc. In addition, there are cyclic irregularities, such as wavy track, rail corrugation, out-of-round wheels, polygonal wheels, and so on. Particularly, at rail joints, impulse irregularities such as hogging rails, dipped rails, large rail gaps, and weld protrusions are even more prevalent. Furthermore, defects exist in the track structure underneath the rail, such as fastener failure, voided sleepers, ballast hardening, fragmentation and voids of CA mortar underneath track slabs. Other defects include the abrupt change of the supporting stiffness at the transition of a bridge approach as well as the transition zone between different types of track structures. All these defects can cause dynamic irregularities which are related to the uneven supporting stiffness of the foundation. These dynamic irregularities can lead to vertical dynamic interactions in the wheel–rail system. The dynamic interactions at the wheel–rail interface can be transmitted upwards to the vehicle subsystems and downwards to the track subsystem. The dynamic interactions further induce coupled vibrations and impacts between the vehicle and track structure, which result in performance deterioration of the wheel–rail system in terms of safety and ride quality. These dynamic interactions also have a direct impact on the daily maintenance requirements of the vehicle–track system. In the high-speed and heavy-haul operational environment, this type of interaction will be further intensified and its hazards will become even more prominent. Under this scenario, several topics become the important issues to be solved by the application of vertical vehicle–track coupled dynamics. These are clarifying the characteristics of vertical wheel–rail dynamic interactions in different forms and their influencing factors, and seeking a mitigation strategy accordingly, such as exploring approaches to reduce vehicle's dynamic loads on the track and adopting vibration reduction techniques for the wheel–rail system.

The lateral vehicle–track coupled dynamics is basically related to the studies of the laterally coupled dynamic behaviors of the vehicle and track. More importantly, the lateral vehicle–track coupled dynamics is to be used in investigating issues of operational safety of the vehicle running on an elastic and damped track structure. First, it is critical to understand what the differences are between the analysis results when considering the elasticity and damping of the track structure and when using the traditional "rigid track" assumption in the study of the vehicle lateral stability, i.e., the hunting problem [14]. The second important topic, as described previously in Sect. 1.2, is the curve negotiation problem of the vehicle. It is inevitable that, when the vehicle negotiates the curved track, elastic lateral movements of the rail will occur and the track gauge will be transiently widened. This factor has a non-negligible influence on the dynamic contact relationship of the wheel–rail system. However, classic vehicle dynamics theory does not consider this influencing factor. Therefore, the lateral vehicle–track coupled dynamics should be utilized in the studies of the dynamic safety problems (including potential derailment) related to the elastic lateral movement of the rail during curve negotiation (especially for small radius curves). The third topic of lateral vehicle–track coupled dynamics is the safety threshold of track geometry irregularities. This is also a research topic that needs to be undertaken for improving track maintenance standards. The vertical, alignment, cross-level and gauge irregularities have important impacts on operational safety. It is very difficult to analyze these dynamic operational safety issues by investigating a single system (vehicle or track) as these problems are jointly determined by the vehicle–track interactive system. Given this situation, vehicle–track coupled dynamics can provide an appropriate theoretical platform for comprehensive studies of the safety thresholds for different types of irregularities. When multiple types of irregularities exist at the same location of the track, the problems for operational safety are even more serious and the dynamic wheel–rail interactions become more complicated. Only by the use of vehicle–track coupled dynamics, where the entire range of interactive factors have been taken into consideration, can the safety thresholds for the operational conditions of multiple irregularities be obtained. In addition, the running safety (especially for high-speed operations) of the vehicle when passing through switches and turnouts and negotiating combined horizontal and vertical curves are also within the research scope of the lateral vehicle–track coupled dynamics.

The longitudinal vehicle–track coupled dynamics is mainly related to the studies of wheel–rail stick–slip oscillation, the wheel–rail abrasion mechanism, the cause of rail corrugation, train longitudinal impact under traction/braking conditions and its interaction with the railway track, the longitudinal dynamic effect of the powertrain system on the vehicle–track system, etc. First of all, the wheel–rail system plays fundamental roles in supporting the train on the track structure vertically as well as in guiding and constraining the wheelset movement laterally. Furthermore, longitudinal wheel–rail creep also has a key role in transforming traction or braking torque into the longitudinal wheel–rail force to realize acceleration or deceleration of the train. Longitudinal wheel–rail stick–slip oscillation, wheel and rail abrasion, and rail corrugation are the main problems of the wheel–rail system during traction or braking. New breakthroughs in investigating these traditional problems are more likely to emerge if the viewpoint of the longitudinal vehicle–track coupled dynamics is adopted. Second, with the increase of train running speed and hauling mass, especially for long heavy-haul trains, longitudinal impacts between adjacent vehicles are evident when starting, braking and correcting speed. Under these circumstances, the longitudinal impacts applied on couplers are aggravated, which could lead to serious incidents such as decoupling, coupler breakage and even derailment under certain conditions [15]. In curved track sections, large coupler lateral forces can greatly exacerbate lateral dynamic interaction between vehicle and track, causing overturn or breakage of the rail. Therefore, it is an important task to explore the characteristic of the longitudinal impact in long heavy-haul trains and its effect on the track. It is also important to seek effective train handling strategies for alleviating the related problems by using longitudinal vehicle–track coupled dynamics. Third, with the rapid development of high-speed and high-powered motorized vehicles, the dynamic coupled effect of motor traction and gear transmission of the powertrain on the vehicle–track system is significantly increased [16], deteriorating the working environment of related components. In fact, heat failure of traction motor bearings, bearing cage fractures, gear tooth breakages, gearbox cracks, oil leaks, and other serious failures could occur during the operations of the vehicle. Therefore, the investigation of the dynamic mechanism of the key components in the powertrain and its coupled effect with the vehicle–track system are also within the scope of longitudinal vehicle–track coupled dynamics.

The above discussion is more related to deterministic disturbances. For nondeterministic disturbances, stochastic vehicle–track coupled dynamics is dedicated to this issue [17]. The stochastic vehicle–track coupled dynamics is mainly related to the studies regarding the vibration response characteristics and evolutionary behaviors of the vehicle–track coupled system under the excitation of stochastic track irregularities, as well as the investigation of characteristics of the vehicle system, track structure system and wheel–rail interaction in the frequency domain, and the cause of the excitation sources inducing the vehicle–track coupled vibrations (i.e., the cause of track irregularity formation). As a result, the study of stochastic vehicle–track coupled dynamics provides the feasibility to restrain and mitigate the detrimental vibrations of vehicle–track system in different situations with a targeted approach, to improve the ride quality and passenger comfort during train operations, and to reduce the fatigue damage of vehicle and track components to minimize the maintenance costs.

As a new research system distinguished from the classic theories of vehicle dynamics and track dynamics, vehicle–track coupled dynamics has a very broad application prospect in the field of railway vehicle and track system dynamics as well as in wheel–rail interaction. Over the past 20 years, the railway speedup strategy was successfully implemented in the Chinese railway and many remarkable results were achieved. Many speedup projects were carried out on existing railway lines. However, the structures of the existing lines, which were not constructed in accordance with the current high design standards, actually cannot be comprehensively upgraded in a large scale. The dynamic influences of trains on the infrastructure are greatly intensified as the operational speed increases. Therefore, the issues of how to reduce the dynamic wheel–rail interaction with an increased operational speed and how to avoid serious deformation and deterioration of the track structure in order to guarantee the operational safety have become major concerns in the Chinese railway. It is necessary to carry out systematic studies and propose appropriate corresponding countermeasures from the perspective of the overall vehicle–track system. In this regard, vehicle–track coupled dynamics has provided appropriate theoretical analysis tools [5].

High-speed and heavy-haul are the two symbolic icons in today's railway industry. However, both high-speed and heavy-haul transportation scenarios aggravate the dynamic wheel–rail interactions, which means the traditional railway systems are not well suited to these new developments. The mitigation of dynamic wheel–rail interactions has played a key role in developing modern railway transportation systems, which has contributed to economic development. In order to achieve low dynamic interactions for wheel–rail systems, optimal integrated solutions for wheel–rail systems and parameters must be sought from the perspective of system engineering where comprehensive factors of vehicle, track, and wheel–rail interface are taken into account. This aim can be achieved by the application of vehicle–track coupled dynamics theory. With detailed parameter and sensitivity analyses of the overall vehicle–track system, the basic approaches in mitigating wheel–rail interactions and the corresponding technical countermeasures can be identified. Meanwhile, the principle of optimal integrated design and the criterion of parameter selection for new types of rolling stocks and track structures can be proposed to provide theoretical guidance for the designs of high-speed and heavy-haul vehicles as well as track systems. In addition, computer simulation systems of vehicle–track coupled dynamics can be used to predict and evaluate the dynamic performance of new or existing designs of vehicle or track. In this case, simulation systems can provide a critical technical platform to evaluate the safety issues of high-speed railway design and reconstruction for existing railway speedup projects. The simulation systems can also be used to optimize vehicle design to achieve better dynamic performance. They can also be used for the analysis of rolling stock overturn, derailment and other major accidents, especially for the study of derailments caused by the wheel–rail interaction and track damage. Using the vehicle–track coupled dynamics in derailment analysis can overcome the bias of the classic theoretical methods, which use single vehicle or track system.

1.4 Research Methodology of Vehicle–Track Coupled Dynamics

Vehicle–track coupled dynamics is a highly focused technical discipline for engineering applications. It is an interdisciplinary field involving many research areas such as mechanical engineering, civil engineering, vibration mechanics, numerical analysis methods, and computer simulation technologies. To study such a complicated issue, it is generally necessary from theoretical models, numerical simulations, and field tests. The research methods should incorporate mathematical models and related experiments. The theoretical analysis should be regarded as the principal part of the research. Validations of theoretical models and simulation systems can be achieved via essential field tests. In this sense, it is also possible to reduce many expensive railway field tests, since many problems can be investigated by using validated mathematical models. However, it has to be pointed out that, for practical engineering applications, theoretical analysis results must ultimately be validated using field tests.

Due to the complexity of the wheel–rail system, the analysis of vehicle–track coupled dynamics is far beyond the scope of theoretical analysis and must be solved numerically by using computers. For large and complex systems, numerical simulations have great advantages and are widely used in modern engineering. First, numerical simulations provide detailed numerical solutions of the complex problems with less investment. Compared with the limited experimental data from field tests, numerical simulations can effectively reduce the reliance on costly experiments. For the study of vehicle–track coupled dynamics, the significance is considerable because real-vehicle tests on railway lines are usually extremely expensive and also affect normal operation of the railway. In the event of any serious accident such as an overturn or derailment during the tests, the economic loss is even greater. In contrast, it is almost always feasible to investigate various extreme conditions such as exceeding the speed limit and/or the axle load, or running over severely damaged track without any risk by utilizing numerical simulation tools. Second, the system parameters or sensitivity analysis can be carried out via numerical simulations within a short time, which enables optimal parameters of new rail transport systems to be found. Therefore, once the simulation system is tested and verified, many intermediate experiments could be largely reduced or even completely replaced by simulations. This saves considerable expense and shortens the engineering design process.

There are three essential points to realize numerical simulations of vehicle–track coupled dynamics. The first is to establish a reasonable mathematical model to describe the physical essence of the vehicle–track coupled system. The second is to choose an efficient numerical simulation algorithm that is suitable for the nonlinear solution of this complicated large-scale dynamic system. The third is to correctly determine the model parameters of the vehicle–track system.

Mathematical models are the foundation of numerical simulations. To be able to handle the complicated factors in the large-scale vehicle–track coupled system, the general influences of these factors first need to be defined. Second, the modeling method of each component in the system needs to be analyzed and then the corresponding modeling principle should be proposed via analysis and comparison. Third, based on this principle, the vertical, lateral and longitudinal interactive vehicle–track coupled model can be established. The model includes differential equations of vehicle motions and track structure vibrations, dynamic wheel–rail coupled relationships, etc. Finally, various disturbances of the wheel–rail system need to be modeled correspondingly in order to provide the excitation inputs to the vehicle–track coupled model.

Stability and accuracy are the most important requirements for all numerical simulation algorithms. The mathematical model of vehicle–track coupled dynamics can eventually be expressed as second-order differential equation sets with quite high degrees of freedom, e.g., hundreds or even thousands. Furthermore, the frequency of the wheel–rail contact vibration is very high (some components have frequencies higher than 500 Hz). Consequently, only a very small calculation step size can be adopted in the numerical simulations. Therefore, the numerical calculation speed has become a key issue of concern in the simulations. Among the existing numerical integration methods, Newmark-β method, Wilson-θ method, and Runge–Kutta method are the most commonly used ones. However, these methods often require a large amount of computation time in the simulations of large-scale engineering dynamics problems. Especially in the early studies conducted by the author (in the early 1990s), the conflict between required computing speed and available computing capacity was very prominent. Therefore, the development of a fast and practical numerical integration method, at that time, was the priority for the implementation of vehicle–track coupled dynamics simulations on ordinary microcomputers. Fortunately, a new fast explicit numerical integration method and a new prediction–correction integration method were constructed by the author [18]. The two methods have obvious advantages in numerical solutions for large-scale dynamic problems. As they do not need to solve large-scale algebraic equation sets at each time step, they are expected to be able to successfully solve the issues mentioned above.

Correct selection of model parameters plays a key role in ensuring the accuracy of numerical simulation results. For vehicle–track coupled systems, it is not very easy to determine or identify all physical parameters accurately due to the complexity of the system, especially for the track structure. This is a common problem for many engineering calculations and analyses. However, it is not always necessary to obtain completely detailed parameters for modeling. The more important thing is to specify the key parameters that characterize the system behaviors. For the vehicle system, many detailed and accurate methods for determination of vehicle parameters are widely adopted. For instance, the dimensions, mass, and inertias of the vehicle components, as well as the stiffness and damping parameters of the suspension systems can be determined or calculated using design drawings or technical documents or by means of relevant laboratory bench tests, etc. For the track system, the situation is much more complicated. However, after years of simulations and experiments, large amounts of relevant data have been collected and the systemic parameter identification methods have been gradually developed. First of all, the physical parameters and profiles of the rail can be determined for specific railway lines. Second, sleeper parameters can also be accurately determined via design standards. Third, the dynamic parameters of rail pads (stiffness and damping) can be obtained from product documents or be identified from loading tests in a laboratory. Fourth, for ballastless track which is widely used in high-speed railways, the structural parameters and physical parameters of track slab, CA mortar, and other components can be determined in more straightforward ways as they are clearly specified in high-speed railway standards. In addition, the parameters of ballasted tracks, such as ballast thickness and ballast density, can also be measured accurately or determined from design specifications. The difficult task is the identification of the stiffness and damping parameters of the ballast and subgrade. The classic axle-dropping test technique [19] and the continuous measurement techniques for track stiffness developed in recent years (especially, the track elasticity test vehicles) [20–23] have provided effective approaches to measure the overall stiffness (and even damping) of the track. It is feasible to identify accurate stiffness parameters of ballast and subgrade from the overall track stiffness, even though this work is not easy. The author believes that an effective indirect method to determine ballast and subgrade stiffness is to measure the elastic modulus of ballast and subgrade (both are easy to measure) and then calculate their stiffness parameters inversely according to the dimensions of the supporting body. This method will be introduced later in this book.

Admittedly, any simulation analysis must be based on certain experimental results. Simulation results should be validated by the corresponding tests. Focusing on practical applications of railway engineering, the simulation results of vehicle–track coupled dynamics must be verified by railway field tests. In fact, it is possible to simulate many typical large-scale wheel–rail dynamics tests and compare the simulation results with field test results. In contrast, by systemically organizing typical dynamics tests, such as vehicles passing over rail joints or negotiating curved tracks, the displacements and accelerations of key components of the wheel–rail system, dynamic wheel–rail contact forces, and other major indexes can be measured and compared with the calculation results so as to validate the vehicle–track coupled dynamics model and improve the simulation software. In addition, based on a series of full-scale field tests including the Chinese speedup tests on existing railway lines, high-speed train running tests on newly constructed high-speed railways and several derailment tests organized by the Chinese railway during last decades, a wide range of test results under various operating conditions could be utilized to validate the simulation models. These validations also enable more functions to be incorporated into the simulation system. Having gone through these systematic and extensive validation processes, the vehicle–track coupled dynamics simulation system is expected to demonstrate significant value in more extensive engineering applications.

References

1. Garg VK, Dukkipati RV. Dynamics of railway vehicle systems. Ontario: Academic Press Canada; 1984.
2. Wickens AH. Fundamentals of rail vehicle dynamics: guidance and stability. Lisse: Swets & Zeitlinger Publishers; 2003.
3. Knothe K. Gleisdynamik. Berlin: Ernst & Sohn; 2001.
4. Lian SL. Track dynamics. Shanghai: Tongji University Press; 2003 (in Chinese).
5. Zhai WM, Cai CB, Wang QC, et al. Dynamic effects of vehicles on tracks in the case of raising train speed. Proc Inst Mech Eng Part F: J Rail Rapid Transit. 2001;215(2):125–35.
6. Zhai WM. Vertical vehicle–track coupled dynamics. PhD thesis. Chengdu, China: Southwest Jiaotong University; 1991 (in Chinese).
7. Zhai WM. The vertical model of vehicle–track system and its coupling dynamics. J China Railw Soc. 1992;14(3):10–21 (in Chinese).
8. Zhai WM, Sun X. A detailed model for investigating vertical interaction between railway vehicle and track. Veh Syst Dyn. 1994;23(Suppl.):603–15.
9. Zhai WM, Cai CB, Guo SZ. Coupling model of vertical and lateral vehicle/track interactions. Veh Syst Dyn. 1996;26(1):61–79.
10. Zhai WM. Vehicle–track coupled dynamics. 1st ed. Beijing: China Railway Publishing House; 1997.
11. Zhai WM, Wang KY, Cai CB. Fundamentals of vehicle–track coupled dynamics. Veh Syst Dyn. 2009;47(11):1349–76.
12. Zhai WM. Vehicle–track coupled dynamics. 4th ed. Beijing: Science Press; 2015.

13. Zhai WM, Wang KY. Lateral interactions of trains and tracks on small-radius curves: simulation and experiment. Veh Syst Dyn. 2006;44(Suppl.):520–30.
14. Zhai WM, Wang KY. Lateral hunting stability of railway vehicles running on elastic track structures. J Comput Nonlinear Dyn ASME. 2010;5(4):041009-1-9.
15. Liu PF, Zhai WM, Wang KY. Establishment and verification of three-dimensional dynamic model for heavy-haul train–track coupled system. Veh Syst Dyn. 2016;54(11):1511–37.
16. Chen ZG, Zhai WM, Wang KY. Dynamic investigation of a locomotive with effect of gear transmissions under tractive conditions. J Sound Vib. 2017;408:220–33.
17. Xu L, Zhai WM. A new model for temporal–spatial stochastic analysis of vehicle–track coupled systems. Veh Syst Dyn. 2017;55(3):427–48.
18. Zhai WM. Two simple fast integration methods for large-scale dynamic problems in engineering. Int J Numer Meth Eng. 1996;39(24):4199–214.
19. Wang MZ, Cai CB, Zhu SY, Zhai WM. Experimental study on dynamic performance of typical non-ballasted track systems using a full-scale test rig. Proc Inst Mech Eng Part F: J Rail Rapid Transit. 2017;231(4):470–81.
20. Wu WQ, Zhang GM, Zhu KM, Luo L. Development of inspection car for measuring railway track elasticity. In: Proceedings of 6th international heavy haul conference, Cape Town. 1997.
21. Li D, Thompson R, Kalay S. Development of continuous lateral and vertical track stiffness measurement techniques. In: Proceedings of railway engineering conference, London. 2002.
22. Berggren E, Jahlénius Å, Bengtsson BE. Continuous track stiffness measurement: an effective method to investigate the structural conditions of the track. In: Proceedings of railway engineering conference, London. 2002.
23. Norman C, Farritor S, Arnold R, et al. Design of a system to measure track modulus from a moving railcar. In: Proceedings of railway engineering conference, London. 2004.

Chapter 2
Vehicle–Track Coupled Dynamics Models

Abstract Theoretical model is the base for the study of vehicle–track coupled dynamics problems. In this chapter, the principle and methodology for modeling of vehicle–track coupled systems are discussed at first. And then, three types of theoretical models are established: the vehicle–track vertically coupled dynamics model, the vehicle–track spatially coupled dynamics model, and the train–track spatially coupled dynamics model, in which typical passenger coaches, freight wagons, and locomotives as well as typical ballasted and ballastless tracks are included. A new dynamic wheel–rail coupling model is also established to connect the vehicle subsystem and track subsystem. Equations of motion of the vehicle and track subsystems are deduced and given in detail.

2.1 On Modeling of Vehicle–Track Coupled System

2.1.1 Evolution of Wheel–Rail Dynamics Analysis Model

The earliest involvement of wheel–rail dynamic analysis dates back to 1867, when Winkler proposed the theory of elastic foundation beam, which was quickly used for track modeling and the deformation analysis of track under static load. In 1926, Timoshenko applied the elastic foundation beam model to first study the dynamic stress of the rail under vehicle loading, which is a classical method still widely used today. In 1943, Dörr proposed that better track models should be developed to accommodate the growth speed of the train. However, few models have been developed to solve practical problems of wheel–rail contact. In the meantime, the railway researchers are more concerned about the dynamic stability of moving loads (due to rolling stock) on the beam (i.e., rail). An important reason is that in 1954, the French National Railways (SNCF) experienced severe track sinusoidal alignment irregularities in the high-speed train test with a maximum speed of 330 km/h, resulting in lateral damage to the track structure. This experience shifted the research focus in the late 50s and early 60s to the lateral running stability of rolling

W. Zhai, *Vehicle–Track Coupled Dynamics*,
https://doi.org/10.1007/978-981-32-9283-3_2

stocks. As a result, the Prud'homme limit was developed based on the experiments carried out by SNCF between 1960 and 1965 [1].

In the 1970s, the rapid development of the railway transportation industry greatly promoted the development of wheel–rail interaction research, especially in the application of mathematical mechanics models to solve the practical problems of railways. The experimental and analytical research activities on rail joint forces carried out by the Derby Railway Technology Research Centre in the UK [2, 3] led the wheel–rail dynamics analysis into a substantive stage. At that time, in order to prevent and remediate the damage in the rail joint area, the British Railways took the lead in carrying out the wheel–rail dynamic test of the vehicle passing through the rail dipped joints, and thus defined two types of wheel–rail forces that existed during the wheel–rail impact process: high-frequency impact force P_1 and low-frequency force P_2.

Lyon [2] and Jenkins et al. [3] developed a fundamental model to analyze the wheel–rail dynamic interaction (Fig. 2.1), and were the first to study of the effect of some principal parameters of the vehicle and track system (such as the unsprung mass, track stiffness) on the wheel–rail force. This model described the track as an Euler beam supported by a continuous elastic foundation, in which the vehicle was simplified to unsprung mass with consideration of primary suspension characteristics. The wheel–rail contact was modeled using a nonlinear Hertzian contact model. In 1979, Newton et al. [4] conducted track dynamics tests to study the dynamic effect of wheel flat on the track, and made a partial improvement on the model developed in [2, 3]. The track was modeled using Timoshenko beam, so that the calculated rail shear strain can be directly compared against the experiments, and the theoretical and experimental results have achieved a good consistency. In

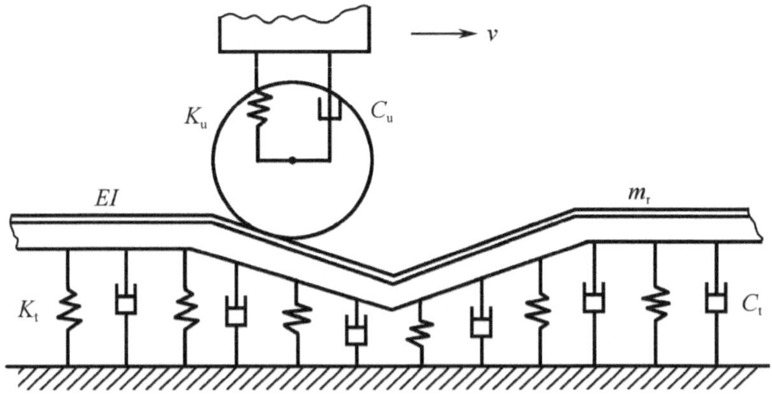

Fig. 2.1 The most basic model for the analysis of wheel–rail dynamic interaction (K_u, C_u—vehicle suspension stiffness and damping; K_t, C_t—track support stiffness and damping; *EI*—rail bending stiffness; m_r—unit length rail mass; *v*—vehicle speed)

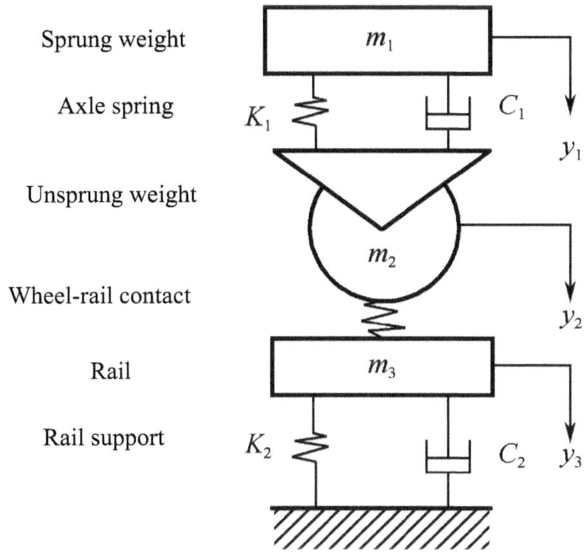

Fig. 2.2 The simplest lumped parameter model of a wheel–rail system

1982, Clark et al. [5] studied the dynamic effects of vehicles traveling on corrugated track, and adopted a continuous track model with discrete elastic supports, and considered the influence of sleeper vibration, so that the simulation was close to the actual track structure.

During the same period, Sato, Ahlbeck, Birmann, Gent, etc. developed a more simplified lumped parameter model to simulate the track structure with distributed parameter characteristics for the study of wheel–rail dynamics [6, 7]. Figure 2.2 is the simplest lumped parameter model of a wheel–rail system in which the rail was simplified as a concentrated mass and the substructure was simplified to a spring and damping element.

Since the 1990s, with the rapid development of heavy-haul and high-speed railways, especially in the booming Chinese railway industry, wheel–rail system dynamics research was significantly active. To meet the needs of railway development, the author of this book began to consider the vehicle and the track as a coupled system in 1990. The author first proposed and carried out the theoretical study of vehicle–track coupled dynamics, and successively established a series of vehicle–track coupled models [8–14]. The vehicle–track coupled model is characterized by considering various major dynamic factors of the vehicle and the track system in detail, which is able to investigate the wheel–rail dynamic interaction from the overall vehicle–track system. The vehicle–track lateral dynamics model was also developed. Subsequently, researchers from many railway research institutes had carried out a large number of theoretical and applied research projects in the field of vehicle–track coupled dynamics [15–35], especially in the application of the vehicle–track coupled dynamics models to study the practical dynamics problems of various wheel–rail systems. There were many improvements and extensions developed in the model, including:

- modeling the rail using Timoshenko beam to achieve high accurate of rail force [15–17];
- modeling the track using Finite Element method [18, 19];
- considering the vibration of subgrade [20, 21];
- considering the effect of stiffness transition between the subgrade and bridge [22, 23];
- considering the interaction between the vehicle and the switch [24–28].

It could be concluded on the research in this period that, the wheel–rail interaction was analyzed from a bogie–track lumped parameter model [29, 30] to a complete vehicle–track interaction model [31–35], and the general trend was being to consider the factors influencing the vehicle and track system more comprehensively.

The theoretical and analysis framework of the vehicle–track coupled dynamics has matured in early twenty-first century after a long time of developments [36]. The theory of vehicle–track coupled dynamics has become a fundamental method for the dynamic analysis of vehicle–track interactions. Early vehicle–track vertical interaction analysis model has been widely used in the field of railway dynamics analysis [37–57], and recent studies had begun to adopt vehicle–track spatially coupled interaction models [58–81]. The research methods started to incorporate with the finite element method [39, 41–44, 62, 78], the boundary element method [55, 67], the discrete element method [75, 78], the flexible multibody system dynamics methods [58, 67, 72, 79], etc. The researched topics are more and more complex, including:

- Vehicle–track–bridge interactive dynamics [69, 70, 73, 80],
- Vehicle–track–tunnel coupled dynamics [55],
- Vehicle–turnout interaction [62, 63],
- Wheel–rail dynamics at various rail joints [41, 42, 54, 76],
- Wheel–rail lateral interaction on curves [60, 61, 66],
- Wheel–rail wear and RCF [43–45, 48, 50, 52, 71],
- Vehicle hunting stability on elastic track structure [65, 68, 72],
- Dynamic effect of track settlement [38, 56, 57],
- Vehicle–track long-term performance degradation [40, 53, 81],
- Environmental vibration induced by vehicle operation [46, 47, 49],
- Wheel–rail noise [37, 59, 64, 80].

These research activities covered the high-, medium-, and low-frequency domain of the wheel–rail dynamics, and started to consider the stochastic characteristics of the structure parameters of the track [77], the nonlinear material characteristics of the track structure [51], the long-term evolution of track structure properties, and the dynamic characteristics of subgrade materials [56, 74], and so on.

2.1.2 Modeling of Track Structure

The track structure is usually modeled using numerical methods (in time domain) and analytical methods (in frequency domain). The time-domain numerical methods include the equivalent lumped parameter method, continuous beam-based modal superposition method, finite element method, boundary element method, discrete element method, etc. [12, 16, 39, 55, 67, 75, 78]. The equivalent lumped parameter method simplifies the track structure into a mass–spring–damper system with a few degrees of freedom; therefore, the calculation speed is very fast, but less accurate. The continuous beam method includes the models of a continuous beam (rail) supported by a continuous elastic foundation and a continuous rail beam with discrete elastic point supports. The continuous elastic foundation beam model simplifies the underlying foundation into a uniformly distributed foundation, while the discrete elastic support beam models the sleeper (or slab) and foundation individually. The continuous beam model generally uses the modal superposition method to solve the vibration differential equations; this method is simple and computational efficient. The finite element method discretizes the track structure into a finite number of elements and assumes a displacement function to obtain a unit matrix. The advantage of the finite element method is that it can model complex track structures in detail, and consider geometric nonlinearity of the track structure; however, this method is time consuming. The boundary element method is mainly used for the vibration transmission and noise problem simulation of the track structure, while the discrete element method is mainly used for modeling of the ballast of the track system. The analytical methods for track structure modeling, such as wavenumber finite element and Green's function method, are mainly used for frequency-domain calculation. The calculation efficiency is high and covers a wide range of frequency domain. However, the disadvantage is that the nonlinear characteristics of the track structure cannot be considered. The following sections discuss the effect of track modeling on the interaction between the vehicle and track based on the time domain continuous beam model and summarize the fundamental principles of track modeling.

1. *The equivalent lumped parameter model*

The equivalent lumped parameter model is based on a certain equivalence principle, transforming a track structure with a complex decentralized parameter system into a simplified model of mass–spring–damping lumped parameters with a few degrees of freedom. Due to its efficient calculation speed, this simplified model is widely used by vehicle dynamics analysis software such as SIMPACK, GENSYS, VAMPIRE, and others.

There are two common equivalent transformation principles for the equivalent lumped parameter track model. One is to derive the equivalent mass and the equivalent spring stiffness from the measured natural vibration frequency of the track structure, and derive the equivalent damping coefficient from the logarithmic decay rate of the measured amplitude–frequency response of the track structure; another transformation principle is to determine the equivalent mass by keeping the kinetic energy of the distribution quality of the elastic foundation beam equal to the kinetic energy of the lumped mass. When determining the equivalent spring stiffness, the static deflection of the elastic foundation beam at the load point is required to be equal to the static deflection of the lumped parameter model.

Obviously, the lumped parameter model has significant limitations compared with the distributed parameter model. Generally, it can only analyze the wheel–rail dynamics problem under the condition that the track parameters are evenly distributed (this is the premise of the equivalent transformation). Because the flexibility of the track structure and the longitudinal transferring of the track vibration are not considered, the lumped parameter model can only be used to qualitatively analyze the wheel–rail system dynamics problems [82], which generally covers the frequency range below 20 Hz [68].

2. *Comparison between beam model with continuous elastic foundation and beam model with discrete elastic point supports for track modeling*

The beam model with continuous elastic foundation (Fig. 2.3) differs from the beam model with discrete elastic point supports (Fig. 2.4). The former model considers the rail substructure as a uniformly distributed overall foundation, and the foundation characteristics are consistent with the Winkler assumption; the latter model considers the rail substructure as a series of discrete spring–damper point support systems separated by the sleeper spacing. Obviously, the former model focuses on the fundamental characteristics of the track system as well as the overall behavior of the track; the latter model can consider the local behavior at each sleeper, such as the impact of sleeper quality on vibration, which can better reflect

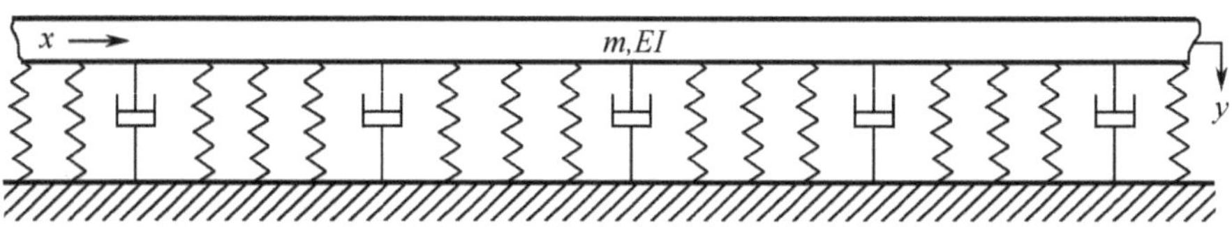

Fig. 2.3 Beam model with continuous elastic foundation (m—track mass per unit length; EI—bending stiffness of rail)

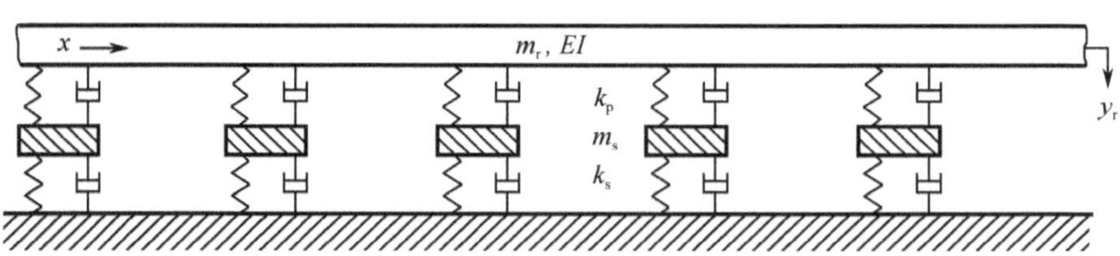

Fig. 2.4 Beam model with discrete elastic point supports (m_r—rail mass per unit length; m_s—sleeper mass; k_p—rail pad stiffness; k_s—under-sleeper stiffness)

the fact that the track is supported by the sleepers and track foundation. Moreover, the beam model with discrete elastic point supports can consider the case where the track system parameters are nonuniformly distributed in the longitudinal direction more conveniently. For example, it can reflect the situation where the sleeper is not uniformly distributed at track joints, or the situation where the track supporting stiffness is nonuniform (such as when under-track high stiffness rubber pads are laid to reduce the force at rail joints). In addition, the beam model with discrete elastic point supports can also consider the defects in the track structure such as loose fasteners, hanging sleepers and track slabs.

British Railway compared the simulation results of the beam model with continuous elastic foundation and the beam model with discrete elastic point supports. The prediction results of the two models were considered to be not much different in the low-speed range. At high speed, the beam model with continuous elastic foundation will overestimate the wheel–rail force, which was also supported by the measured results (Fig. 2.5) [4]. The simulation results using the beam model with discrete elastic point supports had good agreement with the experimental results within the entire speed range.

3. *Comparison between Euler beam model and Timoshenko beam model for rail modeling*

Two types of beam models are commonly used to describe the rail as a continuous elastic body, namely the Bernoulli–Euler beam (Euler beam for short) model and the Rayleigh–Timoshenko beam (Timoshenko beam for short) model. The Euler

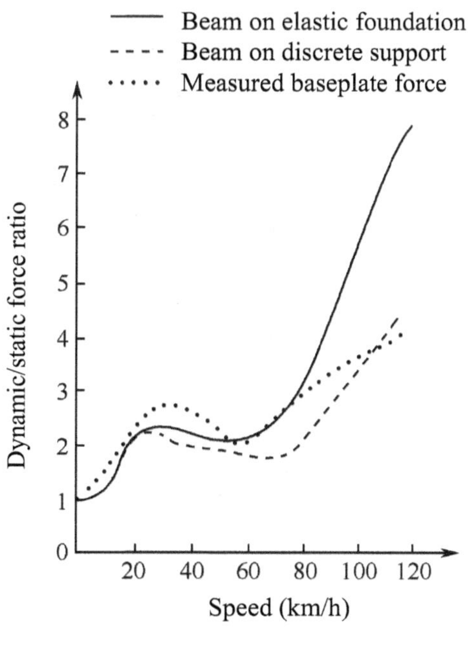

Fig. 2.5 Comparison of the results using the beam models with continuous elastic foundation and with discrete elastic point supports

beam model considers the bending deformation of the rail regardless of its shear deformation. The Timoshenko beam model introduces the shear strain of the beam and considers the rotational inertia of the beam, so that the force analysis of the beam is more complete and the analysis frequency is higher. The shear strain parameters of the Timoshenko beam model are also convenient to be compared to the measured parameters in the field. Generally, the Euler beam model could cover a frequency range up to 500 Hz, while the Timoshenko beam model could cover a frequency range up to 3000 Hz. However, the theory of the Timoshenko beam model is more complex, so the computational efficiency is slightly lower; the Euler beam model is simpler and computationally efficient, and is hence widely used in practice.

The numerical analysis results showed that the P_2 forces calculated by the Timoshenko beam model and by the Euler beam model were almost identical, and only the high-frequency P_1 forces are slightly different. The former is 7–11% larger than the latter (see Fig. 2.6a). The wheel–rail force comparison between the Euler and Timoshenko beam models supported by continuous elastic foundation done by the Derby Railway Research Centre in the UK [4] showed the difference is small at middle to low-frequency while the difference is larger at higher frequency (see Fig. 2.6b). Therefore, for the low-frequency wheel–rail dynamics problem, the Euler beam rail model is generally used, which does not make the calculation process too complicated, and can meet the accuracy requirements of the general wheel–rail dynamic analysis. When it is necessary to pay attention to the high-frequency wheel–rail vibration characteristics (such as the wheel–rail noise problems), the Timoshenko beam model is used to reflect higher frequency behavior of the track.

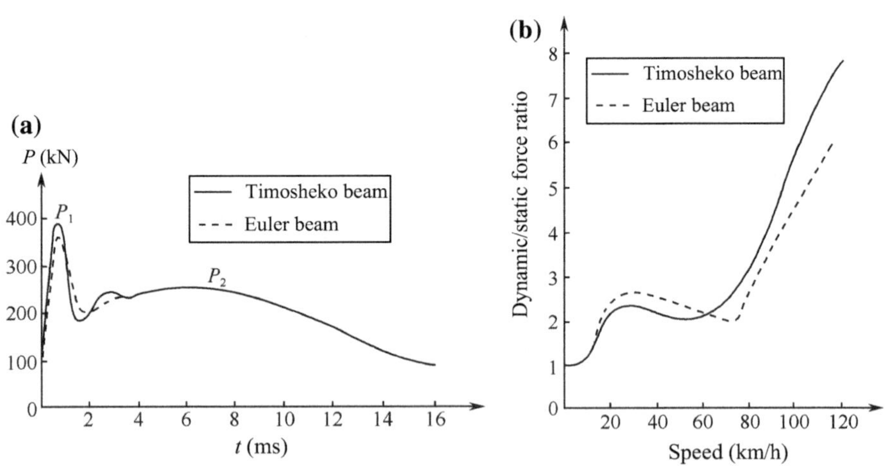

Fig. 2.6 Comparison of results from the Euler beam model and the Timoshenko beam model: **a** dynamic responses of wheel–rail force and **b** ratio of dynamic/static wheel loads

4. *Comparison between single-layer point support beam model and multi-layer point support beam model*

The beam model with discrete elastic point supports can be divided into single-layer, double-layer, and multi-layer (three or more) models according to the detailed degree of description of the track substructure. The single-layer model considers the entire under-track foundation as equivalent spring–damper supports in the vertical direction, without considering the effects of the sleeper/slab and ballast separately. The double-layer model takes into account the role of the sleeper or slab, including the mass of the sleeper or slab and the effect of fastening system. The triple-layer model further includes the dynamic effect of the ballast, which fully considers the mass and elasticity of the track ballasted bed, and even the dynamic effect of the subgrade. The multi-layer model separates the ballast into several separate layers, which is inconsistent with the fact that the ballast acts as an integrated body during vibration [83]. The vibration parameter of each layer of the ballast cannot be reasonably determined, so the multi-layer models are considered to have no obvious advantage and rarely used.

Since the components of the rail infrastructures (rail pad, sleeper, track slab, ballast, and subgrade) have different roles in achieving the track function, and different effects on the wheel–rail dynamic interaction, the model will be more realistic if these components are considered separately. In this sense, the triple-layer model is better than the double-layer model and hence the single-layer model. From the perspective of dynamic simulation, only by considering the rail, sleeper, track slab, ballast, and subgrade separately, can the respective vibration responses of the components be obtained, hence allow a comprehensive understanding of the vibration behavior of the track structure. For example, the ballast plays an important role during the process of track deteriorating, and the deterioration of the ballast is related to the ballast acceleration [83]. It is essential to understand the value and variation of the ballast acceleration, as it is a key indicator in wheel–rail dynamics analysis, therefore, the track dynamics model should be able to predict the ballast acceleration. In this sense, the triple-layer model that is able to predict the ballast vibration is superior to the two-layer and single-layer models. In addition, the triple-layer model can simulate special types of dynamic problems such as the voided sleepers or hardened ballast bed.

The numerical results of the ballasted track modeled with the single-layer, double-layer, and triple-layer discrete elastic point support Euler beams were compared by the author. The dynamic responses when a Chinese C_{62A} freight wagon passing a dipped rail joint on a traditional ballasted track at a speed of 80 km/h are listed in Table 2.1.

It can be seen from Table 2.1 that the calculation results of the double-layer model were almost the same as those of the triple-layer model except that the P_2 force was slightly increased and the ballast acceleration cannot be obtained. The results of the single-layer model had significant errors. Compared with the triple-layer model, both the sleeper reaction force and the P_2 force had an

Table 2.1 Comparison of numerical results for different layered track models

Indicators	Triple-layer model	Double-layer model	Single-layer model
High-frequency force P_1 (kN)	306.49	306.49	313.14
Low-frequency force P_2 (kN)	226.87	253.60	288.52
Sleeper reaction force (kN)	54.43	54.46	68.19
Rail acceleration (m/s^2)	1063.85	1063.83	1003.01
Sleeper acceleration (m/s^2)	351.16	349.79	–[a]
Ballast acceleration (m/s^2)	69.45	–[a]	–[a]

[a]This indicator cannot be calculated in the model

approximately 25% increment. This result was consistent with the conclusion in [4] that the impact force could be overestimated using the single-layer model without considering the effect of the sleeper and rail pad.

It is worth noting that for the slab track, or the track with sleeper or ballast based vibration isolation (such as elastic–supporting–block track, ladder-shaped sleeper track, and floating slab track), the vibration of the track system mainly exists in the rail and sleeper/ballast. The two-layer discrete elastic point support beam model is suitable for the wheel–rail dynamics analysis, which can ensure sufficient accuracy. For the fastener vibration isolation type (or rail pad vibration isolation type) track, since the track system vibration mainly exists on the rails, a single-layer discrete elastic point support beam model can be used to reflect the dynamic behavior of the rail–substructure system, which meets the requirements for engineering analysis.

2.1.3 Modeling of Vehicle

For a long time, in the wheel–rail dynamics analysis model, more attention was paid to the detailed description of the track structure, but less on the part of the vehicle system. In fact, the vehicle and the track models are equally important due to the strong coupling effect between the vehicle and the track, which has a significant impact on the wheel–rail dynamics. Various simplifications of the vehicle model can greatly reduce the workload of the analysis and calculation, but will inevitably lead to different degrees of analysis error. Kisilowski and Knothe [84] specifically discussed the dynamic coupling effect between adjacent wheels of a railway vehicle, using the analysis model as shown in Fig. 2.7. The analysis results showed that in the case of good track elasticity, the wheel displacement can be significantly affected (approx. 37%) by the adjacent wheel; significant forced vibration around 200 Hz can be observed due to the transmission of the wheel–rail force along the rail. Therefore, it is not reasonable to simplify the vehicle into a single wheelset model.

The numerical analysis performed by the author in [82] showed the analysis error caused by neglecting the influence of the adjacent wheels. The wheel–rail

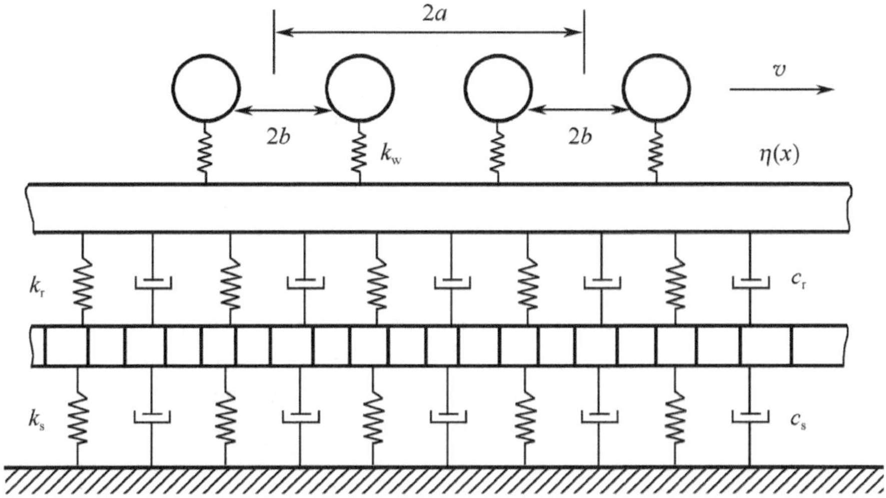

Fig. 2.7 Model for studying the interaction of adjacent wheels on mutual dynamics

force error was around 7%, and the ballast acceleration is about 11% for high-frequency wheel–rail impact vibration. The error was greater in the case of sinusoidal vibration excitation. Table 2.2 lists the wheel–rail dynamic responses of three vehicle models (single wheelset, single bogie, and complete vehicle) under continuous sinusoidal excitation on rail surface (wavelength 250 mm, wave depth 1 mm). It can be seen from Table 2.2 that, compared with the complete vehicle model, the single wheelset–track model without considering the interaction among the wheelsets of the vehicle had 13% lower wheel–rail force, and 35.7% lower ballast acceleration. The single bogie–track model considering the interaction of two adjacent wheelsets had 9% lower wheel–rail force, and 19% lower ballast acceleration. The main reason for the calculation error was that the dynamic action of one wheelset and the track was transmitted to the adjacent wheelsets through the

Table 2.2 Comparison of the predicted results of different vehicle models

Model type	Complete vehicle–track model	Single bogie–track model		Single wheelset–track model	
	Numerical value	Numerical value	Deviation (%)	Numerical value	Deviation (%)
Wheel–rail force (kN)	376.3	342.4	9	327.8	13
Ballast acceleration (m/s^2)	181.2	146.9	19	116.6	35.7

vehicle (bogie frame and body) and the rail, hence creating a dynamic coupling effect of the wheelsets. The track vibration had the largest effect in the range of three sleeper spans before and after the excitation point [85], which is close to the bogie wheelbase. Therefore, the calculation error was greatly reduced after considering the coupling effect of the two wheelsets on a bogie. Thus, the bogie (or half-car) model was obviously superior to the single wheelset model.

Under the excitation of continuous sinusoidal track irregularity, the four wheelsets of the vehicle would be simultaneously excited, which led to coupling and superposition of the dynamic actions of all the four wheelsets of the vehicle. However, the single wheelset and the bogie models could not reflect (or not fully reflect) this coupling effect, hence, the calculation results had large errors compared to the complete vehicle model.

Another topic should be discussed on vehicle modeling: is it necessary to consider several vehicles or even a whole train? The traction and braking actions cause large longitudinal forces in the inter-vehicle coupling system for the long and heavy-haul trains, which has a significant effect on the vehicle curving performance and running stability [86, 87]. In addition, the inter-vehicle couplers and dampers used in high-speed trains can greatly reduce the longitudinal impulse between adjacent vehicles in a train, and also greatly improve the ride comfort and running safety of the train [88]. It is evident that under certain conditions, the interaction between adjacent vehicles in a train has an important influence on the wheel–rail dynamics. Therefore, for the long heavy-haul trains, articulated trains, and trains using compact inter-vehicle coupling, the train–track coupled dynamics model considering multiple vehicles is necessary, which may achieve higher calculation accuracy than the model with only one vehicle. In general, the train–track coupled dynamics model of a 3–5 cars can well reflect the dynamic performance and the inter-vehicle interaction of a long train [87, 88]. For the investigation of conventional vehicle dynamics problems, excluding the aforementioned three types of trains, a single-vehicle model is sufficient.

2.1.4 General Principles for Vehicle–Track Coupled System Modeling

Through the above comparative analysis, it can be found that the simplification of the vehicle and the track structure in the modeling of the vehicle–track coupled system may lead to the loss of the model function or the reduction of the analysis accuracy. Therefore, the ideal model should adequately consider various factors influencing the dynamic performance of the vehicle and track system, especially the wheel–rail interaction characteristic. Meanwhile, the model could not be too complicated to allow rapid calculation.

In sum, the dynamics model of the vehicle–track coupled system should comply with the following basic principles:

(1) A continuous track model should be used instead of a simplified equivalent lumped parameter track model. The lumped parameter model is mainly suitable for qualitative analysis, but it cannot be used for quantitative research of complex problems.
(2) The beam model with discrete elastic point supports should be used instead of the beam model with continuous elastic foundation, which better conforms to the actual track structure, and can solve special types of dynamic problems such as nonuniform track support stiffness along the longitudinal direction.
(3) For analysis of the middle- and low-frequency wheel–rail dynamics problems, the Euler beam rail model could be generally used, which makes the model less complicated and also meets the general wheel–rail dynamics analysis requirements. When it is necessary to consider the high-frequency vibration characteristics of the wheel–rail interaction (especially for wheel–rail noise problems), the Timoshenko beam rail model that reflects higher frequencies should be used.
(4) For the ballasted track, a triple-layer support beam model should be used to completely reflect the function and the interaction of the rail–sleeper–ballast–subgrade system; for the ballastless slab track, a two-layer support beam model should be used to reflect the interaction of the rail–slab–subgrade system; for the rail fastener damping type track, a single-layer support beam model could suffice the requirement for wheel–rail dynamics analysis.
(5) The complete vehicle–track model should be used to consider the effects of car body, bogies, wheelsets, and the primary and secondary suspensions, which could include the superimposed vibrations on the track components due to the mutual influences of all wheelsets. To study dynamics problems of long heavy-haul trains and articulated trains, the vehicle model should be extended to train model including at least 3–5 cars, which could reflect the coupled dynamics between adjacent vehicles, and between the train and the track in longitudinal, lateral and vertical directions.

The vehicle–track coupled dynamics models presented in the following sections of this book will be all based on these principles.

2.2 Vehicle–Track Vertically Coupled Dynamics Model

The vehicle–track vertically coupled dynamics model is the theoretical foundation for analyzing the vertical vehicle–track dynamic interaction, which is a theoretical tool for studying the dynamic effect of vehicles on track structures as well. It is also the necessary foundation for further development of three-dimensional coupled models of the vehicle–track system.

2.2.1 Physical Model

According to the vehicle category, the vehicle–track vertically coupled dynamics model can be classified into three types: the passenger vehicle–track vertically coupled dynamics model, the freight wagon–track vertically coupled dynamics model, and the locomotive–track vertically coupled dynamics model. Based on the modeling principles given in Sect. 2.1.4, we had built the typical vertical interaction models for traditional passenger vehicle–track system (Fig. 2.8), freight wagon–track system (Fig. 2.9), and locomotive–track system (Fig. 2.10).

In the vehicle models, the passenger vehicle, freight wagon, and locomotive are modeled as multi-rigid-body systems that move in the track longitudinal direction at train speed v. The vehicle models consider the mass (M_c) and the pitching moment of inertia (J_c) of the car body, the mass (M_t) and the pitching moment of inertia (J_t) of the lead and rear bogie frames, the mass (M_w) of the four wheelsets, the vertical stiffness (K_{pz}) and damping (C_{pz}) of the primary suspension, and the vertical stiffness (K_{sz}) and damping (C_{sz}) of the secondary suspension.

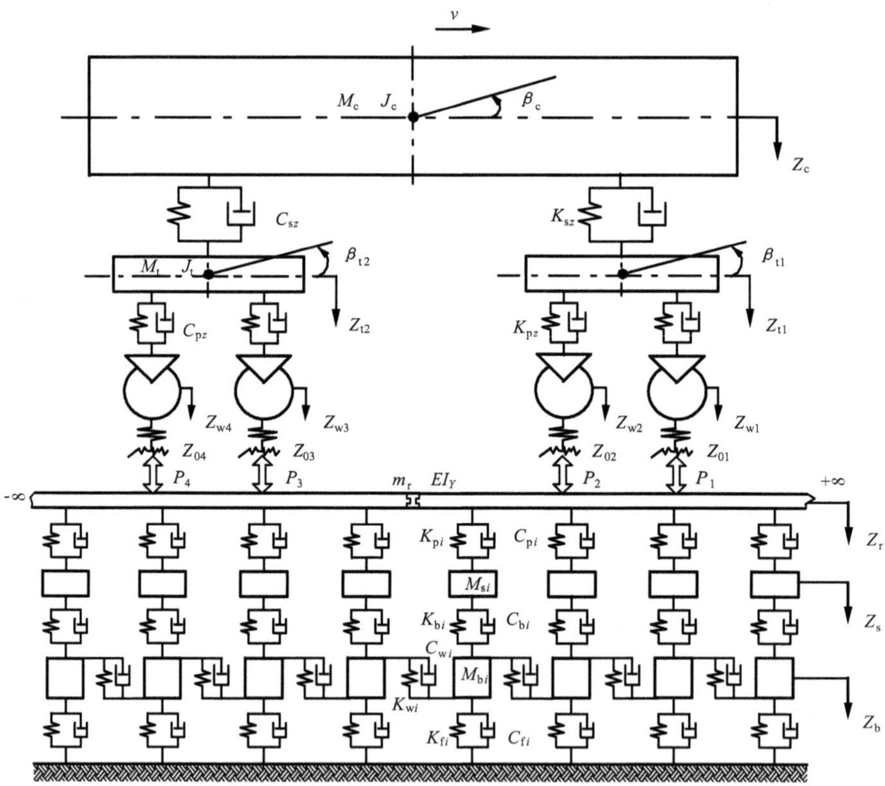

Fig. 2.8 Passenger vehicle–track vertically coupled dynamics model

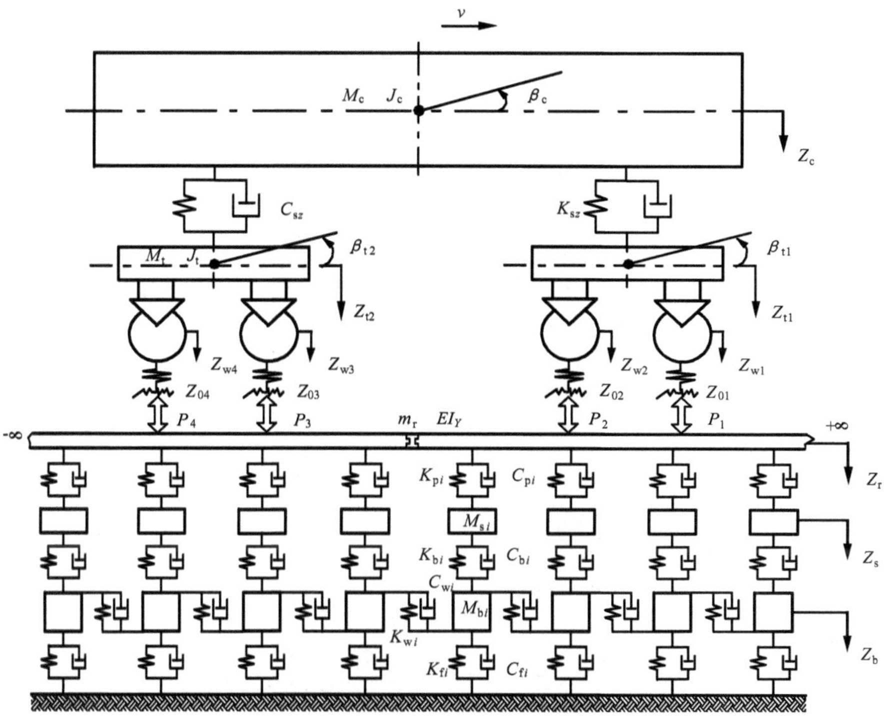

Fig. 2.9 Freight wagon–track vertically coupled dynamics model

In the passenger vehicle model, the car body has the vertical (Z_c) and pitch (β_c) motions, the lead and rear bogie frames also have the vertical (Z_{t1}, Z_{t2}) and pitch (β_{t1}, β_{t2}) motions, and the four wheelsets only have the vertical motion ($Z_{wi}, i = 1\text{--}4$). Thus, each passenger vehicle has 10 Degrees of Freedom (DOFs). In the freight wagon model, both the car body and bogie frames have the vertical and pitch motions. Since the wagon has no primary suspension, the motions of the wheelset coincide with the motions of the bogie frame. Therefore, each wagon has six DOFs. It is noted that the wagon model shown in Fig. 2.9 was built for a main kind of Chinese freight wagon equipped with three-piece bogies without primary suspension. For the other wagons equipped with primary suspension, the modeling method can be found in Ref. [12]. And for the freight wagons with two suspensions, its vertical coupled model should be the same as that shown in Fig. 2.8.

The double-axle locomotive model is similar to the passenger vehicle model. The only difference is that the locomotive model has to consider the traction motor. Figure 2.10 shows a locomotive–track vertically coupled dynamics model that considers the dynamic effect of the traction motor system. In the model, a traction motor (M_m, J_m) is mounted on each wheel axle via axle-hung bearings at one end while elastically linked (K_m) with the bogie frame at the other end. Each rigid motor has only the rotational motion (β_{mi}). Thus, each locomotive has 14 DOFs. It is

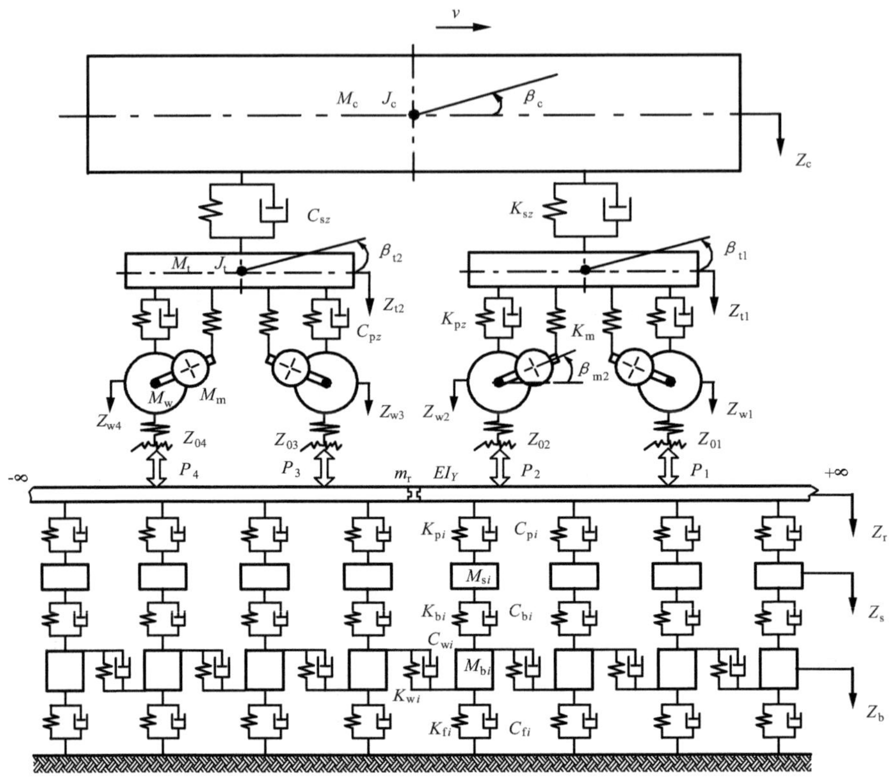

Fig. 2.10 Locomotive–track vertically coupled dynamics model

noted that the current locomotive model can be easily improved when the motor suspension mode is changed. For example, when the motor is linked with the bearing by elastic suspension, the vertical DOF of each motor should be considered, such a model can be referred to [13].

The track sub-model shown in Figs. 2.8, 2.9, and 2.10 represent the conventional ballasted track structure, consisting of the rails, the rail pads, the sleepers, the ballast, and the subgrade. Both the left and the right rails are treated as continuous beams (the Euler or Timoshenko beams), which are discretely supported at rail–sleeper junctions by three layers of springs and dampers representing the elasticity and damping of the rail pad, the ballast, and the subgrade, respectively. In order to account for the continuity and the coupling effects of the interlocking ballast granules, a couple of shear stiffness (K_{wi}) and shear damping (C_{wi}) is introduced between adjacent ballast masses. In the model, m_r and EI_Y are the mass per unit longitudinal length and the bending stiffness of the rails, K_{pi}, K_{bi}, K_{fi} and C_{pi}, C_{bi}, C_{fi} are the stiffness and damping of the rail pad, the ballast and the subgrade in each rail–sleeper junction.

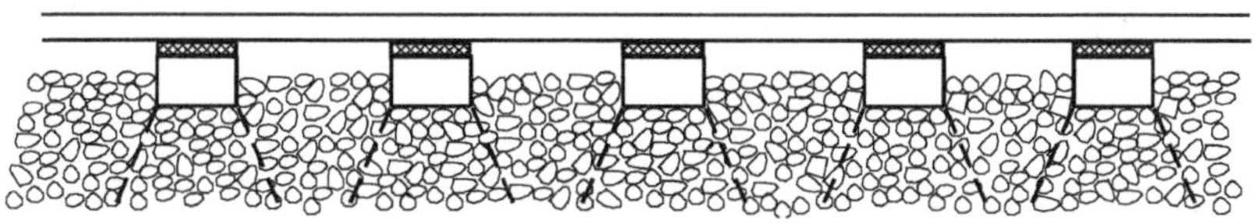

Fig. 2.11 Load distribution region in continuous granular ballast (Reprinted from Ref. [90], Copyright 2003, with permission from Elsevier.)

It is difficult to analyze the vibration of ballast because of its granular configuration and special mechanism of action. So far, little research work has been done on the dynamic modeling of the ballast layer. The representative theoretical work on the ballast vibration modeling is the hypothesis of the load being transmitted within a cone region in the ballast by Ahlbeck [7]. He assumed that the load transmitting from a sleeper to the ballast approximately coincides with the cone distribution. That is to say, the stresses of the ballast are uniformly distributed over the cone region and zero outside the cone. The inclination of the cone is just the ballast stress pervasion angle α corresponding to the Poisson's ratio. Thus, the effective acting region of the ballast under each sleeper can be determined, as shown in Fig. 2.11.

On the basis of this assumption, the vibrating part of the ballast under each sleeper is just the cone region as shown in Fig. 2.12. Therefore, the continuous granular ballast could be modeled as a series of separate vibrating masses when analyzing the track dynamics, by which the analytical process of the ballast vibration is greatly simplified. In fact, Rücker et al. [89] have already concluded that the theoretical investigation on ballast vibrations is extremely difficult and impossible to be applied in engineering if a three-dimensional half-space model is adopted for ballast modeling.

Fig. 2.12 Model of the ballast under one rail support point (Reprinted from Ref. [90], Copyright 2003, with permission from Elsevier.)

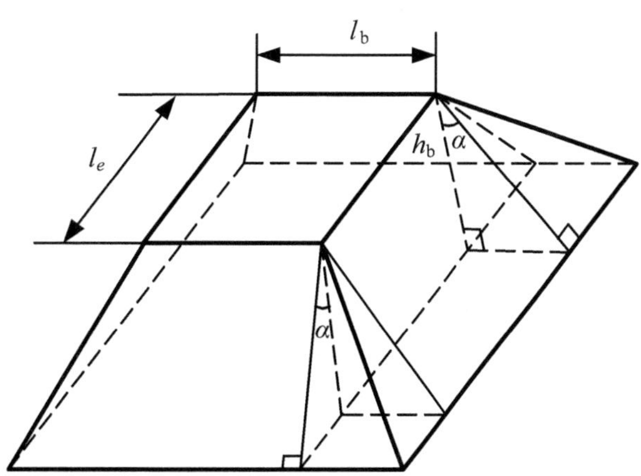

According to the ballast model shown in Fig. 2.12, the vibrating mass of ballast under a sleeper support point could be evaluated as

$$M_b = \rho_b h_b \left[l_e l_b + (l_e + l_b) h_b \tan\alpha + \frac{4}{3} h_b^2 \tan^2\alpha \right] \tag{2.1}$$

where ρ_b is the density of ballast (kg/m^3), h_b is the depth of ballast (m), l_e is the effective supporting length of half sleeper (m), l_b is the width of sleeper underside (m), and α is the ballast stress pervasion angle.

The supporting stiffness of a ballast mass can be determined as

$$K_b = \frac{2(l_e - l_b)\tan\alpha}{\ln\left[\frac{l_e}{l_b} \cdot \frac{(l_b + 2h_b \tan\alpha)}{(l_e + 2h_b \tan\alpha)}\right]} E_b \tag{2.2}$$

where E_b is the Young's modulus of the ballast (N/m^2).

Correspondingly, the subgrade stiffness under one supporting point equals to the product of the cone underside area and the modulus of the subgrade

$$K_f = (l_e + 2h_b \tan\alpha)(l_b + 2h_b \tan\alpha) E_f \tag{2.3}$$

where E_f is the K_{30} modulus of subgrade (Pa/m), which means the force acting on unit area that leads to unit deformation.

The above-proposed ballast model is based on the assumption that there is no overlapping of adjacent cone regions of ballasts. In the case of thick ballast layer, small sleeper spacing, or big ballast stress pervasion angle, an overlapping of adjacent ballast masses may occur, see Fig. 2.13. The above ballast model should be modified appropriately [90]. In this case, the vibrating mass of ballast under a rail support point could be defined as the shadowed region as shown in Fig. 2.13.

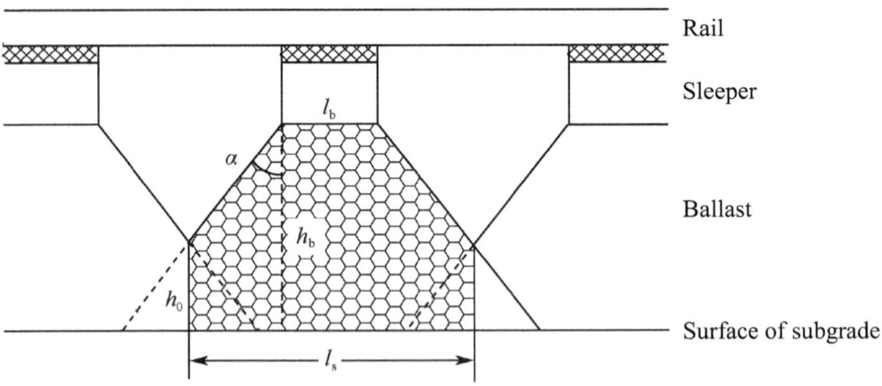

Fig. 2.13 The modified model of ballast (Reprinted from Ref. [90], Copyright 2003, with permission from Elsevier.)

According to the geometry shown in Fig. 2.13, the height of the overlapping regions is calculated by

$$h_0 = h_b - \frac{l_s - l_b}{2\tan\alpha} \tag{2.4}$$

where l_s is the sleeper spacing.

Therefore, it can be seen that there is no overlapping between adjacent ballast masses when $h_0 \leq 0$. But when $h_0 > 0$, the overlapping of adjacent ballast masses occurs, then the ballast vibrating mass is changed into

$$M_b' = \rho_b\left[l_b h_b(l_e + h_b\tan\alpha) + l_e\left(h_b^2 - h_0^2\right)\tan\alpha + \frac{4}{3}\left(h_b^3 - h_0^3\right)\tan^2\alpha\right] \tag{2.5}$$

The ballast supporting stiffness is the combined stiffness of two parts in series

$$K_b' = \frac{K_{b1}K_{b2}}{K_{b1} + K_{b2}} \tag{2.6}$$

where

$$K_{b1} = \frac{2(l_e - l_b)\tan\alpha}{\ln\left[\frac{l_e l_s}{l_b(l_e + l_s - l_b)}\right]}E_b \tag{2.7}$$

and

$$K_{b2} = \frac{l_s(l_s - l_b + 2l_e + 2h_b\tan\alpha)\tan\alpha}{l_b - l_s + 2h_b\tan\alpha}E_b \tag{2.8}$$

And the subgrade stiffness becomes

$$K_f' = l_s(l_e + 2h_b\tan\alpha)E_f \tag{2.9}$$

To consider the continuity and the coupling effects of the interlocking ballast granules, a couple of shear stiffness K_w and shear damping C_w should be introduced between the adjacent ballast masses in the ballast model. The authors had studied the coupling effects of the interlocking ballast granules on the ballast vibration in Ref. [90]. The results show that the model will overestimate the ballast vibration level if the ballast shearing effect is not considered—usually the acceleration of the ballast will be at least 10% higher. The reason is that the effect of friction and impact of ballast stones induces a counteracting motion of adjacent ballast blocks, so that the vibration level of one ballast block will be attenuated by the adjacent blocks. If the shearing effect is not considered, this attenuating effect is absent.

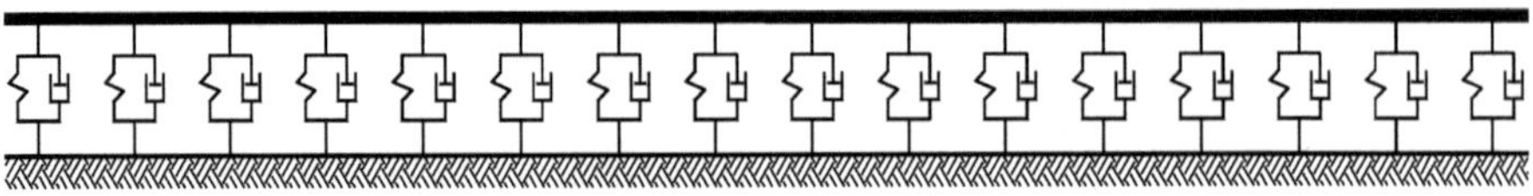

Fig. 2.14 The vertical dynamics model for the long-sleeper embedded track

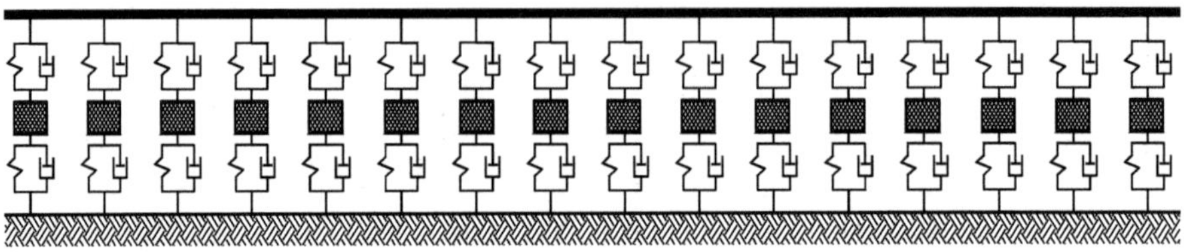

Fig. 2.15 The vertical dynamics model for the elastic supporting block track

Thus, the ballast mass can vibrate more freely and its vibration level will be overestimated.

If the analyzed track structure is a ballastless track, the above ballasted track dynamics model can be improved correspondingly. Figures 2.14, 2.15, and 2.16 establish the vertical dynamics models of three typical ballastless track structures, including the long-sleeper embedded track, the elastic supporting block track, and the slab track.

The long-sleeper embedded track (or double-block track) consists of rails, fastenings and rail pads, concrete sleepers, concrete slabs, and concrete base. For this type of ballastless track, the track structure can be simply modeled as two continuous rail beams discretely supported by fastenings and rail pads (Fig. 2.14). This is because the sleeper blocks are precast into the slab directly and there is no elasticity between the slab and the concrete base. Thus only the vibration of the rails is important for the wheel–rail interaction.

The elastic supporting block track consists of rails, fastenings and rail pads, concrete blocks, block pads and rubber boots, concrete slabs, and concrete base.

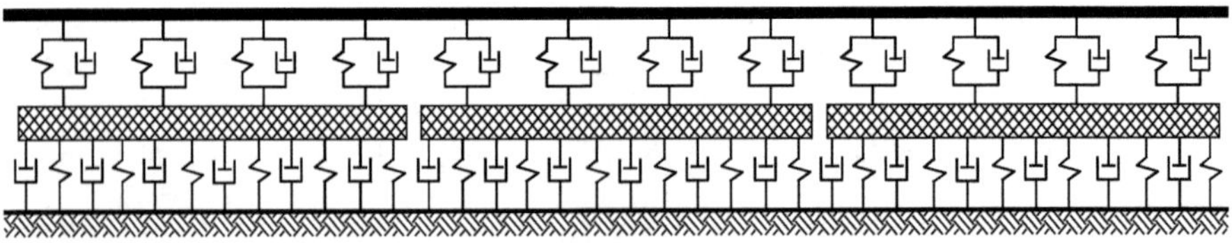

Fig. 2.16 The vertical dynamics model for the typical slab track (Reprinted from Ref. [36], Copyright 2009, with permission from Taylor & Francis.)

The pads provide the vertical supporting stiffness and damping for the blocks under rails, and the rubber boots provide the lateral stiffness and damping for the blocks. Thus, only the vibrations of rails and concrete blocks are important for the wheel–rail interaction. The elastic supporting block track model is shown in Fig. 2.15, where the rails modeled as continuous beams discretely supported by fastenings and rail pads, and the concrete blocks are modeled as rigid bodies.

The slab track consists of rails, fastenings and rail pads, concrete slabs, cement asphalt mortar (CAM) layer, and concrete base. The CAM layer has soft stiffness, which may reduce the vibration of the subgrade. Therefore, both the vertical vibrations of the rails and slabs have a significant effect on the vehicle–track vertical interaction. In the model, the rails are modeled as continuous beams discretely supported by fastenings and rail pads, while the track slabs are simplified as finite length free beams (no constraint at two ends) continuously supported by stiffness and damping of the CAM layer, as shown in Fig. 2.16.

The wheel–rail contact is an essential element that couples the vehicle with the track. The key issue is the contact forces between the wheel and the rail. The wheel–rail vertical contact force can be easily calculated using the nonlinear Hertzian elastic contact theory. It is able to take into account the separation between the wheel and rail.

2.2.2 Equations of Motion

1. *Equations of motion of the passenger vehicle subsystem (refer to* Fig. 2.8*)*

(1) Vertical motion of the car body

$$\begin{aligned} & M_c\ddot{Z}_c + 2C_{sz}\dot{Z}_c + 2K_{sz}Z_c - C_{sz}\dot{Z}_{t1} \\ & \quad - K_{sz}Z_{t1} - C_{sz}\dot{Z}_{t2} - K_{sz}Z_{t2} = M_c g \end{aligned} \tag{2.10}$$

(2) Pitch motion of the car body

$$\begin{aligned} & J_c\ddot{\beta}_c + 2C_{sz}l_c^2\dot{\beta}_c + 2K_{sz}l_c^2\beta_c + C_{sz}l_c\dot{Z}_{t1} - C_{sz}l_c\dot{Z}_{t2} \\ & \quad + K_{sz}l_cZ_{t1} - K_{sz}l_cZ_{t2} = 0 \end{aligned} \tag{2.11}$$

(3) Vertical motion of the lead bogie frame

$$\begin{aligned} & M_t\ddot{Z}_{t1} + (2C_{pz} + C_{sz})\dot{Z}_{t1} + (2K_{pz} + K_{sz})Z_{t1} - C_{sz}\dot{Z}_c \\ & \quad - K_{sz}Z_c - C_{pz}\dot{Z}_{w1} - C_{pz}\dot{Z}_{w2} - K_{pz}Z_{w1} - K_{pz}Z_{w2} \\ & \quad + C_{sz}l_c\dot{\beta}_c + K_{sz}l_c\beta_c = M_t g \end{aligned} \tag{2.12}$$

(4) Pitch motion of the lead bogie frame

$$J_t\ddot{\beta}_{t1} + 2C_{pz}l_t^2\dot{\beta}_{t1} + 2K_{pz}l_t^2\beta_{t1} + C_{pz}l_t\dot{Z}_{w1} - C_{pz}l_t\dot{Z}_{w2} + K_{pz}l_tZ_{w1} - K_{pz}l_tZ_{w2} = 0 \tag{2.13}$$

(5) Vertical motion of the rear bogie frame

$$M_t\ddot{Z}_{t2} + (2C_{pz} + C_{sz})\dot{Z}_{t2} + (2K_{pz} + K_{sz})Z_{t2} - C_{sz}\dot{Z}_c - K_{sz}Z_c - C_{pz}\dot{Z}_{w3} - C_{pz}\dot{Z}_{w4} - K_{pz}Z_{w3} - K_{pz}Z_{w4} - C_{sz}l_c\dot{\beta}_c - K_{sz}l_c\beta_c = M_tg \tag{2.14}$$

(6) Pitch motion of the rear bogie frame

$$J_t\ddot{\beta}_{t2} + 2C_{pz}l_t^2\dot{\beta}_{t2} + 2K_{pz}l_t^2\beta_{t2} + C_{pz}l_t\dot{Z}_{w3} - C_{pz}l_t\dot{Z}_{w4} + K_{pz}l_tZ_{w3} - K_{pz}l_tZ_{w4} = 0 \tag{2.15}$$

(7) Vertical motion of the first wheelset

$$M_w\ddot{Z}_{w1} + C_{pz}\dot{Z}_{w1} + K_{pz}Z_{w1} - C_{pz}\dot{Z}_{t1} - K_{pz}Z_{t1} + C_{pz}l_t\dot{\beta}_{t1} + K_{pz}l_t\beta_{t1} + 2p_1(t) - M_wg = F_{01}(t) \tag{2.16}$$

(8) Vertical motion of the second wheelset

$$M_w\ddot{Z}_{w2} + C_{pz}\dot{Z}_{w2} + K_{pz}Z_{w2} - C_{pz}\dot{Z}_{t1} - K_{pz}Z_{t1} - C_{pz}l_t\dot{\beta}_{t1} - K_{pz}l_t\beta_{t1} + 2p_2(t) - M_wg = F_{02}(t) \tag{2.17}$$

(9) Vertical motion of the third wheelset

$$M_w\ddot{Z}_{w3} + C_{pz}\dot{Z}_{w3} + K_{pz}Z_{w3} - C_{pz}\dot{Z}_{t2} - K_{pz}Z_{t2} + C_{pz}l_t\dot{\beta}_{t2} + K_{pz}l_t\beta_{t2} + 2p_3(t) - M_wg = F_{03}(t) \tag{2.18}$$

(10) Vertical motion of the fourth wheelset

$$M_w\ddot{Z}_{w4} + C_{pz}\dot{Z}_{w4} + K_{pz}Z_{w4} - C_{pz}\dot{Z}_{t2} - K_{pz}Z_{t2} - C_{pz}l_t\dot{\beta}_{t2} - K_{pz}l_t\beta_{t2} + 2p_4(t) - M_wg = F_{04}(t) \tag{2.19}$$

where l_c is half of the distance between two bogie centers of a vehicle (m), l_t is half of the distance between the two axles of a bogie (m); $p_i(t)$ is the vertical wheel–rail force at the ith wheelset ($i = 1-4$); $F_{0i}(t)$ is the self-excitation force at the ith wheelset ($i = 1-4$) if there is, such as the centrifugal force caused by the wheel eccentricity.

2. *Equations of motion of the wagon subsystem (refer to* Fig. 2.9*)*

(1) Vertical motion of the car body

$$\begin{aligned} &M_c\ddot{Z}_c + 2C_{sz}\dot{Z}_c + 2K_{sz}Z_c - C_{sz}\dot{Z}_{t1} \\ &- K_{sz}Z_{t1} - C_{sz}\dot{Z}_{t2} - K_{sz}Z_{t2} = M_c g \end{aligned} \tag{2.20}$$

(2) Pitch motion of the car body

$$\begin{aligned} &J_c\ddot{\beta}_c + 2C_{sz}l_c^2\dot{\beta}_c + 2K_{sz}l_c^2\beta_c + C_{sz}l_c\dot{Z}_{t1} - C_{sz}l_c\dot{Z}_{t2} \\ &+ K_{sz}l_cZ_{t1} - K_{sz}l_cZ_{t2} = 0 \end{aligned} \tag{2.21}$$

(3) Vertical motion of the lead bogie frame

$$\begin{aligned} &(M_t + 2M_w)\ddot{Z}_{t1} + C_{sz}\dot{Z}_{t1} + K_{sz}Z_{t1} - C_{sz}\dot{Z}_c - K_{sz}Z_c \\ &+ C_{sz}l_c\dot{\beta}_c + K_{sz}l_c\beta_c + 2p_1(t) + 2p_2(t) - (M_t + 2M_w)g \\ &= F_{01}(t) + F_{02}(t) \end{aligned} \tag{2.22}$$

(4) Pitch motion of the lead bogie frame

$$(J_t + 2M_wl_t^2)\ddot{\beta}_{t1} - 2[p_1(t) - p_2(t)]l_t + [F_{01}(t) - F_{02}(t)]l_t = 0 \tag{2.23}$$

(5) Vertical motion of the rear bogie frame

$$\begin{aligned} &(M_t + 2M_w)\ddot{Z}_{t2} + C_{sz}\dot{Z}_{t2} + K_{sz}Z_{t2} - C_{sz}\dot{Z}_c - K_{sz}Z_c \\ &- C_{sz}l_c\dot{\beta}_c - K_{sz}l_c\beta_c + 2p_3(t) + 2p_4(t) - (M_t + 2M_w)g \\ &= F_{03}(t) + F_{04}(t) \end{aligned} \tag{2.24}$$

(6) Pitch motion of the rear bogie frame

$$(J_t + 2M_wl_t^2)\ddot{\beta}_{t2} - 2[p_3(t) - p_4(t)]l_t + [F_{03}(t) - F_{04}(t)]l_t = 0 \tag{2.25}$$

For the wagon system with the three-piece bogie considered here, the equations of motion of the wheelsets are depended on the bogie frames, which can be written as

$$\begin{aligned} &Z_{w1} = Z_{t1} - l_t\beta_{t1}, \quad \dot{Z}_{w1} = \dot{Z}_{t1} - l_t\dot{\beta}_{t1}, \quad \ddot{Z}_{w1} = \ddot{Z}_{t1} - l_t\ddot{\beta}_{t1} \\ &Z_{w2} = Z_{t1} + l_t\beta_{t1}, \quad \dot{Z}_{w2} = \dot{Z}_{t1} + l_t\dot{\beta}_{t1}, \quad \ddot{Z}_{w2} = \ddot{Z}_{t1} + l_t\ddot{\beta}_{t1} \\ &Z_{w3} = Z_{t2} - l_t\beta_{t2}, \quad \dot{Z}_{w3} = \dot{Z}_{t2} - l_t\dot{\beta}_{t2}, \quad \ddot{Z}_{w3} = \ddot{Z}_{t2} - l_t\ddot{\beta}_{t2} \\ &Z_{w4} = Z_{t2} + l_t\beta_{t2}, \quad \dot{Z}_{w4} = \dot{Z}_{t2} + l_t\dot{\beta}_{t2}, \quad \ddot{Z}_{w4} = \ddot{Z}_{t2} + l_t\ddot{\beta}_{t2} \end{aligned}$$

3. *Equations of motion of the locomotive subsystem (refer to Fig. 2.10)*

(1) Vertical motion of the car body

$$M_c\ddot{Z}_c + 2C_{sz}\dot{Z}_c + 2K_{sz}Z_c - C_{sz}(\dot{Z}_{t1} + \dot{Z}_{t2}) - K_{sz}(Z_{t1} + Z_{t2}) = M_c g \quad (2.26)$$

(2) Pitch motion of the car body

$$J_c\ddot{\beta}_c + 2C_{sz}l_c^2\dot{\beta}_c + 2K_{sz}l_c^2\beta_c + C_{sz}l_c(\dot{Z}_{t1} - \dot{Z}_{t2}) + K_{sz}l_c(Z_{t1} - Z_{t2}) = 0 \quad (2.27)$$

(3) Vertical motion of the lead bogie frame

$$\begin{aligned} &M_t\ddot{Z}_{t1} + (C_{sz} + 2C_{pz})\dot{Z}_{t1} + (K_{sz} + 2K_{pz} + 2K_m)Z_{t1} - C_{sz}\dot{Z}_c - K_{sz}Z_c \\ &\quad - C_{pz}(\dot{Z}_{w1} + \dot{Z}_{w2}) - (K_{pz} + K_m)(Z_{w1} + Z_{w2}) + C_{sz}l_c\dot{\beta}_c + K_{sz}l_c\beta_c \\ &\quad - K_m(l_1 + l_2)(\beta_{m1} - \beta_{m2}) = M_t g \end{aligned} \quad (2.28)$$

(4) Pitch motion of the lead bogie frame

$$\begin{aligned} &J_t\ddot{\beta}_{t1} + 2C_{pz}l_t^2\dot{\beta}_{t1} + 2\left[K_{pz}l_t^2 + K_m(l_t - l_1 - l_2)^2\right]\beta_{t1} - C_{pz}l_t(\dot{Z}_{w2} - \dot{Z}_{w1}) \\ &\quad - \left[K_{pz}l_t + K_m(l_t - l_1 - l_2)\right](Z_{w2} - Z_{w1}) = 0 \end{aligned} \quad (2.29)$$

(5) Vertical motion of the rear bogie frame

$$\begin{aligned} &M_t\ddot{Z}_{t2} + (C_{sz} + 2C_{pz})\dot{Z}_{t2} + (K_{sz} + 2K_{pz} + 2K_m)Z_{t2} - C_{sz}\dot{Z}_c - K_{sz}Z_c \\ &\quad - C_{pz}(\dot{Z}_{w3} + \dot{Z}_{w4}) - (K_{pz} + K_m)(Z_{w3} + Z_{w4}) - C_{sz}l_c\dot{\beta}_c - K_{sz}l_c\beta_c \\ &\quad - K_m(l_1 + l_2)(\beta_{m3} - \beta_{m4}) = M_t g \end{aligned} \quad (2.30)$$

(6) Pitch motion of the rear bogie frame

$$\begin{aligned} &J_t\ddot{\beta}_{t2} + 2C_{pz}l_t^2\dot{\beta}_{t2} + 2\left[K_{pz}l_t^2 + K_m(l_t - l_1 - l_2)^2\right]\beta_{t2} - C_{pz}l_t(\dot{Z}_{w4} - \dot{Z}_{w3}) \\ &\quad - \left[K_{pz}l_t + K_m(l_t - l_1 - l_2)\right](Z_{w4} - Z_{w3}) = 0 \end{aligned} \quad (2.31)$$

(7) Vertical motion of the first wheelset

$$\begin{aligned} &(M_w + M_m)\ddot{Z}_{w1} + M_m l_1\ddot{\beta}_{m1} + C_{pz}\dot{Z}_{w1} + (K_{pz} + K_m)Z_{w1} - C_{pz}\dot{Z}_{t1} \\ &\quad - (K_{pz} + K_m)Z_{t1} + C_{pz}l_t\dot{\beta}_{t1} + \left[K_{pz}l_t + K_m(l_t - l_1 - l_2)\right]\beta_{t1} \\ &\quad + K_m(l_1 + l_2)\beta_{m1} + 2p_1(t) - (M_w + M_m)g = F_{01}(t) \end{aligned} \quad (2.32)$$

(8) Vertical motion of the second wheelset

$$
\begin{aligned}
&(M_{\mathrm{w}}+M_{\mathrm{m}})\ddot{Z}_{\mathrm{w2}}-M_{\mathrm{m}}l_1\ddot{\beta}_{\mathrm{m2}}+C_{\mathrm{pz}}\dot{Z}_{\mathrm{w2}}+(K_{\mathrm{pz}}+K_{\mathrm{m}})Z_{\mathrm{w2}}-C_{\mathrm{pz}}\dot{Z}_{\mathrm{t1}}\\
&\quad-(K_{\mathrm{pz}}+K_{\mathrm{m}})Z_{\mathrm{t1}}-C_{\mathrm{pz}}l_{\mathrm{t}}\dot{\beta}_{\mathrm{t1}}-\left[K_{\mathrm{pz}}l_{\mathrm{t}}+K_{\mathrm{m}}(l_{\mathrm{t}}-l_1-l_2)\right]\beta_{\mathrm{t1}}\\
&\quad-K_{\mathrm{m}}(l_1+l_2)\beta_{\mathrm{m2}}+2p_2(t)-(M_{\mathrm{w}}+M_{\mathrm{m}})g=F_{02}(t)
\end{aligned}
\tag{2.33}
$$

(9) Vertical motion of the third wheelset

$$
\begin{aligned}
&(M_{\mathrm{w}}+M_{\mathrm{m}})\ddot{Z}_{\mathrm{w3}}+M_{\mathrm{m}}l_1\ddot{\beta}_{\mathrm{m3}}+C_{\mathrm{pz}}\dot{Z}_{\mathrm{w3}}+(K_{\mathrm{pz}}+K_{\mathrm{m}})Z_{\mathrm{w3}}-C_{\mathrm{pz}}\dot{Z}_{\mathrm{t2}}\\
&\quad-(K_{\mathrm{pz}}+K_{\mathrm{m}})Z_{\mathrm{t2}}+C_{\mathrm{pz}}l_{\mathrm{t}}\dot{\beta}_{\mathrm{t2}}+\left[K_{\mathrm{pz}}l_{\mathrm{t}}+K_{\mathrm{m}}(l_{\mathrm{t}}-l_1-l_2)\right]\beta_{\mathrm{t2}}\\
&\quad+K_{\mathrm{m}}(l_1+l_2)\beta_{\mathrm{m3}}+2p_3(t)-(M_{\mathrm{w}}+M_{\mathrm{m}})g=F_{03}(t)
\end{aligned}
\tag{2.34}
$$

(10) Vertical motion of the fourth wheelset

$$
\begin{aligned}
&(M_{\mathrm{w}}+M_{\mathrm{m}})\ddot{Z}_{\mathrm{w4}}-M_{\mathrm{m}}l_1\ddot{\beta}_{\mathrm{m4}}+C_{\mathrm{pz}}\dot{Z}_{\mathrm{w4}}+(K_{\mathrm{pz}}+K_{\mathrm{m}})Z_{\mathrm{w4}}-C_{\mathrm{pz}}\dot{Z}_{\mathrm{t2}}\\
&\quad-(K_{\mathrm{pz}}+K_{\mathrm{m}})Z_{\mathrm{t2}}-C_{\mathrm{pz}}l_{\mathrm{t}}\dot{\beta}_{\mathrm{t2}}-\left[K_{\mathrm{pz}}l_{\mathrm{t}}+K_{\mathrm{m}}(l_{\mathrm{t}}-l_1-l_2)\right]\beta_{\mathrm{t2}}\\
&\quad-K_{\mathrm{m}}(l_1+l_2)\beta_{\mathrm{m4}}+2p_4(t)-(M_{\mathrm{w}}+M_{\mathrm{m}})g=F_{04}(t)
\end{aligned}
\tag{2.35}
$$

(11) Rotation motion of the first motor

$$
\begin{aligned}
&(J_{\mathrm{m}}+M_{\mathrm{m}}l_1^2)\ddot{\beta}_{\mathrm{m1}}+M_{\mathrm{m}}l_1\ddot{Z}_{\mathrm{w1}}+K_{\mathrm{m}}(l_1+l_2)^2\beta_{\mathrm{m1}}\\
&\quad+K_{\mathrm{m}}(l_1+l_2)(l_{\mathrm{t}}-l_1-l_2)\beta_{\mathrm{t1}}-K_{\mathrm{m}}(l_1+l_2)(Z_{\mathrm{t1}}-Z_{\mathrm{w1}})=0
\end{aligned}
\tag{2.36}
$$

(12) Rotation motion of the second motor

$$
\begin{aligned}
&(J_{\mathrm{m}}+M_{\mathrm{m}}l_1^2)\ddot{\beta}_{\mathrm{m2}}-M_{\mathrm{m}}l_1\ddot{Z}_{\mathrm{w2}}+K_{\mathrm{m}}(l_1+l_2)^2\beta_{\mathrm{m2}}\\
&\quad+K_{\mathrm{m}}(l_1+l_2)(l_{\mathrm{t}}-l_1-l_2)\beta_{\mathrm{t1}}+K_{\mathrm{m}}(l_1+l_2)(Z_{\mathrm{t1}}-Z_{\mathrm{w2}})=0
\end{aligned}
\tag{2.37}
$$

(13) Rotation motion of the third motor

$$
\begin{aligned}
&(J_{\mathrm{m}}+M_{\mathrm{m}}l_1^2)\ddot{\beta}_{\mathrm{m3}}+M_{\mathrm{m}}l_1\ddot{Z}_{\mathrm{w3}}+K_{\mathrm{m}}(l_1+l_2)^2\beta_{\mathrm{m3}}\\
&\quad+K_{\mathrm{m}}(l_1+l_2)(l_{\mathrm{t}}-l_1-l_2)\beta_{\mathrm{t2}}-K_{\mathrm{m}}(l_1+l_2)(Z_{\mathrm{t2}}-Z_{\mathrm{w3}})=0
\end{aligned}
\tag{2.38}
$$

(14) Rotation motion of the fourth motor

$$
\begin{aligned}
&(J_{\mathrm{m}}+M_{\mathrm{m}}l_1^2)\ddot{\beta}_{\mathrm{m4}}-M_{\mathrm{m}}l_1\ddot{Z}_{\mathrm{w4}}+K_{\mathrm{m}}(l_1+l_2)^2\beta_{\mathrm{m4}}\\
&\quad+K_{\mathrm{m}}(l_1+l_2)(l_{\mathrm{t}}-l_1-l_2)\beta_{\mathrm{t2}}+K_{\mathrm{m}}(l_1+l_2)(Z_{\mathrm{t2}}-Z_{\mathrm{w4}})=0
\end{aligned}
\tag{2.39}
$$

where l_{c} is half of the distance between bogie centers of the locomotive (m), l_1 is the distance between the motor and the axle-hung bearings (m), l_2 is the distance between the motor and the mounted point on the bogie frame (m).

It is noted that the vertical motion of the traction motor is not independent, its displacement and velocity can be obtained by

$$\begin{cases} Z_{mi} = Z_{wi} \mp l_1 \beta_{mi} \\ \ddot{Z}_{mi} = \ddot{Z}_{wi} \mp l_1 \ddot{\beta}_{mi} \end{cases} \tag{2.40}$$

where symbol in front of l_1 is + when i = 1 or 3, and – when i = 2 or 4.

It can be observed the mass matrix of the above locomotive system equations is off-diagonal, which is different from those of passenger and freight vehicles. It is necessary to diagonalize its mass matrix so as to be convenient for further numerical solution in Chap. 4. Consequently, the equations of motion of the wheelset (2.32)–(2.35) and the motor (2.36)–(2.39) subsystems are normalized as the following form:

① The first wheelset

$$\begin{aligned} &(M_w + \eta_m M_m)\ddot{Z}_{w1} + C_{pz}\dot{Z}_{w1} + (K_{pz} + \xi_m K_m)Z_{w1} - C_{pz}\dot{Z}_{t1} \\ &\quad - (K_{pz} + \xi_m K_m)Z_{t1} + C_{pz}l_t\dot{\beta}_{t1} + \left[K_{pz}l_t + \xi_m K_m(l_t - l_1 - l_2)\right]\beta_{t1} \\ &\quad + \xi_m K_m(l_1 + l_2)\beta_{m1} + 2p_1(t) - (M_w + M_m)g = F_{01}(t) \end{aligned} \tag{2.41}$$

② The second wheelset

$$\begin{aligned} &(M_w + \eta_m M_m)\ddot{Z}_{w2} + C_{pz}\dot{Z}_{w2} + (K_{pz} + \xi_m K_m)Z_{w2} - C_{pz}\dot{Z}_{t1} \\ &\quad - (K_{pz} + \xi_m K_m)Z_{t1} - C_{pz}l_t\dot{\beta}_{t1} - \left[K_{pz}l_t + \xi_m K_m(l_t - l_1 - l_2)\right]\beta_{t1} \\ &\quad - \xi_m K_m(l_1 + l_2)\beta_{m2} + 2p_2(t) - (M_w + M_m)g = F_{02}(t) \end{aligned} \tag{2.42}$$

③ The third wheelset

$$\begin{aligned} &(M_w + \eta_m M_m)\ddot{Z}_{w3} + C_{pz}\dot{Z}_{w3} + (K_{pz} + \xi_m K_m)Z_{w3} - C_{pz}\dot{Z}_{t2} \\ &\quad - (K_{pz} + \xi_m K_m)Z_{t2} + C_{pz}l_t\dot{\beta}_{t2} + \left[K_{pz}l_t + \xi_m K_m(l_t - l_1 - l_2)\right]\beta_{t2} \\ &\quad + \xi_m K_m(l_1 + l_2)\beta_{m3} + 2p_3(t) - (M_w + M_m)g = F_{03}(t) \end{aligned} \tag{2.43}$$

④ The fourth wheelset

$$\begin{aligned} &(M_w + \eta_m M_m)\ddot{Z}_{w4} + C_{pz}\dot{Z}_{w4} + (K_{pz} + \xi_m K_m)Z_{w4} - C_{pz}\dot{Z}_{t2} \\ &\quad - (K_{pz} + \xi_m K_m)Z_{t2} - C_{pz}l_t\dot{\beta}_{t2} - \left[K_{pz}l_t + \xi_m K_m(l_t - l_1 - l_2)\right]\beta_{t2} \\ &\quad - \xi_m K_m(l_1 + l_2)\beta_{m4} + 2p_4(t) - (M_w + M_m)g = F_{04}(t) \end{aligned} \tag{2.44}$$

⑤ The first motor

$$\begin{aligned}&(J_m+\mu_m M_w l_1^2)\ddot{\beta}_{m1}-\mu_m C_{pz} l_1(\dot{Z}_{w1}-\dot{Z}_{t1}+l_t\dot{\beta}_{t1})-[K_m(l_1+l_2-\mu_m l_1)\\&\quad-\mu_m K_{pz} l_1](Z_{t1}-Z_{w1})+[K_m(l_t-l_1-l_2)(l_1+l_2-\mu_m l_1)\\&\quad-\mu_m K_{pz} l_1 l_t]\beta_{t1}+K_m(l_1+l_2)(l_1+l_2-\mu_m l_1)\beta_{m1}\\&\quad-\mu_m l_1[2p_1(t)-(M_w+M_m)g-F_{01}(t)]=0\end{aligned} \tag{2.45}$$

⑥ The second motor

$$\begin{aligned}&(J_m+\mu_m M_w l_1^2)\ddot{\beta}_{m2}+\mu_m C_{pz} l_1(\dot{Z}_{w2}-\dot{Z}_{t1}-l_t\dot{\beta}_{t1})+[K_m(l_1+l_2-\mu_m l_1)\\&\quad-\mu_m K_{pz} l_1](Z_{t1}-Z_{w2})+[K_m(l_t-l_1-l_2)(l_1+l_2-\mu_m l_1)\\&\quad-\mu_m K_{pz} l_1 l_t]\beta_{t1}+K_m(l_1+l_2)(l_1+l_2-\mu_m l_1)\beta_{m2}\\&\quad+\mu_m l_1[2p_2(t)-(M_w+M_m)g-F_{02}(t)]=0\end{aligned} \tag{2.46}$$

⑦ The third motor

$$\begin{aligned}&(J_m+\mu_m M_w l_1^2)\ddot{\beta}_{m3}-\mu_m C_{pz} l_1(\dot{Z}_{w3}-\dot{Z}_{t2}+l_t\dot{\beta}_{t2})-[K_m(l_1+l_2-\mu_m l_1)\\&\quad-\mu_m K_{pz} l_1](Z_{t2}-Z_{w3})+[K_m(l_t-l_1-l_2)(l_1+l_2-\mu_m l_1)\\&\quad-\mu_m K_{pz} l_1 l_t]\beta_{t2}+K_m(l_1+l_2)(l_1+l_2-\mu_m l_1)\beta_{m3}\\&\quad-\mu_m l_1[2p_3(t)-(M_w+M_m)g-F_{03}(t)]=0\end{aligned} \tag{2.47}$$

⑧ The fourth motor

$$\begin{aligned}&(J_m+\mu_m M_w l_1^2)\ddot{\beta}_{m4}+\mu_m C_{pz} l_1(\dot{Z}_{w4}-\dot{Z}_{t2}-l_t\dot{\beta}_{t2})+[K_m(l_1+l_2-\mu_m l_1)\\&\quad-\mu_m K_{pz} l_1](Z_{t2}-Z_{w4})+[K_m(l_t-l_1-l_2)(l_1+l_2-\mu_m l_1)\\&\quad-\mu_m K_{pz} l_1 l_t]\beta_{t2}+K_m(l_1+l_2)(l_1+l_2-\mu_m l_1)\beta_{m4}\\&\quad+\mu_m l_1[2p_4(t)-(M_w+M_m)g-F_{04}(t)]=0\end{aligned} \tag{2.48}$$

where

$$\eta_m=\frac{J_m}{J_m+M_m l_1^2} \tag{2.49}$$

$$\xi_m=\frac{J_m-M_m l_1 l_2}{J_m+M_m l_1^2} \tag{2.50}$$

$$\mu_m=\frac{M_m}{M_w+M_m} \tag{2.51}$$

$\eta_{\rm m}$ is the equivalent mass of the motor that added to the unsprung mass of the locomotive, here it is defined as the unsprung mass contribution coefficient of the motor system. $\xi_{\rm m}$ is the equivalent suspension stiffness of the motor that added to the primary suspension of the locomotive, it is defined as the primary suspension influence coefficient of the motor system. $\eta_{\rm m}$ and $\xi_{\rm m}$ can clearly show the effect of the motor system on the locomotive wheel–rail interaction [13].

4. *Equations of motion of the ballasted track subsystem*

The rails are usually treated as the simply supported Euler beam or Timoshenko beam with finite length. In fact, the result by the finite length beam model is close to that by the infinite beam model when the calculation length of the rail is sufficient long. The criterion for selecting the calculation length l of the rail as the simply supported beam will be proposed by numerical trials as shown in Sect. 4.4.

(1) Differential equations of rail modeled as Euler beam

Figure 2.17 shows the force analysis of a rail when it is modeled as a simply supported Euler beam. In the figure, p_i is the wheel–rail force ($i = 1{-}4$), which travels on the beam at train speed v, $F_{{\rm rs}i}$ $(i = 1{-}N)$ is the sleeper support force, N is the total number of the sleepers within the calculation length l. ox is the coordinate system fixed on the track, $o'x'$ is the moving coordinate system fixed on the vehicle. The conversion relation between the two coordinate systems is

$$x = x' + x_0 + vt \tag{2.52}$$

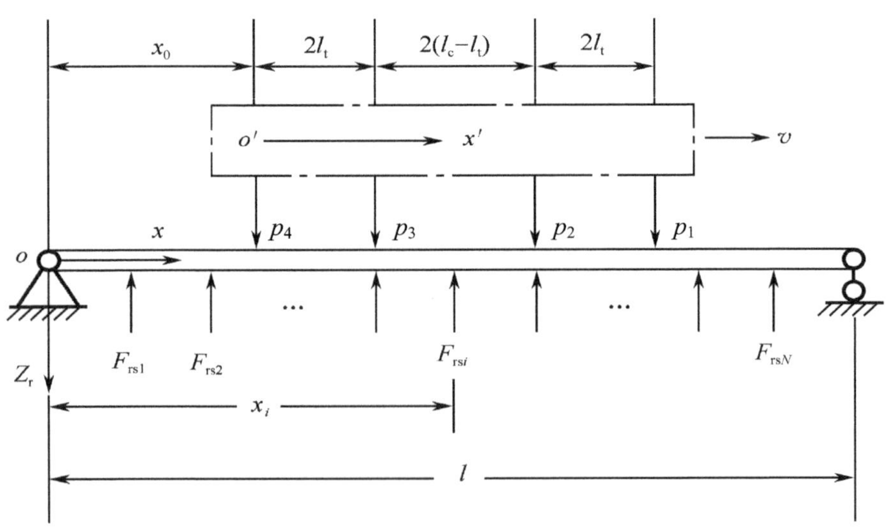

Fig. 2.17 Force analysis of the rail model

where x_0 is the initial coordinate of the fourth wheel in the track coordinate system, t is the time.

The rail deflection $Z_r(x,t)$ can be described as the following differential equation

$$EI_Y\frac{\partial^4 Z_r(x,t)}{\partial x^4}+m_r\frac{\partial^2 Z_r(x,t)}{\partial t^2}=-\sum_{i=1}^{N}F_{rsi}(t)\delta(x-x_i)+\sum_{j=1}^{4}p_j\delta(x-x_{wj}) \quad (2.53)$$

where EI_y is the rail beam bending stiffness (N m^2), δ denotes the Dirac function, and

$$F_{rsi}(t)=K_{pi}[Z_r(x_i,t)-Z_{si}(t)]+C_{pi}\left[\dot{Z}_r(x_i,t)-\dot{Z}_{si}(t)\right] \quad (2.54)$$

in which, $Z_{si}(t)$ is the vertical displacement of the sleeper (m).

The coordinate of each four wheel x_{wj} ($j=1-4$) can be written as

$$\begin{cases} x_{w1}(t)=x_0+2(l_c+l_t)+vt \\ x_{w2}(t)=x_0+2l_c+vt \\ x_{w3}(t)=x_0+2l_t+vt \\ x_{w4}(t)=x_0+vt \end{cases} \quad (2.55)$$

And the coordinate of the sleeper i can be written as

$$x_i=il_s \quad (i=1-N) \quad (2.56)$$

where l_s is the sleeper spacing (m).

By using the Ritz method [91], the fourth-order partial differential equation of the rail, Eq. (2.53), can be converted into the second-order ordinary differential equation. Considering the normalized coordinate $q_k(t)$ of the rail, the modal shape of the rail can be obtain by using the normalized modal function of a simply supported beam

$$Z_k(x)=\sqrt{\frac{2}{m_r l}}\sin\frac{k\pi x}{l} \quad (2.57)$$

The vertical displacement of the rail can then be expressed as

$$Z_r(x,t)=\sum_{k=1}^{NM}Z_k(x)q_k(t) \quad (2.58)$$

where NM is the mode number of the simply supported Euler beam.

Substituting Eq. (2.58) into Eq. (2.53), we can get

$$\sum_{k=1}^{NM} EI_Y \frac{\mathrm{d}^4 Z_k(x)}{\mathrm{d}x^4} q_k(t) + \sum_{k=1}^{NM} m_\mathrm{r} Z_k(x) \ddot{q}_k(t) = -\sum_{i=1}^{N} F_{\mathrm{rs}i}(t)\delta(x - x_i) + \sum_{j=1}^{4} p_j(t)\delta(x - x_{\mathrm{w}j}) \tag{2.59}$$

Multiplying Eq. (2.59) by $Z_h(x)$ $(h = 1, 2, \ldots, NM)$, and integrating it for x from 0 to l, yield

$$\begin{aligned} &\int_0^l EI_Y \frac{\mathrm{d}^4 Z_k(x)}{\mathrm{d}x^4} Z_k(x) q_k(t)\mathrm{d}x + \int_0^l m_\mathrm{r} Z_k(x) Z_k(x) \ddot{q}_k(t)\mathrm{d}x \\ &= -\sum_{i=1}^{N} \int_0^l F_{\mathrm{rs}i}(t) Z_k(x)\delta(x - x_i)\mathrm{d}x \\ &+ \sum_{j=1}^{4} \int_0^l p_j(t) Z_k(x)\delta(x - x_{\mathrm{w}j})\mathrm{d}x \quad (k = 1\text{–}NM) \end{aligned} \tag{2.60}$$

It is noted that the following property due to the modal orthogonality is applied in the derivation of Eq. (2.61):

$$\int_0^l Z_h(x) Z_k(x)\mathrm{d}x = 0 \quad (h \neq k) \tag{2.61}$$

By considering the character of the Dirac δ function, Eq. (2.60) can be expressed as

$$\begin{aligned} &m_\mathrm{r} \ddot{q}_k(t) \int_0^l Z_k^2(x)\mathrm{d}x + EI_Y q_k(t) \int_0^l Z_k(x) \frac{\mathrm{d}^4 Z_k(x)}{\mathrm{d}x^4}\mathrm{d}x \\ &= -\sum_{i=1}^{N} F_{\mathrm{rs}i}(t) Z_k(x_i) + \sum_{j=1}^{4} p_j(t) Z_k(x_{\mathrm{w}j}) \quad (k = 1\text{–}NM) \end{aligned} \tag{2.62}$$

As

$$\int_0^l Z_k^2(x)\mathrm{d}x = \frac{1}{m_\mathrm{r}} \tag{2.63}$$

$$\begin{aligned} \int_0^l Z_k(x) \frac{\mathrm{d}^4 Z_k(x)}{\mathrm{d}x^4}\mathrm{d}x &= \int_0^l \frac{2}{m_\mathrm{r} l}\left(\frac{k\pi}{l}\right)^4 \sin^2\frac{k\pi x}{l}\mathrm{d}x \\ &= \left(\frac{k\pi}{l}\right)^4 \int_0^l Z_k^2(x)\mathrm{d}x \\ &= \frac{1}{m_\mathrm{r}}\left(\frac{k\pi}{l}\right)^4 \end{aligned} \tag{2.64}$$

Thus, Eq. (2.62) can be written as

$$\ddot{q}_k(t) + \frac{EI_Y}{m_r}\left(\frac{k\pi}{l}\right)^4 q_k(t) = -\sum_{i=1}^{N} F_{rsi}(t)Z_k(x_i) + \sum_{j=1}^{4} p_j(t)Z_k(x_{wj}) \quad (k = 1\text{–}NM) \tag{2.65}$$

The above equation is the basic form of the rail second-order ordinary differential equation of the rail.

Substituting Eq. (2.58) into Eq. (2.54), yields

$$F_{rsi}(t) = C_{pi}\sum_{h=1}^{NM} Z_h(x_i)\dot{q}_h(t) + K_{pi}\sum_{h=1}^{NM} Z_h(x_i)q_h(t) - C_{pi}\dot{Z}_{si}(t) - K_{pi}Z_{si}(t) \tag{2.66}$$

Then substituting Eq. (2.66) into Eq. (2.65), we can obtain

$$\begin{aligned}
&\ddot{q}_k(t) + \sum_{i=1}^{N} C_{pi}Z_k(x_i)\sum_{h=1}^{NM} Z_h(x_i)\dot{q}_h(t) + \frac{EI_Y}{m_r}\left(\frac{k\pi}{l}\right)^4 q_k(t) \\
&+ \sum_{i=1}^{N} K_{pi}Z_k(x_i)\sum_{h=1}^{NM} Z_h(x_i)q_h(t) - \sum_{i=1}^{N} C_{pi}Z_k(x_i)\dot{Z}_{si}(t) \\
&- \sum_{i=1}^{N} K_{pi}Z_k(x_i)Z_{si}(t) = \sum_{j=1}^{4} p_j(t)Z_k(x_{wj}) \quad (k = 1\text{–}NM)
\end{aligned} \tag{2.67}$$

This is the final form of the second-order ordinary differential equation of the rail described as the Euler beam.

It should be noted that the selection of the mode number *NM* must comply with the criterion that the intercepted highest frequency relating to *NM* is higher than the analyzed frequency of the rail. We can also determine the *NM* by using the numerical trials, which will be given in Sect. 4.4.

(2) Differential equations of rail modeled as Timoshenko beam

The Timoshenko beam model can consider the rotatory inertia of the beam cross section and beam deformation due to the shear force. Figure 2.18 shows the free-body diagram of a rail element dx when it is modeled as a Timoshenko beam. In the figure, $M(x, t)$ and $Q(x, t)$ are the bending moment and vertical shear force applied on the rail element, ψ is the rotational angle of the cross section due to the bending moment; β is the shear angle at the neutral axis of the same cross section. Thus, the total rotational angle of the cross section can be described as

$$\frac{\partial Z_r(x,t)}{\partial x} = \psi + \beta \tag{2.68}$$

Based on the Timoshenko beam theory, the differential equations for the vertical rail deflection $Z_r(x,t)$ and the shear deformation $\psi(x, t)$ are written as

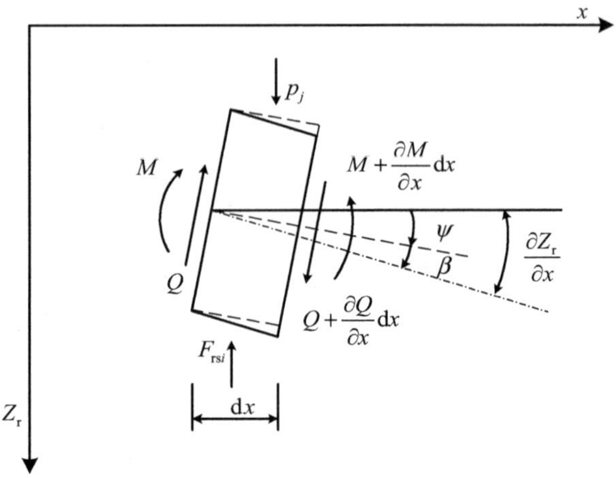

Fig. 2.18 Free-body diagram of a rail element

$$m_r \frac{\partial^2 Z_r(x,t)}{\partial t^2} + \kappa A_r G_r \left[\frac{\partial \psi(x,t)}{\partial x} - \frac{\partial^2 Z_r(x,t)}{\partial x^2} \right] = -\sum_{i=1}^{N} F_{rsi}(t)\delta(x - x_i) + \sum_{j=1}^{4} p_j(t)\delta(x - x_{wj}) \tag{2.69}$$

$$\rho_r I_Y \frac{\partial^2 \psi(x,t)}{\partial t^2} + \kappa A_r G_r \left[\psi - \frac{\partial Z_r(x,t)}{\partial x} \right] - EI_Y \frac{\partial^2 \psi(x,t)}{\partial x^2} = 0 \tag{2.70}$$

where A_r is the area of the rail cross section, G_r is the rail shear modulus, κ is the shear parameter depending on the shape of the rail cross section, ρ_r is the rail density.

Based on the Ritz method, the fourth-order partial differential equations, Eqs. (2.69) and (2.70), can also be converted into the second-order ordinary differential equations. Considering the normalized coordinate of the rail shear deformation $w_k(t)$, the shear deformation modal shape can be obtain by using the normalized modal function of a simply supported beam

$$\Psi_k(x) = \sqrt{\frac{2}{\rho_r I_Y l}} \cos\left(\frac{k\pi}{l} x\right) \tag{2.71}$$

The shear deformation of the rail can then be written as

$$\psi(x,t) = \sum_{k=1}^{NM} \Psi_k(x) w_k(t) \tag{2.72}$$

Substituting Eqs. (2.58) and (2.72) into Eqs. (2.69) and (2.70), we can get

$$\begin{aligned}&\sum_{k=1}^{NM} m_r Z_k(x)\ddot{q}_k(t) + \kappa A_r G_r\left[\sum_{k=1}^{NM}\frac{d\Psi_k(x)}{dx}w_k(t) - \sum_{k=1}^{NM}\frac{d^2 Z_k(x)}{dx^2}q_k(t)\right]\\&= -\sum_{i=1}^{N} F_{rsi}(t)\delta(x-x_i) + \sum_{j=1}^{4} p_j(t)\delta(x-x_{wj})\end{aligned} \tag{2.73}$$

$$\begin{aligned}&\sum_{k=1}^{NM} \rho_r I_Y \Psi_k(x)\ddot{w}_k(t) + \kappa A_r G_r\left[\sum_{k=1}^{NM}\Psi_k(x)w_k(t) - \sum_{k=1}^{NM}\frac{dZ_k(x)}{dx}q_k(t)\right]\\&\quad-\sum_{k=1}^{NM} EI_Y\frac{d^2\Psi_k(x)}{dx^2}w_k(t) = 0\end{aligned} \tag{2.74}$$

Multiplying Eq. (2.73) by $Z_h(x)$ $(h = 1, 2, \ldots, NM)$, and integrating it for x from 0 to l; also multiplying Eq. (2.74) by $\Psi_h(x)$ $(h = 1, 2, \ldots, NM)$, and integrating it for x from 0 to l, we can get

$$\begin{aligned}&\int_0^l m_r Z_k(x)Z_k(x)\ddot{q}_k(t)dx + \int_0^l \kappa A_r G_r\frac{d\Psi_k(x)}{dx}Z_k(x)w_k(t)dx - \int_0^l \kappa A_r G_r\frac{d^2 Z_k(x)}{dx^2}Z_k(x)q_k(t)dx\\&= -\sum_{i=1}^{N}\int_0^l F_{rsi}(t)Z_k(x)\delta(x-x_i)dx + \sum_{j=1}^{4}\int_0^l p_j(t)Z_k(x)\delta(x-x_{wj})dx \quad (k = 1\text{–}NM)\end{aligned} \tag{2.75}$$

$$\begin{aligned}&\int_0^l \rho_r I_Y\Psi_k(x)\Psi_k(x)\ddot{w}_k(t)dx + \int_0^l \kappa A_r G_r\Psi_k(x)\Psi_k(x)w_k(t)dx - \int_0^l \kappa A_r G_r\frac{dZ_k(x)}{dx}\Psi_k(x)q_k(t)dx\\&-\int_0^l EI_Y\frac{d^2\Psi_k(x)}{dx^2}\Psi_k(x)w_k(t)dx = 0 \quad (k = 1\text{–}NM)\end{aligned} \tag{2.76}$$

Here is applied the following properties due to the modal orthogonality in the derivation of Eqs. (2.75) and (2.76):

$$\int_0^l \Psi_h(x)\Psi_k(x)dx = 0 \quad (h \neq k) \tag{2.77}$$

$$\int_0^l Z_h(x)\Psi_k(x)dx = 0 \tag{2.78}$$

Considering the character of the Dirac δ function, Eqs. (2.75) and (2.76) can be rewritten as

$$
\begin{aligned}
& m_{\mathrm{r}}\ddot{q}_k(t)\int_0^l Z_k^2(x)\mathrm{d}x+\kappa A_{\mathrm{r}}G_{\mathrm{r}}w_k(t)\int_0^l \frac{\mathrm{d}\Psi_k(x)}{\mathrm{d}x}Z_k(x)\mathrm{d}x-\kappa A_{\mathrm{r}}G_{\mathrm{r}}q_k(t)\int_0^l \frac{\mathrm{d}^2 Z_k(x)}{\mathrm{d}x^2}Z_k(x)\mathrm{d}x \\
& =-\sum_{i=1}^{N}F_{\mathrm{rs}i}(t)Z_k(x_i)+\sum_{j=1}^{4}p_j(t)Z_k(x_{\mathrm{w}j}) \quad (k=1\text{–}NM)
\end{aligned}
\tag{2.79}
$$

$$
\begin{aligned}
& \rho_{\mathrm{r}}I_Y\ddot{w}_k(t)\int_0^l \Psi_k^2(x)\mathrm{d}x+\kappa A_{\mathrm{r}}G_{\mathrm{r}}w_k(t)\int_0^l \Psi_k^2(x)\mathrm{d}x-\kappa A_{\mathrm{r}}G_{\mathrm{r}}q_k(t)\int_0^l \frac{\mathrm{d}Z_k(x)}{\mathrm{d}x}\Psi_k(x)\mathrm{d}x \\
& \quad -EI_Yw_k(t)\int_0^l \frac{\mathrm{d}^2\Psi_k(x)}{\mathrm{d}x^2}\Psi_k(x)\mathrm{d}x=0 \quad (k=1\text{–}NM)
\end{aligned}
\tag{2.80}
$$

As

$$\int_0^l \frac{\mathrm{d}\Psi_k(x)}{\mathrm{d}x}Z_k(x)\mathrm{d}x=-\frac{k\pi}{l}\sqrt{\frac{1}{m_{\mathrm{r}}\rho_{\mathrm{r}}I_Y}} \tag{2.81}$$

$$\int_0^l \frac{\mathrm{d}^2 Z_k(x)}{\mathrm{d}x^2}Z_k(x)\mathrm{d}x=-\frac{1}{m_{\mathrm{r}}}\left(\frac{k\pi}{l}\right)^2 \tag{2.82}$$

$$\int_0^l \Psi_k^2(x)\mathrm{d}x=\frac{1}{\rho_{\mathrm{r}}I_Y} \tag{2.83}$$

$$\int_0^l \frac{\mathrm{d}Z_k(x)}{\mathrm{d}x}\Psi_k(x)\mathrm{d}x=\frac{k\pi}{l}\sqrt{\frac{1}{m_{\mathrm{r}}\rho_{\mathrm{r}}I_Y}} \tag{2.84}$$

$$\int_0^l \frac{\mathrm{d}^2\Psi_k(x)}{\mathrm{d}x^2}\Psi_k(x)\mathrm{d}x=-\left(\frac{k\pi}{l}\right)^2\frac{1}{\rho_{\mathrm{r}}I_Y} \tag{2.85}$$

Equations (2.79) and (2.80) can be simplified as

$$
\begin{aligned}
& \ddot{q}_k(t)+\kappa A_{\mathrm{r}}G_{\mathrm{r}}\frac{1}{m_{\mathrm{r}}}\left(\frac{k\pi}{l}\right)^2 q_k(t)-\kappa A_{\mathrm{r}}G_{\mathrm{r}}\sqrt{\frac{1}{m_{\mathrm{r}}\rho_{\mathrm{r}}I_Y}}\left(\frac{k\pi}{l}\right)w_k(t) \\
& =-\sum_{i=1}^{N}F_{\mathrm{rs}i}(t)Z_k(x_i)+\sum_{j=1}^{4}p_j(t)Z_k(x_{\mathrm{w}j}) \quad (k=1\text{–}NM)
\end{aligned}
\tag{2.86}
$$

$$\ddot{w}_k(t)+\kappa A_{\mathrm{r}}G_{\mathrm{r}}\frac{1}{\rho_{\mathrm{r}}I_Y}w_k(t)+\frac{E}{\rho_{\mathrm{r}}}\left(\frac{k\pi}{l}\right)^2 w_k(t)-\kappa A_{\mathrm{r}}G_{\mathrm{r}}\frac{k\pi}{l}\sqrt{\frac{1}{m_{\mathrm{r}}\rho_{\mathrm{r}}I_Y}}q_k(t)=0 \quad (k=1\text{–}NM) \tag{2.87}$$

The above two equations are the basic form of the second-order ordinary differential equations of the rail modeled as the Timoshenko beam.

Substituting Eq. (2.66) into Eq. (2.86), we can get the detailed forms of the rail differential equations as follows:

$$\begin{aligned}&\ddot{q}_k(t) + \sum_{i=1}^{N} C_{\mathrm{p}i} Z_k(x_i) \sum_{h=1}^{NM} Z_h(x_i)\dot{q}_h(t) + \sum_{i=1}^{N} K_{\mathrm{p}i} Z_k(x_i) \sum_{h=1}^{NM} Z_h(x_i) q_h(t) \\ &\quad + \frac{\kappa A_{\mathrm{r}} G_{\mathrm{r}}}{m_{\mathrm{r}}}\left(\frac{k\pi}{l}\right)^2 q_k(t) - \kappa A_{\mathrm{r}} G_{\mathrm{r}} \frac{k\pi}{l}\sqrt{\frac{1}{m_{\mathrm{r}}\rho_{\mathrm{r}} I_Y}} w_k(t) - \sum_{i=1}^{N} C_{\mathrm{p}i} Z_k(x_i)\dot{Z}_{\mathrm{s}i}(t) \\ &\quad - \sum_{i=1}^{N} K_{\mathrm{p}i} Z_k(x_i) Z_{\mathrm{s}i}(t) = \sum_{j=1}^{4} p_j(t) Z_k(x_{\mathrm{w}j}) \quad (k = 1\text{–}NM)\end{aligned} \tag{2.88}$$

$$\begin{aligned}&\ddot{w}_k(t) + \frac{1}{\rho_{\mathrm{r}} I_Y}[\kappa A_{\mathrm{r}} G_{\mathrm{r}} + EI_Y(\frac{k\pi}{l})^2] w_k(t) - \kappa A_{\mathrm{r}} G_{\mathrm{r}} \frac{k\pi}{l}\sqrt{\frac{1}{m_{\mathrm{r}}\rho_{\mathrm{r}} I_Y}} q_k(t) = 0 \\ &(k = 1\text{–}NM)\end{aligned} \tag{2.89}$$

(3) Equation of motion of the sleeper

The equation of vertical motion of the sleeper i reads

$$\begin{aligned}&K_{\mathrm{p}i}[Z_{\mathrm{r}}(x_i, t) - Z_{\mathrm{s}i}(t)] + C_{\mathrm{p}i}\left[\dot{Z}_{\mathrm{r}}(x_i, t) - \dot{Z}_{\mathrm{s}i}(t)\right] \\ &\quad - K_{\mathrm{b}i}[Z_{\mathrm{s}i}(t) - Z_{\mathrm{b}i}(t)] - C_{\mathrm{b}i}\left[\dot{Z}_{\mathrm{s}i}(t) - \dot{Z}_{\mathrm{b}i}(t)\right] = M_{\mathrm{s}i}\ddot{Z}_{\mathrm{s}i}(t)\end{aligned} \tag{2.90}$$

Substituting Eq. (2.58) into Eq. (2.90), yields

$$\begin{aligned}&M_{\mathrm{s}i}\ddot{Z}_{\mathrm{s}i}(t) + (C_{\mathrm{p}i} + C_{\mathrm{b}i})\dot{Z}_{\mathrm{s}i}(t) + (K_{\mathrm{p}i} + K_{\mathrm{b}i}) Z_{\mathrm{s}i}(t) - C_{\mathrm{b}i}\dot{Z}_{\mathrm{b}i}(t) - K_{\mathrm{b}i} Z_{\mathrm{b}i}(t) \\ &\quad - C_{\mathrm{p}i}\sum_{h=1}^{NM} Z_h(x_i)\dot{q}_h(t) - K_{\mathrm{p}i}\sum_{h=1}^{NM} Z_h(x_i) q_h(t) = 0 \quad (i = 1\text{–}N)\end{aligned} \tag{2.91}$$

(4) Equation of motion of the ballast

The forces acting on the ith ballast block (Fig. 2.19) include the sleeper force $F_{\mathrm{bs}i}$, the subgrade force $F_{\mathrm{bf}i}$, and the vertical shear forces between the neighboring ballast bodies ($F_{\mathrm{bbl}i}$, $F_{\mathrm{bbr}i}$). Thus the vertical motion equation of the ith ballast block reads

$$F_{\mathrm{bs}i} - F_{\mathrm{bf}i} - F_{\mathrm{bbl}i} - F_{\mathrm{bbr}i} = M_{\mathrm{b}i}\ddot{Z}_{\mathrm{b}i}(t) \tag{2.92}$$

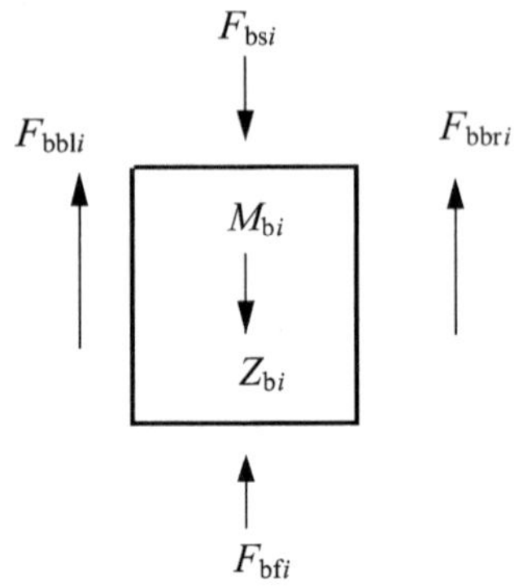

Fig. 2.19 Free-body diagram of the ballast block

The forces acting on the ballast block can be calculated by

$$\begin{cases} F_{\mathrm{bs}i} = K_{\mathrm{b}i}[Z_{\mathrm{s}i}(t) - Z_{\mathrm{b}i}(t)] + C_{\mathrm{b}i}\left[\dot{Z}_{\mathrm{s}i}(t) - \dot{Z}_{\mathrm{b}i}(t)\right] \\ F_{\mathrm{bf}i} = K_{\mathrm{f}i}Z_{\mathrm{b}i}(t) + C_{\mathrm{f}i}\dot{Z}_{\mathrm{b}i}(t) \\ F_{\mathrm{bbl}i} = K_{\mathrm{w}i}\left[Z_{\mathrm{b}i}(t) - Z_{\mathrm{b}(i-1)}(t)\right] + C_{\mathrm{w}i}\left[\dot{Z}_{\mathrm{b}i}(t) - \dot{Z}_{\mathrm{b}(i-1)}(t)\right] \\ F_{\mathrm{bbr}i} = K_{\mathrm{w}i}\left[Z_{\mathrm{b}i}(t) - Z_{\mathrm{b}(i+1)}(t)\right] + C_{\mathrm{w}i}\left[\dot{Z}_{\mathrm{b}i}(t) - \dot{Z}_{\mathrm{b}(i+1)}(t)\right] \end{cases} \tag{2.93}$$

Substituting Eq. (2.93) into Eq. (2.92), yields

$$\begin{aligned} & M_{\mathrm{b}i}\ddot{Z}_{\mathrm{b}i}(t) + (C_{\mathrm{b}i} + C_{\mathrm{f}i} + 2C_{\mathrm{w}i})\dot{Z}_{\mathrm{b}i}(t) + (K_{\mathrm{b}i} + K_{\mathrm{f}i} + 2K_{\mathrm{w}i})Z_{\mathrm{b}i}(t) - C_{\mathrm{b}i}\dot{Z}_{\mathrm{s}i}(t) - K_{\mathrm{b}i}Z_{\mathrm{s}i}(t) \\ & \quad - C_{\mathrm{w}i}\dot{Z}_{\mathrm{b}(i+1)}(t) - K_{\mathrm{w}i}Z_{\mathrm{b}(i+1)}(t) - C_{\mathrm{w}i}\dot{Z}_{\mathrm{b}(i-1)}(t) - K_{\mathrm{w}i}Z_{\mathrm{b}(i-1)}(t) = 0 \quad (i = 1\text{–}N) \end{aligned} \tag{2.94}$$

The boundary conditions are

$$\begin{cases} Z_{\mathrm{b}0} = \dot{Z}_{\mathrm{b}0} = 0 \\ Z_{\mathrm{b}(N+1)} = \dot{Z}_{\mathrm{b}(N+1)} = 0 \end{cases} \tag{2.95}$$

5. *Equations of motion of the ballastless track subsystem*

(1) Equation of motion of the long-sleeper embedded track

The long-sleeper embedded track model only considers the vibrations of the rails, regardless of the vibrations of sleepers. Thus, the equation of motion of the Euler beam model for the rail can be deduced from Eq. (2.67), and the equation of motion of the Timoshenko beam model for the rail can be obtained from Eqs. (2.88) and (2.89), as long as setting $Z_{\mathrm{s}i}(t) = \dot{Z}_{\mathrm{s}i}(t) = 0$. Then Eq. (2.67) becomes

$$\begin{aligned}&\ddot{q}_k(t)+\sum_{i=1}^{N}C_{\mathrm{p}i}Z_k(x_i)\sum_{h=1}^{NM}Z_h(x_i)\dot{q}_h(t)+\frac{EI_Y}{m_{\mathrm{r}}}\left(\frac{k\pi}{l}\right)^4 q_k(t)\\&+\sum_{i=1}^{N}K_{\mathrm{p}i}Z_k(x_i)\sum_{h=1}^{NM}Z_h(x_i)q_h(t)=\sum_{j=1}^{4}p_j(t)Z_k(x_{\mathrm{w}j})\quad(k=1\text{–}NM)\end{aligned}\tag{2.96}$$

Equation (2.88) becomes

$$\begin{aligned}&\ddot{q}_k(t)+\sum_{i=1}^{N}C_{\mathrm{p}i}Z_k(x_i)\sum_{h=1}^{NM}Z_h(x_i)\dot{q}_h(t)+\sum_{i=1}^{N}K_{\mathrm{p}i}Z_k(x_i)\sum_{h=1}^{NM}Z_h(x_i)q_h(t)\\&+\frac{\kappa A_{\mathrm{r}}G_{\mathrm{r}}}{m_{\mathrm{r}}}\left(\frac{k\pi}{l}\right)^2 q_k(t)-\kappa A_{\mathrm{r}}G_{\mathrm{r}}\frac{k\pi}{l}\sqrt{\frac{1}{m_{\mathrm{r}}\rho_{\mathrm{r}}I_Y}}w_k(t)=\sum_{j=1}^{4}p_j(t)Z_k(x_{\mathrm{w}j})\quad(k=1\text{–}NM)\end{aligned}\tag{2.97}$$

(2) Equation of motion of the elastic supporting block track

The elastic supporting block track model considers the vibrations of rails and the supporting blocks. The equation of motion of the Euler beam model for the rail can be seen in Eq. (2.67), and the equation of motion of the Timoshenko beam model for the rail can be seen in Eqs. (2.88) and (2.89).

The equation of motion of the elastic supporting block is similar to the sleeper, which reads

$$\begin{aligned}&M_{\mathrm{s}i}\ddot{Z}_{\mathrm{s}i}(t)+\left(C_{\mathrm{p}i}+C_{\mathrm{b}i}\right)\dot{Z}_{\mathrm{s}i}(t)+\left(K_{\mathrm{p}i}+K_{\mathrm{b}i}\right)Z_{\mathrm{s}i}(t)\\&-C_{\mathrm{p}i}\sum_{h=1}^{NM}Z_h(x_i)\dot{q}_h(t)-K_{\mathrm{p}i}\sum_{h=1}^{NM}Z_h(x_i)q_h(t)=0\quad(i=1\text{–}N)\end{aligned}\tag{2.98}$$

(3) Equation of motion of the slab track

The slab track model considers the vertical vibrations of the rails and the concrete slabs. The equation of motion of the Euler beam model for the rail is the same as Eq. (2.67), and the equation of motion of the Timoshenko beam model for the rail same as Eqs. (2.88) and (2.89).

The track slab is simplified as a finite length free beam on continuous base, as shown in Fig. 2.20. So the equation of vertical motion of the slab can be written as

$$\begin{aligned}&E_{\mathrm{s}}I_{\mathrm{s}}\frac{\partial^4 Z_{\mathrm{s}}(x,t)}{\partial x^4}+\frac{M_{\mathrm{s}}}{L_{\mathrm{s}}}\frac{\partial^2 Z_{\mathrm{s}}(x,t)}{\partial t^2}+c_{\mathrm{s}}\frac{\partial Z_{\mathrm{s}}(x,t)}{\partial t}+k_{\mathrm{s}}Z_{\mathrm{s}}(x,t)\\&=\sum_{j=1}^{n_0}F_{\mathrm{rs}j}(t)\delta\left(x-x_j\right)\end{aligned}\tag{2.99}$$

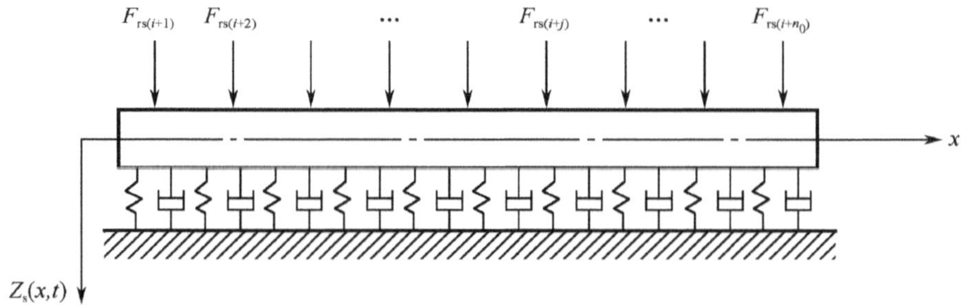

Fig. 2.20 Vertical vibration model for the track slab

where E_sI_s is the vertical bending stiffness of the concrete slab (N m^2), $Z_s(x, t)$ is the slab vertical deflection (m), M_s is the slab mass (kg), L_s is the length of the slab (m), while K_s and C_s are the vertical stiffness and damping of the CAM layer underneath the slab, n_0 is the number of rail fastenings on a track slab.

By using the Ritz method, the fourth-order partial differential equation, Eq. (2.99), can be converted into the second-order ordinary differential equation. The modal functions of a free end beam read [92]

$$\begin{cases} X_1(x) = 1 \\ X_2(x) = \sqrt{3}\left(1 - \frac{2x}{L_s}\right) \\ X_m(x) = (ch\beta_m x + \cos\beta_m x) - C_m(sh\beta_m x + \sin\beta_m x) \quad (m \geq 3) \end{cases} \tag{2.100}$$

where C_m and β_m are, respectively, the frequency coefficient and the function coefficient of a beam with free–free boundary condition. Table 2.3 lists the values of C_m and $\beta_m L_s$.

The vertical deflection of the concrete slab reads

$$Z_s(x,t) = \sum_{n=1}^{NMS} X_n(x)T_n(t) \tag{2.101}$$

Substituting Eq. (2.101) into Eq. (2.99), and multiplying it by $X_p(x)$ (p = 1–NMS), and then integrating it for x from 0 to l. By using the properties of the modal orthogonality and the Dirac δ function, we can get

$$\begin{aligned} &\frac{M_s}{L_s}\ddot{T}_n(t)\int_0^{L_s} X_n^2(x)dx + c_s\dot{T}_n(t)\int_0^{L_s} X_n^2(x)dx + k_sT_n(t)\int_0^{L_s} X_n^2(x)dx \\ &+ E_sI_sT_n(t)\int_0^{L_s} X_n(x)\frac{d^4X_n(x)}{dx^4}dx = \sum_{j=1}^{n0} F_{rsj}(t)X_n(x_j) \end{aligned} \tag{2.102}$$

Table 2.3 The frequency and function coefficient of a beam with free–free boundary condition

m	1	2	3	4	5	≥ 6
C_m	–	–	0.982502	1.000777	0.999966	1.000000
$\beta_m L_s$	0	0	4.73004	7.85320	10.9956	$(2m-3)\pi/2$

As

$$\int_0^{L_s} X_n^2(x)dx = L_s \tag{2.103}$$

$$\int_0^{L_s} X_n(x)\frac{d^4X_n(x)}{dx^4}dx = L_s\beta_n^4 \tag{2.104}$$

Equation (2.102) can be simplified as

$$\ddot{T}_n(t) + \frac{c_sL_s}{M_s}\dot{T}_n(t) + \frac{k_s + E_sI_s\beta_n^4}{M_s}\cdot L_s \cdot T_n(t) = \sum_{j=1}^{n0}\frac{F_{rsj}(t)}{M_s}X_n(x_j) \tag{2.105}$$

This is the basic form of the second-order ordinary differential equations (n = 1–NMS) of the beam-based slab model.

6. *Vehicle–track vertical coupling relation*

In vertical direction, the vehicle and the track is coupled with the wheel–rail contact. The vertical wheel–rail contact force is calculated using the nonlinear Hertzian elastic contact theory. It reads

$$p(t) = \left[\frac{1}{G}\delta Z(t)\right]^{3/2} \tag{2.106}$$

where G is the wheel–rail contact coefficient (m/N$^{2/3}$), $\delta Z(t)$ is the elastic compressing amount at the wheel–rail contact point (m).

For the wheel with cone tread, G can be chosen as

$$G = 4.57R^{-0.149} \times 10^{-8}\left(\mathrm{m/N^{2/3}}\right) \tag{2.107}$$

For the wheel with worn tread, G can be chosen as

$$G = 3.86R^{-0.115} \times 10^{-8}\ (\mathrm{m/N^{2/3}}) \tag{2.108}$$

where R is the wheel radius (m).

The wheel–rail elastic compressing amount is calculated by

$$\delta Z(t) = Z_{wj}(t) - Z_r(x_{wj}, t) \quad (j = 1\text{–}4) \tag{2.109}$$

where $Z_{wj}(t)$ is the vertical displacement of the jth wheel (m), $Z_r(x_{wj}, t)$ is the vertical displacement of the rail under the jth wheel (m).

It is noted that $\delta Z(t) < 0$ indicates the wheel separating from the rail, thus the wheel–rail force $p(t) = 0$.

When the rail irregularity $Z_0(t)$ is considered, the wheel–rail force can be expressed as

$$p_j(t) = \begin{cases} \left\{\frac{1}{G}\left[Z_{wj}(t) - Z_r(x_{wj}, t) - Z_0(t)\right]\right\}^{3/2} \\ 0 \quad \text{(Loss of wheel–rail contact)} \end{cases} \tag{2.110}$$

The wheel–rail contact stress reads

$$\sigma(t) = S[p(t)]^{1/3} \tag{2.111}$$

where S is the stress coefficient ($N^{2/3}/m^2$) determined by the Hertzian elastic contact theory.

If R is in the range of 0.15–0.6 m, S can be chosen by

$$S = \begin{cases} 2.49R^{-0.251} \times 10^7 & \text{(Cone-tread wheel)} \\ 1.49R^{-0.376} \times 10^7 & \text{(Worn-tread wheel)} \end{cases} \tag{2.112}$$

2.3 Vehicle–Track Spatially Coupled Dynamics Model

The vehicle–track vertically coupled dynamics models established in Sect. 2.2 are further extended to the vehicle–track spatially coupled dynamics model by considering the lateral motions of the vehicle–track system, which provide the theoretical tool for studying the vertical and lateral dynamic performances of the vehicle–track coupled system.

2.3.1 Physical Model

This section presents modeling details of the vehicle–track spatially coupled dynamics models for typical passenger vehicle–track system, freight wagon–track system, and locomotive–track system, where the ballasted track model is presented at first, and the ballastless track models will be depicted at the end of this section.

1. *Passenger vehicle–track spatially coupled dynamics model*

Nowadays, various types of railway vehicles with different traveling speeds are designed around the world, while most of the passenger vehicles have two bogies, four wheelsets, and two suspensions. In this section, a vehicle–track spatially coupled dynamics model for a typical passenger vehicle and a ballasted track is established, as shown in Figs. 2.21, 2.22, and 2.23. The notations of the symbols in the figures are given in Tables 2.4 and 2.5.

In the vehicle sub-model, the car body is supported on two double-axle bogies at each end. The bogie frames are linked with the wheelsets through the primary suspensions and linked with the car body through the secondary suspensions. Three-dimensional spring–damper elements are used to represent the primary and the secondary suspensions. Yaw dampers and anti-roll springs are considered in the secondary suspensions. Furthermore, the lateral clearances between the car body and the stopblocks on the bogie frames are also considered in the secondary suspensions. The vehicle is assumed to move along the track with a constant traveling

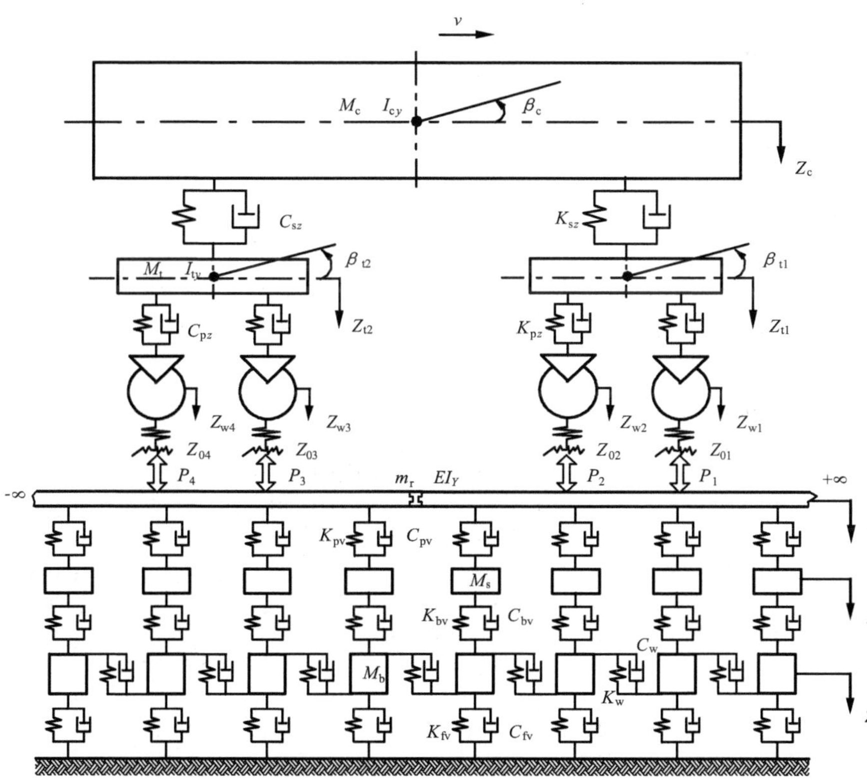

Fig. 2.21 Passenger vehicle–track spatially coupled model (side view) (Reprinted from Ref. [36], Copyright 2009, with permission from Taylor & Francis.)

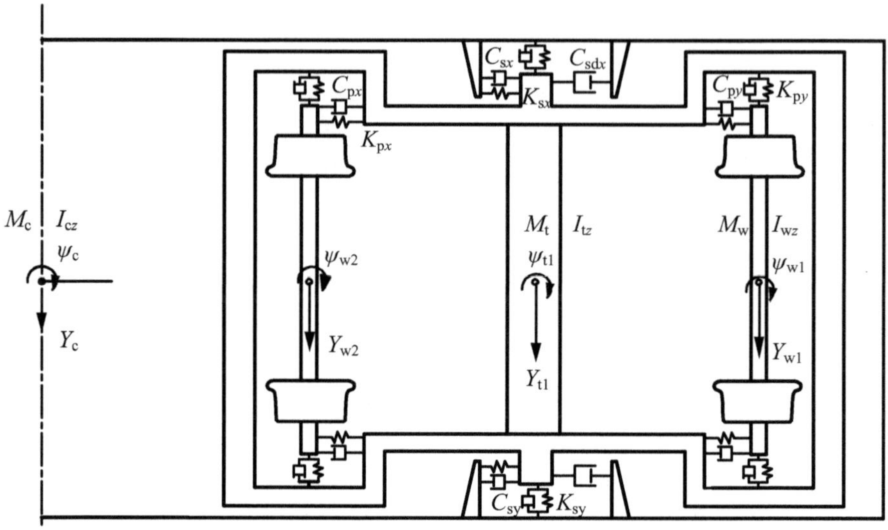

Fig. 2.22 Passenger vehicle–track spatially coupled model (top view) (Reprinted from Ref. [36], Copyright 2009, with permission from Taylor & Francis.)

speed. Each component of the vehicle has five DOFs: the vertical displacement Z, the lateral displacement Y, the roll angle Φ, the yaw angle ψ, and the pitch angle β with respect to its center of mass (the pitch for a wheelset corresponds to the variation of rotation around its mean rotational speed). All angles are assumed to be small, which simplifies the kinematics and the equations of motion. As a result, the total DOFs of the passenger vehicle sub-model are 35, as shown in Table 2.6.

The typical ballasted track model consists of rails, rail pads, sleepers, ballast, and subgrade. Both the left and right rails are treated as continuous Euler or Timoshenko beams, which are discretely supported at rail–sleeper junctions by three layers of springs and dampers representing the elasticity and damping of rail pads and fastenings, ballast and subgrade, respectively. Three kinds of vibrations of the rails are considered: vertical, lateral, and torsional. The sleeper is assumed to be a rigid body with three DOFs of the vertical displacement, the lateral displacement, and the roll angle displacement. Lateral springs and dampers are considered to represent the lateral dynamic properties in the fastener system. Similarly, the lateral springs and dampers are used to represent the elasticity and damping property between the sleeper and the ballast in lateral direction. A five-parameter model of ballast under each rail-supporting point is adopted [90], only the vertical motion of the ballast mass is taken into account, see also Fig. 2.13. In order to account for the continuity and the coupling effects of interlocking ballast granules, a couple of shear stiffness and shear damping is introduced between adjacent ballast masses in the ballast model. Linear springs and dampers are hired to represent the subgrade supporting the ballast.

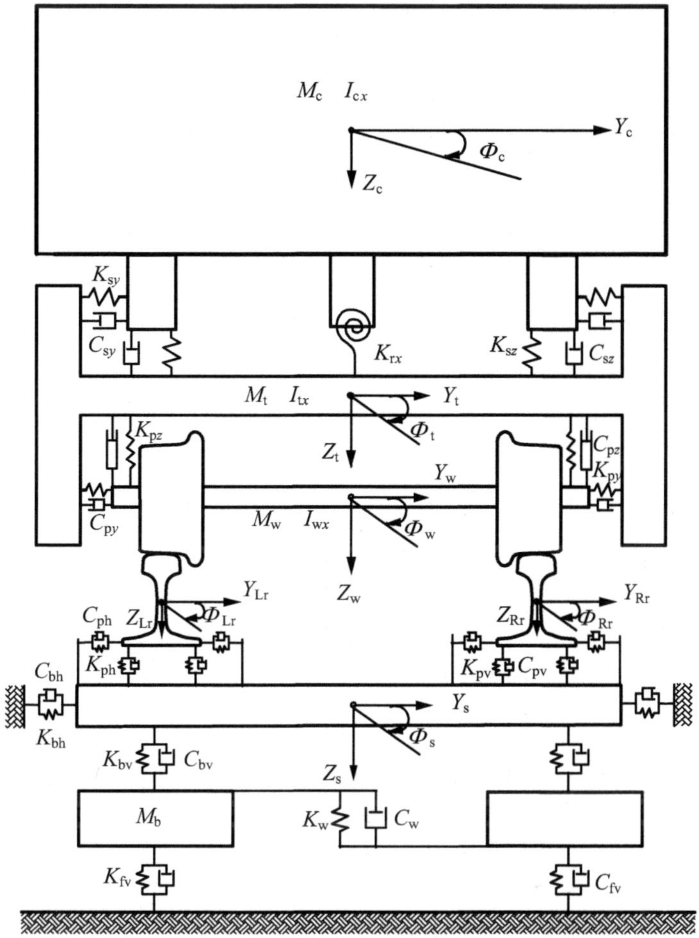

Fig. 2.23 Passenger vehicle–track spatially coupled model (end view) (Reprinted from Ref. [36], Copyright 2009, with permission from Taylor & Francis.)

2. *Freight wagon–track spatially coupled dynamics model*

The freight wagon–track spatially coupled dynamics model is established as shown in Figs. 2.24, 2.25 and 2.26, where a typical freight wagon with the three-piece bogie of Chinese Z8A type is considered. The Z8A type bogie comprises one bolster and two side frames. There are no spring and no viscous damping component in the primary suspensions. The side frames directly contact with the wheel axles in the vertical direction. There are longitudinal and lateral clearances between the axle boxes and the side frames. Therefore, the combination effect of the Coulomb friction and the clearances between the axle boxes and the side frames should be considered. In the secondary suspension, coil steel springs and friction

Table 2.4 Notations for parameters of the passenger vehicle sub-model

Notation	Parameter	Unit
M_c	Car body mass	kg
M_t	Bogie mass	kg
M_w	Wheelset mass	kg
I_{cx}	Mass moment of inertia of car body about *X*-axis	$kg \cdot m^2$
I_{cy}	Mass moment of inertia of car body about *Y*-axis	$kg \cdot m^2$
I_{cz}	Mass moment of inertia of car body about *Z*-axis	$kg \cdot m^2$
I_{tx}	Mass moment of inertia of bogie about *X*-axis	$kg \cdot m^2$
I_{ty}	Mass moment of inertia of bogie about *Y*-axis	$kg \cdot m^2$
I_{tz}	Mass moment of inertia of bogie about *Z*-axis	$kg \cdot m^2$
I_{wx}	Mass moment of inertia of wheelset about *X*-axis	$kg \cdot m^2$
I_{wy}	Mass moment of inertia of wheelset about *Y*-axis	$kg \cdot m^2$
I_{wz}	Mass moment of inertia of wheelset about *Z*-axis	$kg \cdot m^2$
K_{px}	Stiffness coefficient of primary suspension along *X*-axis	MN/m
K_{py}	Stiffness coefficient of primary suspension along *Y*-axis	MN/m
K_{pz}	Stiffness coefficient of primary suspension along *Z*-axis	MN/m
K_{sx}	Stiffness coefficient of secondary suspension along *X*-axis	MN/m
K_{sy}	Stiffness coefficient of secondary suspension along *Y*-axis	MN/m
K_{sz}	Stiffness coefficient of secondary suspension along *Z*-axis	MN/m
C_{pz}	Damping coefficient of primary suspension along *Z*-axis	$kN \cdot s/m$
C_{sy}	Damping coefficient of secondary suspension along *Y*-axis	$kN \cdot s/m$
C_{sz}	Damping coefficient of secondary suspension along *Z*-axis	$kN \cdot s/m$
l_c	Semi-longitudinal distance between bogies	m
l_t	Semi-longitudinal distance between wheelsets in bogie	m
R_0	Wheel radius	m

wedges are used between the side frames and two ends of the bolsters, which are modeled as linear springs and the Coulomb friction elements in both the vertical and the lateral directions. The nonlinear characteristics of these suspension components are taken into account in the models. The notations of the symbols in Figs. 2.24, 2.25, and 2.26 are given in Table 2.7.

The freight wagon is modeled as a multibody system. Both the car body and the wheelset have five DOFs: the vertical displacement *Z*, the lateral displacement *Y*, the roll angle Φ, the yaw angle ψ, and the pitch angle β with respect to its center of mass. The bolster only has one DOF: the yaw angle describing the rotation motion, because the bolster is connected with the car body at its center bowl allowing rotation about its vertical centerline. Three DOFs are considered to describe the movements of each side frame, i.e., the longitudinal displacement, the lateral displacement, and the yaw angle. The vertical displacement and the pitch angle of the side frame are depended on the movements of two relating wheelsets due to the direct contact of them in the vertical direction. Thus, the total DOFs of the wagon model are 39, as shown in Table 2.8.

Table 2.5 Notations for parameters of the ballasted track sub-model

Notation	Parameter	Unit
E	Elastic modulus of rail	N/m^2
ρ	Density of rail	kg/m^3
I_0	Torsional inertia of rail	m^4
I_y	Rail second moment of area about *Y*-axis	m^4
I_z	Rail second moment of area about *Z*-axis	m^4
GK	Rail torsional stiffness	N · m/rad
m_r	Rail mass per unit length	kg/m
M_s	Sleeper mass (half)	kg
K_{pv}	Fastener stiffness in vertical direction	N/m
K_{ph}	Fastener stiffness in lateral direction	N/m
C_{pv}	Fastener damping in vertical direction	N · s/m
C_{ph}	Fastener damping in lateral direction	N · s/m
l_s	Sleeper spacing	m
l_e	Effective support length of half sleeper	m
l_b	Sleeper width	m
ρ_b	Ballast density	kg/m^3
E_b	Elastic modulus of ballast	Pa
C_b	Ballast damping	N · s/m
K_w	Ballast shear stiffness	N/m
C_w	Ballast shear damping	N · s/m
α	Ballast stress distribution angle	°
h_b	Ballast thickness	m
E_f	Subgrade K_{30} modulus	Pa/m
C_f	Subgrade damping	N · s/m

Table 2.6 Degrees of freedom of passenger vehicle dynamics model

Vehicle component	Lateral motion	Vertical motion	Roll motion	Yaw motion	Pitch motion
Car body	Y_c	Z_c	ϕ_c	ψ_c	β_c
Front bogie frame	Y_{t1}	Z_{t1}	ϕ_{t1}	ψ_{t1}	β_{t1}
Rear bogie frame	Y_{t2}	Z_{t2}	ϕ_{t2}	ψ_{t2}	β_{t2}
First wheelset	Y_{w1}	Z_{w1}	ϕ_{w1}	ψ_{w1}	β_{w1}
Second wheelset	Y_{w2}	Z_{w2}	ϕ_{w2}	ψ_{w2}	β_{w2}
Third wheelset	Y_{w3}	Z_{w3}	ϕ_{w3}	ψ_{w3}	β_{w3}
Fourth wheelset	Y_{w4}	Z_{w4}	ϕ_{w4}	ψ_{w4}	β_{w4}

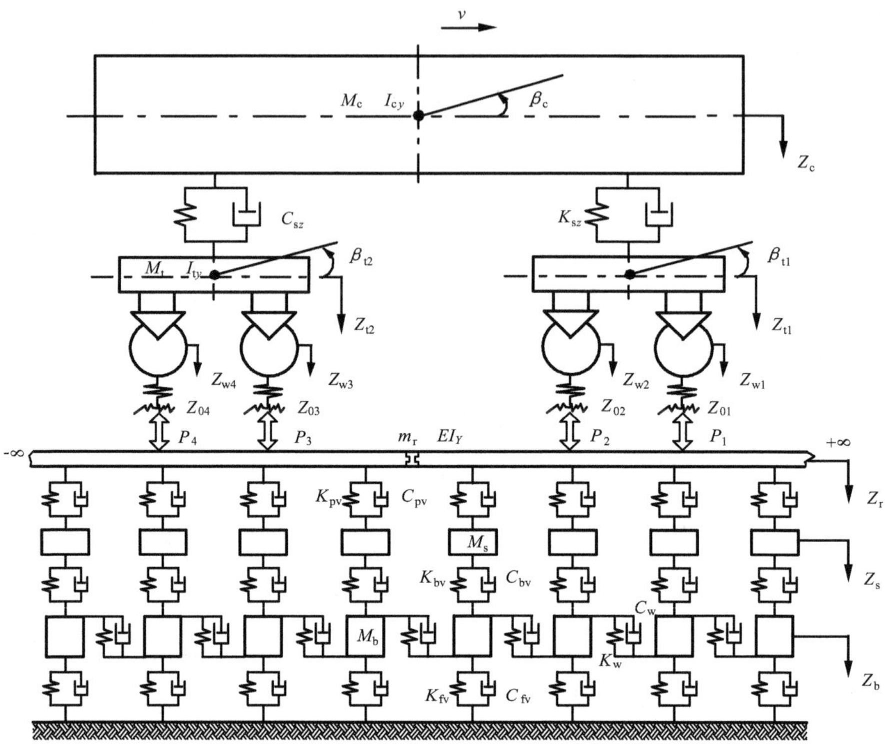

Fig. 2.24 Wagon–track spatially coupled dynamics model (side view)

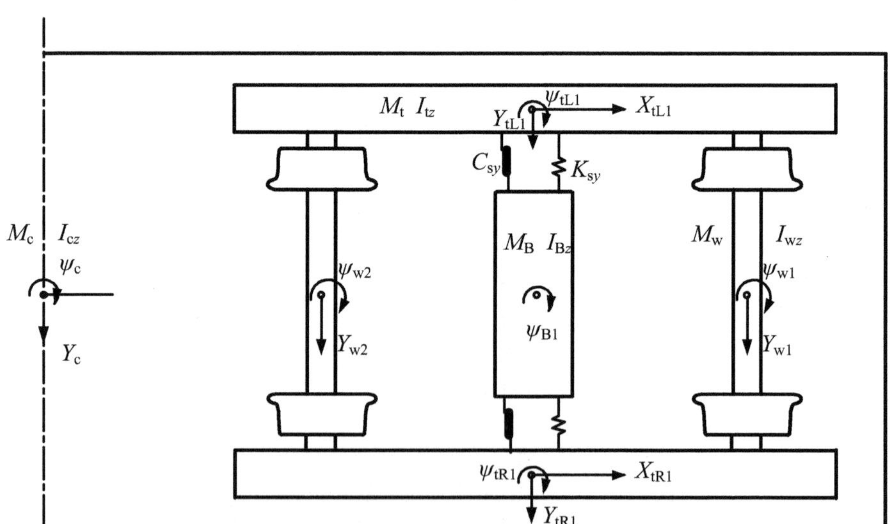

Fig. 2.25 Wagon–track spatially coupled dynamics model (top view)

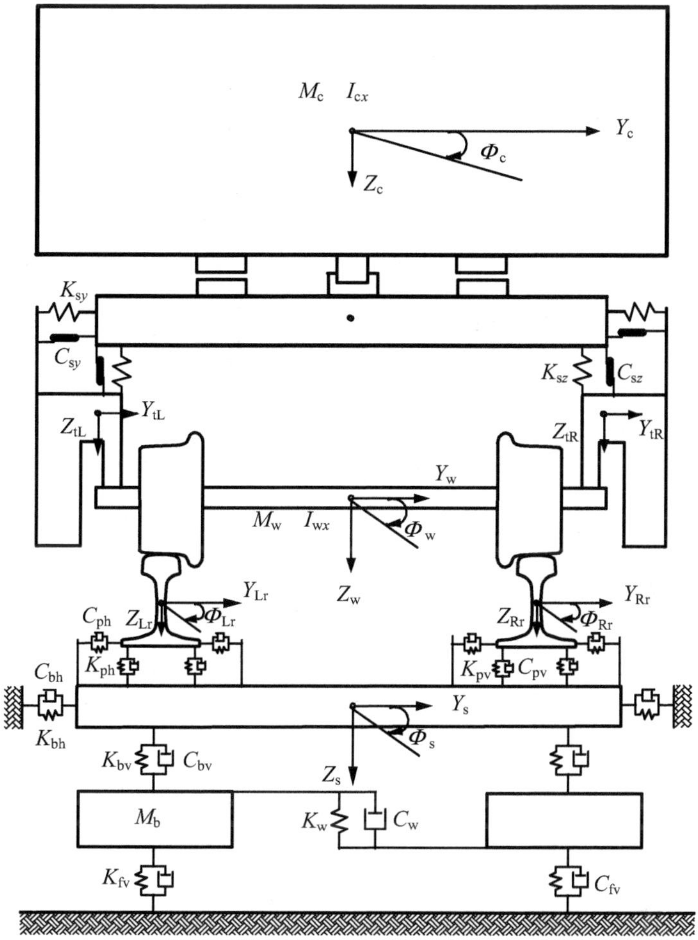

Fig. 2.26 Wagon–track spatially coupled dynamics model (end view)

3. *Locomotive–track spatially coupled dynamics model*

In traditional locomotive–track coupled dynamics models, the gear transmission subsystem was usually neglected. However, it plays a very important role in the traction or braking power transmission between the motor and the wheel–rail contact interface. In recent years, the author's research group has conducted a series of work on the dynamic investigation of the locomotive–track coupled dynamics system, where the dynamic effects of the gear transmission subsystem are considered [81, 93–97]. Compared with the traditional B_0–B_0 locomotive dynamics model, the most obvious feature of this locomotive dynamics model is the detailed consideration of the gear transmission structure which consists of traction motor, motor rotor, pinion, and gear. The established locomotive–track spatially coupled

Table 2.7 Notations for parameters of the wagon sub-model

Notation	Parameter	Unit
M_c	Car body mass	kg
M_t	Side frame mass	kg
M_B	Bolster mass	kg
M_w	Wheelset mass	kg
I_{cx}	Mass moment of inertia of car body about *X*-axis	$kg \cdot m^2$
I_{cy}	Mass moment of inertia of car body about *Y*-axis	$kg \cdot m^2$
I_{cz}	Mass moment of inertia of car body about *Z*-axis	$kg \cdot m^2$
I_{Bz}	Mass moment of inertia of bolster about *Z*-axis	$kg \cdot m^2$
I_{ty}	Mass moment of inertia of side frame about *Y*-axis	$kg \cdot m^2$
I_{tz}	Mass moment of inertia of side frame about *Z*-axis	$kg \cdot m^2$
I_{wx}	Mass moment of inertia of wheelset about *X*-axis	$kg \cdot m^2$
I_{wy}	Mass moment of inertia of wheelset about *Y*-axis	$kg \cdot m^2$
I_{wz}	Mass moment of inertia of wheelset about *Z*-axis	$kg \cdot m^2$
K_{sx}	Stiffness coefficient of secondary suspension along *X*-axis	MN/m
K_{sy}	Stiffness coefficient of secondary suspension along *Y*-axis	MN/m
K_{sz}	Stiffness coefficient of secondary suspension along *Z*-axis	MN/m
K_{sz1}	Stiffness coefficient of a wedge along *Z*-axis	MN/m

Table 2.8 Degrees of freedom of the wagon model

Component	Type of motion					
	Longitudinal	Lateral	Vertical	Roll	Pitch	Yaw
Car body	–	Y_c	Z_c	ϕ_c	β_c	ψ_c
Bolster (i = 2)	–	–	–	–	–	ψ_{Bi}
Side frame (i = 1, 4)	$X_{t(L,R)i}$	$Y_{t(L,R)i}$	–	–	$\beta_{t(L,R)i}$	$\psi_{t(L,R)i}$
Wheelset (i = 1, 4)	–	Y_{wi}	Z_{wi}	ϕ_{wi}	β_{wi}	ψ_{wi}

dynamics model with consideration of the gear transmission subsystem is shown in Figs. 2.27, 2.28, and 2.29 [93, 94]. The notations of the symbols in the figures are given in Table 2.9. In this dynamics model, the coupling effect between the locomotive–track system and the gear transmission subsystem is revealed from a spatial perspective, which is more consistent with the actual situation. This dynamics model could be not only adopted to study the vibration characteristics of locomotive–track system more comprehensively, but also to assess the dynamic characteristics of the gear transmission system in the entire vehicle dynamics system.

The established locomotive dynamics model consists of 23 rigid bodies with a total of 78 DOFs. Except for the rotor, the pinion and the gear, each of the rigid bodies has 6 DOFs, namely longitudinal (*X*), lateral (*Y*), bounce (*Z*), roll (Φ), pitch (β), and yaw (ψ) motions. While for the rotor, the pinion, and the gear, only the rotational DOF is considered in the present dynamics model. The symbols

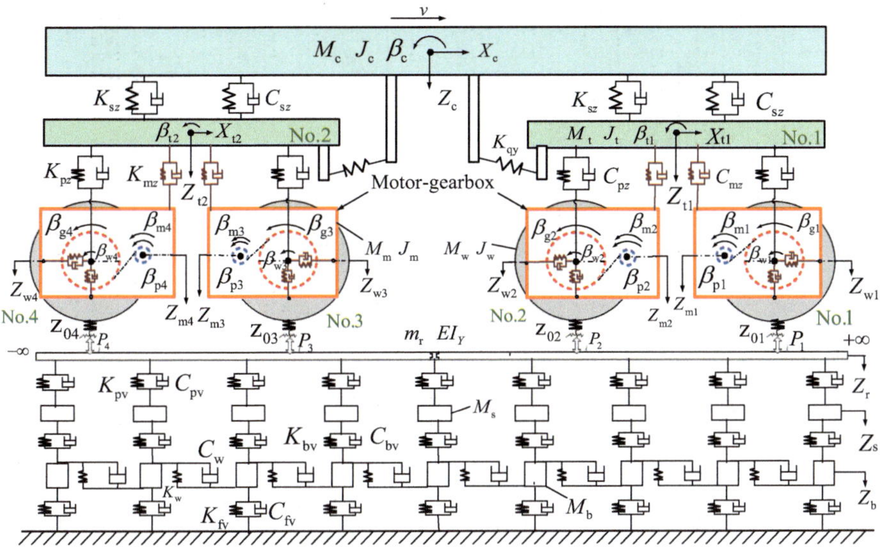

Fig. 2.27 Locomotive–track spatially coupled dynamics model (side view)

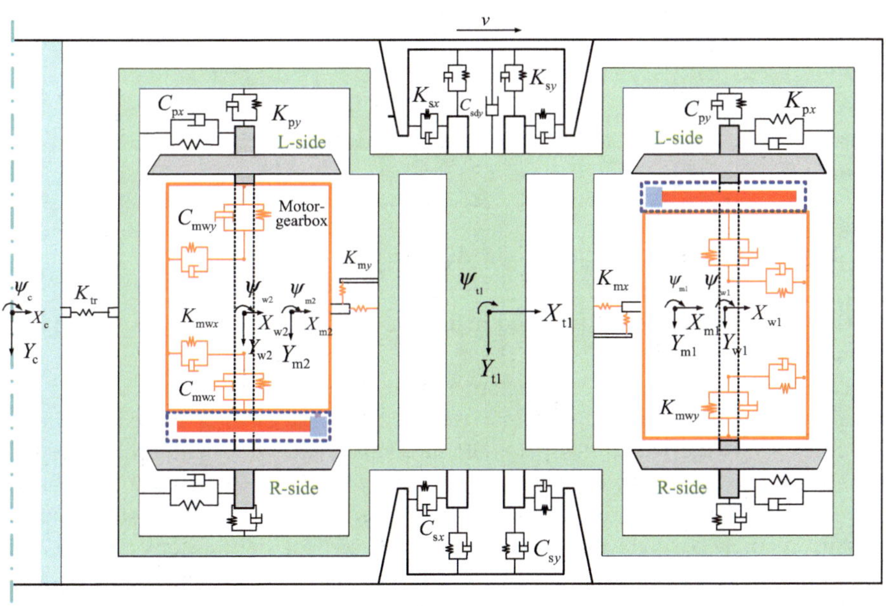

Fig. 2.28 Locomotive–track spatially coupled dynamics model (top view)

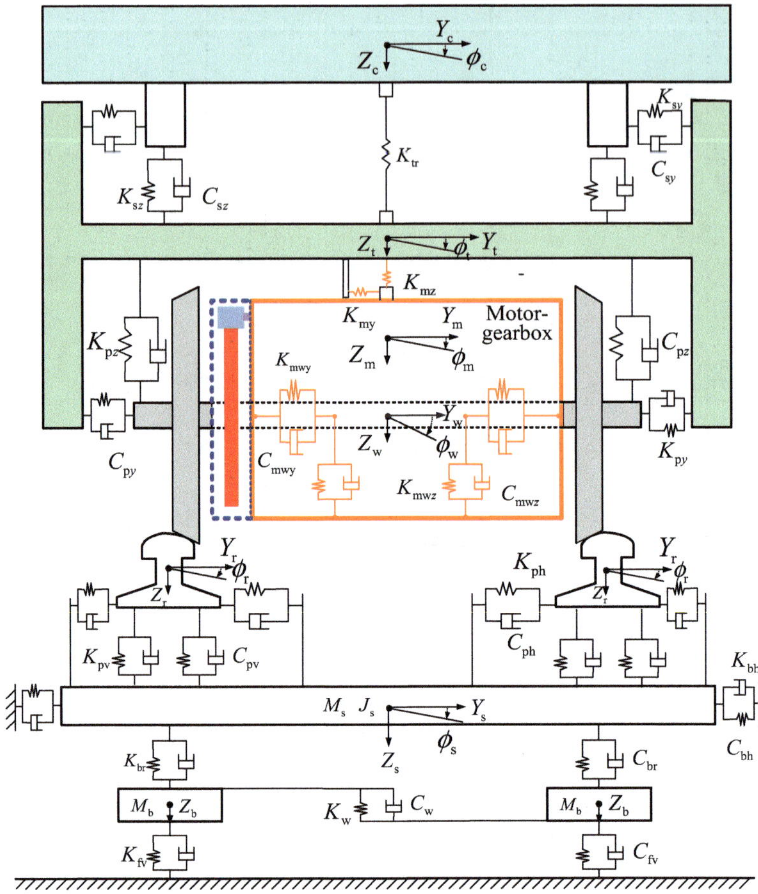

Fig. 2.29 Locomotive–track spatially coupled dynamics model (end view)

representing the motions of the rigid bodies in the locomotive dynamics system are shown in Table 2.10.

The major components of the locomotive dynamics system are usually connected via bogie suspension systems. The secondary suspension system, including the coil springs and lateral dampers, lateral bump–stops and traction rod, is usually used to connect the car body to the bogie frames and is able to mobilize the required rotational stiffness. The primary suspension system in each bogie, including the coil springs, vertical viscous dampers, lateral and vertical bump–stops and axlebox bushings, connects the bogie frames to the four wheelsets. Nonlinear 3D spring–damper elements were used to build the suspension model. The viscous dampers were represented as Maxwell elements consisting of a damper and a spring in series, in which the damping coefficient has piecewise nonlinear characteristics relating the velocity to the force generated. The bump–stops were modeled as bilinear spring elements.

Table 2.9 Notations for parameters of the locomotive sub-model

Notation	Parameter	Unit
M_c	Car body mass	kg
M_t	Bogie mass	kg
M_m	Motor mass	kg
M_w	Wheelset mass	kg
I_{cx}	Mass moment of inertia of car body about *X*-axis	$kg \cdot m^2$
I_{cy}	Mass moment of inertia of car body about *Y*-axis	$kg \cdot m^2$
I_{cz}	Mass moment of inertia of car body about *Z*-axis	$kg \cdot m^2$
I_{tx}	Mass moment of inertia of bogie about *X*-axis	$kg \cdot m^2$
I_{ty}	Mass moment of inertia of bogie about *Y*-axis	$kg \cdot m^2$
I_{tz}	Mass moment of inertia of bogie about *Z*-axis	$kg \cdot m^2$
I_{mx}	Mass moment of inertia of motor about *X*-axis	$kg \cdot m^2$
I_{my}	Mass moment of inertia of motor about *Y*-axis	$kg \cdot m^2$
I_{mz}	Mass moment of inertia of motor about *Z*-axis	$kg \cdot m^2$
I_{wx}	Mass moment of inertia of wheelset about *X*-axis	$kg \cdot m^2$
I_{wy}	Mass moment of inertia of wheelset about *Y*-axis	$kg \cdot m^2$
I_{wz}	Mass moment of inertia of wheelset about *Z*-axis	$kg \cdot m^2$
K_{px}	Stiffness coefficient of primary suspension along *X*-axis	MN/m
K_{py}	Stiffness coefficient of primary suspension along *Y*-axis	MN/m
K_{pz}	Stiffness coefficient of primary suspension along *Z*-axis	MN/m
K_{sx}	Stiffness coefficient of secondary suspension along *X*-axis	MN/m
K_{sy}	Stiffness coefficient of secondary suspension along *Y*-axis	MN/m
K_{sz}	Stiffness coefficient of secondary suspension along *Z*-axis	MN/m
K_{mx}	Stiffness coefficient of the traction motor suspension along *X*-axis	MN/m
K_{my}	Stiffness coefficient of the traction motor suspension along *Y*-axis	MN/m
K_{mz}	Stiffness coefficient of the traction motor suspension along *Z*-axis	MN/m
K_{mwx}	Stiffness coefficient of the axle-hung bearing along *X*-axis	MN/m
K_{mwy}	Stiffness coefficient of the axle-hung bearing along *Y*-axis	MN/m
K_{mwz}	Stiffness coefficient of the axle-hung bearing along *Z*-axis	MN/m
C_{mx}	Damping coefficient of the traction motor suspension along *X*-axis	$kN \cdot s/m$
C_{my}	Damping coefficient of the traction motor suspension along *Y*-axis	$kN \cdot s/m$
C_{mz}	Damping coefficient of the traction motor suspension along *Z*-axis	$kN \cdot s/m$
C_{mwx}	Damping coefficient of the axle-hung bearing along *X*-axis	$kN \cdot s/m$
C_{mwy}	Damping coefficient of the axle-hung bearing along *Y*-axis	$kN \cdot s/m$
C_{mwz}	Damping coefficient of the axle-hung bearing along *Z*-axis	$kN \cdot s/m$
C_{px}	Damping coefficient of primary suspension along *X*-axis	$kN \cdot s/m$
C_{py}	Damping coefficient of primary suspension along *Y*-axis	$kN \cdot s/m$
C_{pz}	Damping coefficient of primary suspension along *Z*-axis	$kN \cdot s/m$
C_{sx}	Damping coefficient of secondary suspension along *X*-axis	$kN \cdot s/m$
C_{sy}	Damping coefficient of secondary suspension along *Y*-axis	$kN \cdot s/m$
C_{sz}	Damping coefficient of secondary suspension along *Z*-axis	$kN \cdot s/m$

Table 2.10 DOFs of the locomotive dynamics model with gear transmissions

Component	Type of motion					
	Longitudinal	Lateral	Vertical	Roll	Pitch	Yaw
Car body	X_c	Y_c	Z_c	ϕ_c	β_c	ψ_c
Bogie frame (i = 1, 2)	X_{ti}	Y_{ti}	Z_{ti}	ϕ_{ti}	β_{ti}	ψ_{ti}
Wheelset (i = 1–4)	X_{wi}	Y_{wi}	Z_{wi}	ϕ_{wi}	β_{wi}	ψ_{wi}
Motor (i = 1–4)	X_{mi}	Y_{mi}	Z_{mi}	ϕ_{mi}	β_{mi}	ψ_{mi}
Rotor (i = 1–4)	–	–	–	–	β_{ri}	–
Pinion (i = 1–4)	–	–	–	–	β_{pi}	–
Gear (i = 1–4)	–	–	–	–	β_{gi}	–

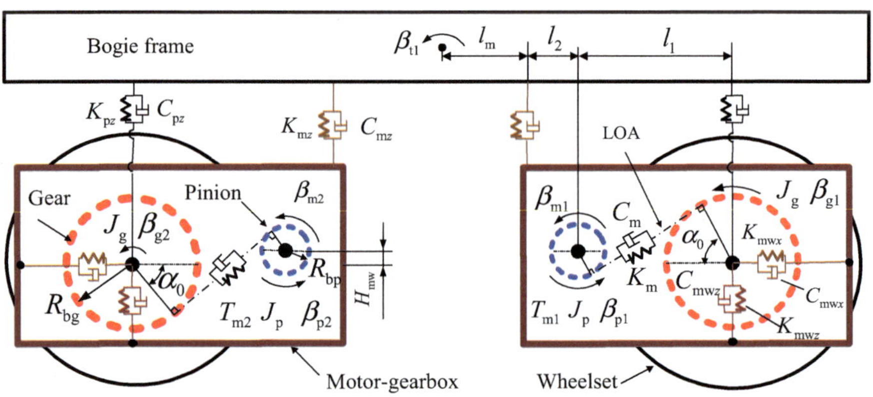

Fig. 2.30 Zoom-in plot of a bogie equipped with gear transmissions

Since the traction motor and the gearbox are bolted together, the mass of the gearbox and the lubricating oil are lumped to the traction motor and symmetrically distributed along the transverse direction. One end of the traction motor is elastically suspended on the bogie frame through a suspender and a ball-type rubber joint, and the other end is supported on the wheel axle by two axle-hung bearings. Zoom-in plot of the lead bogie is depicted in Fig. 2.30, where the dash circles in a gear transmission denote the base circles of the gears in engagement. The power transmission path from the rotor to the pinion, then to the gear and finally to the wheel is shown in Fig. 2.31a, and the corresponding dynamics model of the power transmission subsystem is shown in Fig. 2.31b. The output power of the motor is transmitted to the pinion through a torsional spring–damper element (K_{rp}, C_{rp}) representing the flexibility of the shaft between the motor rotor and the pinion. Except for the power consumed by the damping of the connecting shaft and the pinion, the left power is transmitted from the pinion to the gear by gear teeth engagement. Finally, the remaining power is transmitted from the gear to the left

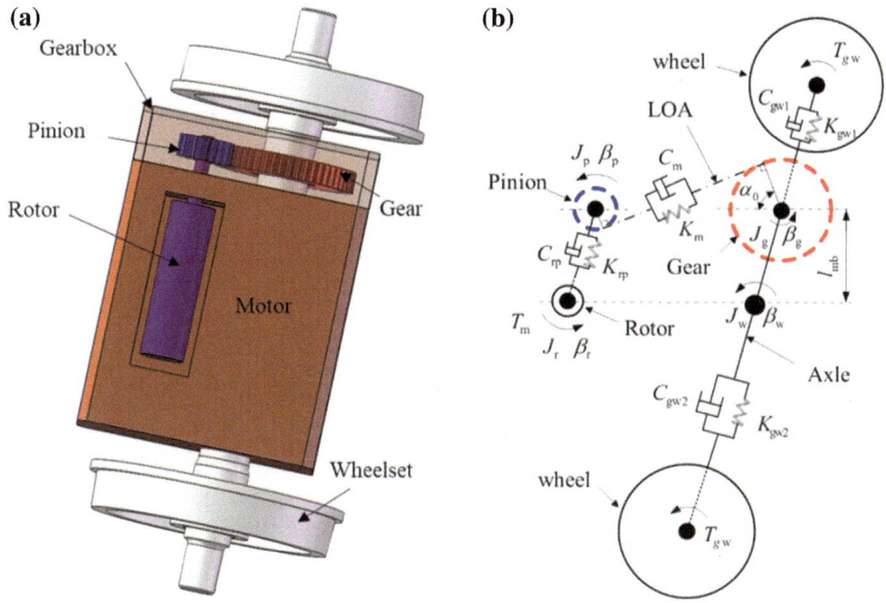

Fig. 2.31 The model of the power transmission: **a** geometrical model, and **b** dynamics model

and right wheels via torsional spring–damper elements (K_{gw}, C_{gw}) representing the flexibility of the wheel axle to generate the driving forces for the locomotive running by wheel–rail adhesion. The elastic deformations of the connecting shaft/coupling and the elastic deformations of the wheel axle are modeled by the torsional spring–damping unit (K_{gw}, C_{gw}), and the compressive deformations of the engaged gear teeth along line of action (LOA) are represented by the spring–damping element (K_m, C_m) through which the time-varying mesh stiffness and/or the gear tooth error excitations could be considered.

4. *The ballastless track model*

The vehicle–track spatially coupled dynamics models shown in Figs. 2.21, 2.22, 2.23, 2.24, 2.25, 2.26, 2.27, 2.28, and 2.29 only consider the ballasted track although the railways can contain either the ballasted track or the ballastless track. If the analyzed track structure is a ballastless track, the above-ballasted track dynamics model can be changed correspondingly. Figures 2.32, 2.33, 2.34, and 2.35 show the dynamics models of four typical ballastless track structures, including the long-sleeper embedded track (systems with sleepers firmly poured into an in situ concrete track slab), the elastic supporting block track (systems with elastically encased supporting blocks between rails and concrete slabs), the slab track (systems with track slabs supported by elastic cement asphalt mortar layer), and the floating slab track (systems with the concrete floating slab supported by the steel springs).

Fig. 2.32 Vertical and lateral dynamics model for the long-sleeper embedded ballastless track

The long-sleeper embedded track (or double-block ballastless track) consists of the rails, the fastenings and rail pads, the concrete sleepers, the concrete slab, and the concrete base. For this type of ballastless track, the track structure can be simply modeled as two Euler or Timoshenko beams discretely supported by fastenings and rail pads, while the lateral and the vertical bending deformations and the torsional rotation of the rails are considered (Fig. 2.32). This is because the sleeper blocks are precast into the slab directly and there is no elasticity between the slab and the concrete base. Thus, only the vibration of the rails is prominent in the wheel–rail dynamic interaction.

The elastic supporting block track consists of the rails, the fastenings and rail pads, the concrete blocks, the block pads and the rubber boots, the concrete slab, and the concrete base. The pads provide the vertical supporting stiffness and damping (K_{bv}, C_{bv}) for the blocks under the rails, and the rubber boots provide the lateral stiffness and damping (K_{bh}, C_{bh}) for the blocks. Thus both the vibrations of the rails and the concrete blocks are important for the wheel–rail interaction. The vertical and lateral coupled model for the elastic supporting block ballastless track is shown in Fig. 2.33, where the rails are modeled as the Euler or Timoshenko beams with considering the vertical, lateral and torsional motions. The concrete blocks are modeled as rigid bodies, which have the vertical and lateral DOFs.

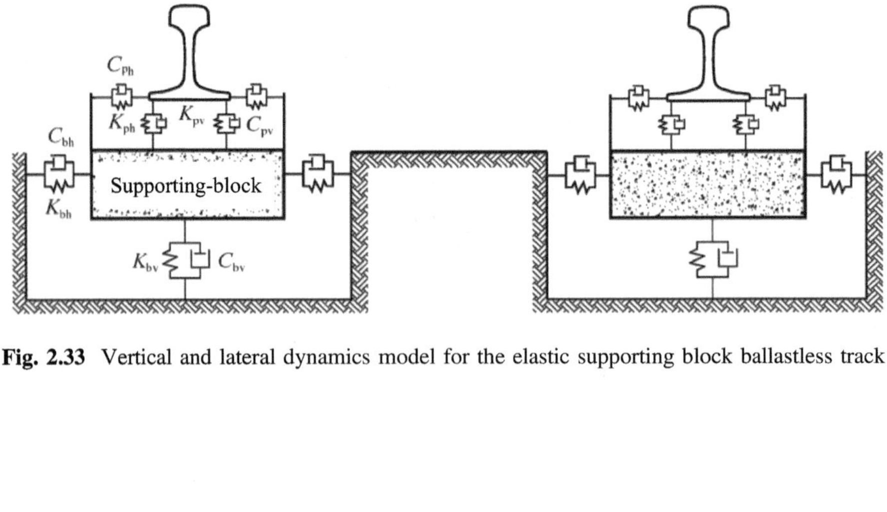

Fig. 2.33 Vertical and lateral dynamics model for the elastic supporting block ballastless track

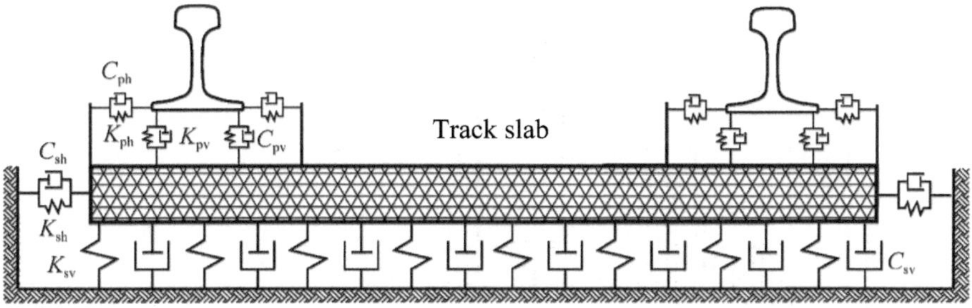

Fig. 2.34 Vertical and lateral dynamics model for the typical slab track (Reprinted from Ref. [36], Copyright 2009, with permission from Taylor & Francis.)

The slab track widely used in high-speed railways consists of the rails, the fastenings and rail pads, the concrete slabs, the cement asphalt mortar (CAM) layer, and the concrete base. In the slab track model, the rails are modeled as the Euler or Timoshenko beams with consideration of the vertical, lateral and torsional motions, while the track slabs are described as elastic rectangle plates supported on viscoelastic foundation, as shown in Fig. 2.34. In the figure, K_{sv} and C_{sv} are the vertical stiffness and damping of the CAM layer, and K_{sh} and C_{sh} are the lateral stiffness and damping of the CAM layer, respectively. Because the lateral bending stiffness of the slab is very large, it is sufficient to consider the rigid mode of the slab vibration in the lateral direction.

The floating slab track that is commonly used in metro systems consists of the rails, the fasteners and rail pads, the concrete floating slab, and the steel springs. In the floating slab track model, the rails are modeled as the Euler or Timoshenko beams with considering the vertical, lateral and torsional motions, while the track slabs are described as elastic rectangle thick plates supported on steel springs, as shown in Fig. 2.35. In the figure, K_{sv} and C_{sv} are the vertical stiffness and damping of the steel spring, respectively; and K_{sh} and C_{sh} are the lateral stiffness and

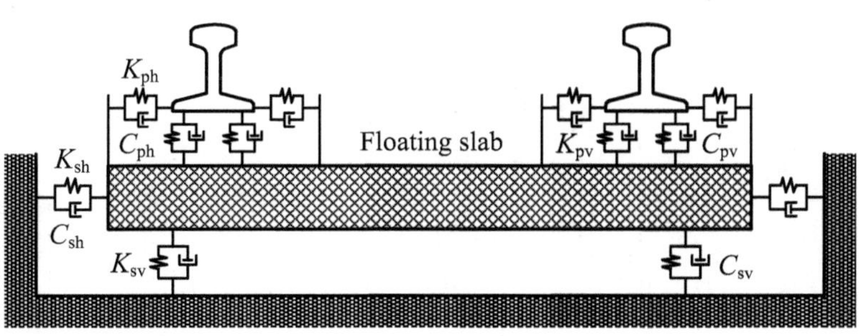

Fig. 2.35 Vertical and lateral dynamics model for the floating slab track

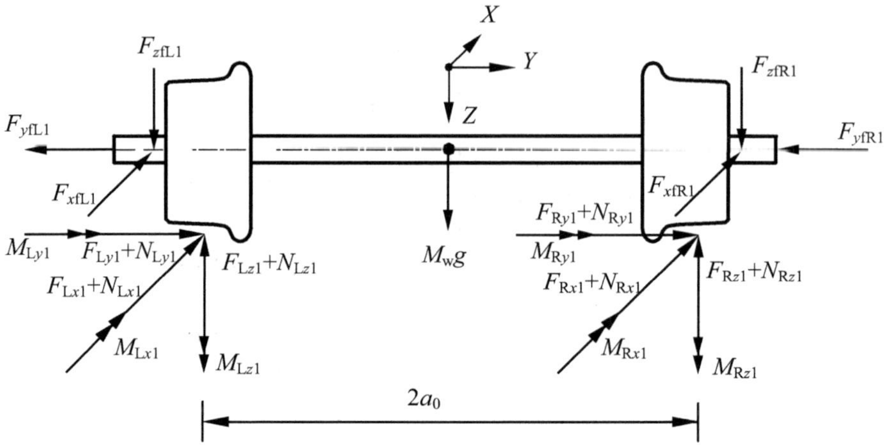

Fig. 2.36 Free-body diagram of the first wheelset of a passenger car

damping of the steel spring. Because the lateral bending stiffness of the slab is rather large, it is reasonable to consider the rigid mode of the slab vibration in the lateral direction.

2.3.2 Equations of Motion

1. *Equations of motion of the passenger vehicle subsystem (refer to* Figs. 2.36, 2.37 and 2.38*)*

To formulate the equations of motion of a railway vehicle, the forces applied on the components should be made clear. The forces between the components of a railway passenger vehicle system include the primary and secondary suspension forces, the wheel–rail normal contact forces and tangent creep forces, which are shown in Figs. 2.36, 2.37, and 2.38 and Table 2.11.

The suspension forces can be calculated with the displacements and velocities of the vehicle components, which are given below.

① Longitudinal forces of the primary suspension (i = 1–4)[1]

[1]① When i = 1 or 2, n = 1; when i = 3 or 4, n = 2.
② When the subscript on the left side of the equal sign is L, the symbol, $\pm$ and $\mp$, takes the symbol above; while when the subscript is R, the symbol, $\pm$ and $\mp$, takes the symbol below.
③ Similar to the situation is so appointed.

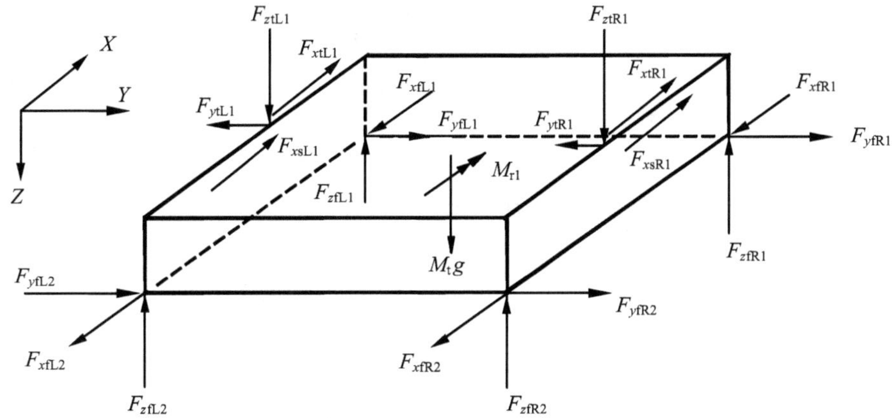

Fig. 2.37 Free-body diagram of the lead bogie frame of a passenger car

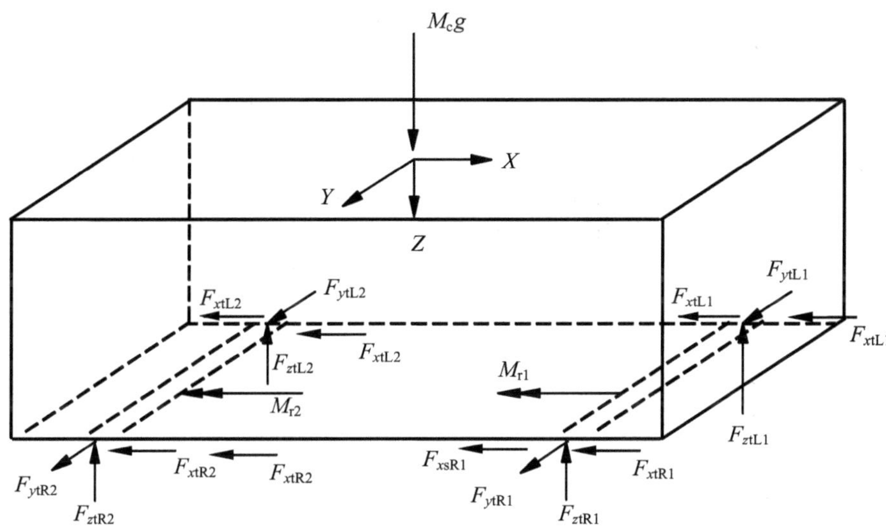

Fig. 2.38 Free-body diagram of the car body of a passenger car

$$\begin{aligned} F_{xf(\mathrm{L,R})i} = {} & K_{\mathrm{p}x}\left[\pm d_{\mathrm{w}}\psi_{\mathrm{t}n} + H_{\mathrm{tw}}\beta_{\mathrm{t}n} \mp d_{\mathrm{w}}\psi_{\mathrm{w}i} \mp (-1)^{i-1} d_{\mathrm{w}}\left(\frac{l_{\mathrm{t}}}{R_{\mathrm{t}n}}\right)\right] \\ & + C_{\mathrm{p}x}\left[\pm d_{\mathrm{w}}\dot{\psi}_{\mathrm{t}n} + H_{\mathrm{tw}}\dot{\beta}_{\mathrm{t}n} \mp d_{\mathrm{w}}\dot{\psi}_{\mathrm{w}i} \mp (-1)^{i-1} d_{\mathrm{w}}\frac{\mathrm{d}}{\mathrm{d}t}\left(\frac{l_{\mathrm{t}}}{R_{\mathrm{t}n}}\right)\right] \end{aligned} \tag{2.113}$$

② Lateral forces of the primary suspension (i = 1–4)

Table 2.11 Notations for the forces in the passenger vehicle subsystem

Notation	Physical meaning
F_{Lxi}, F_{Lyi}, and F_{Lzi}	The *x*-, *y*-, and *z*-direction creep forces of the left wheel of the *i*th wheelset
F_{Rxi}, F_{Ryi}, and F_{Rzi}	The *x*-, *y*-, and *z*-direction creep forces of the right wheel of the *i*th wheelset
N_{Lxi}, N_{Lyi}, and N_{Lzi}	The *x*-, *y*-, and *z*-direction contact forces of the left wheel of the *i*th wheelset
N_{Rxi}, N_{Ryi}, and N_{Rzi}	The *x*-, *y*-, and *z*-direction contact forces of the right wheel of the *i*th wheelset
M_{Lxi}, M_{Lyi}, and M_{Lzi}	The *x*-, *y*-, and *z*-direction spin creep torque of the left wheel of the *i*th wheelset
M_{Rxi}, M_{Ryi}, and M_{Rzi}	The *x*-, *y*-, and *z*-direction spin creep torque of the right wheel of the *i*th wheelset
F_{xfLi} and F_{xfRi}	The left and right longitudinal forces at primary suspension of the *i*th wheelset
F_{yfLi} and F_{yfRi}	The left and right lateral forces at primary suspension of the *i*th wheelset
F_{zfKLi} and F_{zfKRi}	The left and right vertical forces at primary suspension of the *i*th wheelset
F_{zfCLi} and F_{zfCRi}	The left and right vertical damping forces at primary suspension of the *i*th wheelset
F_{xtLi} and F_{xtRi}	The left and right longitudinal forces at secondary suspension of the *i*th bogie
F_{ytLi} and F_{ytRi}	The left and right lateral forces at secondary suspension of the *i*th bogie
F_{ztKLi} and F_{ztKRi}	The left and right vertical forces at secondary suspension of the *i*th bogie
F_{ztCLi} and F_{ztCRi}	The left and right vertical damping forces at secondary suspension of the *i*th bogie
F_{xsLi} and F_{xsRi}	The left and right longitudinal forces at anti-hunting dampers of the *i*th bogie
F_{yRi}	The lateral forces of the stopblock on the ith bogie
M_{Ri}	Anti-roll torque of the *i*th bogie

$$
\begin{aligned}
F_{\mathrm{yf(L,R)}i} = {} & K_{\mathrm{py}}\left[Y_{\mathrm{w}i} - Y_{\mathrm{t}n} + H_{\mathrm{tw}}\phi_{\mathrm{t}n} + (-1)^i l_{\mathrm{t}}\psi_{\mathrm{t}n} + \frac{l_{\mathrm{t}}^2}{2R_{\mathrm{t}n}}\right] \\
& + C_{\mathrm{py}}\left[\dot{Y}_{\mathrm{w}i} - \dot{Y}_{\mathrm{t}n} + H_{\mathrm{tw}}\dot{\phi}_{\mathrm{t}n} + (-1)^i l_{\mathrm{t}}\dot{\psi}_{\mathrm{t}n} + \frac{l_{\mathrm{t}}^2}{2}\frac{\mathrm{d}}{\mathrm{d}t}\left(\frac{1}{R_{\mathrm{t}n}}\right)\right]
\end{aligned} \tag{2.114}
$$

③ Vertical forces of the primary suspension (i = 1–4)

$$
\begin{aligned}
F_{\mathrm{zf(L,R)}i} = {} & K_{\mathrm{pz}}\left[Z_{\mathrm{t}n} - Z_{\mathrm{w}i} + (-1)^i l_{\mathrm{t}}\beta_{\mathrm{t}n} \pm d_{\mathrm{w}}\phi_{\mathrm{w}i} \mp d_{\mathrm{w}}\phi_{\mathrm{t}n}\right] \\
& + C_{\mathrm{pz}}\left[\dot{Z}_{\mathrm{t}n} - \dot{Z}_{\mathrm{w}i} + (-1)^i l_{\mathrm{t}}\dot{\beta}_{\mathrm{t}n} \pm d_{\mathrm{w}}\dot{\phi}_{\mathrm{w}i} \mp d_{\mathrm{w}}\dot{\phi}_{\mathrm{t}n}\right]
\end{aligned} \tag{2.115}
$$

④ Longitudinal forces of the secondary suspension (i = 1–2)

$$\begin{aligned} F_{xt(L,R)i} = & K_{sx}\left[H_{cB}\beta_c + H_{Bt}\beta_{ti} \pm d_s\psi_c \mp d_s\psi_{ti} \mp (-1)^{i-1}d_s\left(\frac{l_c}{R_c}\right)\right] \\ & + C_{sx}\left[H_{cB}\dot{\beta}_c + H_{Bt}\dot{\beta}_{ti} \pm d_s\dot{\psi}_c \mp d_s\dot{\psi}_{ti} \mp (-1)^{i-1}d_s\frac{d}{dt}\left(\frac{l_c}{R_c}\right)\right] \end{aligned} \tag{2.116}$$

If the yaw dampers are equipped in the secondary suspension, the longitudinal forces due to the yaw dampers read

$$F_{xs(L,R)i} = \begin{cases} F_{max}v_{xct(L,R)i}/v_0 & \left(\left|v_{xct(L,R)i}\right| < v_0\right) \\ F_{max}\mathrm{sign}\left(v_{xct(L,R)i}\right) & \left(\left|v_{xct(L,R)i}\right| \geq v_0\right) \end{cases} \tag{2.117}$$

where F_{max} is the saturation force of the damper; v_0 is the unloading velocity of the damper; v_{xct} is the relative velocity between two ends of the damper connecting the car body and the bogie frame in the longitudinal direction, which can be calculated by

$$v_{xct(L,R)i} = \pm d_{sc}\dot{\psi}_c \mp d_{sc}\dot{\psi}_{ti} + H_{cB}\dot{\beta}_c + H_{Bt}\dot{\beta}_{ti} \mp (-1)^{i-1}d_{sc}\frac{d}{dt}\left(\frac{l_c}{R_c}\right) \tag{2.118}$$

⑤ Lateral forces of the secondary suspension (i = 1–2)

$$\begin{aligned} F_{yt(L,R)i} = & K_{sy}\left[Y_{ti} - Y_c + H_{Bt}\phi_{ti} + H_{cB}\phi_c + (-1)^i l_c\psi_c + \frac{l_c^2}{2R_c}\right] \\ & + C_{sy}\left[\dot{Y}_{ti} - \dot{Y}_c + H_{Bt}\dot{\phi}_{ti} + H_{cB}\dot{\phi}_c + (-1)^i l_c\dot{\psi}_c + \frac{l_c^2}{2}\frac{d}{dt}\left(\frac{1}{R_c}\right)\right] \end{aligned} \tag{2.119}$$

⑥ Vertical forces of the secondary suspension (i = 1–2)

$$\begin{aligned} F_{zt(L,R)i} = & K_{sz}\left[Z_c - Z_{ti} \pm d_s\phi_{ti} \mp d_s\phi_c + (-1)^i l_c\beta_c\right] \\ & + C_{sz}\left[\dot{Z}_c - \dot{Z}_{ti} \pm d_s\dot{\phi}_{ti} \mp d_s\dot{\phi}_c + (-1)^i l_c\dot{\beta}_c\right] \end{aligned} \tag{2.120}$$

⑦ Anti-roll torque of the secondary suspension (i = 1–2)

$$M_{ri} = K_{rx}(\phi_c - \phi_{ti}) \tag{2.121}$$

Based on the calculated suspension forces and the Newton's second law, the equations of motion of the vehicle components can be formulated.

(1) Equations of motion of wheelset (i = 1–4)

Lateral motion:

$$M_{\rm w}\left(\ddot{Y}_{{\rm w}i}+\frac{v^2}{R_{{\rm w}i}}+r_0\ddot{\phi}_{{\rm sew}i}\right)=-F_{{\rm yfL}i}-F_{{\rm yfR}i}+F_{{\rm L}yi}+F_{{\rm R}yi}+N_{{\rm L}yi}+N_{{\rm R}yi}+M_{\rm w}g\phi_{{\rm sew}i} \tag{2.122}$$

Vertical motion:

$$\begin{aligned}M_{\rm w}\left(\ddot{Z}_{{\rm w}i}-a_0\ddot{\phi}_{{\rm sew}i}-\frac{v^2}{R_{{\rm w}i}}\phi_{{\rm sew}i}\right)&=-F_{{\rm L}zi}-F_{{\rm R}zi}-N_{{\rm L}zi}\\&\quad-N_{{\rm R}zi}+F_{z{\rm fL}i}+F_{z{\rm fR}i}+M_{\rm w}g\end{aligned} \tag{2.123}$$

Roll motion:

$$\begin{aligned}&I_{{\rm w}x}\left(\ddot{\phi}_{{\rm sew}i}+\ddot{\phi}_{{\rm w}i}\right)-I_{{\rm w}y}\left(\dot{\beta}_{{\rm w}i}-\Omega\right)\left(\dot{\psi}_{{\rm w}i}+\frac{v}{R_{{\rm w}i}}\right)=a_0(F_{{\rm L}zi}+N_{{\rm L}zi}-F_{{\rm R}zi}-N_{{\rm R}zi})\\&-r_{{\rm L}i}\left(F_{{\rm L}yi}+N_{{\rm L}yi}\right)-r_{{\rm R}i}\left(F_{{\rm R}yi}+N_{{\rm R}yi}\right)+d_{\rm w}(F_{z{\rm fR}i}-F_{z{\rm fL}i})\end{aligned} \tag{2.124}$$

Yaw motion:

$$\begin{aligned}&I_{{\rm w}z}\left[\ddot{\psi}_{{\rm w}i}+v\frac{{\rm d}}{{\rm d}t}\left(\frac{1}{R_{{\rm w}i}}\right)\right]-I_{{\rm w}y}\left(\dot{\phi}_{{\rm sew}i}+\dot{\phi}_{{\rm w}i}\right)\left(\dot{\beta}_{{\rm w}i}-\Omega\right)\\&=a_0(F_{{\rm L}xi}-F_{{\rm R}xi})+a_0\psi_{{\rm w}i}\left(F_{{\rm L}yi}+N_{{\rm L}yi}-F_{{\rm R}yi}-N_{{\rm R}yi}\right)\\&\quad+M_{{\rm L}zi}+M_{{\rm R}zi}+d_{\rm w}(F_{x{\rm fL}i}-F_{x{\rm fR}i})+a_0(N_{{\rm L}xi}-N_{{\rm R}xi})\end{aligned} \tag{2.125}$$

Rotation motion:

$$\begin{aligned}I_{{\rm w}y}\ddot{\beta}_{{\rm w}i}&=r_{{\rm R}i}F_{{\rm R}xi}+r_{{\rm L}i}F_{{\rm L}xi}+r_{{\rm R}i}\psi_{{\rm w}i}\left(F_{{\rm R}yi}+N_{{\rm R}yi}\right)\\&\quad+r_{{\rm L}i}\psi_{{\rm w}i}\left(F_{{\rm L}yi}+N_{{\rm L}yi}\right)+M_{{\rm L}yi}+M_{{\rm R}yi}+N_{{\rm L}xi}r_{{\rm L}i}+N_{{\rm R}xi}r_{{\rm R}i}\end{aligned} \tag{2.126}$$

(2) Equations of motion of bogie frame (i = 1, 2)

Lateral motion:

$$\begin{aligned}&M_{\rm t}\left[\ddot{Y}_{{\rm t}i}+\frac{v^2}{R_{{\rm t}i}}+(r_0+H_{\rm tw})\ddot{\phi}_{{\rm set}i}\right]=F_{{\rm yfL}(2i-1)}+F_{{\rm yfL}(2i)}-F_{{\rm ytL}i}\\&+F_{{\rm yfR}(2i-1)}+F_{{\rm yfR}(2i)}-F_{{\rm ytR}i}+M_{\rm t}g\phi_{{\rm set}i}\end{aligned} \tag{2.127}$$

Vertical motion:

$$M_{\mathrm{t}}\left[\ddot{Z}_{\mathrm{t}i}-a_0\ddot{\phi}_{\mathrm{set}i}-\frac{v^2}{R_{\mathrm{t}i}}\phi_{\mathrm{set}i}\right]=F_{z\mathrm{tL}i}-F_{z\mathrm{fL}(2i-1)}-F_{z\mathrm{fL}(2i)} \\ +F_{z\mathrm{tR}i}-F_{z\mathrm{fR}(2i-1)}-F_{z\mathrm{fR}(2i)}+M_{\mathrm{t}}g \tag{2.128}$$

Roll motion:

$$\begin{aligned} I_{\mathrm{tx}}\left(\ddot{\phi}_{\mathrm{t}i}+\ddot{\phi}_{\mathrm{set}i}\right)=&-\left[F_{y\mathrm{fL}(2i-1)}+F_{y\mathrm{fR}(2i-1)}+F_{y\mathrm{fL}(2i)}+F_{y\mathrm{fR}(2i)}\right]H_{\mathrm{tw}} \\ &+\left[F_{z\mathrm{fL}(2i-1)}+F_{z\mathrm{fL}(2i)}-F_{z\mathrm{fR}(2i-1)}-F_{z\mathrm{fR}(2i)}\right]d_{\mathrm{w}} \\ &+(F_{z\mathrm{tR}i}-F_{z\mathrm{tL}i})d_{\mathrm{s}}-\left(F_{y\mathrm{tL}i}+F_{y\mathrm{tR}i}\right)H_{\mathrm{Bt}}+M_{\mathrm{r}i} \end{aligned} \tag{2.129}$$

Pitch motion:

$$\begin{aligned} I_{\mathrm{tz}}\left[\ddot{\psi}_{\mathrm{t}i}+v\frac{\mathrm{d}}{\mathrm{d}t}\left(\frac{1}{R_{\mathrm{t}i}}\right)\right]=&\left[F_{y\mathrm{fL}(2i-1)}+F_{y\mathrm{fR}(2i-1)}-F_{y\mathrm{fL}(2i)}-F_{y\mathrm{fR}(2i)}\right]l_{\mathrm{t}} \\ &+\left[F_{x\mathrm{fR}(2i-1)}+F_{x\mathrm{fR}(2i)}-F_{x\mathrm{fL}(2i-1)}-F_{x\mathrm{fL}(2i)}\right]d_{\mathrm{w}} \\ &+(F_{x\mathrm{tL}i}-F_{x\mathrm{tR}i})d_{\mathrm{s}}+(F_{x\mathrm{sL}i}-F_{x\mathrm{sR}i})d_{\mathrm{sc}} \end{aligned} \tag{2.130}$$

Yaw motion:

$$\begin{aligned} I_{\mathrm{ty}}\ddot{\beta}_{\mathrm{t}i}=&\left[F_{z\mathrm{fL}(2i-1)}+F_{z\mathrm{fR}(2i-1)}-F_{z\mathrm{fL}(2i)}-F_{z\mathrm{fR}(2i)}\right]l_{\mathrm{t}} \\ &-\left[F_{x\mathrm{fL}(2i-1)}+F_{x\mathrm{fR}(2i-1)}+F_{x\mathrm{fL}(2i)}+F_{x\mathrm{fR}(2i)}\right]H_{\mathrm{tw}} \\ &-(F_{x\mathrm{tL}i}+F_{x\mathrm{tR}i})H_{\mathrm{Bt}}-(F_{x\mathrm{sL}i}+F_{x\mathrm{sR}i})H_{\mathrm{Bt}} \end{aligned} \tag{2.131}$$

(3) Equations of motion of car body

Lateral motion:

$$\begin{aligned} &M_{\mathrm{c}}\left[\ddot{Y}_{\mathrm{c}}+\frac{v^2}{R_{\mathrm{c}}}+(r_0+H_{\mathrm{tw}}+H_{\mathrm{Bt}}+H_{\mathrm{cB}})\ddot{\phi}_{\mathrm{sec}}\right] \\ &=F_{y\mathrm{tL}1}+F_{y\mathrm{tL}2}+F_{y\mathrm{tR}1}+F_{y\mathrm{tR}2}+M_{\mathrm{c}}g\phi_{\mathrm{sec}} \end{aligned} \tag{2.132}$$

Vertical motion:

$$M_{\mathrm{c}}\left[\ddot{Z}_{\mathrm{c}}-a_0\ddot{\phi}_{\mathrm{sec}}-\frac{v^2}{R_{\mathrm{c}}}\phi_{\mathrm{sec}}\right]=-F_{z\mathrm{tL}1}-F_{z\mathrm{tR}1}-F_{z\mathrm{tL}2}-F_{z\mathrm{tR}2}+M_{\mathrm{c}}g \tag{2.133}$$

Roll motion:

$$I_{cx}[\ddot{\phi}_{c} + \ddot{\phi}_{sec}] = -(F_{ytL1} + F_{ytR1} + F_{ytL2} + F_{ytR2})H_{cB} + (F_{ztL1} + F_{ztL2} - F_{ztR1} - F_{ztR2})d_{s} - M_{r1} - M_{r2} \tag{2.134}$$

Pitch motion:

$$\begin{aligned} I_{cy}\ddot{\beta}_{c} = &(F_{ztL1} + F_{ztR1} - F_{ztL2} - F_{ztR2})\, l_{c} \\ &- (F_{xtL1} + F_{xtR1} + F_{xtL2} + F_{xtR2})H_{cB} \\ &- (F_{xsL1} + F_{xsR1} + F_{xsL2} + F_{xsR2})H_{cB} \end{aligned} \tag{2.135}$$

Yaw motion:

$$\begin{aligned} I_{cz}\left[\ddot{\psi}_{c} + v\frac{\mathrm{d}}{\mathrm{d}t}\left(\frac{1}{R_{c}}\right)\right] = &(F_{ytL1} + F_{ytR1} - F_{ytL2} - F_{ytR2})\, l_{c} \\ &+ (F_{xtR1} + F_{xtR2} - F_{xtL1} - F_{xtL2})\, d_{s} \\ &+ (F_{xsR1} + F_{xsR2} - F_{xsL1} - F_{xsL2})\, d_{sc} \end{aligned} \tag{2.136}$$

The denotations of undefined notations in Eqs. (2.113)–(2.136) are given in Table 2.12.

Table 2.12 Physical meaning of notations used in equations of motion of vehicle subsystem

Notation	Physical meaning
ϕ_{sewi}	The super elevation angle of the curve high rail where the *i*th wheelset locates
ϕ_{seti}	The super elevation angle of the curve high rail where the *i*th bogie center locates
ϕ_{sec}	The super elevation angle of the curve high rail where the car body center locates
R_{wi}	The curvature radius of the track where the *i*th wheelset locates
R_{ti}	The curvature radius of the track where the *i*th bogie frame locates
R_{c}	The curvature radius of the track where the car body locates
v	Train speed
r_{0}	Nominal contact rolling radius of the wheel
r_{Li} and r_{Ri}	The left and right contact rolling radii of the *i*th wheel
Ω	Nominal rolling angular velocity of the wheel
g	Gravity acceleration
δsc	Half-distance between the yaw dampers on the two sides of the bogie
δs	Half-distance between the secondary suspension of the two sides of the bogie
δw	Half-distance between the primary suspension of the two sides of the bogie
l_{c}	Half-distance between bogie centers
l_{t}	Half-distance between the two axles of the bogie
H_{tw}	Height of the bogie center from the wheelset center
H_{Bt}	Height of the secondary suspension from the bogie center
H_{cB}	Height of the car body center from the secondary suspension location

2. *Equations of motion of the freight wagon subsystem (refer to Figs.* 2.39, 2.40, 2.41, and 2.42*)*

The forces between the components of a railway wagon equipped with the three-piece bogies include the spring forces, the friction forces, the wheel–rail normal contact forces, and tangent creep forces, which are shown in Figs. 2.39, 2.40, 2.41, and 2.42 and Table 2.13. Noted that the free-body diagram of a wagon wheelset is as the same as shown in Fig. 2.36, and not given here again.

① Longitudinal and lateral forces of the primary suspension (i = 1–4)

The three-piece bogie Z8A comprises one bolster and two side frames. There are no spring and no viscous damping component in the primary suspensions. The side frames directly contact with the wheel axles in the vertical direction. There are

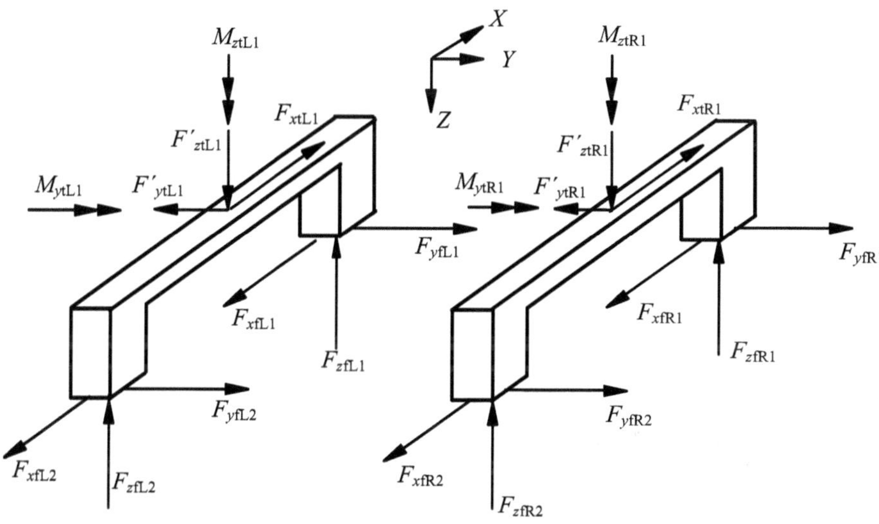

Fig. 2.39 Free-body diagram of the side frames of a wagon

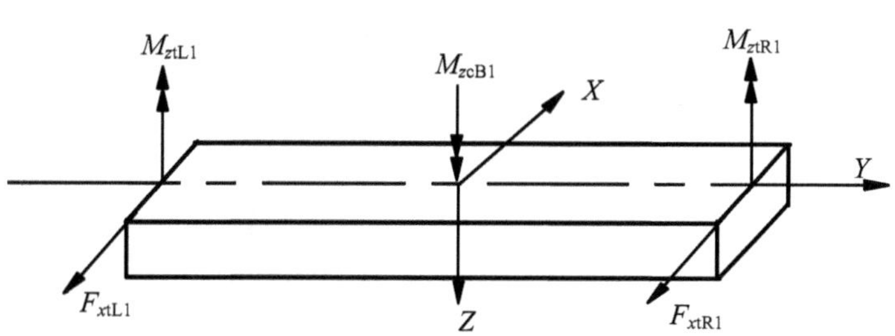

Fig. 2.40 Free-body diagram of the bolster of a wagon

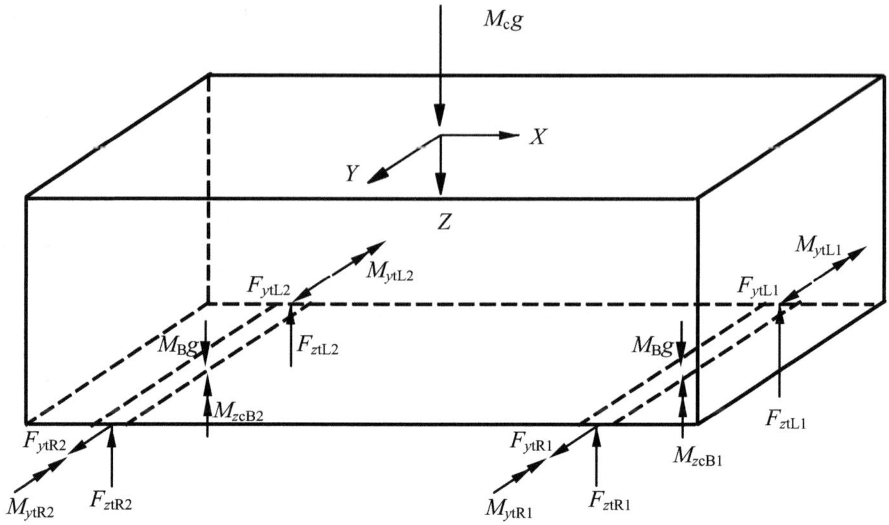

Fig. 2.41 Free-body diagram of the car body of a freight wagon

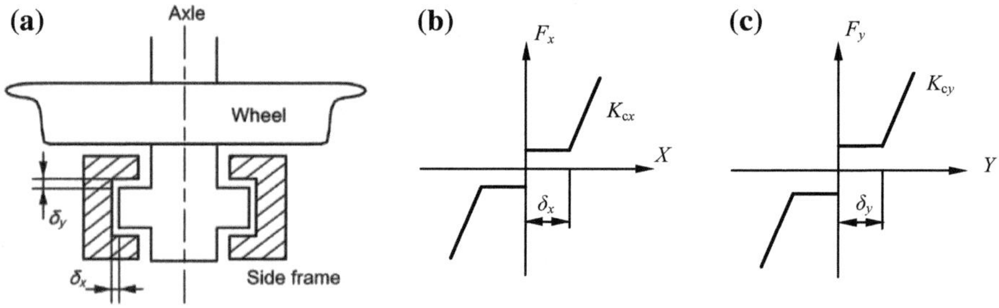

Fig. 2.42 Modeling of the axle box connection between side frame and wheelset: **a** the clearances in axle box, **b** the longitudinal force characteristic, and **c** the lateral force characteristic

Table 2.13 Notations for the forces in the wagon subsystem

Notation	Physical meaning
F_{xtLi} and F_{xtRi}	The left and right longitudinal forces at side frames of the *i*th bogie
F_{ytLi} and F_{ytRi}	The left and right lateral forces of the *i*th bolster
F'_{ytLi} and F'_{ytRi}	The left and right lateral forces at side frames of the *i*th bogie
F_{ztLi} and F_{ztRi}	The left and right vertical forces of the *i*th bolster
F'_{ztLi} and F'_{ztRi}	The left and right vertical forces at side frames of the *i*th bogie
M_{ytLi} and M_{ytRi}	The left and right torques around *y*-axis of the *i*th bolster
M_{ztLi} and M_{ztRi}	The left and right torques around *z*-axis of the *i*th bolster
M_{zcBi}	The friction torque around *x*-axis of the *i*th bolster

longitudinal clearance δ_x and lateral clearance δ_y between the axle box and the side frame, see Fig. 2.42a. In consideration of the combination effect of the Coulomb friction and the clearances between the axle box and the side frame, the force–displacement characteristics in these places are described in Fig. 2.42b and c.

According to Fig. 2.42b, the longitudinal forces at the axle box can be expressed as

$$F_{x\mathrm{f(L,R)}i} = \begin{cases} \mu_s F_{z\mathrm{f(L,R)}i}\mathrm{sign}\left(v_{x\mathrm{tw(L,R)}i}\right) & \left(\left|X_{x\mathrm{tw(L,R)}i}\right| < \delta_x\right) \\ K_{\mathrm{c}x}\left[X_{x\mathrm{tw(L,R)}i} - \mathrm{sign}\left(X_{x\mathrm{tw(L,R)}i}\right)\delta_x\right] \\ \quad + \mu_s F_{z\mathrm{f(L,R)}i}\mathrm{sign}\left(v_{x\mathrm{tw(L,R)}i}\right) & \left(\left|X_{x\mathrm{tw(L,R)}i}\right| \geq \delta_x\right) \end{cases} \tag{2.137}$$

where μ_s is the Coulomb friction coefficient between the axle and the side frame, $X_{x\mathrm{tw(L,R)}}$ is the longitudinal relative displacement between the axle and the side frame, and $V_{x\mathrm{tw(L,R)}i}$ is the longitudinal relative velocity between the axle and the side frame, which are given by

$$X_{x\mathrm{tw(L,R)}i} = X_{\mathrm{t(L,R)}n} + H_{\mathrm{tw}}\beta_{\mathrm{t(L,R)}n} \mp d_{\mathrm{w}}\psi_{\mathrm{w}i} \pm (-1)^i d_{\mathrm{w}}\frac{l_{\mathrm{t}}}{R_{\mathrm{t}n}} \tag{2.138}$$

$$v_{x\mathrm{tw(L,R)}i} = \dot{X}_{\mathrm{t(L,R)}n} + H_{\mathrm{tw}}\dot{\beta}_{\mathrm{t(L,R)}n} \mp d_{\mathrm{w}}\dot{\psi}_{\mathrm{w}i} \pm (-1)^i d_{\mathrm{w}}\frac{\mathrm{d}}{\mathrm{d}t}\left(\frac{l_{\mathrm{t}}}{R_{\mathrm{t}n}}\right) \tag{2.139}$$

According to Fig. 2.42c, the lateral forces at the axle box can be expressed as

$$F_{y\mathrm{f(L,R)}i} = \begin{cases} \mu_s F_{z\mathrm{f(L,R)}i}\mathrm{sign}\left(v_{y\mathrm{tw(L,R)}i}\right) & \left(\left|X_{y\mathrm{tw(L,R)}i}\right| < \delta_y\right) \\ K_{\mathrm{c}y}\left[X_{y\mathrm{tw(L,R)}i} - \mathrm{sign}\left(X_{y\mathrm{tw(L,R)}i}\right)\delta_y\right] \\ \quad + \mu_s F_{z\mathrm{f(L,R)}i}\mathrm{sign}\left(v_{y\mathrm{tw(L,R)}i}\right) & \left(\left|X_{y\mathrm{tw(L,R)}i}\right| \geq \delta_y\right) \end{cases} \tag{2.140}$$

where $X_{y\mathrm{tw(L,R)}i}$ is the lateral relative displacements between the axle and the side frame and $V_{y\mathrm{tw(L,R)}i}$ is the lateral relative velocities between the axle and the side frame, which can be obtained by

$$X_{y\mathrm{tw(L,R)}i} = Y_{\mathrm{w}i} - Y_{\mathrm{t(L,R)}n} + (-1)^i l_{\mathrm{t}}\psi_{\mathrm{t(L,R)}n} + \frac{l_{\mathrm{t}}^2}{2R_{\mathrm{t}n}} \tag{2.141}$$

$$v_{y\mathrm{tw(L,R)}i} = \dot{Y}_{\mathrm{w}i} - \dot{Y}_{\mathrm{t(L,R)}n} + (-1)^i l_{\mathrm{t}}\dot{\psi}_{\mathrm{t(L,R)}n} + \frac{l_{\mathrm{t}}^2}{2}\frac{\mathrm{d}}{\mathrm{d}t}\left(\frac{1}{R_{\mathrm{t}n}}\right) \tag{2.142}$$

② Vertical forces of the primary suspension (i = 1–2)

To calculate the vertical forces of the primary suspension, the vertical motion $Z_{\mathrm{t(L,R)}i}$ and pitch motion $\beta_{\mathrm{t(L,R)}i}$ of the side frames should be first obtained, which are dependent to the motions of wheelsets due to the fact that the side frames rigidly contact with the wheel axles in the vertical direction. The equations of the vertical motion and pitch motion of the ith side frame read

$$M_t\left[\ddot{Z}_{tLi} - a_0\ddot{\phi}_{seti} - \frac{v^2}{R_{ti}}\phi_{seti}\right] = -F_{zfL(2i-1)} - F_{zfL(2i)} + F'_{ztLi} + M_t g \tag{2.143}$$

$$I_{ty}\ddot{\beta}_{tLi} = \left[F_{zfL(2i-1)} - F_{zfL(2i)}\right] l_t - \left[F_{xfL(2i-1)} + F_{xfL(2i)}\right] H_{tw} - F_{xtLi} H_{Bt} + M_{ytLi} \tag{2.144}$$

where H_{Bt} is height of the side frame center from the bolster center.

From Eqs. (2.143) and (2.144), the vertical forces of the left primary suspension can be deduced:

$$\begin{cases} F_{zfL(2i-1)} = (A_1 + B_1)/2 \\ F_{zfL(2i)} = (A_1 - B_1)/2 \end{cases} \tag{2.145}$$

where

$$\begin{cases} A_1 = F'_{ztLi} + M_t g - M_t\left(\ddot{Z}_{tLi} - a_0\ddot{\phi}_{seti} - \frac{v^2}{R_{ti}}\phi_{seti}\right) \\ B_1 = \left[I_{ty}\ddot{\beta}_{tLi} + \left(F_{xfL(2i-1)} + F_{xfL(2i)}\right)H_{tw} + F_{xtLi}H_{Bt} - M_{ytLi}\right]/l_t \end{cases} \tag{2.146}$$

The vertical and pitch displacements of the left side frames can be calculated by

$$\begin{cases} Z_{tL1} = [Z_{w1} + Z_{w2} - d_w(\phi_{w1} + \phi_{w2})]/2 \\ Z_{tL2} = [Z_{w3} + Z_{w4} - d_w(\phi_{w3} + \phi_{w4})]/2 \end{cases} \tag{2.147}$$

$$\begin{cases} \beta_{tL1} = [Z_{w2} - Z_{w1} + d_w(\phi_{w1} - \phi_{w2})]/(2l_t) \\ \beta_{tL2} = [Z_{w4} - Z_{w3} + d_w(\phi_{w3} - \phi_{w4})]/(2l_t) \end{cases} \tag{2.148}$$

The vertical forces of the right primary suspension read

$$\begin{cases} F_{zfR(2i-1)} = (A_2 + B_2)/2 \\ F_{zfR(2i)} = (A_2 - B_2)/2 \end{cases} \tag{2.149}$$

where

$$\begin{cases} A_2 = F'_{ztRi} + M_t g - M_t\left(\ddot{Z}_{tRi} - a_0\ddot{\phi}_{seti} - \frac{v^2}{R_{ti}}\phi_{seti}\right) \\ B_2 = \left[I_{ty}\ddot{\beta}_{tRi} + \left(F_{xfR(2i-1)} + F_{xfR(2i)}\right)H_{tw} + F_{xtRi}H_{Bt} - M_{ytRi}\right]/l_t \end{cases} \tag{2.150}$$

The vertical and pitch displacements of the right side frames can be calculated by

$$\begin{cases} Z_{tR1} = [Z_{w1} + Z_{w2} + d_w(\phi_{w1} + \phi_{w2})]/2 \\ Z_{tR2} = [Z_{w3} + Z_{w4} + d_w(\phi_{w3} + \phi_{w4})]/2 \end{cases} \tag{2.151}$$

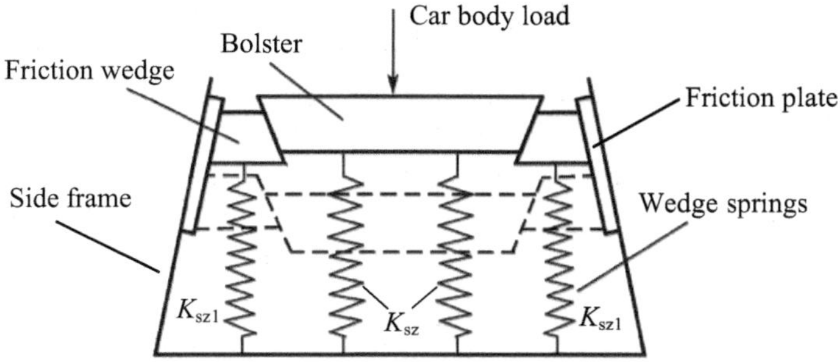

Fig. 2.43 Configuration of the secondary suspension of the three-piece bogie

$$\begin{cases} \beta_{tR1} = [Z_{w2} - Z_{w1} - d_w(\phi_{w1} - \phi_{w2})]/(2l_t) \\ \beta_{tR2} = [Z_{w4} - Z_{w3} - d_w(\phi_{w3} - \phi_{w4})]/(2l_t) \end{cases} \tag{2.152}$$

③ Secondary suspension forces

Figure 2.43 shows the configuration of the secondary suspension of the three-piece bogie, which includes the steel springs and friction wedges. When the wagon operates on railway track, the vertical forces from the coil steel springs and friction wedges will support the wagon body via the bolster. The vibrations of the wagon system induce the relative motions between the friction plates on the side frames, the friction wedges, and the bolsters, which generate the dry friction forces in the secondary suspension, damping the vibration of the whole wagon system.

Figure 2.44 shows the free-body diagram of the left friction wedge, where α is the contact angle between the wedge and the bolster; β is the contact angle between the wedge and the side frame; P_a is the support force from the steel springs; μ_1 and μ_2 are the friction coefficients of the interfaces between the wedge and the friction plate on the side frame and between the wedge and the bolster, respectively. N_1 and

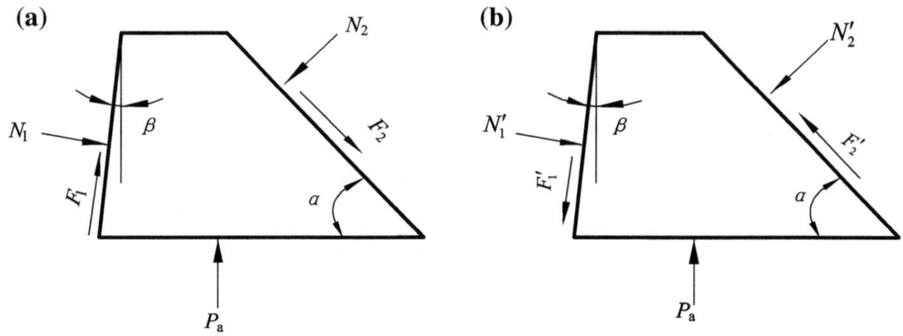

Fig. 2.44 Free-body diagram of the friction wedge: **a** bolster and side frame move face to face, and **b** bolster and side frame move in direction of separation

N_2 are the normal forces applied on the friction wedge when the bolster and side frame move face to face, N_1' and N_2' are the normal forces applied on the friction wedge when the bolster and the side frame move in the direction of separation. F_1 and F_2 are the friction forces of the wedge when the bolster and side frame move face to face, F_1' and F_2' are the friction forces of the wedge when the bolster and the side frame move in the direction of separation. These forces are calculated as follows:

$$\begin{cases} N_1 = \frac{\sin\alpha - \mu_2 \cos\alpha}{\Delta A} P_a \\ N_2 = \frac{\cos\beta + \mu_1 \sin\beta}{\Delta A} P_a \end{cases} \tag{2.153}$$

$$\begin{cases} F_1 = \mu_1 N_1 \\ F_2 = \mu_2 N_2 \end{cases} \tag{2.154}$$

$$\begin{cases} N_1' = \frac{\sin\alpha + \mu_2 \cos\alpha}{\Delta B} P_a \\ N_2' = \frac{\cos\beta - \mu_1 \sin\beta}{\Delta B} P_a \end{cases} \tag{2.155}$$

$$\begin{cases} F_1' = \mu_1 N_1' \\ F_2' = \mu_2 N_2' \end{cases} \tag{2.156}$$

$$\begin{cases} \Delta A = (1 + \mu_1 u_2)\cos(\alpha - \beta) + (\mu_2 - \mu_1)\sin(\alpha - \beta) \\ \Delta B = (1 + \mu_1 \mu_2)\cos(\alpha - \beta) - (\mu_2 - \mu_1)\sin(\alpha - \beta) \end{cases} \tag{2.157}$$

$$P_a = K_{sz1} Z_1 \tag{2.158}$$

where K_{sz1} is the stiffness of the steel spring supporting the friction wedge, Z_1 is the compression of the wedge springs.

If the relative vertical displacement between the bolster and the side frame is ΔZ_{Bt}, Z_1 can then be expressed as

$$Z_1 = (\Delta Z_{Bt} + \Delta Z_0)/(1 + \tan\alpha\tan\beta) \tag{2.159}$$

where ΔZ_0 is the initial compression of the coil steel spring, which can be obtained by

$$\Delta Z_0 = \frac{1}{4}(M_c + 2M_B)g/[K_{sz} + 2K_{sz1}/(1 + \tan\alpha\tan\beta)] \tag{2.160}$$

where K_{sz} is the stiffness of the steel spring supporting the bolster, i.e., the secondary suspension stiffness, i.e., the secondary suspension stiffness.

Consequently, when the bolster and side frame move face to face, the secondary suspension forces (i = 1, 2) can be calculated as follows:

Vertical forces

$$\begin{cases} F_{zt(L,R)i} = K_{sz}[Z_c + (-1)^i l_c \beta_c \mp d_s \phi_c - Z_{t(L,R)i} + \Delta Z_0] + 2N_2 \cos\alpha + 2F_2 \sin\alpha \\ F'_{zt(L,R)i} = K_{sz}[Z_c + (-1)^i l_c \beta_c \mp d_s \phi_c - Z_{t(L,R)i} + \Delta Z_0] + 2P_a + 2F_1 \cos\beta - 2N_1 \sin\beta \end{cases} \tag{2.161}$$

Lateral friction forces

$$\begin{cases} S_{yt(L,R)i} = 2\mu_2 N_2 \mathrm{sign}\left[\dot{Y}_c - \dot{Y}_{t(L,R)i} + (-1)^{i-1}\dot{\psi}_c l_c - \dot{\phi}_c H_{cB} - \frac{l_c^2}{2}\frac{\mathrm{d}}{\mathrm{d}t}\left(\frac{1}{R_c}\right)\right] \\ S'_{yt(L,R)i} = 2\mu_1 N_1 \mathrm{sign}\left[\dot{Y}_c - \dot{Y}_{t(L,R)i} + (-1)^{i-1}\dot{\psi}_c l_c - \dot{\phi}_c H_{cB} - \frac{l_c^2}{2}\frac{\mathrm{d}}{\mathrm{d}t}\left(\frac{1}{R_c}\right)\right] \end{cases} \tag{2.162}$$

Friction moments

$$\begin{cases} M_{yt(L,R)i} = \mu_1 N_1 l_{rs} \mathrm{sign}\left[\dot{\beta}_c - \dot{\beta}_{t(L,R)i}\right] \\ M_{zB(L,R)i} = \mu_1 N_1 l_{rs} \mathrm{sign}\left[\dot{\psi}_{Bi} - \dot{\psi}_{t(L,R)i}\right] \end{cases} \tag{2.163}$$

where l_{rs} is the distance between the center positions of two friction plates (see Fig. 2.41) on the side frame.

When the bolster and side frame move in the direction of separation, the secondary suspension forces (i = 1, 2) becomes:

Vertical forces

$$\begin{cases} F_{zt(L,R)i} = K_{sz}[Z_c + (-1)^i l_c \beta_c \mp d_s \phi_c - Z_{t(L,R)i} + \Delta Z_0] + 2N'_2 \cos\alpha - 2F'_2 \sin\alpha \\ F'_{zt(L,R)i} = K_{sz}[Z_c + (-1)^i l_c \beta_c \mp d_s \phi_c - Z_{t(L,R)i} + \Delta Z_0] + 2P_a - 2F'_1 \cos\beta - 2N'_1 \sin\beta \end{cases} \tag{2.164}$$

Lateral friction forces

$$\begin{cases} S_{yt(L,R)i} = 2\mu_2 N'_2 \mathrm{sign}\left[\dot{Y}_c - \dot{Y}_{t(L,R)i} + (-1)^{i-1}\dot{\psi}_c l_c - \dot{\phi}_c H_{cB} - \frac{l_c^2}{2}\frac{\mathrm{d}}{\mathrm{d}t}\left(\frac{1}{R_c}\right)\right] \\ S'_{yt(L,R)i} = 2\mu_1 N'_1 \mathrm{sign}\left[\dot{Y}_c - \dot{Y}_{t(L,R)i} + (-1)^{i-1}\dot{\psi}_c l_c - \dot{\phi}_c H_{cB} - \frac{l_c^2}{2}\frac{\mathrm{d}}{\mathrm{d}t}\left(\frac{1}{R_c}\right)\right] \end{cases} \tag{2.165}$$

Friction moments

$$\begin{cases} M_{\mathrm{yt(L,R)}i} = \mu_1 N_1' l_{\mathrm{rs}} \mathrm{sign}\left[\dot{\beta}_{\mathrm{c}} - \dot{\beta}_{\mathrm{t(L,R)}i}\right] \\ M_{z\mathrm{B(L,R)}i} = \mu_1 N_1' l_{\mathrm{rs}} \mathrm{sign}\left[\dot{\psi}_{\mathrm{B}i} - \dot{\psi}_{\mathrm{t(L,R)}i}\right] \end{cases} \tag{2.166}$$

The longitudinal and lateral forces due to the coil steel springs can be calculated by (i = 1, 2)

$$\begin{cases} F_{\mathrm{ys(L,R)}i} = K_{\mathrm{sy}}[-Y_{\mathrm{c}} + Y_{\mathrm{t(L,R)}i} + (-1)^i \psi_{\mathrm{c}} l_{\mathrm{c}} + \phi_{\mathrm{c}} H_{\mathrm{cB}} + \frac{l_{\mathrm{c}}^2}{2R_{\mathrm{c}}}] \\ F_{x\mathrm{s(L,R)}i} = K_{\mathrm{s}x}[\pm \psi_{\mathrm{B}i} d_{\mathrm{s}} + \beta_{\mathrm{c}} H_{\mathrm{cB}} - X_{\mathrm{t(L,R)}i} + \beta_{\mathrm{t(L,R)}i} H_{\mathrm{Bt}}] \end{cases} \tag{2.167}$$

The anti-warp moment due to the bottom frame brace system can be expressed as (i = 1, 2)

$$M_{\psi\mathrm{(L,R)}i} = K_{\psi\mathrm{t}}\left[\psi_{\mathrm{B}i} - \psi_{\mathrm{t(L,R)}i}\right] \tag{2.168}$$

where $K_{\psi\mathrm{t}}$ is the anti-warp stiffness of the three-piece bogie, which can be obtained from Table 2.14.

The friction moments resulting from the center plate and the side bearings on the bolster can be calculated by

$$M_{z\mathrm{cB}i} = \left[F_{\mathrm{p}} \mu_{\mathrm{s}} l_{\mathrm{rp}} + \frac{1}{2}(M_{\mathrm{c}} g - 2F_{\mathrm{p}}) \mu_{\mathrm{s}} l_{\mathrm{r0}}\right] \mathrm{sign}\left(\dot{\psi}_{\mathrm{c}} - \dot{\psi}_{\mathrm{B}i}\right) \tag{2.169}$$

where F_{p} is the preload of the side bearing on the bolster, μ_{s} is the friction coefficient of the side bearing, l_{r0} is the radius of the friction moment of the center plate, l_{rp} is the half distance of the two the side bearings on the bolster.

To sum up, the secondary suspension forces include (i = 1, 2)

Table 2.14 Anti-warp stiffness of the three-piece bogie

Warp angle (°)	$K_{\psi\mathrm{t}} \times 10^5$ (m · N/°)
0.1	0.87800000
0.2	0.42702103
0.3	0.27595789
0.4	0.21081081
0.5	0.17729730
0.6	0.15405405
0.7	0.13243243
0.8	0.11621622
0.9	0.10000000
1.0	0.08918109

Longitudinal: F_{xsLi}, F_{xsRi}
Lateral: F_{ysLi}, F_{ysRi}, S_{ytLi}, S'_{ytLi}, S_{ytRi}, S'_{ytRi}
Vertical: F_{ztLi}, F_{ztRi}, F'_{ztLi}, F'_{ztRi}

The secondary suspension moments include (i = 1, 2)
Around y-axis; M_{ytLi}, M_{ytRi}
Around z-axis: M_{zBLi}, M_{zBRi}, $M_{\psi Li}$, $M_{\psi Ri}$, M_{zcBi}

And the combined forces in the secondary suspension of the three-piece bogie are expressed as

$$\begin{cases} F_{ytLi} = F_{ysLi} - S_{ytLi} \\ F_{ytRi} = F_{ysRi} - S_{ytRi} \\ F'_{ytLi} = F_{ysLi} - S'_{ytLi} \\ F'_{ytRi} = F_{ysRi} - S'_{ytRi} \\ M_{ztLi} = M_{zBLi} + M_{\psi Li} \\ M_{ztRi} = M_{zBRi} + M_{\psi Ri} \end{cases} \tag{2.170}$$

The equations of motion of the wagon components can then be formulated when the abovementioned suspension forces are all determined.

(1) Equations of motion of wheelset (i = 1–4)

The motion equations of the wheelsets of the wagon are similar to those of the passenger vehicle (see Eqs. (2.122)–(2.126)), which are omitted here.

(2) Equations of motion of side frame (i = 1, 2)

Longitudinal motion

$$M_t \ddot{X}_{t(L,R)i} = F_{xt(L,R)i} - F_{xf(L,R)(2i-1)} - F_{xf(L,R)(2i)} \tag{2.171}$$

Lateral motion

$$M_t \left[\ddot{Y}_{t(L,R)i} + \frac{v^2}{R_{ti}} + (r_0 + H_{tw})\ddot{\phi}_{seti} \right] = -F'_{yt(L,R)i} + F_{yf(L,R)(2i-1)} + F_{yf(L,R)(2i)} + M_t g \phi_{seti} \tag{2.172}$$

Yaw motion

$$I_{tz}\left[\ddot{\psi}_{t(L,R)i} + v\frac{d}{dt}\left(\frac{1}{R_{ti}}\right) \right] = \left[F_{yf(L,R)(2i-1)} - F_{yf(L,R)(2i)} \right] l_t + M_{zt(L,R)i} \tag{2.173}$$

(3) Equations of motion of bolster (i = 1, 2)

$$I_{Bz}\left[\ddot{\psi}_{Bi}+v\frac{d}{dt}\left(\frac{1}{R_{ti}}\right)\right]=M_{zcBi}-M_{ztLi}-M_{ztRi}-d_s(F_{xtLi}-F_{xtRi}) \tag{2.174}$$

(4) Equations of motion of car body

Lateral motion

$$\begin{aligned}&M_c\left[\ddot{Y}_c+\frac{v^2}{R_c}+(r_0+H_{tw}+H_{Bt}+H_{cB})\ddot{\phi}_{sec}\right]\\&+2M_B\left[\ddot{Y}_c-H_{cB}\ddot{\phi}_c+\frac{v^2}{R_c}+(r_0+H_{tw}+H_{Bt})\ddot{\phi}_{sec}\right]\\&=F_{ytL1}+F_{ytL2}+F_{ytR1}+F_{ytR2}+(M_c+2M_B)g\phi_{sec}\end{aligned} \tag{2.175}$$

where H_{cB} is the height of the car body center from the bolster center.

Vertical motion

$$\begin{aligned}(M_c+2M_B)\left[\ddot{Z}_c-a_0\ddot{\phi}_{sec}-\frac{v^2}{R_c}\phi_{sec}\right]=&-F_{ztL1}-F_{ztR1}-F_{ztL2}\\&-F_{ztR2}+(M_c+2M_B)g\end{aligned} \tag{2.176}$$

Roll motion

$$\begin{aligned}&\left(I_{cx}+2I_{Bx}+2M_BH_{cB}^2\right)\left[\ddot{\phi}_c+\ddot{\phi}_{sec}\right]=-\left(F_{ytL1}+F_{ytR1}+F_{ytL2}+F_{ytR2}\right)H_{cB}\\&+(F_{ztL1}+F_{ztL2}-F_{ztR1}-F_{ztR2})d_s\end{aligned} \tag{2.177}$$

Pitch motion

$$\begin{aligned}\left(I_{cy}+2M_Bl_c^2\right)\ddot{\beta}_c=&(F_{ztL1}+F_{ztR1}-F_{ztL2}-F_{ztR2})l_c\\&-(F_{xtL1}+F_{xtR1}+F_{xtL2}+F_{xtR2})H_{cB}\\&-M_{ytL1}-M_{ytL2}-M_{ytR1}-M_{ytR2}\end{aligned} \tag{2.178}$$

Yaw motion

$$I_{cz}\left[\ddot{\psi}_c+v\frac{d}{dt}\left(\frac{1}{R_c}\right)\right]=\left(F_{ytL1}+F_{ytR1}-F_{ytL2}-F_{ytR2}\right)l_c-M_{zcB1}-M_{zcB2} \tag{2.179}$$

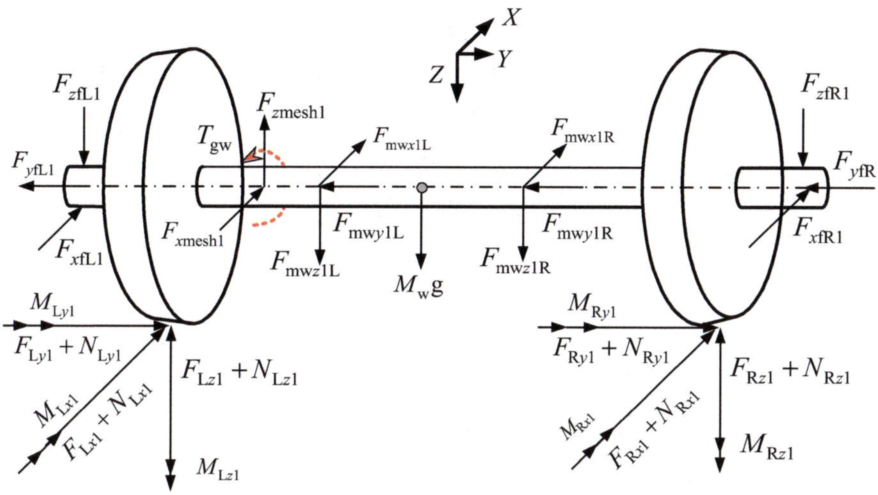

Fig. 2.45 Free-body diagram of the first wheelset of a locomotive

3. *Equations of motion of the locomotive subsystem (shown in* Figs. 2.45, 2.46, 2.47, and 2.48*)*

The acting forces between the major components of a locomotive system usually include the primary and secondary suspension forces, traction rod forces, motor suspension forces, bearing forces for axle-hung motor, the wheel–rail normal forces and tangent forces, and the gear mesh forces, which are shown in Figs. 2.45, 2.46, 2.47 and 2.48, Tables 2.11 and 2.15. Calculation of these forces is going to be introduced in the following contents:

① Primary suspension forces (i = 1–4)

Longitudinal forces

$$\begin{aligned}F_{xf(\mathrm{L,R})i} = {} & K_{px}\left[X_{tn} - X_{wi} + H_{tw}\beta_{tn} \pm d_{w}\psi_{tn} \mp d_{w}\psi_{wi} \mp (-1)^{i-1} d_{w}\left(\frac{l_t}{R_{tn}}\right)\right] \\ & + C_{px}\left[\dot{X}_{tn} - \dot{X}_{wi} + H_{tw}\dot{\beta}_{tn} \pm d_{w}\dot{\psi}_{tn} \mp d_{w}\dot{\psi}_{wi} \mp (-1)^{i-1} d_{w}\frac{d}{dt}\left(\frac{l_t}{R_{tn}}\right)\right]\end{aligned} \tag{2.180}$$

Lateral forces

$$\begin{aligned}F_{yf(\mathrm{L,R})i} = {} & K_{py}\left[Y_{wi} - Y_{tn} + H_{tw}\phi_{tn} + (-1)^{i-1} l_{t}\psi_{tn} + \frac{l_t^2}{2R_{tn}}\right] \\ & + C_{py}\left[\dot{Y}_{wi} - \dot{Y}_{tn} + H_{tw}\dot{\phi}_{tn} + (-1)^{i-1} l_{t}\dot{\psi}_{tn} + \frac{l_t^2}{2}\frac{d}{d_t}\left(\frac{1}{R_{tn}}\right)\right]\end{aligned} \tag{2.181}$$

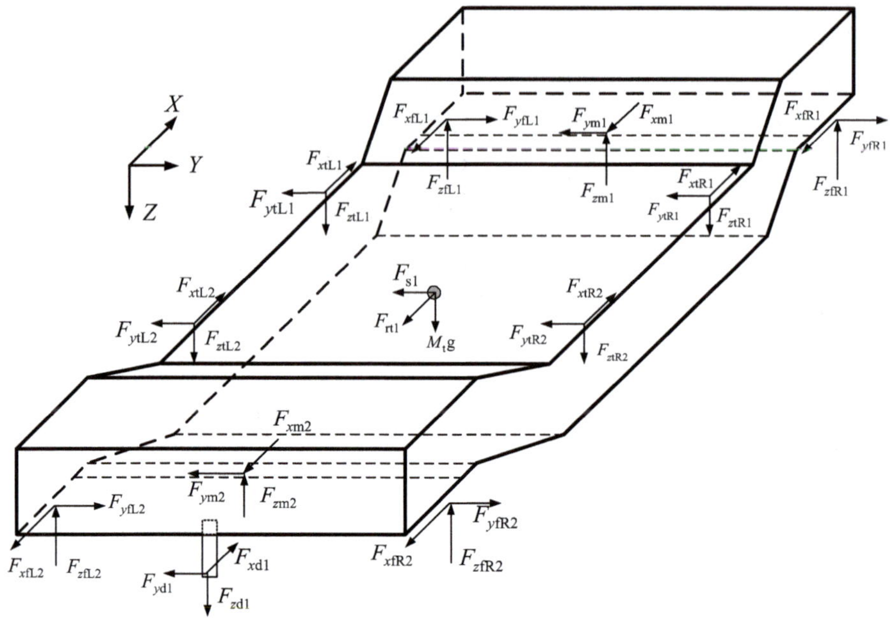

Fig. 2.46 Free-body diagram of the lead bogie frame of a locomotive

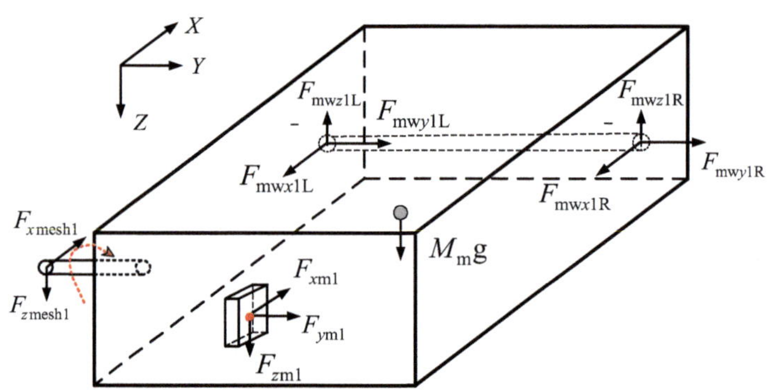

Fig. 2.47 Free-body diagram of the first traction motor of a locomotive

Vertical forces

$$\begin{aligned} F_{zf(L,R)i} &= K_{pz}\left[Z_{tn} - Z_{wi} + (-1)^i l_t \beta_{tn} \pm d_w \phi_{wi} \mp d_w \phi_{tn}\right] \\ &\quad + C_{pz}\left[\dot{Z}_{tn} - \dot{Z}_{wi} + (-1)^i l_t \dot{\beta}_{tn} \pm d_w \dot{\phi}_{wi} \mp d_w \dot{\phi}_{tn}\right] \end{aligned} \tag{2.182}$$

② Secondary suspension forces (i = 1–4)

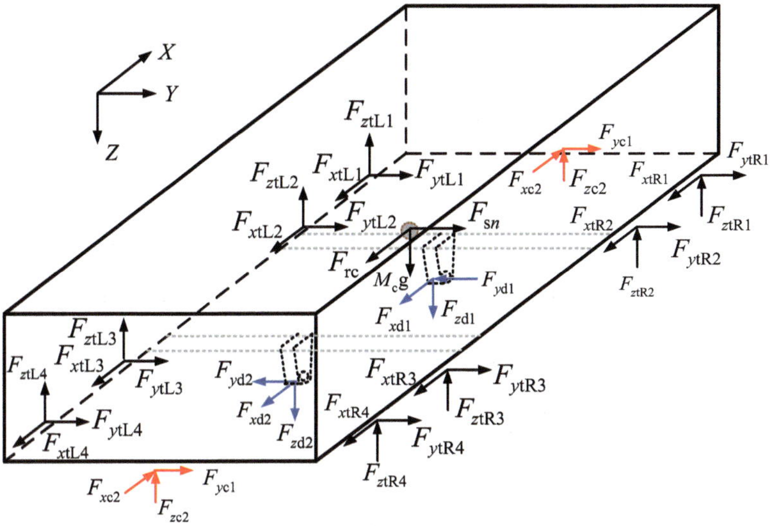

Fig. 2.48 Free-body diagram of the car body of a locomotive

Longitudinal forces

$$
\begin{aligned}
F_{xt(\mathrm{L,R})i} = {} & K_{sx}\left[X_{\mathrm{c}} - X_{tn} + H_{\mathrm{cB}}\beta_{\mathrm{c}} + H_{\mathrm{Bt}}\beta_{tn} \pm d_{\mathrm{s}}\psi_{\mathrm{c}} \mp d_{\mathrm{s}}\psi_{tn} \mp (-1)^{i-1} d_{\mathrm{s}}\left(\frac{L}{R_{\mathrm{c}}}\right)\right] \\
& + C_{sx}\left[\dot{X}_{\mathrm{c}} - \dot{X}_{tn} + H_{\mathrm{cB}}\dot{\beta}_{\mathrm{c}} + H_{\mathrm{Bt}}\dot{\beta}_{tn} \pm d_{\mathrm{s}}\dot{\psi}_{\mathrm{c}} \mp d_{\mathrm{s}}\dot{\psi}_{tn} \mp (-1)^{i-1} d_{\mathrm{s}}\frac{d}{d_{\mathrm{t}}}\left(\frac{L}{R_{\mathrm{c}}}\right)\right]
\end{aligned}
\tag{2.183}
$$

where

$$
L = \begin{cases} l_{\mathrm{c}} + l_{\mathrm{sb}} & (i = 1, 4) \\ l_{\mathrm{c}} - l_{\mathrm{sb}} & (i = 2, 3) \end{cases}
\tag{2.184}
$$

where l_{sb} is the longitudinal distance between the secondary suspension and the bogie frame center.

Lateral forces

$$
\begin{aligned}
F_{yt(\mathrm{L,R})i} = {} & K_{\mathrm{sy}}\begin{bmatrix} Y_{tn} - Y_{\mathrm{c}} + H_{\mathrm{cB}}\phi_{\mathrm{c}} + H_{\mathrm{Bt}}\phi_{tn} + (-1)^{i-1} l_{\mathrm{sb}}\psi_{tn} \\ + (-1)^{n} l_{\mathrm{c}}\psi_{\mathrm{c}} + (-1)^{i-1} l_{\mathrm{sb}}\psi_{\mathrm{c}} + \frac{L^2}{2R_{\mathrm{c}}} \end{bmatrix} \\
& + C_{\mathrm{sy}}\begin{bmatrix} \dot{Y}_{tn} - \dot{Y}_{\mathrm{c}} + H_{\mathrm{cB}}\dot{\phi}_{\mathrm{c}} + H_{\mathrm{Bt}}\dot{\phi}_{tn} + (-1)^{i-1} l_{\mathrm{sb}}\dot{\psi}_{tn} \\ + (-1)^{n} l_{\mathrm{c}}\dot{\psi}_{\mathrm{c}} + (-1)^{i-1} l_{\mathrm{sb}}\dot{\psi}_{\mathrm{c}} + \frac{L^2}{2}\frac{d}{d_{\mathrm{t}}}\left(\frac{1}{R_{\mathrm{c}}}\right) \end{bmatrix}
\end{aligned}
\tag{2.185}
$$

Table 2.15 Notations for the forces in the locomotive subsystem

Notation	Physical meaning
F_{xmeshi} and F_{zmeshi}	The *x*- and *z*-direction mesh forces of the *i*th gear pairs
F_{xmi}, F_{ymi}, and F_{zmi}	The *x*-, *y*-, and *z*-direction suspension forces of the *i*th traction motor
F_{mwxLi}, F_{mwyLi}, and F_{mwzLi}	The *x*-, *y*-, and *z*-direction the left bearing forces of the *i*th axle-hung motor
F_{mwxRi}, F_{mwyRi}, and F_{mwzRi}	The *x*-, *y*-, and *z*-direction the right bearing forces of the *i*th axle-hung motor
F_{xdn}, F_{ydn}, and F_{zdn}	The *x*-, *y*-, and *z*-direction traction forces of the *n*th traction rod
F_{xcn}, F_{ycn}, and F_{zcn}	The *x*-, *y*-, and *z*-direction coupler forces of the *n*th coupler
F_{rtn}	The suspension forces of the *n*th lateral dampers
F_{sn}	The lateral forces of the stopblock on the *n*th bogie
F_{rc}	The basic resistance of the car body

In addition, the lateral forces generated by the lateral dampers can be calculated as

$$F_{yctn} = \begin{cases} F_{\max} v_{yctn}/v_0 \dot{Y}_{tn} & (|v_{yctn}| < v_0) \\ F_{\max} \operatorname{sign}(v_{yctn}) & (|v_{yctn}| \geq v_0) \end{cases} \tag{2.186}$$

where $F_{\max}$ is the saturation force of the damper; v_0 is the unloading velocity of the damper; v_{yct} is the relative velocity between two ends of the damper connecting the car body and the bogie frame in the lateral direction, which can be obtained by

$$v_{yctn} = \dot{Y}_{tn} - \dot{Y}_c + H_{cB}\dot{\phi}_c + H_{Bt}\dot{\phi}_{tn} + (-1)^{n-1} l_c \dot{\psi}_{tn} + \frac{l_c^2}{2}\frac{d}{d_t}\left(\frac{1}{R_c}\right) \tag{2.187}$$

Vertical forces

$$\begin{aligned} F_{zt(L,R)i} = {} & K_{sz}\begin{bmatrix} Z_c - Z_{tn} \pm d_s\phi_{tn} \mp d_s\phi_c + (-1)^{i-1} l_{sb}\beta_{tn} \\ + (-1)^n l_c\beta_c + (-1)^i l_{sb}\beta_c \end{bmatrix} \\ & + C_{sz}\begin{bmatrix} \dot{Z}_c - \dot{Z}_{tn} \pm d_s\dot{\phi}_{tn} \mp d_s\dot{\phi}_c + (-1)^{i-1} l_{sb}\dot{\beta}_{tn} \\ + (-1)^n l_c\dot{\beta}_c + (-1)^i l_{sb}\dot{\beta}_c \end{bmatrix} \end{aligned} \tag{2.188}$$

③ Suspension forces of the traction motor (i = 1–4)

The specific structures of traction motor are shown in Figs. 2.30 and 2.31. The suspension stiffness of the traction motor in all directions is determined by the swing motion of the suspender and the stiffness of the rubber joint. The radial stiffness and axial stiffness of axle-hung bearing are selected according to the

empirical value. The equations for the traction motor suspension forces and the axle-hung bearing forces can be calculated as follows:
Longitudinal forces

$$\begin{aligned} F_{xmi} = & K_{mx}\left[X_{tn} - X_{mi} + (-1)^{i-1}H_{mt}\beta_{tn} + (-1)^{i-1}H_{mw}\beta_{mi}\right] \\ & + C_{mx}\left[\dot{X}_{tn} - \dot{X}_{mi} + (-1)^{i-1}H_{mt}\dot{\beta}_{tn} + (-1)^{i-1}H_{mw}\dot{\beta}_{mi}\right] \end{aligned} \tag{2.189}$$

where H_{mw} is the vertical distance between the gravity center of the motor and that of the wheelset, while H_{mt} is the vertical distance between the gravity center of the motor and that of the bogie frame.
Lateral forces

$$\begin{aligned} F_{ymi} = & K_{my}\left[Y_{tn} - Y_{mi} + (-1)^{i-1}l_m\psi_{tn} + (-1)^{i}l_2\psi_{mi}\right] \\ & + C_{my}\left[\dot{Y}_{tn} - \dot{Y}_{mi} + (-1)^{i-1}l_m\dot{\psi}_{tn} + (-1)^{i}l_2\dot{\psi}_{mi}\right] \end{aligned} \tag{2.190}$$

where l_m represents the longitudinal distance between the gravity center of the bogie frame and the motor suspension position on the frame, and l_2 the longitudinal distance between the gravity center of the motor and the motor suspension position on the frame.
Vertical forces

$$\begin{aligned} F_{zmi} = & K_{mz}\left[Z_{tn} - Z_{mi} + (-1)^{i}l_m\beta_{tn} + (-1)^{i}l_m\beta_{mi}\right] \\ & + C_{mz}\left[\dot{Z}_{tn} - \dot{Z}_{mi} + (-1)^{i}l_m\dot{\beta}_{tn} + (-1)^{i}l_m\dot{\beta}_{mi}\right] \end{aligned} \tag{2.191}$$

④ Bearing forces of the axle-hung motor (i = 1–4)

Longitudinal forces

$$\begin{aligned} F_{mwx(L,R)i} = & K_{mwx}\left[X_{mi} - X_{wi} + H_{mw}\beta_{mi} \mp l_{mb}(\psi_{mi} + \psi_{wi}) \mp (-1)^{i-1}l_{mb}\left(\frac{l_1}{R_{mi}}\right)\right] \\ & + C_{mwx}\left[\dot{X}_{mi} - \dot{X}_{wi} + H_{mw}\dot{\beta}_{mi} \mp l_{mb}\left(\dot{\psi}_{mi} + \dot{\psi}_{wi}\right) \mp (-1)^{i-1}l_{mb}\frac{d}{dt}\left(\frac{l_1}{R_{mi}}\right)\right] \end{aligned} \tag{2.192}$$

where l_{mb} the horizontal distance between the left and right bearing installation location and the gravity center of the wheelset. It should be noted that the left and right bearings are symmetrically mounted about the gravity center of the wheelset. R_{mi} represents the curvature radius of the track where the ith motor locates.

Lateral forces

$$\begin{aligned} F_{\mathrm{mwy(L,R)}i} = K_{\mathrm{mwy}} \left[Y_{\mathrm{w}i} - Y_{\mathrm{m}i} + (-1)^i l_1 \psi_{\mathrm{m}i} + H_{\mathrm{mw}} \phi_{\mathrm{m}i} + \frac{l_1^2}{2R_{\mathrm{m}i}} \right] \\ + C_{\mathrm{mwy}} \left[\dot{Y}_{\mathrm{w}i} - \dot{Y}_{\mathrm{m}i} + (-1)^i l_1 \dot{\psi}_{\mathrm{m}i} + H_{\mathrm{mw}} \dot{\phi}_{\mathrm{m}i} + \frac{l_1^2}{2} \frac{d}{dt} \left(\frac{1}{R_{\mathrm{m}i}} \right) \right] \end{aligned} \tag{2.193}$$

where l_1 denotes the longitudinal distance between the gravity center of the motor and the center of the wheelset.

Vertical forces

$$\begin{aligned} F_{\mathrm{mwz(L,R)}i} = K_{\mathrm{mwz}} \left[Z_{\mathrm{m}i} - Z_{\mathrm{w}i} + (-1)^i l_1 \beta_{\mathrm{m}i} \mp l_{\mathrm{m}} \phi_{\mathrm{m}i} \pm l_{\mathrm{mb}} \phi_{\mathrm{w}i} \right] \\ + C_{\mathrm{mwz}} \left[\dot{Z}_{\mathrm{m}i} - \dot{Z}_{\mathrm{w}i} + (-1)^i l_1 \dot{\beta}_{\mathrm{m}i} \mp l_{\mathrm{m}} \dot{\phi}_{\mathrm{m}i} \pm l_{\mathrm{mb}} \dot{\phi}_{\mathrm{w}i} \right] \end{aligned} \tag{2.194}$$

⑤ Forces of the traction rod (n = 1–2)

There are many kinds of traction devices in locomotives. The central push–pull low-position traction rods are adopted in this model. Figure 2.49 show the decomposition diagrams of the forces acting on the car body by the rear traction rod in the vertical–longitudinal plane and the longitudinal–lateral plane, respectively. The actual force transmitted by the traction rod can be obtained by the stiffness of traction rod K_{tr} and the dynamic variation of space distance $\Delta l_{\mathrm{tr}n}$ between the traction points of the car body and bogie frame.

Assuming the initial equilibrium state, the longitudinal and vertical distances between the car body traction point and the bogie frame traction point are l_{O} and h_{O}, respectively. Based on the abovementioned assumptions and the relative displacement between the car body and the bogie frame $l_{x\mathrm{tr}n}$, $l_{y\mathrm{tr}n}$, and $l_{z\mathrm{tr}n}$, which denote the dynamic distances from the traction point of the bogie frame to that of the car body in x-, y-, and z-direction, can be calculated as follows:

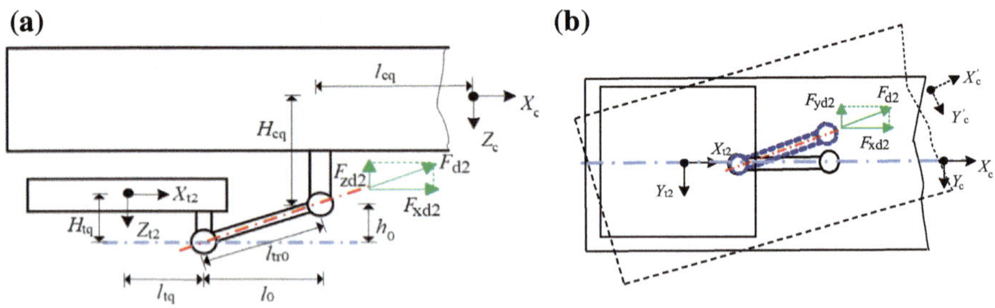

Fig. 2.49 Decomposition diagrams of the forces acting on the car body by the rear traction rod: **a** vertical–longitudinal plane and **b** longitudinal–lateral plane

$$\begin{cases} l_{xtrn} = l_0 + (-1)^n (X_c - X_{tn} + H_{cq}\beta_c - H_{tq}\beta_{tn}) \\ l_{ytrn} = Y_c - Y_{tn} + (-1)^{n-1} l_{cq}\psi_c + (-1)^{n-1} l_{tq}\psi_{tn} \\ l_{ztrn} = h_0 - \left[Z_c - Z_{tn} + (-1)^n l_{cq} \cdot \beta_c + (-1)^{n-1} l_{tq}\beta_{tn}\right] \end{cases} \tag{2.195}$$

where H_{cq} and l_{cq}, respectively, denote the vertical and longitudinal distance between the traction point of the car body and the car body center, while H_{tq} and l_{tq} represent the vertical and longitudinal distance between the traction point of the bogie frame and the bogie frame center, respectively.

Therefore, the space distance between the traction points of the frame and the car body l_{trn} and the dynamic variation of space distance of the traction rod Δl_{trn} are as follows:

$$l_{trn} = \sqrt{l_{xtrn}^2 + l_{ytrn}^2 + l_{ztrn}^2} \tag{2.196}$$

$$\Delta l_{trn} = l_{trn} - \sqrt{l_0^2 + h_0^2} \tag{2.197}$$

The force transmitted by the traction rod is as follows:

$$F_{dn} = K_{tr} \cdot \Delta l_{trn} \tag{2.198}$$

Combined with the dynamic geometric position relationship of the traction rod, the components of the traction rod force in the x-, y-, and z-direction can be calculated as follows:

$$\begin{cases} F_{xdn} = (-1)^n F_{dn} \cdot l_{xtrn}/l_{trn} \\ F_{ydn} = -F_{dn} \cdot l_{ytrn}/l_{trn} \\ F_{zdn} = -F_{dn} \cdot l_{ztrn}/l_{trn} \end{cases} \tag{2.199}$$

⑥ Mesh forces of gear pairs (i = 1–4)

Gear transmissions are usually employed to transmit forces and motions in a mechanical transmission system by the teeth elastic deformations of the gear pairs in engagement. The flexible deformations of the gear pairs can be lumped to a time-varying spring with damping in the LOA direction. The mesh force is calculated as follows:

$$F_{mesh} = K_m\delta + C_m\dot{\delta} \tag{2.200}$$

where F_{mesh}, K_m and C_m represents the mesh force, mesh stiffness, and mesh damping, respectively; δ denotes the Dynamic Transmission Error (DTE) indicating the relative compressive displacement of the engaged gear pairs along LOA.

During the mesh process, the gear mesh stiffness varies with time periodically due to the variations in the number of tooth pairs in mesh and the variations of the contact position along the tooth profile. The time-varying mesh stiffness is calculated as [98, 101]

$$K_{\mathrm{m}}(t)=\frac{\sum_{j=1}^{N}K_j(t)}{1+\sum_{j=1}^{N}K_j(t)E_{ij}(t)/F(t)} \tag{2.201}$$

where N represents the number of tooth pairs in mesh, E denotes the errors of tooth profile. The subscripts i and j denote the number of the tooth pairs, F is the total mesh force of the gear pairs, K is the single-tooth mesh stiffness, which can be calculated as follows [81, 99–101]:

$$\frac{1}{K_j(t)}=\frac{1}{K_{\mathrm{to1}}(t)}+\frac{1}{K_{\mathrm{ff1}}(t)}+\frac{1}{K_{\mathrm{to2}}(t)}+\frac{1}{K_{\mathrm{ff2}}(t)}+\frac{1}{K_{\mathrm{h}}(t)} \tag{2.202}$$

where K_{to}, K_{ff}, and K_{h} indicates the tooth stiffness, the fillet-foundation stiffness, and the Hertz contact stiffness, respectively. The subscripts 1 and 2 represent the pinion and the gear, respectively. It should be noted that the gear fillet-foundation stiffness will be altered in the presence of gear tooth root crack fault, which could be referenced to Ref. [102].

According to the rotational direction of the pinion shown in Fig. 2.30, its counterclockwise rotation is defined as positive, and the corresponding relative displacement of the pinion and the gear teeth along LOA can be calculated as

$$\delta=R_{\mathrm{p}}\beta_{\mathrm{p}}+R_{\mathrm{g}}\beta_{\mathrm{g}}-e \tag{2.203}$$

where R_{p} and R_{g} are the base circle radius of the pinion and the gear, respectively.

However, during the actual operation of the locomotive, the pinion vibrates with the motor, and the gear vibrates with the wheelset. Thus, gear teeth contact loss or even double-sided contacts may happen to the engaged gear teeth due to the drastic speed variations [103]. Consequently, Eq. (2.203) should be revised as follows:

$$\delta_i=\begin{cases}R_{\mathrm{p}}\beta_{\mathrm{p}i}+R_{\mathrm{g}}\beta_{\mathrm{g}i}+(-1)^i\Delta Z_{\mathrm{mw}i}\cos\alpha_0+(-1)^{i-1}\Delta X_{\mathrm{mw}i}\sin\alpha_0-e_i & \text{(for drive-side contact)}\\ 0 & \text{(for teeth contact loss)}\\ R_{\mathrm{p}}\beta_{\mathrm{p}i}+R_{\mathrm{g}}\beta_{\mathrm{g}i}+(-1)^i\Delta Z_{\mathrm{mw}i}\cos\alpha_0-(-1)^{i-1}\Delta X_{\mathrm{mw}i}\sin\alpha_0+e_i & \text{(for coast-side contact)}\end{cases} \tag{2.204}$$

where α_0 denotes the pressure angle of the gear pair, and the symbol e is half of the gear tooth backlash. ΔZ_{mw} and ΔX_{mw} are the vertical and longitudinal relative displacements of the motor and the wheelset, respectively, including the effect of roll and yaw motions of the motor and wheelset. They are calculated as follows:

$$\Delta Z_{mwi} = \Delta Z_{mi} - \Delta Z_{wi} + (-1)^{i} l_{gy}[\sin(\phi_{mi}) - \sin(\phi_{wi})] \tag{2.205}$$

$$\Delta X_{mwi} = \Delta X_{mi} - \Delta X_{wi} + (-1)^{i-1} l_{gy}[\sin(\psi_{mi}) - \sin(\psi_{wi})] \tag{2.206}$$

Based on the calculated suspension forces and the Newton's second law, the equations of motion of the locomotive dynamics system can be derived.

(1) Equations of motion of the wheelset (i = 1–4)

Longitudinal motion:

$$\begin{aligned} M_w \ddot{X}_w = {} & F_{xfLi} + F_{xfRi} + F_{Lxi} + F_{Rxi} + N_{Lxi} \\ & + N_{Rxi} + F_{mwxLi} + F_{mwxRi} + (-1)^{i-1} |F_{xmeshi}| \end{aligned} \tag{2.207}$$

Lateral motion:

$$\begin{aligned} & M_w \left(\ddot{Y}_w + \frac{v^2}{R_{wi}} + r_0 \ddot{\phi}_{sewi} \right) = F_{Lyi} + F_{Ryi} - F_{yfLi} - F_{yfRi} + N_{Lyi} \\ & + N_{Ryi} - F_{mwyRi} - F_{mwyLi} + M_w g \phi_{sewi} \end{aligned} \tag{2.208}$$

Bounce motion:

$$\begin{aligned} & M_w \left(\ddot{Z}_{wi} - a_{0i} \ddot{\phi}_{sewi} - \frac{v^2}{R_{wi}} \phi_{sewi} \right) = -F_{Lzi} - F_{Rzi} - N_{Lzi} - N_{Rzi} \\ & + F_{zfRi} + F_{zfLi} + F_{mwzRi} + F_{mwzLi} + (-1)^{i} F_{zmeshi} + M_w g \end{aligned} \tag{2.209}$$

Roll motion:

$$\begin{aligned} & I_{wx} \left(\ddot{\phi}_{sewi} + \ddot{\phi}_{wi} \right) - I_{wy} \left(\dot{\beta}_{wi} - \Omega \right) \left(\dot{\psi}_{wi} + \frac{v}{R_{wi}} \right) = a_0 (F_{Lzi} + N_{Lzi}) \\ & - a_0 (F_{Rzi} + N_{Rzi}) - r_{Li} (F_{Lyi} + N_{Lyi}) + d_w (F_{zfRi} - F_{zfLi}) + M_{Lxi} \\ & + M_{Rxi} + (F_{mwzLi} - F_{mwzRi}) l_{mb} - r_{Ri} (F_{Ryi} + N_{Ryi}) + l_{gy} F_{zmeshi} \end{aligned} \tag{2.210}$$

where l_{gy} represents the horizontal distance between installation location of the pinion and the gravity center of the motor.

Pitch motion:

$$\begin{aligned} & I_{wz} \left[\ddot{\psi}_{wi} + v \frac{d}{dt} \left(\frac{1}{R_{wi}} \right) \right] - I_{wy} \left(\dot{\phi}_{sewi} + \dot{\phi}_{wi} \right) \left(\dot{\beta}_{wi} - \Omega \right) = M_{Lzi} + M_{Rzi} \\ & + a_0 (F_{Lxi} + N_{Lxi} - F_{Rxi} - N_{Rxi}) + a_0 \psi_{wi} (F_{Lyi} + N_{Lyi} - F_{Ryi} - N_{Ryi}) \\ & + d_w (F_{xfLi} - F_{xfRi}) - (F_{mwxLi} - F_{mwxRi}) l_{mb} + l_{gy} |F_{xmeshi}| \end{aligned} \tag{2.211}$$

Rotational motion:

$$
\begin{aligned}
I_{\mathrm{wy}}\ddot{\beta}_{\mathrm{w}i} = {} & r_{\mathrm{L}i}(F_{\mathrm{L}xi}+N_{\mathrm{L}xi})+r_{\mathrm{R}i}(F_{\mathrm{R}xi}+N_{\mathrm{R}xi})+M_{\mathrm{R}yi}+M_{\mathrm{L}yi} \\
& +r_{\mathrm{R}i}\psi_{\mathrm{w}i}(F_{\mathrm{R}yi}+N_{\mathrm{R}yi})+r_{\mathrm{L}i}\psi_{\mathrm{w}i}(F_{\mathrm{L}yi}+N_{\mathrm{L}yi})+T_{\mathrm{gw}i}
\end{aligned} \tag{2.212}
$$

where T_{gw} denotes the torque generated in the axle between the gear and the wheels.

(2) Equations of motion of the bogie frame (n = 1–2)

Longitudinal motion:

$$
\begin{aligned}
M_{\mathrm{t}}\ddot{X}_{\mathrm{t}n} = {} & (-1)^{n-1}F_{x\mathrm{d}n}-\sum_{i=2n-1}^{2n}(F_{x\mathrm{fL}i}+F_{x\mathrm{fR}i}) \\
& +\sum_{i=2n-1}^{2n}(F_{x\mathrm{tL}i}+F_{x\mathrm{tR}i})-\sum_{i=2n-1}^{2n}F_{x\mathrm{m}i}-F_{\mathrm{rt}n}
\end{aligned} \tag{2.213}
$$

Lateral motion:

$$
\begin{aligned}
& M_{\mathrm{t}}\left[\ddot{Y}_{\mathrm{t}n}+\frac{v^2}{R_{\mathrm{t}n}}+(r_0+H_{\mathrm{tw}})\ddot{\phi}_{\mathrm{set}n}\right]=F_{y\mathrm{d}n}+\sum_{j=2n-1}^{2n}(F_{y\mathrm{fL}j}+F_{y\mathrm{fR}j}) \\
& -\sum_{j=2n-1}^{2n}(F_{y\mathrm{tL}j}+F_{y\mathrm{tR}j})-\sum_{j=2n-1}^{2n}F_{y\mathrm{m}j}+M_{\mathrm{t}}g\phi_{\mathrm{set}i}-F_{\mathrm{s}n}-F_{y\mathrm{ct}n}
\end{aligned} \tag{2.214}
$$

Bounce motion:

$$
\begin{aligned}
& M_{\mathrm{t}}\left[\ddot{Z}_{\mathrm{t}n}-a_0\ddot{\phi}_{\mathrm{set}n}-\frac{v^2}{R_{\mathrm{t}i}}\phi_{\mathrm{set}n}\right]=F_{z\mathrm{d}n}-\sum_{j=2n-1}^{2n}(F_{z\mathrm{fL}j}+F_{z\mathrm{fR}j}) \\
& +\sum_{j=2n-1}^{2n}(F_{z\mathrm{tL}j}+F_{z\mathrm{tR}j})-\sum_{j=2n-1}^{2n}F_{z\mathrm{m}j}+M_{\mathrm{t}}g
\end{aligned} \tag{2.215}
$$

Roll motion:

$$
\begin{aligned}
I_{\mathrm{tx}}\left(\ddot{\phi}_{\mathrm{t}n}+\ddot{\phi}_{\mathrm{set}n}\right) = {} & -H_{\mathrm{tw}}\times\sum_{j=2n-1}^{2n}(F_{y\mathrm{fL}j}+F_{y\mathrm{fR}j})+d_{\mathrm{w}}\times\sum_{j=2n-1}^{2n}(F_{z\mathrm{fL}j}-F_{z\mathrm{fR}j}) \\
& -d_{\mathrm{s}}\times\sum_{j=2n-1}^{2n}(F_{z\mathrm{tL}j}-F_{z\mathrm{tR}j})-H_{\mathrm{mt}}\times\sum_{j=2n-1}^{2n}F_{y\mathrm{m}j}-H_{\mathrm{tq}}F_{y\mathrm{d}n} \\
& +H_{\mathrm{tld}}F_{y\mathrm{ct}n}+H_{\mathrm{Bt}}\times\sum_{j=2n-1}^{2n}(F_{y\mathrm{tL}j}+F_{y\mathrm{tR}j})+l_{\mathrm{dy}}\times\sum_{j=2n-1}^{2n}(-1)^{j-1}F_{z\mathrm{m}j}
\end{aligned} \tag{2.216}
$$

where l_{dy} represents the horizontal distance between suspension location of the motor on bogie frame and the bogie frame center. And H_{tld} denotes the vertical distance between installation location of the lateral damper on bogie frame and the bogie frame center.
Yaw motion:

$$\begin{aligned} I_{\text{tz}}\left(\ddot{\psi}_{\text{t}n} + v\frac{\text{d}}{\text{d}t}\left(\frac{1}{R_{\text{t}n}}\right)\right) &= l_{\text{t}} \times \sum_{j=2n-1}^{2n}\left[(-1)^{j-1}F_{\text{yfL}j} + (-1)^{j-1}F_{\text{yfR}j}\right] \\ &+ (-1)^{n} l_{\text{tq}} F_{\text{yd}n} + l_{\text{sb}} \times \sum_{j=2n-1}^{2n}\left[(-1)^{j}F_{\text{ytL}j} + (-1)^{j}F_{\text{ytR}j}\right] - l_{\text{m}} \times \sum_{j=2n-1}^{2n}(-1)^{j-1}F_{\text{ym}j} \\ &- d_{\text{w}} \times \sum_{j=2n-1}^{2n}(F_{x\text{fL}j} - F_{x\text{fR}j}) + d_{\text{s}} \times \sum_{j=2n-1}^{2n}(F_{x\text{tL}j} - F_{x\text{tR}j}) - l_{\text{dy}} \times \sum_{j=2n-1}^{2n}(-1)^{j-1}F_{x\text{m}j} \end{aligned} \tag{2.217}$$

Pitch motion:

$$\begin{aligned} I_{\text{ty}}\ddot{\beta}_{\text{t}n} &= l_{\text{t}} \times \sum_{j=2n-1}^{2n}\left[(-1)^{j-1}F_{\text{zfL}j} + (-1)^{j-1}F_{\text{zfR}j}\right] + (-1)^{n-1} l_{\text{tq}} F_{\text{zd}n} + (-1)^{n-1} H_{\text{tq}} F_{\text{xd}n} \\ &- H_{\text{tw}} \times \sum_{j=2n-1}^{2n}(F_{x\text{fL}j} + F_{x\text{fR}j}) - H_{\text{Bt}} \times \sum_{j=2n-1}^{2n}(F_{x\text{tL}j} + F_{x\text{tR}j}) - H_{\text{mt}} \times \sum_{j=2n-1}^{2n}F_{x\text{m}j} \\ &+ l_{\text{sb}} \times \sum_{j=2n-1}^{2n}\left[(-1)^{j}F_{\text{ztL}j} + (-1)^{j}F_{\text{ztR}j}\right] + l_{\text{m}} \times \sum_{j=2n-1}^{2n}(-1)^{j-1}F_{\text{zm}j} \end{aligned} \tag{2.218}$$

(3) Equations of motion of the traction motor (i = 1–4)

Longitudinal motion:

$$M_{\text{m}}\ddot{X}_{\text{m}i} = F_{x\text{m}i} - F_{\text{mw}x\text{R}i} - F_{\text{mw}x\text{L}i} + (-1)^{i}\left|F_{x\text{mesh}i}\right| \tag{2.219}$$

Lateral motion:

$$M_{\text{m}}\left[\ddot{Y}_{\text{m}i} + \frac{v^2}{R_{\text{m}i}} + (r_0 + H_{\text{mw}})\ddot{\phi}_{\text{sem}i}\right] = F_{y\text{m}i} + F_{\text{mwyR}i} + F_{\text{mwyL}i} + M_{\text{m}}g\phi_{\text{sem}i} \tag{2.220}$$

where $\phi_{\text{sem}i}$ denotes the super elevation angle of the curve high rail where the ith motor center locates.

Bounce motion:

$$M_{\rm m}\left[\ddot{Z}_{{\rm m}i} - a_0\ddot{\phi}_{{\rm sem}i} - \frac{v^2}{R_{{\rm m}i}}\phi_{{\rm sem}i}\right] = F_{z{\rm m}i} - F_{{\rm mw}z{\rm R}i} - F_{{\rm mw}z{\rm L}i} + (-1)^{i-1}F_{z{\rm mesh}i} + M_{\rm m}g \tag{2.221}$$

Roll motion:

$$\begin{aligned} I_{{\rm m}x}\left(\ddot{\phi}_{{\rm m}i} + \ddot{\phi}_{{\rm sem}i}\right) &= (F_{{\rm mw}z{\rm L}i} - F_{{\rm mw}z{\rm R}i})l_{\rm mb} \\ &- (F_{{\rm mw}y{\rm R}i} + F_{{\rm mw}y{\rm L}i} + F_{y{\rm m}i})H_{\rm mw} - l_{\rm gy}F_{z{\rm mesh}i} \end{aligned} \tag{2.222}$$

Pitch motion:

$$\begin{aligned} I_{{\rm m}z}\left[\ddot{\psi}_{{\rm m}i} + v\frac{{\rm d}}{{\rm d}t}\left(\frac{1}{R_{{\rm m}i}}\right)\right] &= (-1)^{i-1}(F_{{\rm mw}y{\rm R}i} + F_{{\rm mw}y{\rm L}i})l_1 \\ + (-1)^i F_{y{\rm m}i}l_2 - l_{\rm gy}F_{x{\rm mesh}i} &- (F_{{\rm mw}x{\rm L}i} - F_{{\rm mw}x{\rm R}i})l_{\rm mb} \end{aligned} \tag{2.223}$$

Yaw motion:

$$\begin{aligned} I_{{\rm m}y}\ddot{\beta}_{{\rm m}i} &= -(-1)^i(F_{{\rm mw}z{\rm R}i} + F_{{\rm mw}z{\rm L}i})l_1 - (-1)^i F_{z{\rm m}i}l_2 \\ &+ F_{x{\rm m}i}H_{\rm mw} - (F_{{\rm mw}x{\rm L}i} - F_{{\rm mw}x{\rm R}i})H_{\rm mw} - T_{{\rm m}i} \end{aligned} \tag{2.224}$$

where $T_{\rm m}$ denotes the electromagnetic torque applied to the rotor.

(4) Equations of motion of the car body (i = 1–4)

Longitudinal motion:

$$M_{\rm c}\ddot{X}_{\rm c} = \sum_{n=1}^{2}(-1)^{{\rm n}-1}F_{x{\rm c}n} - \sum_{n=1}^{2}F_{x{\rm d}n} - \sum_{j=1}^{4}(F_{x{\rm tL}j} + F_{x{\rm tR}j}) - F_{\rm rc} \tag{2.225}$$

Lateral motion:

$$\begin{aligned} &M_{\rm c}\left[\ddot{Y}_{\rm c} + \frac{v^2}{R_{\rm c}} + (r_0 + H_{\rm tw} + H_{\rm Bt} + H_{\rm cB})\ddot{\phi}_{\rm sec}\right] \\ &= \sum_{n=1}^{2}(F_{y{\rm c}n} - F_{y{\rm d}n} + F_{{\rm s}n} + F_{y{\rm ct}n}) + \sum_{j=1}^{4}(F_{y{\rm tL}j} + F_{y{\rm tR}j}) + M_{\rm c}g\phi_{\rm sec} \end{aligned} \tag{2.226}$$

Bounce motion:

$$M_{\rm c}\left(\ddot{Z}_{\rm c}-a_0\ddot{\phi}_{\rm sec}-\frac{v^2}{R_{\rm c}}\phi_{\rm sec}\right)=M_{\rm c}g-\sum_{n=1}^{2}(F_{zdn}-F_{zcn})-\sum_{j=1}^{4}(F_{z{\rm tL}j}+F_{z{\rm tR}j}) \tag{2.227}$$

Roll motion:

$$\begin{aligned}I_{cx}\left(\ddot{\phi}_{\rm c}+\ddot{\phi}_{\rm sec}\right)&=-H_{\rm cB}\times\sum_{j=1}^{4}(F_{y{\rm tL}j}+F_{y{\rm tR}j})-d_{\rm s}\times\sum_{j=1}^{4}(F_{z{\rm tL}j}-F_{z{\rm tR}j})\\&\quad-\sum_{n=1}^{2}(H_{\rm cg}F_{ycn}-H_{\rm cq}F_{ydn}+H_{\rm cld}F_{yctn}+H_{\rm s}F_{sn})\end{aligned} \tag{2.228}$$

where $H_{\rm cg}$ represents the height of the coupler from the car body center; $H_{\rm cld}$ denotes the height of installation location of the lateral damper on car body from the car body center; $H_{\rm s}$ represents the height of the secondary lateral stopper from the car body center.

Pitch motion:

$$\begin{aligned}I_{cy}\ddot{\beta}_{\rm c}&=l_{\rm sb}\times\sum_{n=1}^{2}\left[(-1)^{n-1}\left(F_{z{\rm tL}n}+F_{z{\rm tR}n}+F_{z{\rm tL}(n+2)}+F_{z{\rm tR}(n+2)}\right)\right]\\&\quad+l_{\rm c}\times\sum_{n=1}^{2}\left[F_{z{\rm tL}n}+F_{z{\rm tR}n}-F_{z{\rm tL}(n+2)}-F_{z{\rm tR}(n+2)}\right]-H_{\rm cB}\times\sum_{j=1}^{4}(F_{x{\rm tL}j}+F_{x{\rm tR}j})\\&\quad+H_{\rm cg}\times\sum_{n=1}^{2}(-1)^{n-1}F_{xcn}+\sum_{n=1}^{2}\left[(-1)^{n}H_{\rm cq}F_{xdn}+(-1)^{n-1}l_{\rm cg}F_{zcn}+(-1)^{n-1}l_{\rm cq}F_{zdn}\right]\end{aligned} \tag{2.229}$$

where $l_{\rm cg}$ denotes longitudinal distance between the coupler and the car body center.

Yaw motion:

$$\begin{aligned}I_{cz}\left[\ddot{\psi}_{\rm c}+v\frac{\rm d}{{\rm d}t}\left(\frac{1}{R_{\rm c}}\right)\right]&=l_{\rm sb}\times\sum_{n=1}^{2}\left[(-1)^{n-1}\left(F_{y{\rm tL}n}+F_{y{\rm tR}n}+F_{y{\rm tL}(n+2)}+F_{y{\rm tR}(n+2)}\right)\right]\\&\quad+l_{\rm c}\times\sum_{n=1}^{2}\left[F_{y{\rm tL}n}+F_{y{\rm tR}n}-F_{y{\rm tL}(n+2)}-F_{y{\rm tR}(n+2)}\right]+\sum_{n=1}^{2}\left[(-1)^{n}l_{\rm cg}F_{ycn}\right]\\&\quad+d_{\rm s}\times\sum_{j=1}^{4}(F_{x{\rm tL}j}-F_{x{\rm tR}j})+\sum_{n=1}^{2}\left[(-1)^{n-1}l_{\rm cq}F_{ydn}+(-1)^{n-1}l_{\rm c}F_{sn}+(-1)^{n-1}l_{\rm c}F_{yctn}\right]\end{aligned} \tag{2.230}$$

(5) Rotational motion of the motor rotor (i = 1–4)

$$J_{ri}\ddot{\beta}_{ri} = T_{mi} - T_{rpi} \tag{2.231}$$

where T_{rp} represents the torque generated in the shaft connecting the rotor and the pinion due to their relative rotational displacements. It can be calculated as

$$T_{rpi} = K_{rpi}(\beta_{ri} - \beta_{pi}) + C_{rpi}(\dot{\beta}_{ri} - \dot{\beta}_{pi}) \tag{2.232}$$

(6) Rotational motion of the pinion (i = 1–4)

$$J_{ri}\ddot{\beta}_{pi} = T_{rpi} - F_{mi}R_p \tag{2.233}$$

(7) Rotational motion of the gear (i = 1–4)

$$J_{gi}\ddot{\beta}_{gi} = -T_{gwi} - F_{mi}R_g \tag{2.234}$$

where

$$T_{gwi} = K_{gwi}(\beta_{gi} - \beta_{wi}) + C_{gwi}(\dot{\beta}_{gi} - \dot{\beta}_{wi}) \tag{2.235}$$

4. *Equations of motion of the ballasted track subsystem*

(1) Differential equations of rail modeled as Euler beam

When the rail is modeled as an Euler beam, the equations of the vertical, lateral, and torsional motions are

$$EI_Y\frac{\partial^4 Z_r(x,t)}{\partial x^4} + m_r\frac{\partial^2 Z_r(x,t)}{\partial t^2} = -\sum_{i=1}^{N} F_{Vi}(t)\delta(x - x_i) + \sum_{j=1}^{4} P_j\delta(x - x_{wj}) \tag{2.236}$$

$$EI_Z\frac{\partial^4 Y_r(x,t)}{\partial x^4} + m_r\frac{\partial^2 Y_r(x,t)}{\partial t^2} = -\sum_{i=1}^{N} F_{Li}(t)\delta(x - x_i) + \sum_{j=1}^{4} Q_j\delta(x - x_{wj}) \tag{2.237}$$

$$\rho_{\rm r} I_0 \frac{\partial^2 \phi_{\rm r}(x,t)}{\partial t^2} - G_{\rm r} I_{\rm t} \frac{\partial^2 \phi_{\rm r}(x,t)}{\partial x^2} = -\sum_{i=1}^{N} M_{{\rm s}i}\delta(x - x_i) + \sum_{j=1}^{4} M_{{\rm w}j}\delta(x - x_{{\rm w}j}) \tag{2.238}$$

in Eqs. (2.236)–(2.238), $EI_{\rm y}$ and $EI_{\rm z}$ are the rail bending stiffness to the y-axle and to the z-axle, respectively; I_0 is the torsional inertia of the rail, $G_{\rm r}$ is the rail shear modulus; $Z_{\rm r}(x, t)$, $Y_{\rm r}(x, t)$, and $\Phi_{\rm r}(x, t)$ are the vertical, lateral, and torsional displacements of the rail, respectively; $F_{{\rm V}i}(t)$ and $F_{{\rm L}i}(t)$ are the vertical and lateral dynamic forces at the ith rail-supporting point; $m_{\rm r}$ is the rail mass per unit length, ρ is the rail density; $P_j(t)$ and $Q_j(t)$ are the vertical and lateral forces of the jth wheel acting on the rail; $M_{{\rm s}i}(t)$ and $M_{{\rm w}j}(t)$ are the moments acting on the rails due to the forces $F_{{\rm V}i}(t)$ and $F_{{\rm L}i}(t)$ and due to the forces $P_j(t)$ and $Q_j(t)$, respectively; $\delta(x)$ is the Dirac delta function; $x_{{\rm s}i}$ is the coordinate of the ith sleeper, $x_{{\rm w}j}$ is the coordinate of the jth wheel, and N is the number of sleepers under the rail.

To solve the fourth-order partial differential equations of rails with time-stepping integration methods, it is necessary to transform Eqs. (2.236)–(2.238) into a series of second-order ordinary differential equations in terms of the generalized coordinates. This could be done by means of the Ritz's method and results are given as follows:

$$\begin{cases} \ddot{q}_{{\rm V}k}(t) + \frac{EI_Y}{m_{\rm r}}\left(\frac{k\pi}{l}\right)^4 q_{{\rm V}k}(t) = -\sum\limits_{i=1}^{N} F_{{\rm V}i} Z_k(x_{{\rm s}i}) + \sum\limits_{j=1}^{4} P_j Z_k(x_{{\rm w}j}) & (k = 1\text{–}N_{\rm V}) \\ \ddot{q}_{{\rm L}k}(t) + \frac{EI_Z}{m_{\rm r}}\left(\frac{k\pi}{l}\right)^4 q_{{\rm L}k}(t) = -\sum\limits_{i=1}^{N} F_{{\rm L}i} Y_k(x_{{\rm s}i}) + \sum\limits_{j=1}^{4} Q_j Y_k(x_{{\rm w}j}) & (k = 1\text{–}N_{\rm L}) \\ \ddot{q}_{{\rm T}k}(t) + \frac{G_{\rm r} I_{\rm t}}{\rho_{\rm r} I_0}\left(\frac{k\pi}{l}\right)^2 q_{{\rm T}k}(t) = -\sum\limits_{i=1}^{N} M_{{\rm s}i} \Theta_k(x_{{\rm s}i}) + \sum\limits_{j=1}^{4} M_{{\rm w}j} \Theta_k(x_{{\rm w}j}) & (k = 1\text{–}N_{\rm T}) \end{cases} \tag{2.239}$$

where $q_{zk}(t)$, $q_{yk}(t)$, and $q_{tk}(t)$ are the kth vertical, lateral, and torsional mode time coordinates, respectively; l is the calculated length of the rail; and Z_k, Y_k, and Φ_k are the rail vertical, lateral, and torsional mode functions, described as

$$\begin{cases} Z_k(x) = \sqrt{\frac{2}{m_{\rm r} l}} \sin\frac{k\pi x}{l} \\ Y_k(x) = \sqrt{\frac{2}{m_{\rm r} l}} \sin\frac{k\pi x}{l} \\ \Theta_k(x) = \sqrt{\frac{2}{\rho_{\rm r} I_0 l}} \sin\frac{k\pi x}{l} \end{cases} \tag{2.240}$$

Figure 2.50 illustrates the free-body force diagram of the right rail, where $O_{\rm r}$ is the twisting center of the rail, e is the distance from the contact point to the central

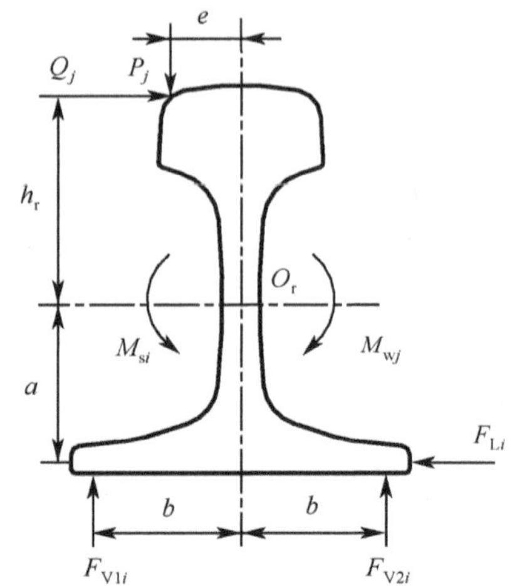

Fig. 2.50 Free-body diagram of rail (Reprinted from Ref. [36], Copyright 2009, with permission from Taylor & Francis.)

line of the rail, h_r is the height from the contact point to the central of rail torsion, a is the height from the rail bottom to the central of rail torsion, b is half distance of two supporting forces under the rail bottom, F_{V1i} and F_{V2i} are the vertical forces between the rail and the sleeper.

F_{V1i} and F_{V2i} can be expressed as

$$\begin{cases} F_{V1i} = \frac{1}{2}K_{pv}[Z_r - b\phi_r - Z_s - (d-b)\phi_s] + \frac{1}{2}C_{pv}\left[\dot{Z}_r - b\dot{\phi}_r - \dot{Z}_s - (d-b)\dot{\phi}_s\right] \\ F_{V2i} = \frac{1}{2}K_{pv}[Z_r + b\phi_r - Z_s - (d+b)\phi_s] + \frac{1}{2}C_{pv}\left[\dot{Z}_r + b\dot{\phi}_r - \dot{Z}_s - (d+b)\dot{\phi}_s\right] \end{cases} \tag{2.241}$$

where Z_r and ϕ_r are the vertical and torsional displacements of the rail; Z_s and ϕ_s are the vertical and torsional displacements of the sleeper.

$$\begin{cases} F_{Vi} = F_{V1i} + F_{V2i} \\ F_{Li} = K_{ph}(Y_r - Y_s - a\phi_r) + C_{ph}(\dot{Y}_r - \dot{Y}_s - a\dot{\phi}_r) \end{cases} \tag{2.242}$$

where Y_r and Y_s are the lateral displacements of rail and sleeper.

$$\begin{cases} M_{si} = (F_{V2i} - F_{V1i})b - F_{Li}a \\ M_{wj} = Q_j h_r - P_j e \end{cases} \tag{2.243}$$

The rail vertical, lateral, and torsional displacements at the time t can then be expressed as

$$\begin{cases} Z_r(x,t) = \sum_{k=1}^{N_V} Z_k(x) q_{Vk}(t) \\ Y_r(x,t) = \sum_{k=1}^{N_L} Y_k(x) q_{Lk}(t) \\ \phi_r(x,t) = \sum_{k=1}^{N_T} \Theta_k(x) q_{Tk}(t) \end{cases} \tag{2.244}$$

where N_V, N_L and N_T are the total mode numbers of the vertical, lateral and torsional mode functions of the rail selected in the calculation.

(2) Differential equations of rail modeled as Timoshenko beam

When the rail is modeled as a Timoshenko beam, the differential equations of the vertical motion are the same as shown in Eqs. (2.69) and (2.70), while the differential equations of the lateral motion are similar to those of the vertical motion, and the differential equation of the torsional motion is the same of that as shown in Eq. (2.238).

The fourth-order partial differential equations of the Timoshenko beam rail can be transformed into a series of second-order ordinary differential equations in terms of the generalized coordinates by using the Ritz's method, which are given as follows:

Vertical motion (k = 1–N_V)

$$\begin{cases} \ddot{q}_{Vk}(t) + \kappa_z A_r G_r \frac{1}{m_r}\left(\frac{k\pi}{l}\right)^2 q_{Vk}(t) - \kappa_z A_r G_r \frac{k\pi}{l}\sqrt{\frac{1}{m_r \rho_r I_Y}} w_{Vk}(t) = -\sum_{i=1}^{N} F_{Vi} Z_k(x_{si}) + \sum_{j=1}^{4} P_j Z_k(x_{wj}) \\ \ddot{w}_{Vk}(t) + \left[\frac{\kappa_z A_r G_r}{\rho_r I_y} + \frac{E}{\rho_r}\left(\frac{k\pi}{l}\right)^2\right] w_{Vk}(t) - \kappa_z A_r G_r \frac{k\pi}{l}\sqrt{\frac{1}{m_r \rho_r I_Y}} q_{Vk}(t) = 0 \end{cases} \tag{2.245}$$

Lateral motion (k = 1–N_L)

$$\begin{cases} \ddot{q}_{Lk}(t) + \kappa_y A_r G_r \frac{1}{m_r}\left(\frac{k\pi}{l}\right)^2 q_{Lk}(t) - \kappa_y A_r G_r \frac{k\pi}{l}\sqrt{\frac{1}{m_r \rho_r I_Z}} w_{Lk}(t) = -\sum_{i=1}^{N} F_{Li} Y_k(x_{si}) + \sum_{j=1}^{4} Q_j Y_k(x_{wj}) \\ \ddot{w}_{Lk}(t) + \left[\frac{\kappa_y A_r G_r}{\rho_r I_Z} + \frac{E}{\rho_r}\left(\frac{k\pi}{l}\right)^2\right] w_{Lk}(t) - \kappa_y A_r G_r \frac{k\pi}{l}\sqrt{\frac{1}{m_r \rho_r I_Z}} q_{Lk}(t) = 0 \end{cases} \tag{2.246}$$

Torsional motion (k = 1–N_T)

$$\ddot{q}_{Tk}(t) + \frac{G_r I_t}{\rho_r I_0}\left(\frac{k\pi}{l}\right)^2 q_{Tk}(t) = -\sum_{i=1}^{N} M_{si}\Theta_k(x_{si}) + \sum_{j=1}^{4} M_{wj}\Theta_k(x_{wj}) \tag{2.247}$$

where κ_z and κ_y are the vertical and lateral shear parameters for the rail cross section.

Based on the modal superposition principle, the vertical and lateral displacements (Z_r, Y_r) and rotation displacements (ψ_{zr}, ψ_{yr}) of the rail can be written as

Vertical

$$\begin{cases} Z_r(x,t) - \sum_{k=1}^{N_V} Z_k(x) q_{Vk}(t) \\ \psi_{zr}(x,t) = \sum_{k=1}^{N_V} \Psi_{zk}(x) w_{Vk}(t) \end{cases} \tag{2.248}$$

Lateral

$$\begin{cases} Y_r(x,t) = \sum_{k=1}^{N_L} Y_k(x) q_{Lk}(t) \\ \psi_{yr}(x,t) = \sum_{k=1}^{N_L} \Psi_{yk}(x) w_{Lk}(t) \end{cases} \tag{2.249}$$

Torsional

$$\phi_r(x,t) = \sum_{k=1}^{N_T} \Theta_k(x) q_{Tk}(t) \tag{2.250}$$

where the normalized shape functions of a simply supported beam are given by

Vertical

$$\begin{cases} Z_k(x) = \sqrt{\frac{2}{m_r l}} \sin(\frac{k\pi}{l} x) \\ \Psi_{zk}(x) = \sqrt{\frac{2}{\rho_r I_Y l}} \cos(\frac{k\pi}{l} x) \end{cases} \tag{2.251}$$

Lateral

$$\begin{cases} Y_k(x) = \sqrt{\frac{2}{m_r l}} \sin(\frac{k\pi}{l} x) \\ \Psi_{yk}(x) = \sqrt{\frac{2}{\rho_r I_Z l}} \cos(\frac{k\pi}{l} x) \end{cases} \tag{2.252}$$

Torsional

$$\Theta_k(x) = \sqrt{\frac{2}{\rho_r I_0 l}} \sin(\frac{k\pi}{l} x) \tag{2.253}$$

(3) Equations of motion of the sleeper

The ith (i = 1, …, N_s) sleeper is represented by a rectangular rigid body, as shown in Fig. 2.51. In the figure, F_{LrV} and F_{RrVi} are the vertical forces between the sleeper

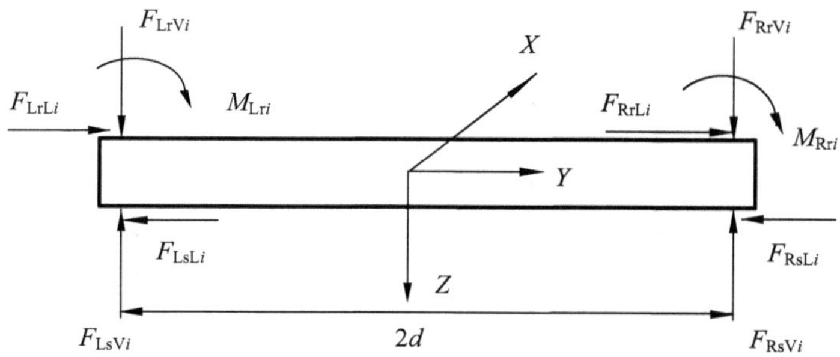

Fig. 2.51 Free-body diagram of sleeper

i and the left and right rails; F_{LrLi} and F_{RrLi} are the lateral forces between the sleeper i and the left and right rails; M_{Lri} and M_{Rri} are the moments acting on the sleeper i from the left and right rails; F_{LsVi} and F_{RsVi} are the vertical supporting forces due to the equivalent ballast bodies; F_{LsLi} and F_{RsLi} are the lateral forces acting on the sleeper from the ballast; d is half length of the sleeper.

The equations of vertical, lateral and rotational motions of the sleeper i read

Vertical motion

$$M_s\ddot{Z}_s = F_{LrVi} + F_{RrVi} - F_{LsVi} - F_{RsVi} \tag{2.254}$$

Lateral motion

$$M_s\ddot{Y}_s = F_{LrLi} + F_{RrLi} - F_{LsLi} - F_{RsLi} \tag{2.255}$$

Rotational motion

$$J_s\ddot{\phi}_s = M_{Lri} + M_{Rri} + (F_{RrVi} - F_{RsVi})d - (F_{LrVi} - F_{LsVi})d \tag{2.256}$$

where J_s is the moment of inertia of the sleeper in roll, $J_s = M_sL_s^2/12$.

The forces applied on the sleeper can be calculated by

$$\begin{cases} F_{LsVi} = K_{bv}(Z_s - Z_{Lb} - d\phi_s) + C_{bv}\left(\dot{Z}_s - \dot{Z}_{Lb} - d\dot{\phi}_s\right) \\ F_{RsVi} = K_{bv}(Z_s - Z_{Rb} + d\phi_s) + C_{bv}\left(\dot{Z}_s - \dot{Z}_{Rb} + d\dot{\phi}_s\right) \end{cases} \tag{2.257}$$

$$\begin{cases} M_{Lri} = -b(F_{LrV1i} - F_{LrV2i}) \\ M_{Rri} = -b(F_{RrV1i} - F_{RrV2i}) \end{cases} \tag{2.258}$$

$$\begin{cases} F_{LsLi} = K_{bh}Y_s + C_{bh}\dot{Y}_s \\ F_{RsLi} = K_{bh}Y_s + C_{bh}\dot{Y}_s \end{cases} \tag{2.259}$$

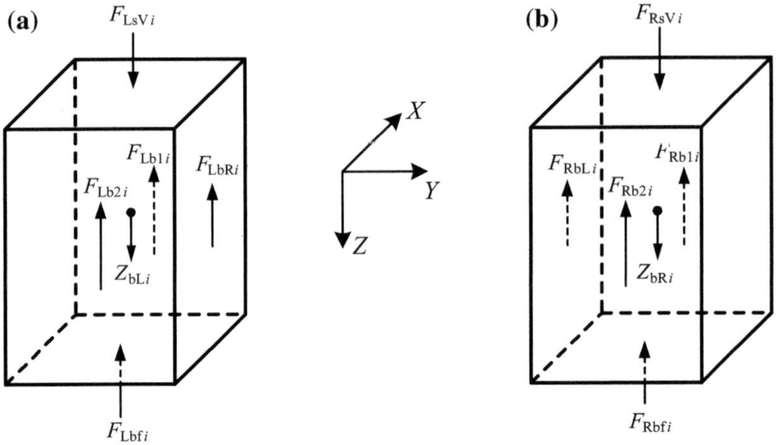

Fig. 2.52 Force diagram of equivalent ballast bodies under a sleeper: **a** left side, and **b** right side

In Eq. (2.257), Z_{Lb} and Z_{Rb} are the vertical displacements of the left and right ballast blocks.

(4) Equations of motion of the ballast

The ballast bed is modeled as equivalent rigid bodies as shown in Fig. 2.52, where F_{Rb1i}, F_{Rb2}, F_{Lb1i}, F_{Lb2i}, F_{LbRi}, and F_{RbLi} are the vertical shear forces between the neighboring ballast bodies, F_{Rbfi} and F_{Lbfi} are the vertical forces between the ballast bodies and the subgrade. The interactive influence of the left and right equivalent ballast bodies are taken into account. Only the vertical motion of each ballast body is considered.

The vertical motion equations of the ballast body i read

Left:

$$M_b\ddot{Z}_{Lbi} = F_{LsVi} - F_{Lbfi} - F_{Lb1i} - F_{Lb2i} - F_{LbRi} \tag{2.260}$$

Right:

$$M_b\ddot{Z}_{Rbi} = F_{RsVi} - F_{Rbfi} - F_{Rb1i} - F_{Rb2i} - F_{RbLi} \tag{2.261}$$

where the forces can be calculated by

$$F_{Rb1i} = K_w\left[Z_{Rbi} - Z_{Rb(i+1)}\right] + C_w\left[\dot{Z}_{Rbi} - \dot{Z}_{Rb(i+1)}\right] \tag{2.262}$$

$$F_{Rb2i} = K_w\left[Z_{Rbi} - Z_{Rb(i-1)}\right] + C_w\left[\dot{Z}_{Rbi} - \dot{Z}_{Rb(i-1)}\right] \tag{2.263}$$

$$F_{\mathrm{Lb1}i} = K_{\mathrm{w}}\left[Z_{\mathrm{Lb}i} - Z_{\mathrm{Lb}(i+1)}\right] + C_{\mathrm{w}}\left[\dot{Z}_{\mathrm{Lb}i} - \dot{Z}_{\mathrm{Lb}(i+1)}\right] \tag{2.264}$$

$$F_{\mathrm{Lb2}i} = K_{\mathrm{w}}\left[Z_{\mathrm{Lb}i} - Z_{\mathrm{Lb}(i-1)}\right] + C_{\mathrm{w}}\left[\dot{Z}_{\mathrm{Lb}i} - \dot{Z}_{\mathrm{Lb}(i-1)}\right] \tag{2.265}$$

$$F_{\mathrm{LbR}i} = K_{\mathrm{w}}(Z_{\mathrm{Lb}i} - Z_{\mathrm{Rb}i}) + C_{\mathrm{w}}\left(\dot{Z}_{\mathrm{Lb}i} - \dot{Z}_{\mathrm{Rb}i}\right) \tag{2.266}$$

$$F_{\mathrm{RbL}i} = -F_{\mathrm{LbR}i} \tag{2.267}$$

$$F_{\mathrm{Rbf}i} = K_{\mathrm{fv}}Z_{\mathrm{Rb}i} + C_{\mathrm{fv}}\dot{Z}_{\mathrm{Rb}i} \tag{2.268}$$

$$F_{\mathrm{Lbf}i} = K_{\mathrm{fv}}Z_{\mathrm{Lb}i} + C_{\mathrm{fv}}\dot{Z}_{\mathrm{Lb}i} \tag{2.269}$$

The boundary conditions are

$$\begin{cases} Z_{\mathrm{Lb0}} = \dot{Z}_{\mathrm{Lb0}} = 0 \\ Z_{\mathrm{Lb}(N+1)} = \dot{Z}_{\mathrm{Lb}(N+1)} = 0 \end{cases} \tag{2.270}$$

$$\begin{cases} Z_{\mathrm{Rb0}} = \dot{Z}_{\mathrm{Rb0}} = 0 \\ Z_{\mathrm{Rb}(N+1)} = \dot{Z}_{\mathrm{Rb}(N+1)} = 0 \end{cases} \tag{2.271}$$

5. Equations of motion of the ballastless track subsystem

(1) Equations of motion of the long-sleeper embedded track

The long-sleeper embedded track only considers the vertical, lateral, and torsional motions of rails. The equations of motion of the Euler beam model for the rail are shown in Eqs. (2.236)–(2.244), and the equation of motion of the Timoshenko beam model for the rail are shown in Eqs. (2.245)–(2.253), where the values of Z_s, ϕ_s, and Y_s in these equations must be set as 0.

(2) Equations of motion of the elastic supporting block track

In this case, the equations of motion of rails are also the same as shown in Eqs. (2.236)–(2.244) for the Euler beam rail model and Eqs. (2.245)–(2.253) for the Timoshenko beam rail model. Based on the free-body diagram of the concrete supporting blocks shown in Fig. 2.53, the equations of motion of the blocks can be easily formulated.

Vertical motion of the left supporting block:

$$M_{\mathrm{b}}\ddot{Z}_{\mathrm{Lb}i} = F_{\mathrm{LrV}i} - F_{\mathrm{LsV}i} \tag{2.272}$$

Lateral motion of the left supporting block:

$$M_{\mathrm{b}}\ddot{Y}_{\mathrm{Lb}i} = F_{\mathrm{LrL}i} - F_{\mathrm{LsL}i} \tag{2.273}$$

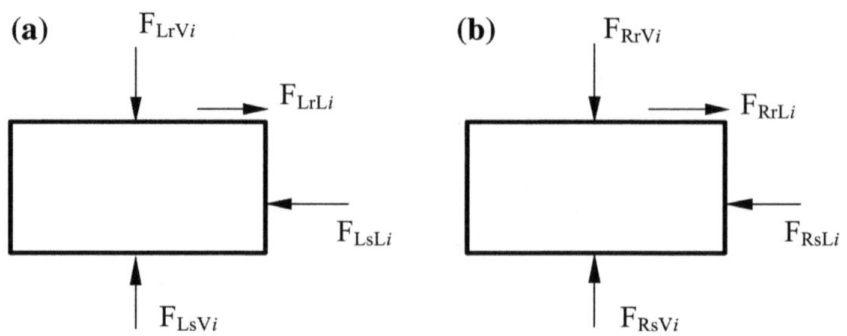

Fig. 2.53 Free-body diagram of concrete supporting blocks: **a** left side, and **b** right side

Vertical motion of the right supporting block:

$$M_{\mathrm{b}}\ddot{Z}_{\mathrm{Rb}i} = F_{\mathrm{RrV}i} - F_{\mathrm{RsV}i} \tag{2.274}$$

Lateral motion of the right supporting block:

$$M_{\mathrm{b}}\ddot{Y}_{\mathrm{Rb}i} = F_{\mathrm{RrL}i} - F_{\mathrm{RsL}i} \tag{2.275}$$

In Eqs. (2.272)–(2.275), $F_{\mathrm{RrV}i}$, $F_{\mathrm{RrL}i}$, $F_{\mathrm{LrV}i}$, $F_{\mathrm{LrL}i}$ can be obtained by Eqs. (2.241) and (2.242), as long as setting $\phi_{\mathrm{s}} = 0$; $F_{\mathrm{RsV}i}$, $F_{\mathrm{RsL}i}$, $F_{\mathrm{LsV}i}$, $F_{\mathrm{LsL}i}$ can be calculated by Eqs. (2.257) and (2.259), as long as setting $Z_{\mathrm{Lb}} = Z_{\mathrm{Rb}} = \phi_{\mathrm{s}} = 0$.

(3) Equation of motion of the slab track in high-speed railways

In the slab track model, the equation of motion of rails are shown in Eqs. (2.236)–(2.244), and Eqs. (2.245)–(2.253). For the track slab in high-speed railways, its thickness is usually much smaller than the length and width, hence, the track slabs are regarded as elastic rectangular thin plates supported on viscoelastic foundation in the vertical direction, while only the rigid mode of the slab vibration in the lateral direction is considered due to the very large lateral bending stiffness. When the damping of the plate is not considered, the differential equation that governs the vibration of classical thin plate can be expressed as

$$D_s\left(\frac{\partial^4 w(x,y,t)}{\partial x^4} + 2\frac{\partial^4 w(x,y,t)}{\partial x^2 \partial y^2} + \frac{\partial^4 w(x,y,t)}{\partial y^4}\right) + \rho_s h_s \frac{\partial^2 w(x,y,t)}{\partial t^2} - q(x,y,t) = 0 \tag{2.276}$$

where

$$D_s = \frac{E_s h_s^3}{12\left(1 - \nu_s^2\right)} \tag{2.277}$$

in Eqs. (2.276)–(2.277), $w(x, y, t)$ is the vertical displacement of the plate; h_s is the thickness of the plate; ρ_s is the plate density; D_s is the vertical bending stiffness of the plate; E_s and ν_s are the Young's elastic modulus and the Poisson's ratio of the plate; $q(x, y, t)$ is the external uniform load acting on the plate.

First, free vibration of the plate ($q = 0$) is studied to obtain the plate natural frequencies. The vertical displacement of the plate takes the following form by introducing the separation variable method

$$w(x, y, t) = W(x, y)\mathrm{e}^{i\omega t} \tag{2.278}$$

where $W(x, y)$ is the mode function of w; i is $\sqrt{-1}$; ω is the angular eigenfrequency of the plate.

The strain energy U and the kinetic energy T of the plate can be expressed as

$$U = \frac{D_s}{2}\iint\limits_S \left[\left(\frac{\partial^2 W(x,y)}{\partial x^2} + \frac{\partial^2 W(x,y)}{\partial y^2}\right)^2 - 2(1-\nu_s)\left(\frac{\partial^2 W(x,y)}{\partial x^2}\cdot\frac{\partial^2 W(x,y)}{\partial y^2} - \left(\frac{\partial^2 W(x,y)}{\partial x\partial y}\right)^2\right)\right] dxdy \tag{2.279}$$

$$T = \frac{\rho_s h_s}{2}\omega^2 \iint\limits_S W^2(x,y)dxdy \tag{2.280}$$

Furthermore, the plate modal function can be can be further equivalent to the product of the Euler–Bernoulli beam functions along the length and width of the plate, respectively.

$$W_{mn}(x, y) = A_{mn}X_m(x)Y_n(y) \tag{2.281}$$

where the A_{mn} is the modal coefficient; $X_m(x)$ and $Y_n(y)$ are Euler–Bernoulli beam functions along the length and width directions, respectively, which are given by [92]

$$\begin{cases} X_1(x) = 1 \\ X_2(x) = \sqrt{3}\left(1 - \frac{2x}{L_s}\right) \\ X_m(x) = \cosh(\alpha_m x) + \cos(\alpha_m x) - \beta_m[\sinh(\alpha_m x) + \sin(\alpha_m x)] \quad (m \geq 3) \end{cases} \tag{2.282a}$$

$$\begin{cases} Y_1(y) = 1 \\ Y_2(y) = \sqrt{3}\left(1 - \frac{2y}{W_s}\right) \\ Y_n(y) = \cosh(\varepsilon_n y) + \cos(\varepsilon_n y) - \xi_n[\sinh(\varepsilon_n y) + \sin(\varepsilon_n y)] \quad (n \geq 3) \end{cases} \tag{2.282b}$$

where L_s and W_s are the length and width of the plate, respectively; α_m and ε_n are the frequency coefficients corresponding to $X_m(x)$ and $Y_n(y)$, which can be calculated by

$$\begin{cases} \alpha_3 = \frac{4.73004}{L_s} \\ \alpha_m = \frac{2m-3}{2L_s}\pi \quad m \geq 4 \end{cases} \tag{2.283a}$$

$$\begin{cases} \varepsilon_3 = \frac{4.73004}{W_s} \\ \varepsilon_n = \frac{2n-3}{2W_s}\pi \quad n \geq 4 \end{cases} \tag{2.283b}$$

β_m and ξ_n are the mode coefficients corresponding to $X_m(x)$ and $Y_n(y)$, which can be calculated by

$$\begin{cases} \beta_3 = 0.982502 \\ \beta_m = \frac{\cosh(\alpha_m L_s) - \cos(\alpha_m L_s)}{\sinh(\alpha_m L_s) - \sin(\alpha_m L_s)} \quad m \geq 4 \end{cases} \tag{2.284a}$$

$$\begin{cases} \xi_3 = 0.982502 \\ \xi_n = \frac{\cosh(\varepsilon_n W_s) - \cos(\varepsilon_n W_s)}{\sinh(\varepsilon_n W_s) - \sin(\varepsilon_n W_s)} \quad n \geq 4 \end{cases} \tag{2.284b}$$

Introduce the Hamilton principle of a conservative system

$$\delta \int_{t_1}^{t_2} (T - U)dt = 0 \tag{2.285}$$

where δ denotes the variation symbol; T and U denote the kinetic and potential energy for a dynamic system, respectively; t_1–t_2 is an arbitrary integration time period.

By substituting Eqs. (2.279)–(2.280) into Eq. (2.285), homogeneous linear equations with respect to the modal coefficient A_{mn} can be arranged in the form

$$\left\{\frac{\rho_s h_s}{2}\omega_{mn}^2 B_{1m}B_{1n} - \frac{D_s}{2}[B_{2m}B_{1n} + B_{1m}B_{2n} + 2(v_s B_{3m}B_{3n} + (1 - v_s)B_{4m}B_{4n})]\right\}A_{mn} = 0 \tag{2.286}$$

where B_{1m}, B_{1n}, B_{2m}, B_{2n}, B_{3m}, B_{3n}, B_{4m}, B_{4n} are the integral constants with respect to the Euler–Bernoulli beam functions, which take the form

$$\begin{aligned} B_{1m} &= \int_0^{L_s} X_m^2(x)dx\, B_{1n} = \int_0^{W_s} Y_n^2(y)dy\, B_{2m} = \int_0^{L_s} X_m''^2(x)dx \\ B_{2n} &= \int_0^{W_s} Y_n''^2(y)dy\, B_{3m} = \int_0^{L_s} X_m''(x)X_m(x)dx\, B_{3n} = \int_0^{W_s} Y_n''(y)Y_n(y)dy \\ B_{4m} &= \int_0^{L_s} X_m'^2(x)dx\, B_{4n} = \int_0^{W_s} Y_n'^2(y)dy \end{aligned} \tag{2.287}$$

Based on Eq. (2.286), the natural frequency of the thin plate is given by

$$\omega_{mn} = \sqrt{\frac{D_s}{\rho_s h_s}} \sqrt{\frac{B_{2m}B_{1n} + B_{1m}B_{2n} + 2[\nu_s B_{3m}B_{3n} + (1-\nu_s)B_{4m}B_{4n}]}{B_{1m}B_{1n}}} \tag{2.288}$$

After obtaining the natural frequency of the thin plate, the classical modal damping can be introduced into the vibration equation of the thin plate, and the forced vibration of the damped thin plate ($q \neq 0$) can be deduced as follows.

According to Ritz method, the solutions of the dynamic vertical displacements can be separately expanded as

$$w(x,y,t) = \sum_{m=1}^{N_x}\sum_{n=1}^{N_y} W_{mn}(x,y)T_{mn}(t) = \sum_{m=1}^{N_x}\sum_{n=1}^{N_y} X_m(x)Y_n(y)T_{mn}(t) \tag{2.289}$$

where $T_{mn}(t)$ is time dependent modal coordinate; N_x or N_y is the truncated mode numbers of $X_m(x)$ and $Y_n(y)$, respectively.

Note that the following mode equations can be established by substituting Eq. (2.278) into the free vibration equation of the thin plate

$$D_s\left(\frac{\partial^4 W(x,y)}{\partial x^4} + 2\frac{\partial^4 W(x,y)}{\partial x^2 \partial y^2} + \frac{\partial^4 W(x,y)}{\partial y^4}\right) - \rho_s h_s \omega^2 W(x,y) = 0 \tag{2.290}$$

Substitution of Eqs. (2.289) and (2.290) into Eq. (2.276) yields

$$\sum_{m=1}^{N_x}\sum_{n=1}^{N_y}\left[\ddot{T}_{mn}(t) + \omega_{mn}^2 T_{mn}(t)\right]\rho_s h_s W_{mn}(x,y) - q(x,y,t) = 0 \tag{2.291}$$

By utilizing orthogonality condition of the mode shapes of the thin plates,

$$\iint\limits_S \rho_s h_s W_{mn} W_{kl}\,dxdy \begin{cases} = 0 & m \neq k \quad \text{or} \quad n \neq l \\ \neq 0 & m = k \quad \text{and} \quad n = l \end{cases} \tag{2.292}$$

the second-order ordinary differential equations of the undamped slab vibration in terms of the generalized coordinate can be first obtained as follows:

$$\ddot{T}_{mn}(t) + \omega_{mn}^2 T_{mn}(t) = \frac{\iint\limits_S q(x,y,t)W_{mn}(x,y)dxdy}{\iint\limits_S \rho_s h_s W_{mn}^2(x,y)dxdy} \tag{2.293}$$

Furthermore, by taking the classical modal damping into account and combining with the actual force distribution of the track slab as shown in Fig. 2.54, the forced vibration equations of the damped concrete slab according to the thin plate theory can be recast as

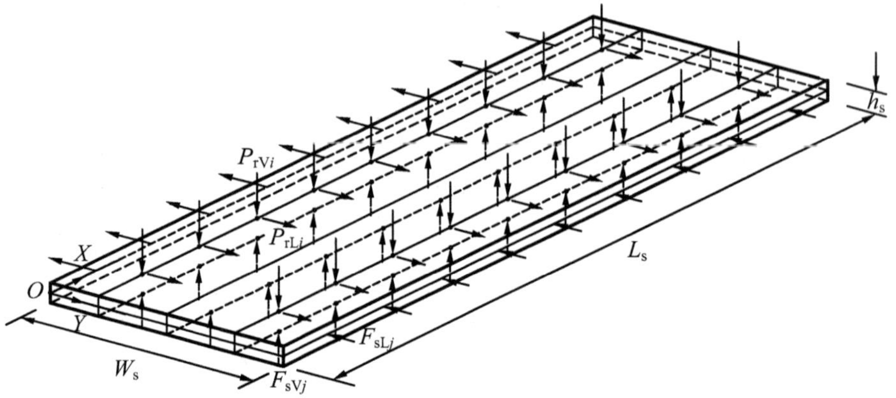

Fig. 2.54 Free-body diagram of the concrete slab

$$\ddot{T}_{mn}(t) + 2\zeta_{mn}\omega_{mn}\dot{T}_{mn}(t) + \omega_{mn}^2 T_{mn}(t) = \frac{\left[\sum_{i=1}^{N_p} P_{rVi}(t)X_m(x_{pi})Y_n(y_{pi}) - \sum_{j=1}^{N_b} F_{sVj}(t)X_m(x_{bj})Y_n(y_{bj})\right]}{\rho_s h_s B_{1m} B_{1n}} \tag{2.294}$$

where $m = 1, 2, \ldots, N_x$, $n = 1, 2, \ldots, N_y$; ζ_{mn} is the damping ratio of the concrete slab; $P_{\mathrm{rV}i}$ is the vertical force at the ith rail fastener, $F_{\mathrm{sV}j}$ is the vertical dynamic force at the jth slab supporting point; N_p and N_b are the total number of the rail fasteners on one slab and the total number of the discrete supporting points under one slab used in the calculation; $x_{\mathrm{p}i}$ and $y_{\mathrm{p}i}$ are the x-coordinate and y-coordinate of the ith rail-supporting points, respectively; $x_{\mathrm{b}j}$ and $y_{\mathrm{b}j}$ are the x-coordinate and y-coordinate of the jth slab supporting point, respectively.

For the lateral motions, the slab is considered as a rigid body. Thus, the lateral vibrations are described by

$$\rho_\mathrm{s} L_\mathrm{s} W_\mathrm{s} h_\mathrm{s} \ddot{y}_\mathrm{s} = \sum_{i=1}^{N_\mathrm{p}} P_{\mathrm{rL}i} - 2\sum_{j=1}^{N_\mathrm{l}} F_{\mathrm{sL}j} \tag{2.295}$$

where $P_{\mathrm{rL}i}$ is the lateral force at the ith rail fastener, $F_{\mathrm{sL}j}$ is the lateral dynamic force at the jth slab supporting point; N_l is the the total node numbers of the slab.

The equation of torsional motion reads

$$J_{\mathrm{sz}} \ddot{\phi}_\mathrm{s} = \sum_{i=1}^{N\mathrm{p}} P_{\mathrm{rL}i} d_{\mathrm{p}i} - 2\sum_{j=1}^{N_\mathrm{l}} F_{\mathrm{sL}j} d_{\mathrm{b}j} \tag{2.296}$$

where $d_{\mathrm{p}i}$ is the longitudinal distance of the ith fastener to the slab center, $d_{\mathrm{b}j}$ is the longitudinal distance of the jth supporting point to the slab center, J_{sz} is the moment of inertia of the sleeper in roll, which has a value of $J_{\mathrm{sz}} = \rho_\mathrm{s} L_\mathrm{s} h_\mathrm{s} W_\mathrm{s}^3/12$.

(4) Equation of motion of floating slab track in metro lines

In the floating slab track model, the equation of motion of rails is shown in Eqs. (2.236)–(2.244), and Eqs. (2.245)–(2.253). For the floating slab in metro lines, the track slab is usually relatively thicker and its thickness can often reach 0.5 m or more. When the thickness of the track slab is relatively large compared to the length and width, shear effects of the plate will be significant and it is more reliable to regard the track slab as an elastic thick plate than adopt the classical thin plate model. A dynamic model for the floating slab based on Mindlin plate theory [104] will be introduced in this section. The floating slabs are described as elastic rectangle thick plates supported on steel springs in the vertical direction, for the lateral motion and torsional motions in the x-y plane, rigid behavior is still assumed as the last section stated. The modal functions of the Mindlin plate are approximated by the bidirectional beam functions, while deflection and shear deformation of the beams need to be considered at the same time.

To obtain vibration equations for the Mindlin plate, the Timoshenko beam functions with free–free boundary conditions has to be deduced first. When the effect of transverse–shear deformation of a beam is considered, a pair of coupled differential equations for the deflection w and the bending slope ψ are given by Timoshenko as

$$EI\frac{\partial^2\psi}{\partial x^2} + G\kappa_r A\left(\frac{\partial w}{\partial x} - \psi\right) - I\rho\frac{\partial^2\psi}{\partial t^2} = 0 \tag{2.297a}$$

$$\rho A\frac{\partial^2 w}{\partial t^2} - G\kappa_r A\left(\frac{\partial^2 w}{\partial x^2} - \frac{\partial\psi}{\partial x}\right) = 0 \tag{2.297b}$$

where E is the modulus of elasticity, G is the shear modulus, I is the area moment of inertia of cross section, ρ is the density, A is the cross-sectional area, and κ_r is the shear factor.

After some algebraic manipulations, the bending slope ψ or the deflection w can be eliminated from Eqs. (2.297a) and (2.297b), respectively.

$$EI\frac{\partial^4 w}{\partial x^4} + \rho A\frac{\partial^2 w}{\partial t^2} - \left(I\rho + \frac{EI\rho}{G\kappa_r}\right)\frac{\partial^4 w}{\partial x^2\partial t^2} + \frac{I\rho^2}{G\kappa_r}\frac{\partial^4 w}{\partial t^4} = 0 \tag{2.298a}$$

$$EI\frac{\partial^4\psi}{\partial x^4} + \rho A\frac{\partial^2\psi}{\partial t^2} - \left(I\rho + \frac{EI\rho}{G\kappa_r}\right)\frac{\partial^4\psi}{\partial x^2\partial t^2} + \frac{I\rho^2}{G\kappa_r}\frac{\partial^4\psi}{\partial t^4} = 0 \tag{2.298b}$$

Let

$$w = X\mathrm{e}^{i\omega t}\ \psi = \Psi\mathrm{e}^{i\omega t}\ \xi = x/L \tag{2.299}$$

where X is the mode function of w, Ψ is the mode function of ψ, i is $\sqrt{-1}$; ω is the angular eigenfrequency of the beam, L is the length of the beam, and ξ is the nondimensional coordinate.

Substituting the Eq. (2.299) into Eq. (2.298a, 2.298b), the solutions of the mode functions $X(\xi)$ and $\Psi(\xi)$ can be obtained as [105]

$$X = A_1 \cosh b\alpha\xi + A_2 \sinh b\alpha\xi + A_3 \cos b\beta\xi + A_4 \sin b\beta\xi \tag{2.300}$$

$$\begin{aligned}\Psi = & \frac{b\alpha A_1}{L[1 - b^2 s^2(\alpha^2 + r^2)]} \sinh b\alpha\xi + \frac{b\alpha A_2}{L[1 - b^2 s^2(\alpha^2 + r^2)]} \cosh b\alpha\xi \\ & - \frac{b\beta A_3}{L\left[1 + b^2 s^2\left(\beta^2 - r^2\right)\right]} \sin b\beta\xi + \frac{b\beta A_4}{L\left[1 + b^2 s^2\left(\beta^2 - r^2\right)\right]} \cos b\beta\xi\end{aligned} \tag{2.301}$$

where

$$\begin{aligned}&\begin{pmatrix}\alpha \\ \beta\end{pmatrix} = \frac{1}{\sqrt{2}} \sqrt{\mp(r^2 + s^2) + \sqrt{(r^2 - s^2) + \frac{4}{b^2}}} \\ &r^2 = \frac{I}{AL^2} \quad s^2 = \frac{EI}{G\kappa_r AL^2} \quad b^2 = \frac{\rho AL^4}{EI}\omega^2\end{aligned} \tag{2.302}$$

By applying the boundary conditions of a free–free Timoshenko beam

$$\begin{aligned}&\Psi'(0) = 0 \quad \Psi'(1) = 0 \\ &\frac{X'(0)}{L} - \Psi(0) = 0 \quad \frac{X'(1)}{L} - \Psi(1) = 0\end{aligned} \tag{2.303}$$

Nonzero solutions for the integration coefficients A_1–A_4 can be obtained by evaluating the fourth-order coefficient determinant of Eq. (2.303), and thus the frequency equation takes the form

$$\begin{aligned}2 - 2\cosh b\alpha \cos b\beta + \frac{b}{\sqrt{1 - b^2 r^2 s^2}} \left[b^2 r^2 \left(r^2 - s^2\right)^2 + \left(3r^2 - s^2\right)\right] \sinh b\alpha \sin b\beta \\ = 0\end{aligned} \tag{2.304}$$

The roots of the above transcendental equation indicate the values of a series of frequency parameters $b_1, b_2, \ldots b_k, \ldots$, and the corresponding mode functions can be determined accordingly.

$$X = \cosh b\alpha\xi + \lambda\delta \sinh b\alpha\xi + \frac{1}{\zeta} \cos b\beta\xi + \delta \sin b\beta\xi \tag{2.305}$$

$$\begin{aligned}\Psi &= \frac{b\alpha}{L[1 - b^2 s^2(\alpha^2 + r^2)]}\sinh b\alpha\xi + \frac{b\alpha\lambda\delta}{L[1 - b^2 s^2(\alpha^2 + r^2)]}\cosh b\alpha\xi \\ &\quad - \frac{b\beta}{\zeta L\left[1 + b^2 s^2\left(\beta^2 - r^2\right)\right]}\sin b\beta\xi + \frac{b\beta\delta}{L\left[1 + b^2 s^2\left(\beta^2 - r^2\right)\right]}\cos b\beta\xi \\ &= k_1 \sinh b\alpha\xi + k_1\lambda\delta\cosh b\alpha\xi - \frac{k_2}{\zeta}\sin b\beta\xi + k_2\delta\cos b\beta\xi\end{aligned} \tag{2.306}$$

in which

$$\lambda = \frac{\alpha}{\beta} \quad \zeta = \frac{\alpha^2 + r^2}{\alpha^2 + s^2} \quad \delta = -\frac{\cosh b\alpha - \cos b\beta}{\lambda\sinh b\alpha - \zeta\sin b\beta} \tag{2.307}$$

It should be noted that the free–free beam modes given by Eqs. (2.305) and (2.306) do not constitute the complete expressions of the mode functions of the Timoshenko beams having a pair of opposite edges free. The above bending modes provide the third and higher trail functions and can be supplemented by the first two-order rigid mode functions representing the translation and rotation degrees of freedom.

$$\begin{cases} X_1 = 1 \\ X_2 = \sqrt{3}(1 - 2\xi) \\ X_k = \cosh b_{k-2}\alpha\xi + \lambda\delta\sinh b_{k-2}\alpha\xi + \frac{1}{\zeta}\cos b_{k-2}\beta\xi + \delta\sin b_{k-2}\beta\xi \quad k \geq 3 \end{cases} \tag{2.308}$$

$$\begin{cases} \Psi_{x1} = 0 \\ \Psi_{x2} = 1 \\ \Psi_{xk} = k_1\sinh b_{k-2}\alpha\xi + k_1\lambda\delta\cosh b_{k-2}\alpha\xi - \frac{k_2}{\zeta}\sin b_{k-2}\beta\xi + k_2\delta\cos b_{k-2}\beta\xi \quad k \geq 3 \end{cases} \tag{2.309}$$

Now consider the vibration equations of Mindlin plate. Different from the classical thin plate, the inclusion of shear effects in Mindlin plate theory means that the cross-sectional rotations β_x, β_y can no longer be expressed solely in terms of the deflection w of the plate median surface, thus, three independent quantities namely w, β_x, and β_y are introduced to represent the deformations of the plate. The three differential equations that govern the vibration of the undamped Mindlin plate can be expressed as [104]

$$G\kappa_r h_s\left(\frac{\partial^2 w}{\partial x^2} + \frac{\partial^2 w}{\partial y^2} + \frac{\partial \beta_x}{\partial x} + \frac{\partial \beta_y}{\partial y}\right) - \rho_s h_s\frac{\partial^2 w}{\partial t^2} + q(x, y, t) = 0 \tag{2.310a}$$

$$\frac{D_s}{2}\left[(1-\nu_s)\left(\frac{\partial^2\beta_x}{\partial x^2}+\frac{\partial^2\beta_y}{\partial y^2}\right)+(1+\nu_s)\left(\frac{\partial^2\beta_x}{\partial x^2}+\frac{\partial^2\beta_y}{\partial x\partial y}\right)\right]-G\kappa_r h_s\left(\beta_x+\frac{\partial w}{\partial x}\right)-\rho_s J\frac{\partial^2\beta_x}{\partial t^2}=0 \tag{2.310b}$$

$$\frac{D_s}{2}\left[(1-\nu_s)\left(\frac{\partial^2\beta_y}{\partial x^2}+\frac{\partial^2\beta_y}{\partial y^2}\right)+(1+\nu_s)\left(\frac{\partial^2\beta_y}{\partial y^2}+\frac{\partial^2\beta_x}{\partial x\partial y}\right)\right]-G\kappa_r h_s\left(\beta_y+\frac{\partial w}{\partial y}\right)-\rho_s J\frac{\partial^2\beta_y}{\partial t^2}=0 \tag{2.310c}$$

where w is the vertical displacement of the plate; β_x and β_y are the cross-sectional rotations in x- and y-directions, respectively; J is the plate rotary inertia per unit length with the value of $J = h_s^3/12$; and other parameters can be referred to the previous section.

Analogously, free vibration of the plate ($q = 0$) is studied first to obtain the plate natural frequencies. The three independent quantities, namely, w, β_x, and β_y, take the following forms by introducing the separation variable method:

$$w(x,y,t) = W(x,y)e^{i\omega t} \tag{2.311a}$$

$$\beta_x(x,y,t) = \Phi_x(x,y)e^{i\omega t} \tag{2.311b}$$

$$\beta_y(x,y,t) = \Phi_y(x,y)e^{i\omega t} \tag{2.311c}$$

where $W(x, y)$ is the dynamic deflection function of the plate; $\Phi_x(x, y)$ and $\Phi_y(x, y)$ are the dynamic rotation functions; and ω is the angular eigenfrequency of the plate.

Based on Eq. (2.311a–2.311c), the strain energy U and the kinetic energy T of the plate can be expressed as [106]

$$\begin{aligned} U = {} & \frac{D_s}{2}\iint\limits_S\left[\left(\frac{\partial\Phi_x}{\partial x}\right)^2+\left(\frac{\partial\Phi_y}{\partial y}\right)^2+2\nu_s\frac{\partial\Phi_x}{\partial x}\frac{\partial\Phi_y}{\partial y}+\frac{1-\nu_s}{2}\left(\frac{\partial\Phi_x}{\partial y}+\frac{\partial\Phi_y}{\partial x}\right)^2\right]dxdy \\ & +\frac{Gh_s}{2\kappa_r}\iint\limits_S\left[\left(\frac{\partial W}{\partial x}+\Phi_x\right)^2+\left(\frac{\partial W}{\partial y}+\Phi_y\right)^2\right]dxdy \end{aligned} \tag{2.312a}$$

$$T = \frac{\rho_s h_s}{2}\omega^2\iint\limits_S W^2 dxdy + \frac{\rho_s J}{2}\omega^2\iint\limits_S\left(\Phi_x^2+\Phi_y^2\right)dxdy \tag{2.312b}$$

Moreover, the modal functions of the Mindlin plate corresponding to the *mn*th natural frequency ω_{mn} can be further equivalent to the product of the Timoshenko beam functions along the length and width of the plate, respectively.

$$W_{mn}(x,y) = A_{mn}X_m(x)Y_n(y) \tag{2.313a}$$

$$\Phi_{xmn}(x,y) = B_{mn}\Psi_{xm}(x)Y_n(y) \tag{2.313b}$$

$$\Phi_{ymn}(x,y) = C_{mn}X_m(x)\Psi_{yn}(y) \tag{2.313c}$$

where the A_{mn}, B_{mn}, C_{mn} are the modal coefficients representing the arbitrary amplitudes of deflection and rotation, the four unidirectional functions $X_m(x)$, $Y_n(y)$, $\Psi_{xm}(x)$, and $\Psi_{yn}(y)$ are those Timoshenko beam functions derived in Eqs. (2.263)–(2.264), which are appropriate to the vibration of thick plates with free boundary conditions.

The free vibration of the thick plates satisfies the Hamilton principle of conservative systems. Substitution of Eqs. (2.312a, 2.312b) and (2.313a–2.313c) into Eq. (2.285) yields homogeneous linear equations with respect to the modal coefficients A_{mn}, B_{mn}, C_{mn} that, after some algebraic manipulation, can be arranged in the form

$$\begin{bmatrix} G\kappa_r h_s(B_{3m}B_{1n}+B_{1m}B_{3n})-\rho_s h_s\omega_{mn}^2 B_{1m}B_{1n} & G\kappa_r h_s B_{5m}B_{1n} & G\kappa_r h_s B_{1m}B_{5n} \\ G\kappa_r h_s B_{5m}B_{1n} & D_s B_{4m}B_{1n}+G\kappa_r h_s B_{2m}B_{1n}+\frac{D_s(1-\nu_s)}{2}B_{2m}B_{3n}-\rho_s J\omega_{mn}^2 B_{2m}B_{1n} & D_s\nu_s B_{6m}B_{6n}+\frac{D_s(1-\nu_s)}{2}B_{5m}B_{5n} \\ G\kappa_r h_s B_{1m}B_{5n} & D_s\nu_s B_{6m}B_{6n}+\frac{D_s(1-\nu_s)}{2}B_{5m}B_{5n} & D_s B_{1m}B_{4n}+G\kappa_r h_s B_{1m}B_{2n}+\frac{D_s(1-\nu_s)}{2}B_{3m}B_{2n}-\rho_s J\omega_{mn}^2 B_{1m}B_{2n} \end{bmatrix} \begin{bmatrix} A_{mn} \\ B_{mn} \\ C_{mn} \end{bmatrix} = 0 \tag{2.314}$$

where B_{1m}, B_{1n}, B_{2m}, B_{2n}, B_{3m}, B_{3n}, B_{4m}, B_{4n}, B_{5m}, B_{5n}, B_{6m}, and B_{6n} are the integral constants with respect to the Timoshenko beam functions, which take the form

$$\begin{aligned} B_{1m} &= \int_0^{L_s} X_m^2(x)dx\, B_{1n} = \int_0^{W_s} Y_n^2(y)dy\, B_{2m} = \int_0^{L_s} \Psi_{xm}^2(x)dx \\ B_{2n} &= \int_0^{W_s} \Psi_{yn}^2(y)dy\, B_{3m} = \int_0^{L_s} X_m'^2(x)dx\, B_{3n} = \int_0^{W_s} Y_n'^2(y)dy \\ B_{4m} &= \int_0^{L_s} \Psi_{xm}'^2(x)dx\, B_{4n} = \int_0^{W_s} \Psi_{yn}'^2(y)dy\, B_{5m} = \int_0^{L_s} \Psi_{xm}(x)X_m'(x)dx \\ B_{5n} &= \int_0^{W_s} \Psi_{yn}(y)Y_n'(y)dy\, B_{6m} = \int_0^{L_s} \Psi_{xm}'(x)X_m(x)dx\, B_{6n} = \int_0^{W_s} \Psi_{yn}'(y)Y_n(y)dy \end{aligned} \tag{2.315}$$

In order to obtain the nonzero solutions of the modal coefficients, the coefficient determinant of Eq. (2.314) must be zero

$$\begin{vmatrix} D_{mn(11)} & D_{mn(12)} & D_{mn(13)} \\ D_{mn(21)} & D_{mn(22)} & D_{mn(23)} \\ D_{mn(31)} & D_{mn(32)} & D_{mn(33)} \end{vmatrix} = 0 \tag{2.316}$$

Equation (2.316) is an univariate cubic equation related to the ω^2, for any set of (m, n), three sequent frequencies can be solved $\omega_{mn}^{(\lambda)}$ (λ = 1, 2, 3, m = 1, 2, 3, …, n = 1, 2, 3, …), namely low, medium, and high ones in which the low one represents the flexural frequency and the other two represent thickness–shear frequencnics.

Once a series of frequencies are obtained, the proportional relationship between the modal coefficients can be determined by substituting them back to Eq. (2.314)

$$B_{mn} = -\frac{\begin{vmatrix} D_{mn(21)} & D_{mn(23)} \\ D_{mn(31)} & D_{mn(33)} \end{vmatrix}}{\begin{vmatrix} D_{mn(22)} & D_{mn(23)} \\ D_{mn(32)} & D_{mn(33)} \end{vmatrix}} A_{mn} \quad C_{mn} = -\frac{\begin{vmatrix} D_{mn(21)} & D_{mn(23)} \\ D_{mn(31)} & D_{mn(33)} \end{vmatrix}}{\begin{vmatrix} D_{mn(23)} & D_{mn(22)} \\ D_{mn(33)} & D_{mn(32)} \end{vmatrix}} A_{mn} \tag{2.317}$$

Meanwhile, the modal functions in Eq. (2.313a–2.313c) can also be determined according to Eq. (2.317).

Now, to consider the forced vibration of the damped Mindlin plate ($q \neq 0$), the solutions of the dynamic vertical and angular displacements can be expanded separately by Ritz method, given as

$$w(x, y, t) = \sum_{m=1}^{N_x} \sum_{n=1}^{N_y} \sum_{\lambda=1}^{3} W_{mn}^{(\lambda)}(x, y) T_{mn}^{(\lambda)}(t) = \sum_{m=1}^{N_x} \sum_{n=1}^{N_y} \sum_{\lambda=1}^{3} A_{mn}^{(\lambda)} X_m(x) Y_n(y) T_{mn}^{(\lambda)}(t) \tag{2.318a}$$

$$\beta_x(x, y, t) = \sum_{m=1}^{N_x} \sum_{n=1}^{N_y} \sum_{\lambda=1}^{3} \Phi_{xmn}^{(\lambda)}(x, y) T_{mn}^{(\lambda)}(t) = \sum_{m=1}^{N_x} \sum_{n=1}^{N_y} \sum_{\lambda=1}^{3} B_{mn}^{(\lambda)} \Psi_{xm}(x) Y_n(y) T_{mn}^{(\lambda)}(t) \tag{2.318b}$$

$$\beta_y(x, y, t) = \sum_{m=1}^{N_x} \sum_{n=1}^{N_y} \sum_{\lambda=1}^{3} \Phi_{ymn}^{(\lambda)}(x, y) T_{mn}^{(\lambda)}(t) = \sum_{m=1}^{N_x} \sum_{n=1}^{N_y} \sum_{\lambda=1}^{3} C_{mn}^{(\lambda)} X_m(x) \Psi_{yn}(y) T_{mn}^{(\lambda)}(t) \tag{2.318c}$$

where $T_{mn}(t)$ is time dependent modal coordinate; N_x or N_y is the truncated mode numbers of $X_m(x)$ and $\Psi_{xm}(x)$ or $Y_n(y)$ and $\Psi_{yn}(y)$, respectively.

Note that the following mode equations can be established by substituting Eq. (2.311a–2.311c) into the free vibration equation of the Mindlin plate

$$G\kappa_r h_s \left(\frac{\partial^2 W}{\partial x^2} + \frac{\partial^2 W}{\partial y^2} + \frac{\partial \Phi_x}{\partial x} + \frac{\partial \Phi_y}{\partial y} \right) = -\rho_s h_s \omega^2 W \tag{2.319a}$$

$$\frac{D_s}{2}\left[(1-\nu_s)\left(\frac{\partial^2\Phi_x}{\partial x^2}+\frac{\partial^2\Phi_y}{\partial y^2}\right)+(1+\nu_s)\left(\frac{\partial^2\Phi_x}{\partial x^2}+\frac{\partial^2\Phi_y}{\partial x\partial y}\right)\right]-G\kappa_r h_s\left(\Phi_x+\frac{\partial W}{\partial x}\right)$$
$$=-\rho_s J\omega^2\Phi_x \tag{2.319b}$$

$$\frac{D_s}{2}\left[(1-\nu_s)\left(\frac{\partial^2\Phi_y}{\partial x^2}+\frac{\partial^2\Phi_y}{\partial y^2}\right)+(1+\nu_s)\left(\frac{\partial^2\Phi_y}{\partial y^2}+\frac{\partial^2\Phi_x}{\partial x\partial y}\right)\right]-G\kappa_r h_s\left(\Phi_y+\frac{\partial W}{\partial y}\right)$$
$$=-\rho_s J\omega^2\Phi_y \tag{2.319c}$$

Substitution of Eqs. (2.318a–2.318c)–(2.319a–2.319c) into Eq. (2.310a–2.310c) yields

$$\sum_{m=1}^{N_x}\sum_{n=1}^{N_y}\sum_{\lambda=1}^{3}\left[\ddot{T}_{mn}^{(\lambda)}(t)+\omega_{mn}^{(\lambda)2}T_{mn}^{(\lambda)}(t)\right]\rho_s h_s W_{mn}^{(\lambda)}-q(x,y,t)=0 \tag{2.320a}$$

$$\sum_{m=1}^{N_x}\sum_{n=1}^{N_y}\sum_{\lambda=1}^{3}\left[\ddot{T}_{mn}^{(\lambda)}(t)+\omega_{mn}^{(\lambda)2}T_{mn}^{(\lambda)}(t)\right]\rho_s J\Phi_{xmn}^{(\lambda)}=0 \tag{2.320b}$$

$$\sum_{m=1}^{N_x}\sum_{n=1}^{N_y}\sum_{\lambda=1}^{3}\left[\ddot{T}_{mn}^{(\lambda)}(t)+\omega_{mn}^{(\lambda)2}T_{mn}^{(\lambda)}(t)\right]\rho_s J\Phi_{ymn}^{(\lambda)}=0 \tag{2.320c}$$

By utilizing orthogonality condition of the mode shapes of the Mindlin plate

$$\iint_S\left[\rho_s h_s W_{mn}^{(\lambda)}W_{kl}^{(\eta)}+\rho_s J\left(\Phi_{xmn}^{(\lambda)}\Phi_{xkl}^{(\eta)}+\Phi_{ymn}^{(\lambda)}\Phi_{ykl}^{(\eta)}\right)\right]dxdy$$
$$\begin{cases}=0 & m\neq k \quad \text{or} \quad n\neq l \quad \text{or} \quad \lambda\neq\eta\\ \neq 0 & m=k \quad \text{and} \quad n=l \quad \text{and} \quad \lambda=\eta\end{cases} \tag{2.321}$$

the three modal coordinates can be decoupled into the following three independent equations ultimately.

$$\ddot{T}_{mn}^{(\lambda)}(t)+\omega_{mn}^{(\lambda)2}T_{mn}^{(\lambda)}(t)=\frac{\iint_S q(x,y,t)W_{mn}^{(\lambda)}(x,y)dxdy}{\iint_S\left[\rho_s h_s W_{mn}^{(\lambda)2}+\rho_s J\left(\Phi_{xmn}^{(\lambda)2}+\Phi_{ymn}^{(\lambda)2}\right)\right]dxdy}\quad \lambda=1,2,3 \tag{2.322}$$

Furthermore, by taking the classical modal damping into account and combining with the actual force distribution of the floating slab as shown in Fig. 2.55, the forced vibration equations of the damped floating slab according to the Mindlin plate theory can be recast as a series of second-order ordinary differential equations

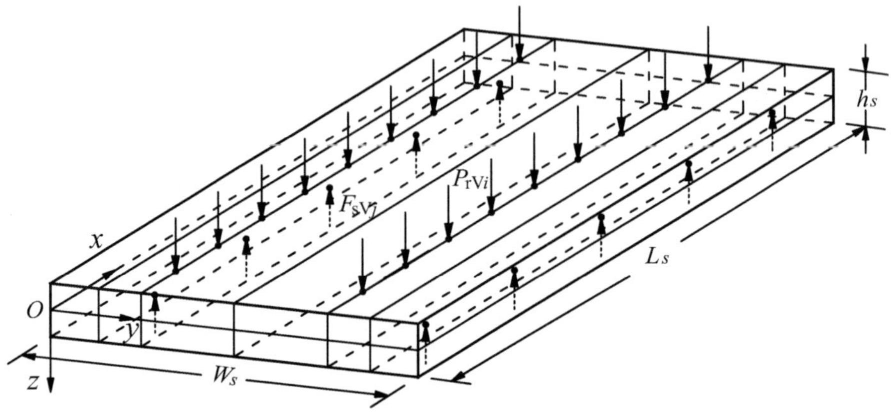

Fig. 2.55 The floating slab subjected to the vertical forces from fasteners and steel springs

$$\ddot{T}_{mn}^{(\lambda)}(t) + 2\zeta_{mn}\omega_{mn}^{(\lambda)}\dot{T}_{mn}^{(\lambda)}(t) + \omega_{mn}^{(\lambda)2}T_{mn}^{(\lambda)}(t) = \frac{\left[\sum_{i=1}^{N_p} P_{rVi}(t)X_m(x_{pi})Y_n(y_{pi}) - \sum_{j=1}^{N_b} F_{sVj}(t)X_m(x_{bj})Y_n(y_{bj})\right]}{\rho_s h_s B_{1m}B_{1n} + \rho_s J\left(B_{mn}^{(\lambda)2}B_{2m}B_{1n} + C_{mn}^{(\lambda)2}B_{1m}B_{2n}\right)} \tag{2.323}$$

where ζ_{mn} denotes the damping ratio of the floating slab; x_{pi} and y_{pi} are the x-coordinate and y-coordinate of the rail-supporting points; x_{bj} and y_{bj} are the x-coordinate and y-coordinate of the steel springs; N_p and N_b are the total number of rail fasteners on the slab and the total number of steel springs under the slab, respectively; P_{rVi} is the vertical force at the ith rail fastener, F_{sVj} is the vertical steel spring force at the jth slab supporting point; and λ = 1, 2, 3 denote the vertical motion, torsional motion in x- and y-directions, respectively.

For the lateral motion and torsional motion of the floating slab in the x-y plane, the equations of motion can be referred to Eqs. (2.295)–(2.296).

2.3.3 *Dynamic Wheel–Rail Coupling Model*

The wheel–rail coupling model is an essential element that couples the vehicle subsystem with the track subsystem at the wheel–rail interfaces. Unlike the early wheel–rail contact model used in the classical vehicle system dynamics, in which the rails are assumed to be fixed without any movement, the proposed dynamic wheel–rail coupling model in this book for analysis of the three-dimensional vehicle–track coupled dynamics problems could consider three kinds of rail motions in the vertical, lateral, and torsional directions.

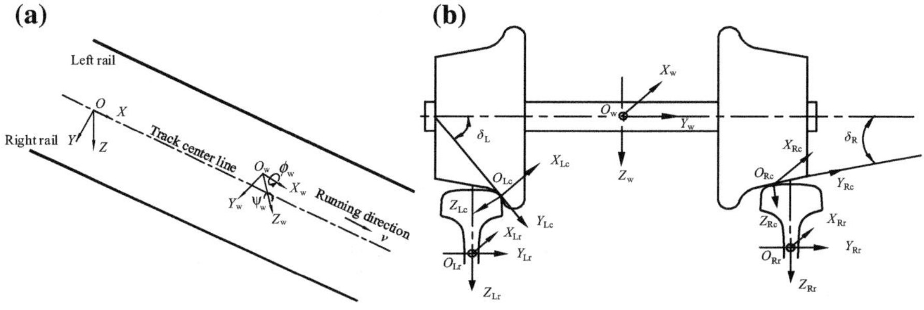

Fig. 2.56 Definitions of coordinate systems: **a** absolute coordinate system and **b** wheelset and rail coordinate systems (Reprinted from Ref. [36], Copyright 2009, with permission from Taylor & Francis.)

1. *Coordinate systems in wheel–rail system*

The wheel–rail coordinate systems are the basis for the calculation of wheel–rail relationship, which should be defined first. Figure 2.56 shows the definitions of the wheelset and rail coordinate systems, which are described in detail as follows:

① O-XYZ is the absolute coordinate system. It is fixed on the centerline of track, which does not change with vehicle moving. Its vector basis is $\boldsymbol{e} = [\boldsymbol{i}, \boldsymbol{j}, \boldsymbol{k}]$.

② O_w-$X_wY_wZ_w$ is the wheelset coordinate system, which is fixed on the mass center of wheelset. It can move in the X, Y, and Z directions, and rotate around the X- and Z-axises of the absolute coordinate system. Its vector basis is $\boldsymbol{e}_w = [\boldsymbol{i}_w, \boldsymbol{j}_w, \boldsymbol{k}_w]$.

③ O_{Lc}-$X_{Lc}Y_{Lc}Z_{Lc}$ and O_{Rc}-$X_{Rc}Y_{Rc}Z_{Rc}$ are the left and right wheel–rail contact point coordinate systems. They are fixed on the left and right contact spots, and change with wheelset moving, their vector basis are $\boldsymbol{e}_{Lc} = [\boldsymbol{i}_{Lc}, \boldsymbol{j}_{Lc}, \boldsymbol{k}_{Lc}]$, $\boldsymbol{e}_{Rc} = [\boldsymbol{i}_{Rc}, \boldsymbol{j}_{Rc}, \boldsymbol{k}_{Rc}]$.

④ O_{Lr}-$X_{Lr}Y_{Lr}Z_{Lr}$ and O_{Rr}-$X_{Rr}Y_{Rr}Z_{Rr}$ are the left and right rail coordinate systems, which are fixed on the mass centers of the left and right rails, and change with rails vibrating. Their vector basis are $\boldsymbol{e}_{Lr} = [\boldsymbol{i}_{Lr}, \boldsymbol{j}_{Lr}, \boldsymbol{k}_{Lr}]$, $\boldsymbol{e}_{Rr} = [\boldsymbol{i}_{Rr}, \boldsymbol{j}_{Rr}, \boldsymbol{k}_{Rr}]$.

The transformation between the above coordinate systems is given by

$$\begin{Bmatrix} \boldsymbol{i}_w \\ \boldsymbol{j}_w \\ \boldsymbol{k}_w \end{Bmatrix} = \begin{bmatrix} \cos\psi_w & \sin\psi_w & 0 \\ -\cos\phi_w \sin\psi_w & \cos\phi_w \cos\psi_w & \sin\phi_w \\ \sin\phi_w \sin\psi_w & -\sin\phi_w \cos\psi_w & \cos\phi_w \end{bmatrix} \begin{Bmatrix} \boldsymbol{i} \\ \boldsymbol{j} \\ \boldsymbol{k} \end{Bmatrix} \tag{2.324}$$

$$\begin{Bmatrix} \boldsymbol{i}_{Lc} \\ \boldsymbol{j}_{Lc} \\ \boldsymbol{k}_{Lc} \end{Bmatrix} = \begin{bmatrix} \cos\psi_w & \sin\psi_w & 0 \\ -\cos(\delta_L+\phi_w) \sin\psi_w & \cos(\delta_L+\phi_w) \cos\psi_w & \sin(\delta_L+\phi_w) \\ \sin(\delta_L+\phi_w) \sin\psi_w & -\sin(\delta_L+\phi_w) \cos\psi_w & \cos(\delta_L+\phi_w) \end{bmatrix} \begin{Bmatrix} \boldsymbol{i} \\ \boldsymbol{j} \\ \boldsymbol{k} \end{Bmatrix} \tag{2.325}$$

$$\begin{Bmatrix} \boldsymbol{i}_{\mathrm{Rc}} \\ \boldsymbol{j}_{\mathrm{Rc}} \\ \boldsymbol{k}_{\mathrm{Rc}} \end{Bmatrix} = \begin{bmatrix} \cos\psi_{\mathrm{w}} & \sin\psi_{\mathrm{w}} & 0 \\ -\cos(\delta_{\mathrm{R}}-\phi_{\mathrm{w}})\sin\psi_{\mathrm{w}} & \cos(\delta_{\mathrm{R}}-\phi_{\mathrm{w}})\cos\psi_{\mathrm{w}} & -\sin(\delta_{\mathrm{R}}-\phi_{\mathrm{w}}) \\ -\sin(\delta_{\mathrm{R}}-\phi_{\mathrm{w}})\sin\psi_{\mathrm{w}} & \sin(\delta_{\mathrm{R}}-\phi_{\mathrm{w}})\cos\psi_{\mathrm{w}} & \cos(\delta_{\mathrm{R}}-\phi_{\mathrm{w}}) \end{bmatrix} \begin{Bmatrix} \boldsymbol{i} \\ \boldsymbol{j} \\ \boldsymbol{k} \end{Bmatrix} \tag{2.326}$$

$$\begin{Bmatrix} \boldsymbol{i}_{\mathrm{Lr}} \\ \boldsymbol{j}_{\mathrm{Lr}} \\ \boldsymbol{k}_{\mathrm{Lr}} \end{Bmatrix} = \begin{bmatrix} 1 & 0 & 0 \\ 0 & \cos(\phi_{\mathrm{rL}}+\phi_0) & \sin(\phi_{\mathrm{rL}}+\phi_0) \\ 0 & -\sin(\phi_{\mathrm{rL}}+\phi_0) & \cos(\phi_{\mathrm{rL}}+\phi_0) \end{bmatrix} \begin{Bmatrix} \boldsymbol{i} \\ \boldsymbol{j} \\ \boldsymbol{k} \end{Bmatrix} \tag{2.327}$$

$$\begin{Bmatrix} \boldsymbol{i}_{\mathrm{Rr}} \\ \boldsymbol{j}_{\mathrm{Rr}} \\ \boldsymbol{k}_{\mathrm{Rr}} \end{Bmatrix} = \begin{bmatrix} 1 & 0 & 0 \\ 0 & \cos(\phi_{\mathrm{rR}}-\phi_0) & \sin(\phi_{\mathrm{rR}}-\phi_0) \\ 0 & -\sin(\phi_{\mathrm{rR}}-\phi_0) & \cos(\phi_{\mathrm{rR}}-\phi_0) \end{bmatrix} \begin{Bmatrix} \boldsymbol{i} \\ \boldsymbol{j} \\ \boldsymbol{k} \end{Bmatrix} \tag{2.328}$$

where δ_{L}, δ_{R} are the left and right wheel–rail contact angles; ϕ_{rL}, ϕ_{rR} are the roll angles of the left and right rails; ϕ_0 is standard rail cant.

2. Dynamic wheel–rail spatially coupling model

(1) Wheel–rail contact geometry calculation

Trace curve method is a common computation method of wheel–rail spatial contact geometry. According to the trace curve method [107], the wheel–rail spatial contact points are only on a spatial curve, which is named trace curve, and thus the contact points could be searched out through seeking a trace curve. Therefore, it becomes a one dimensional scanning problem.

The detailed procedure for solving the wheel–rail contact geometry relationship is described as follows:

Step 1 Obtain the lateral displacement, yaw angle and roll angle of wheelset, as well as the lateral displacements, vertical displacements and the torsion angles of rails by solving the vehicle and track system equations with numerical integration method.

Step 2 Based on the discrete datum of wheel and rail profiles and the instantaneous motions of the wheelsets, calculate the contact trace curve of the wheels in 3D spaces by using the trace curve method. Figure 2.57 schematically shows the wheel–rail spatial contact geometry relationship, where O_{R} is the right wheel–rail contact point, B is the center of the rolling circle, A is the intersection point of the common normal line of the contact surfaces and the wheel axle centerline, O_{w} is the center of the wheelset, O'_{R}, B', A', O'_{w} are, respectively, the projections of O_{R}, B, A, O_{w} on the O-XY plane, $O'_{\mathrm{w}}B'$ is the projections of $O_{\mathrm{w}}B$ on the O-XY plane, O''_{w} is the projection point of O'_{w} on the OY axis.

From Fig. 2.57, the wheel–rail contact point O_{R} in the absolute coordinate system O-XYZ can be calculated as

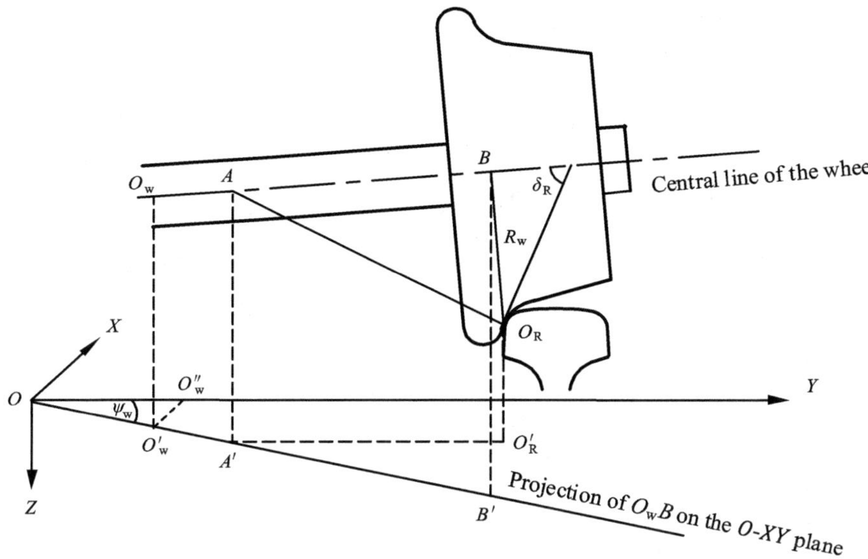

Fig. 2.57 Schematic diagram of wheel–rail spatial contact geometry

$$\begin{cases} x = x_B + l_x R_w \tan\delta_R \\ y = y_B - \frac{R_w}{1-l_x^2}(l_x^2 l_y \tan\delta_R + l_z m) \\ z = z_B - \frac{R_w}{1-l_x^2}(l_x^2 l_z \tan\delta_R - l_y m) \end{cases} \tag{2.329}$$

where $m = \sqrt{1 - l_x^2(1 + \tan^2\delta_R)}$; l_x, l_y, l_z can be calculated by

$$\begin{cases} l_x = -\cos\phi_w \sin\psi_w \\ l_y = \cos\phi_w \cos\psi_w \\ l_z = \sin\phi_w \end{cases} \tag{2.330}$$

The coordinates of the rolling circle center B in the absolute coordinate system are

$$\begin{cases} x_B = d_w l_x \\ y_B = d_w l_y + Y_w \\ z_B = d_w l_z \end{cases} \tag{2.331}$$

where d_w is the lateral distance of the rolling circle from wheelset center.

Thus, the series of the wheel rolling circles can be determined with the changing of d_w, and then the wheel–rail contact trace curve can be determined.

Step 3 Create the discrete datum of profiles of the wheels and rails by dealing with the given profiles (measured or artificially generated) based on the spline interpolation algorithm, here the rail deformation and the surface defects are combined into the rail profile.

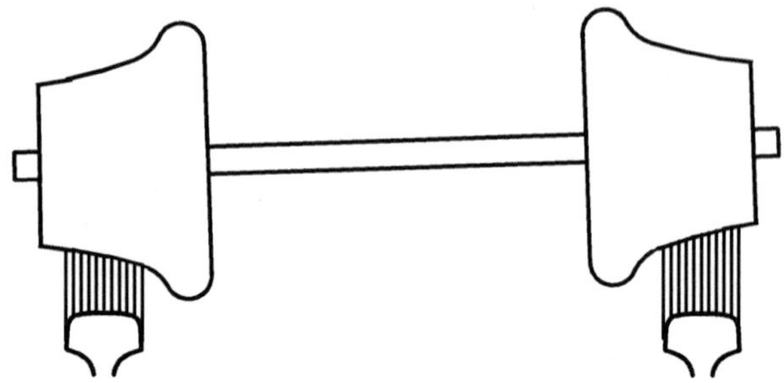

Fig. 2.58 Schematic diagram of minimum distance algorithm

Step 4 Artificially upraise the wheelset to a height enough to ensure no penetration occurring between the wheels and the rails (see Fig. 2.58), and search for the contact points using the minimum distance algorithm. The contact points just locate at the line that has the minimum distance from the wheel to the rail at each side.

Step 5 Using the known locations of the contact points to calculate the wheel–rail contact geometry parameters, including the curvature radii of the wheels and rails at their contact points according to the prescribed wheel and rail profiles, the coordinates of wheel–rail contact points in the wheel and rail local coordinate systems, and the contact angles.

It can be seen that the above wheel–rail contact geometry calculation avoids the iteration solution of the wheelset rolling angle.

(2) Wheel–rail normal force model

The nonlinear Hertzian elastic contact theory is used to calculate the wheel–rail normal contact forces, which is shown in Eq. (2.106). In order to solve the wheel–rail normal force correctly, the key issue is to accurately obtain the wheel–rail normal elastic compressing deformation at each contact point at every moment. When only the vertical vibration is considered, the wheel–rail normal compressing deformation equals to the vertical relative displacement of wheel and rail. But when the lateral and vertical wheel and rail vibrations are considered at the same time, the factors influencing the wheel–rail normal compressing deformation are multitudinous, such as the lateral displacement, vertical displacement, yaw angle, and rolling angle of the wheelset, and the lateral displacement, vertical displacement, and torsional angle of the rail, and also the track geometry irregularities. The current model to solve the wheel–rail normal compressing deformation considers all these factors comprehensively.

By using the wheel–rail contact geometry calculation model, we can obtain the minimum vertical distances from the wheel to the rail, $\Delta Z_{\mathrm{Lw}jt}$ (at the left side) and $\Delta Z_{\mathrm{Rw}jt}$ (at the right side). To obtain the vertical relative displacements of wheel and rail at the left and right sides at moment t, the left and right minimum vertical distances at initial moment, $\Delta Z_{\mathrm{Lw}j0}$ (at the left side) and $\Delta Z_{\mathrm{Rw}j0}$ (at the right side), should be subtracted. Assume that the wheelset moves downwards with $\Delta Z_{\mathrm{w}j0}$, the wheelset just contacts and does not compress the rails. Considering the symmetry relationship, we have $\Delta Z_{Lwj0} = \Delta Z_{Rwj0} = \Delta Z_{wj0}$. Therefore, the left and right wheel–rail vertical relative displacements can be expressed as

$$\begin{cases} \delta Z_{Lj} = Z_{wj}(t) - (\Delta Z_{Lwjt} - \Delta Z_{wj0}) \\ \delta Z_{Rj} = Z_{wj}(t) - (\Delta Z_{Rwjt} - \Delta Z_{wj0}) \end{cases} \tag{2.332}$$

where $Z_{wj}(t)$ is the vertical displacement of jth wheelset.

According to the wheel–rail contact geometry, the left and right wheel–rail normal elastic compression deformations can be described as

$$\begin{cases} \delta Z_{Lcj} = \dfrac{\delta Z_{Lj}}{\cos(\delta_L + \phi_w)} \\ \delta Z_{Rcj} = \dfrac{\delta Z_{Rj}}{\cos(\delta_R - \phi_w)} \end{cases} \tag{2.333}$$

Obviously, when δZ_{Lcj} or δZ_{Rcj} is less than zero, which indicates that the wheel at left or right side separates from the rail, the left or right side wheel–rail normal contact force is then equal to zero.

When vehicles move on an irregular track, three types of wheel–rail contact states as shown in Fig. 2.59 are possible. In some particular conditions, the instantaneous loss of contact between one side wheel of a wheelset and the rail would appear (Fig. 2.59b), and worse still, it might happen to both sides of the wheelset under some extreme circumstances (Fig. 2.59c).

(3) Wheel–rail creep force model

The tangential wheel–rail creep forces are calculated first by use of Kalker's linear creep theory [108] and then modified by Shen–Hedrick–Elkins nonlinear model [109, 110]. Unlike the classical wheel–rail contact model, in which the rails are assumed to be fixed without any movement, the current model is capable of considering the rail motions in vertical, lateral and torsional directions when the creep forces are calculated.

Figure 2.60 shows the schematic diagram of wheel–rail contact in coordinate system O_{Rc}-$X_{Rc}Y_{Rc}Z_{Rc}$, where O_{Rc} is the wheel–rail contact point. Here, the right side is taken as the example. The wheelset moves along the axis $O_{Rc}X_{Rc}$, which is coincident with the axis $O_{Rc}X$. $O_{Rc}Y_{Rc}$ lies in the wheel–rail contact plane, $O_{Rc}Z_{Rc}$ indicates the normal direction of the wheel–rail contact plane. The intersection

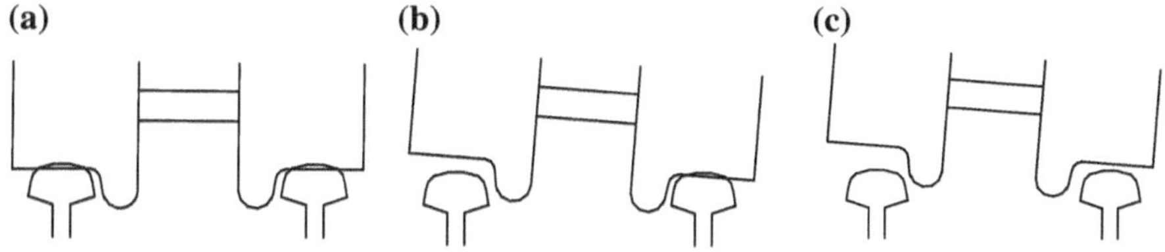

Fig. 2.59 Wheel–rail contact states: **a** constant contact, **b** instantaneous loss of contact at one side, and **c** loss of contact at two sides

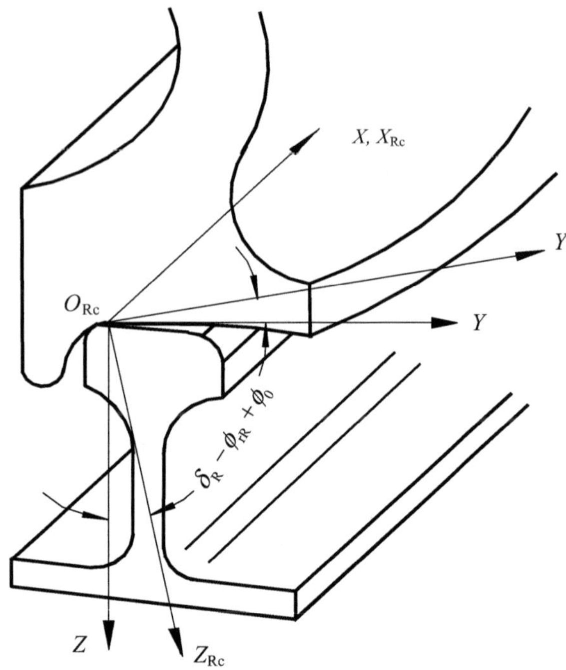

Fig. 2.60 Schematic diagram of wheel–rail contact

angle between the axis $O_{Rc}Y_{Rc}$ and the axis $O_{Rc}Y$ is $\delta_L + \phi_{rL} + \phi_0$ for the left wheel–rail contact point, and δ_R-$\phi_{rR} + \phi_0$ for the right wheel–rail contact point.

In wheel–rail contact spot coordinate O_{Rc}-$X_{Rc}Y_{Rc}Z_{Rc}$, the longitudinal, lateral, and spin creepages are defined as

$$\begin{cases} \xi_x = \frac{v_{w1} - v_{r1}}{v} \\ \xi_y = \frac{v_{w2} - v_{r2}}{v} \\ \xi_\phi = \frac{\Omega_{w3} - \Omega_{r3}}{v} \end{cases} \tag{2.334}$$

where v is the wheelset speed moving on rails, v_{w1}, v_{w2} and Ω_{w3} are the speed of contact ellipse on the wheel along $O_{Rc}X_{Rc}$ axis, $O_{Rc}Y_{Rc}$ axis and around $O_{Rc}Z_{Rc}$ axis, respectively. v_{r1}, v_{r2} and Ω_{r3} are the speed of contact ellipse on the rail along $O_{Rc}X_{Rc}$ axis, $O_{Rc}Y_{Rc}$ axis and around $O_{Rc}Z_{Rc}$ axis, respectively.

(1) Calculation of longitudinal and lateral creepages

The translational motion speed of wheelset center in the absolute coordinate system is

$$\boldsymbol{v}_{\text{ow}} = \dot{X}_{\text{ow}}\boldsymbol{i} + \dot{Y}_{\text{ow}}\boldsymbol{j} + \dot{Z}_{\text{ow}}\boldsymbol{k} \tag{2.335}$$

The absolute angle speed of the wheelset in the absolute coordinate system is expressed as

$$\begin{aligned}\boldsymbol{\omega}_{\mathrm{w}} &= \begin{bmatrix} \dot{\phi}_{\mathrm{w}}\cos\psi_{\mathrm{w}} - (-\Omega+\dot{\beta}_{\mathrm{w}})\cos\phi_{\mathrm{w}}\sin\psi_{\mathrm{w}} \\ \dot{\phi}_{\mathrm{w}}\sin\psi_{\mathrm{w}} + (-\Omega+\dot{\beta}_{\mathrm{w}})\cos\phi_{\mathrm{w}}\cos\psi_{\mathrm{w}} \\ (-\Omega+\dot{\beta}_{\mathrm{w}})\sin\phi_{\mathrm{w}} + \dot{\psi}_{\mathrm{w}} \end{bmatrix}^{\mathrm{T}} \begin{Bmatrix} \boldsymbol{i} \\ \boldsymbol{j} \\ \boldsymbol{k} \end{Bmatrix} \\ &= \begin{bmatrix} \omega_{\mathrm{w}x} \\ \omega_{\mathrm{w}y} \\ \omega_{\mathrm{w}z} \end{bmatrix}^{\mathrm{T}} \begin{Bmatrix} \boldsymbol{i} \\ \boldsymbol{j} \\ \boldsymbol{k} \end{Bmatrix}\end{aligned} \tag{2.336}$$

The vectors from the left and right contact points to the wheelset center $\boldsymbol{R}_{\mathrm{L}}$, $\boldsymbol{R}_{\mathrm{R}}$ can be described as

$$\begin{aligned}\boldsymbol{R}_{\mathrm{L}} &= \begin{bmatrix} R'_{x\mathrm{L}} \\ R'_{y\mathrm{L}} \\ R'_{z\mathrm{L}} \end{bmatrix}^{\mathrm{T}} \begin{Bmatrix} \boldsymbol{i}_{\mathrm{w}} \\ \boldsymbol{j}_{\mathrm{w}} \\ \boldsymbol{k}_{\mathrm{w}} \end{Bmatrix} \\ &= \begin{bmatrix} R'_{x\mathrm{L}}\cos\psi_{\mathrm{w}} - R'_{y\mathrm{L}}\cos\phi_{\mathrm{w}}\sin\psi_{\mathrm{w}} + R'_{z\mathrm{L}}\sin\phi_{\mathrm{w}}\sin\psi_{\mathrm{w}} \\ R'_{x\mathrm{L}}\sin\psi_{\mathrm{w}} + R'_{y\mathrm{L}}\cos\phi_{\mathrm{w}}\cos\psi_{\mathrm{w}} - R'_{z\mathrm{L}}\sin\phi_{\mathrm{w}}\cos\psi_{\mathrm{w}} \\ R'_{y\mathrm{L}}\sin\phi_{\mathrm{w}} + R'_{z\mathrm{L}}\cos\phi_{\mathrm{w}} \end{bmatrix}^{\mathrm{T}} \begin{Bmatrix} \boldsymbol{i} \\ \boldsymbol{j} \\ \boldsymbol{k} \end{Bmatrix} \\ &= \begin{bmatrix} R_{x\mathrm{L}} \\ R_{y\mathrm{L}} \\ R_{z\mathrm{L}} \end{bmatrix}^{\mathrm{T}} \begin{Bmatrix} \boldsymbol{i} \\ \boldsymbol{j} \\ \boldsymbol{k} \end{Bmatrix}\end{aligned} \tag{2.337}$$

$$\begin{aligned}\boldsymbol{R}_{\mathrm{R}} &= \begin{bmatrix} R'_{x\mathrm{R}} \\ R'_{y\mathrm{R}} \\ R'_{z\mathrm{R}} \end{bmatrix}^{\mathrm{T}} \begin{Bmatrix} \boldsymbol{i}_{\mathrm{w}} \\ \boldsymbol{j}_{\mathrm{w}} \\ \boldsymbol{k}_{\mathrm{w}} \end{Bmatrix} \\ &= \begin{bmatrix} R'_{x\mathrm{R}}\cos\psi_{\mathrm{w}} - R'_{y\mathrm{R}}\cos\phi_{\mathrm{w}}\sin\psi_{\mathrm{w}} + R'_{z\mathrm{R}}\sin\phi_{\mathrm{w}}\sin\psi_{\mathrm{w}} \\ R'_{x\mathrm{R}}\sin\psi_{\mathrm{w}} + R'_{y\mathrm{R}}\cos\phi_{\mathrm{w}}\cos\psi_{\mathrm{w}} - R'_{z\mathrm{R}}\sin\phi_{\mathrm{w}}\cos\psi_{\mathrm{w}} \\ R'_{y\mathrm{R}}\sin\phi_{\mathrm{w}} + R'_{z\mathrm{R}}\cos\phi_{\mathrm{w}} \end{bmatrix}^{\mathrm{T}} \begin{Bmatrix} \boldsymbol{i} \\ \boldsymbol{j} \\ \boldsymbol{k} \end{Bmatrix} \\ &= \begin{bmatrix} R_{x\mathrm{R}} \\ R_{y\mathrm{R}} \\ R_{z\mathrm{R}} \end{bmatrix}^{\mathrm{T}} \begin{Bmatrix} \boldsymbol{i} \\ \boldsymbol{j} \\ \boldsymbol{k} \end{Bmatrix}\end{aligned} \tag{2.338}$$

The relative speeds of the contact points with respect to the wheelset center read

$$\begin{aligned}\boldsymbol{v}_{\mathrm{R(L,R)}} &= \boldsymbol{\omega}_{\mathrm{w}} \times \boldsymbol{R}_{\mathrm{(L,R)}} \\ &= \begin{vmatrix} \boldsymbol{i} & \boldsymbol{j} & \boldsymbol{k} \\ \omega_{\mathrm{w}x} & \omega_{\mathrm{w}y} & \omega_{\mathrm{w}z} \\ R_{x\mathrm{(L,R)}} & R_{y\mathrm{(L,R)}} & R_{z\mathrm{(L,R)}} \end{vmatrix} \\ &= \begin{bmatrix} \omega_{\mathrm{w}y}R_{z\mathrm{(L,R)}} - \omega_{\mathrm{w}z}R_{y\mathrm{(L,R)}} \\ \omega_{\mathrm{w}z}R_{x\mathrm{(L,R)}} - \omega_{\mathrm{w}x}R_{z\mathrm{(L,R)}} \\ \omega_{\mathrm{w}x}R_{y\mathrm{(L,R)}} - \omega_{\mathrm{w}y}R_{x\mathrm{(L,R)}} \end{bmatrix}^{\mathrm{T}} \begin{Bmatrix} \boldsymbol{i} \\ \boldsymbol{j} \\ \boldsymbol{k} \end{Bmatrix}\end{aligned} \tag{2.339}$$

According to the speed synthesis theorem, the absolute speeds of the wheelset and the rails at the contact points are obtained:

$$\boldsymbol{v}_{\mathrm{w(L,R)}} = \begin{bmatrix} \dot{X}_{\mathrm{ow}} + \omega_{\mathrm{w}y}R_{z\mathrm{(L,R)}} - \omega_{\mathrm{w}z}R_{y\mathrm{(L,R)}} \\ \dot{Y}_{\mathrm{ow}} + \omega_{\mathrm{w}z}R_{x\mathrm{(L,R)}} - \omega_{\mathrm{w}x}R_{z\mathrm{(L,R)}} \\ \dot{Z}_{\mathrm{ow}} + \omega_{\mathrm{w}x}R_{y\mathrm{(L,R)}} - \omega_{\mathrm{w}y}R_{x\mathrm{(L,R)}} \end{bmatrix}^{\mathrm{T}} \begin{Bmatrix} \boldsymbol{i} \\ \boldsymbol{j} \\ \boldsymbol{k} \end{Bmatrix} \tag{2.340}$$

$$\boldsymbol{v}_{\mathrm{r(L,R)}} = \begin{bmatrix} 0 \\ \delta\dot{Y}_{\mathrm{r(L,R)}} \\ \delta\dot{Z}_{\mathrm{r(L,R)}} \end{bmatrix}^{\mathrm{T}} \begin{Bmatrix} \boldsymbol{i} \\ \boldsymbol{j} \\ \boldsymbol{k} \end{Bmatrix} + \begin{bmatrix} 0 \\ \dot{Y}_{\mathrm{r(L,R)}} + h_{\mathrm{r}}\dot{\phi}_{\mathrm{r(L,R)}} \\ \dot{Z}_{\mathrm{r(L,R)}} \end{bmatrix}^{\mathrm{T}} \begin{Bmatrix} \boldsymbol{i} \\ \boldsymbol{j} \\ \boldsymbol{k} \end{Bmatrix} \tag{2.341}$$

where $\dot{Y}_{\mathrm{r(L,R)}}, \dot{Z}_{\mathrm{r(L,R)}}, \dot{\phi}_{\mathrm{r(L,R)}}$ are the lateral, vertical, and rotational speeds of rails, $\delta\dot{Y}_{\mathrm{r(L,R)}}, \delta\dot{Z}_{\mathrm{r}(L,R)}$ are the lateral and vertical relative speeds induced by track irregularities.

Thus, the differences of absolute speeds of the wheelset and rails at the contact points are calculated as

$$\begin{aligned}\Delta\boldsymbol{v}'_{\mathrm{(L,R)}} &= \begin{bmatrix} \dot{X}_{\mathrm{ow}} + \omega_{\mathrm{w}y}R_{z\mathrm{(L,R)}} - \omega_{\mathrm{w}z}R_{y\mathrm{(L,R)}} \\ \dot{Y}_{\mathrm{ow}} + \omega_{\mathrm{w}z}R_{x\mathrm{(L,R)}} - \omega_{\mathrm{w}x}R_{z\mathrm{(L,R)}} - \dot{Y}_{\mathrm{r(L,R)}} - \delta\dot{Y}_{\mathrm{r(L,R)}} - h_{\mathrm{r}}\dot{\phi}_{\mathrm{r(L,R)}} \\ \dot{Z}_{\mathrm{ow}} + \omega_{\mathrm{w}x}R_{y\mathrm{(L,R)}} - \omega_{\mathrm{w}y}R_{x\mathrm{(L,R)}} - \dot{Z}_{\mathrm{r(L,R)}} - \delta\dot{Z}_{\mathrm{r(L,R)}} \end{bmatrix}^{\mathrm{T}} \begin{Bmatrix} \boldsymbol{i} \\ \boldsymbol{j} \\ \boldsymbol{k} \end{Bmatrix} \\ &= \begin{bmatrix} \Delta v_{x\mathrm{(L,R)}} \\ \Delta v_{y\mathrm{(L,R)}} \\ \Delta v_{z\mathrm{(L,R)}} \end{bmatrix}^{\mathrm{T}} \begin{Bmatrix} \boldsymbol{i} \\ \boldsymbol{j} \\ \boldsymbol{k} \end{Bmatrix}\end{aligned} \tag{2.342}$$

Because the creepage is defined in the wheel–rail contact spot coordinate system, the absolute speed differences should be transformed into the contact spot coordinate system. The transformation between the wheel–rail contact spot coordinate and the absolute coordinate complies with

$$\begin{Bmatrix} \boldsymbol{i} \\ \boldsymbol{j} \\ \boldsymbol{k} \end{Bmatrix} = \begin{bmatrix} \cos\psi_{\mathrm{w}} & \sin\psi_{\mathrm{w}} & 0 \\ -\cos(\delta_{\mathrm{L}}+\phi_{\mathrm{w}})\sin\psi_{\mathrm{w}} & \cos(\delta_{\mathrm{L}}+\phi_{\mathrm{w}})\cos\psi_{\mathrm{w}} & \sin(\delta_{\mathrm{L}}+\phi_{\mathrm{w}}) \\ \sin(\delta_{\mathrm{L}}+\phi_{\mathrm{w}})\sin\psi_{\mathrm{w}} & -\sin(\delta_{\mathrm{L}}+\phi_{\mathrm{w}})\cos\psi_{\mathrm{w}} & \cos(\delta_{\mathrm{L}}+\phi_{\mathrm{w}}) \end{bmatrix}^{-1} \begin{Bmatrix} \boldsymbol{i}_{\mathrm{Lc}} \\ \boldsymbol{j}_{\mathrm{Lc}} \\ \boldsymbol{k}_{\mathrm{Lc}} \end{Bmatrix} \tag{2.343}$$

$$\begin{Bmatrix} \boldsymbol{i} \\ \boldsymbol{j} \\ \boldsymbol{k} \end{Bmatrix} = \begin{bmatrix} \cos\psi_{\mathrm{w}} & \sin\psi_{\mathrm{w}} & 0 \\ -\cos(\delta_{\mathrm{R}}-\phi_{\mathrm{w}})\sin\psi_{\mathrm{w}} & \cos(\delta_{\mathrm{R}}-\phi_{\mathrm{w}})\cos\psi_{\mathrm{w}} & -\sin(\delta_{\mathrm{R}}-\phi_{\mathrm{w}}) \\ -\sin(\delta_{\mathrm{R}}-\phi_{\mathrm{w}})\sin\psi_{\mathrm{w}} & \sin(\delta_{\mathrm{R}}-\phi_{\mathrm{w}})\cos\psi_{\mathrm{w}} & \cos(\delta_{\mathrm{R}}-\phi_{\mathrm{w}}) \end{bmatrix}^{-1} \begin{Bmatrix} \boldsymbol{i}_{\mathrm{Rc}} \\ \boldsymbol{j}_{\mathrm{Rc}} \\ \boldsymbol{k}_{\mathrm{Rc}} \end{Bmatrix} \tag{2.344}$$

Thus, the relative speed differences of the wheelset and rails at the contact points in the wheel–rail contact spot coordinate system are calculated as

$$\begin{aligned} \Delta\boldsymbol{v}_{\mathrm{L}} &= \begin{bmatrix} \Delta v_{x\mathrm{L}} \\ \Delta v_{y\mathrm{L}} \\ \Delta v_{z\mathrm{L}} \end{bmatrix}^{\mathrm{T}} \begin{bmatrix} \cos\psi_{\mathrm{w}} & \sin\psi_{\mathrm{w}} & 0 \\ -\cos(\delta_{\mathrm{L}}+\phi_{\mathrm{w}})\sin\psi_{\mathrm{w}} & \cos(\delta_{\mathrm{L}}+\phi_{\mathrm{w}})\cos\psi_{\mathrm{w}} & \sin(\delta_{\mathrm{L}}+\phi_{\mathrm{w}}) \\ \sin(\delta_{\mathrm{L}}+\phi_{\mathrm{w}})\sin\psi_{\mathrm{w}} & -\sin(\delta_{\mathrm{L}}+\phi_{\mathrm{w}})\cos\psi_{\mathrm{w}} & \cos(\delta_{\mathrm{L}}+\phi_{\mathrm{w}}) \end{bmatrix}^{-1} \begin{Bmatrix} \boldsymbol{i}_{\mathrm{Lc}} \\ \boldsymbol{j}_{\mathrm{Lc}} \\ \boldsymbol{k}_{\mathrm{Lc}} \end{Bmatrix} \\ &= \begin{bmatrix} \Delta v_{1\mathrm{L}} \\ \Delta v_{2\mathrm{L}} \\ \Delta v_{3\mathrm{L}} \end{bmatrix}^{\mathrm{T}} \begin{Bmatrix} \boldsymbol{i}_{\mathrm{Lc}} \\ \boldsymbol{j}_{\mathrm{Lc}} \\ \boldsymbol{k}_{\mathrm{Lc}} \end{Bmatrix} \end{aligned} \tag{2.345}$$

$$\begin{aligned} \Delta\boldsymbol{v}_{\mathrm{R}} &= \begin{bmatrix} \Delta v_{x\mathrm{R}} \\ \Delta v_{y\mathrm{R}} \\ \Delta v_{z\mathrm{R}} \end{bmatrix}^{\mathrm{T}} \begin{bmatrix} \cos\psi_{\mathrm{w}} & \sin\psi_{\mathrm{w}} & 0 \\ -\cos(\delta_{\mathrm{R}}-\phi_{\mathrm{w}})\sin\psi_{\mathrm{w}} & \cos(\delta_{\mathrm{R}}-\phi_{\mathrm{w}})\cos\psi & -\sin(\delta_{\mathrm{R}}-\phi_{\mathrm{w}}) \\ -\sin(\delta_{\mathrm{R}}-\phi_{\mathrm{w}})\sin\psi_{\mathrm{w}} & \sin(\delta_{\mathrm{R}}-\phi_{\mathrm{w}})\cos\psi & \cos(\delta_{\mathrm{R}}-\phi_{\mathrm{w}}) \end{bmatrix}^{-1} \begin{Bmatrix} \boldsymbol{i}_{\mathrm{Rc}} \\ \boldsymbol{j}_{\mathrm{Rc}} \\ \boldsymbol{k}_{\mathrm{Rc}} \end{Bmatrix} \\ &= \begin{bmatrix} \Delta v_{1\mathrm{R}} \\ \Delta v_{2\mathrm{R}} \\ \Delta v_{3\mathrm{R}} \end{bmatrix}^{\mathrm{T}} \begin{Bmatrix} \boldsymbol{i}_{\mathrm{Rc}} \\ \boldsymbol{j}_{\mathrm{Rc}} \\ \boldsymbol{k}_{\mathrm{Rc}} \end{Bmatrix} \end{aligned} \tag{2.346}$$

Therefore, according to the creepage definitions, the longitudinal and lateral creepages at the wheel–rail contact points read

$$\begin{cases} \xi_{x(\mathrm{L,R})} = \dfrac{\Delta v_{1(\mathrm{L,R})}}{v_{(\mathrm{L,R})}} \\ \xi_{y(\mathrm{L,R})} = \dfrac{\Delta v_{2(\mathrm{L,R})}}{v_{(\mathrm{L,R})}} \end{cases} \tag{2.347}$$

where

$$v_{(\mathrm{L,R})} = \frac{1}{2}\left(v + \frac{r_{(\mathrm{L,R})}}{r_0} v\cos\psi_{\mathrm{w}}\right) \tag{2.348}$$

In Eq. (2.348), r_0 is the nominal rolling radius of wheel, while $r_{(\mathrm{L,R})}$ are the actual rolling radii of the left and right wheels.

(2) Calculation of spin creepage

The absolute angular speeds of the left and right rails

$$\boldsymbol{\omega}_{\mathrm{r(L,R)}} = \begin{bmatrix} \dot{\phi}_{\mathrm{r(L,R)}} \\ 0 \\ 0 \end{bmatrix}^{\mathrm{T}} \begin{Bmatrix} \boldsymbol{i} \\ \boldsymbol{j} \\ \boldsymbol{k} \end{Bmatrix} \tag{2.349}$$

The relative differences of the angular speeds of the wheelset and rails in the absolute coordinate system are described as

$$\begin{aligned} \Delta\boldsymbol{\omega}'_{\mathrm{(L,R)}} &= \begin{bmatrix} \dot{\phi}_{\mathrm{w}} \cos\psi_{\mathrm{w}} - \left(-\Omega + \dot{\beta}_{\mathrm{w}}\right) \cos\phi_{\mathrm{w}} \sin\psi_{\mathrm{w}} - \dot{\phi}_{\mathrm{r(L,R)}} \\ \dot{\phi}_{\mathrm{w}} \sin\psi_{\mathrm{w}} + \left(-\Omega + \dot{\beta}_{\mathrm{w}}\right) \cos\phi_{\mathrm{w}} \cos\psi_{\mathrm{w}} \\ \left(-\Omega + \dot{\beta}_{\mathrm{w}}\right) \sin\phi_{\mathrm{w}} + \dot{\psi}_{\mathrm{w}} \end{bmatrix}^{\mathrm{T}} \begin{Bmatrix} \boldsymbol{i} \\ \boldsymbol{j} \\ \boldsymbol{k} \end{Bmatrix} \\ &= \begin{bmatrix} \Delta\omega_{x\mathrm{(L,R)}} \\ \Delta\omega_{y\mathrm{(L,R)}} \\ \Delta\omega_{z\mathrm{(L,R)}} \end{bmatrix}^{\mathrm{T}} \begin{Bmatrix} \boldsymbol{i} \\ \boldsymbol{j} \\ \boldsymbol{k} \end{Bmatrix} \end{aligned} \tag{2.350}$$

When the angular speed differences are transformed into the contact spot coordinate system, yield

$$\begin{aligned} \Delta\boldsymbol{\omega}_{\mathrm{(L,R)}} = & \begin{bmatrix} \Delta\omega_{x\mathrm{(L,R)}} \\ \Delta\omega_{y\mathrm{(L,R)}} \\ \Delta\omega_{z\mathrm{(L,R)}} \end{bmatrix}^{\mathrm{T}} \begin{bmatrix} \cos\psi_{\mathrm{w}} & \sin\psi_{\mathrm{w}} & 0 \\ -\cos\left(\delta_{\mathrm{(L,R)}} \pm \phi_{\mathrm{w}}\right) \sin\psi_{\mathrm{w}} & \cos\left(\delta_{\mathrm{(L,R)}} \pm \phi_{\mathrm{w}}\right) \cos\psi_{\mathrm{w}} & \pm\sin\left(\delta_{\mathrm{(L,R)}} \pm \phi_{\mathrm{w}}\right) \\ \pm\sin\left(\delta_{\mathrm{(L,R)}} \pm \phi_{\mathrm{w}}\right) \sin\psi_{\mathrm{w}} & \mp\sin\left(\delta_{\mathrm{(L,R)}} \pm \phi_{\mathrm{w}}\right) \cos\psi_{\mathrm{w}} & \cos\left(\delta_{\mathrm{(L,R)}} \pm \phi_{\mathrm{w}}\right) \end{bmatrix}^{-1} \\ & \cdot \begin{Bmatrix} \boldsymbol{i}_{\mathrm{(L,R)c}} \\ \boldsymbol{j}_{\mathrm{(L,R)c}} \\ \boldsymbol{k}_{\mathrm{(L,R)c}} \end{Bmatrix} = \begin{bmatrix} \Delta\omega_{1\mathrm{(L,R)}} \\ \Delta\omega_{2\mathrm{(L,R)}} \\ \Delta\omega_{3\mathrm{(L,R)}} \end{bmatrix}^{\mathrm{T}} \begin{Bmatrix} \boldsymbol{i}_{\mathrm{(L,R)c}} \\ \boldsymbol{j}_{\mathrm{(L,R)c}} \\ \boldsymbol{k}_{\mathrm{(L,R)c}} \end{Bmatrix} \end{aligned} \tag{2.351}$$

Thus, the spin creepage at the left and right wheel–rail contact points read

$$\xi_{\phi\mathrm{(L,R)}} = \frac{\Delta\omega_{3\mathrm{(L,R)}}}{V_{\mathrm{(L,R)}}} \tag{2.352}$$

(3) Calculation of wheel–rail creep forces

Based on the Kalker's linear creep theory [108], the wheel–rail longitudinal creep force F_x, the lateral creep force F_y, and the spin creep torque M_z can be described by

$$\begin{cases} F_x = -f_{11}\xi_x \\ F_y = -f_{22}\xi_y - f_{23}\xi_\phi \\ M_z = f_{23}\xi_y - f_{33}\xi_\phi \end{cases} \tag{2.353}$$

where f_{ij} represents the creep coefficients, read

$$\begin{cases} f_{11} = G_{\mathrm{wr}}(ab)C_{11} \\ f_{22} = G_{\mathrm{wr}}(ab)C_{22} \\ f_{23} = G_{\mathrm{wr}}(ab)^{3/2}C_{23} \\ f_{33} = G_{\mathrm{wr}}(ab)^2C_{33} \end{cases} \tag{2.354}$$

in which G_{wr} is the combined shear modulus of the wheel and rail materials, a and b are the semi-axil lengths of the wheel–rail contact patches, C_{ij} are the Kalker's creep coefficients [108], which are depended on the ratios of the semi-axil lengths of the wheel–rail contact patches.

The combined shear modulus G_{wr} and the combined Poisson ratio ν for the wheel and rail materials are given as

$$\begin{cases} G_{\mathrm{wr}} = \frac{2G_{\mathrm{w}}G_{\mathrm{r}}}{G_{\mathrm{w}} + G_{\mathrm{r}}} \\ \nu = \frac{G_{\mathrm{w}}\nu_{\mathrm{r}} + G_{\mathrm{r}}\nu_{\mathrm{w}}}{2G_{\mathrm{w}}G_{\mathrm{r}}} G_{\mathrm{wr}} \end{cases} \tag{2.355}$$

G_{wr} and ν comply with the following relationship:

$$G_{\mathrm{wr}} = \frac{E}{2(1+\nu)} \tag{2.356}$$

where E is the Young's modulus of wheel–rail material, G_{w} and G_{r} are the shear modulus of the wheel and the rail, ν_{w} and ν_{r} are the Poisson ratios of the wheel and the rail.

The semi-axil lengths of the wheel–rail contact patches a and b can be calculated by

$$\begin{cases} a = a_{\mathrm{e}}(NR_{\mathrm{w}})^{1/3} \\ b = b_{\mathrm{e}}(NR_{\mathrm{w}})^{1/3} \\ ab = a_{\mathrm{e}}b_{\mathrm{e}}(NR_{\mathrm{w}})^{2/3} \end{cases} \tag{2.357}$$

where N is the wheel–rail normal force at the contact point, a_{e} and b_{e} are parameters dependent on ρ/R_{w} given in [111].

If $\rho/R_{\mathrm{w}} \leq 2$,

$$\begin{cases} a_{\mathrm{e}} = 0.1506m\left(\frac{\rho}{R_{\mathrm{w}}}\right)^{1/3} \times 10^{-3} \quad (\mathrm{m}^2/\mathrm{N})^{1/3} \\ b_{\mathrm{e}} = 0.1506n\left(\frac{\rho}{R_{\mathrm{w}}}\right)^{1/3} \times 10^{-3} \quad (\mathrm{m}^2/\mathrm{N})^{1/3} \end{cases} \tag{2.358}$$

If $\rho/R_w > 2$,

$$\begin{cases} a_e = 0.1506n\left(\frac{\rho}{R_w}\right)^{1/3} \times 10^{-3} \quad (m^2/N)^{1/3} \\ b_e = 0.1506m\left(\frac{\rho}{R_w}\right)^{1/3} \times 10^{-3} \quad (m^2/N)^{1/3} \end{cases} \tag{2.359}$$

where $\nu = 0.3$ and $E = 2\times10^{11}$ N/m^2 are assumed for the common wheel–rail contact condition, and ρ is given by

$$\frac{1}{\rho} = \frac{1}{4}\left[\frac{1}{R_w} + \left(\frac{1}{r_w} + \frac{1}{r_r}\right)\right] \tag{2.360}$$

where R_w is the rolling radius of wheel, r_w is the curve radius of wheel tread, r_r is the curve radius of rail head.

$$a_e b_e = 22.68mn\left(\frac{\rho}{R_w}\right)^{2/3} \times 10^{-9} \quad (m^2/N)^{2/3} \tag{2.361}$$

According to the Hertzian elastic contact theory, m and n are depended on the coefficient β

$$\beta = \arccos\left[\frac{\rho}{4}\left|\frac{1}{R_w} - \frac{1}{r_w} - \frac{1}{r_r}\right|\right] \tag{2.362}$$

The relationship of the β and m, n are shown in Table 2.16.

It is noted that the linear creep theory by Kalker is only available for small creepage situation, where no slip occurs at wheel–rail interface, and the creepages are in the region of adhesion. For the large creepages situation that is usually the case in real wheel–rail contact patches, the creepages are in the regions of creep and slip, and the relationship between the creep forces and creepages are nonlinear, as shown in Fig. 2.61. In such a situation, linear creep force model is not suitable. Therefore, a nonlinear modification is need. Here, we introduce a nonlinear modification model [109, 110] that is available for the calculation of the wheel–rail creep forces under the condition of large creepages and small spin.

Table 2.16 Relationship of the β and m, n

β	0°	10°	20°	30°	35°	40°	45°	50°
m	∞	6.612	3.778	2.731	2.397	2.130	1.926	1.754
n	0	0.319	0.408	0.493	0.530	0.567	0.604	0.641
β	55°	60°	65°	70°	75°	80°	85°	90°
m	1.611	1.486	1.378	1.284	1.202	1.128	1.061	1.000
n	0.678	0.717	0.759	0.802	0.846	0.893	0.944	1.000

Fig. 2.61 The relationship between creepage and creep force

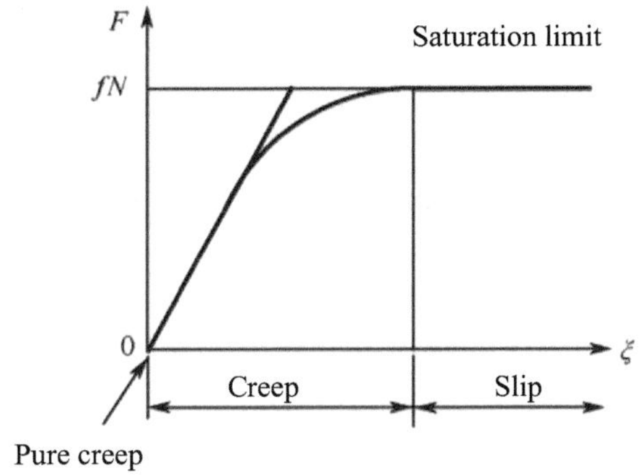

The resultant force of the longitudinal and lateral linear creep forces reads

$$F = \sqrt{F_x^2 + F_y^2} \tag{2.363}$$

Define the nonlinear modified coefficient for the creep forces as

$$\varepsilon = \frac{F'}{F} \tag{2.364}$$

where

$$F' = \begin{cases} fN\left[\frac{F}{fN} - \frac{1}{3}\left(\frac{F}{fN}\right)^2 + \frac{1}{27}\left(\frac{F}{fN}\right)^3\right] & (F \leq 3fN) \\ fN & (F > 3fN) \end{cases} \tag{2.365}$$

In Eq. (2.365), f is the Coulomb friction coefficient between the wheel and the rail.

The modified global tangent creep forces/torque between the wheel and the rail can then be obtained by

$$\begin{cases} F_x' = \varepsilon \cdot F_x \\ F_y' = \varepsilon \cdot F_y \\ M_z' = \varepsilon \cdot M_z \end{cases} \tag{2.366}$$

Unlike the classical wheel–rail contact model, the above new spatial wheel–rail coupling model eliminates the following assumptions: (1) the wheel and the rail are in contact all the time, (2) the rails are assumed to be fixed without any movement, and (3) the wheels and rails are assumed to be rigid bodies. Thus, the current model is capable of considering three kinds of rail motions in vertical, lateral and torsional directions, and dealing with the situation that the wheel loses its contact with the rail existing in practical railway operations. Therefore, it is well suitable for analyzing the dynamic behavior of wheel–rail interactions.

2.4 Train–Track Spatially Coupled Dynamics Model

Based on the established vehicle–track spatially coupled dynamics model presented in Sect. 2.3, a three-dimensional dynamics model for train–track coupled system is developed in this section. Here, the internal dynamic interactions between adjacent vehicles in a train are considered, especially for heavy-haul trains with long formation under the traction/braking conditions where the longitudinal shock and impact are much more intensified, which even threatens the train operation safety.

2.4.1 Basic Principle of Train–Track Dynamic Interaction

The main components of the train and the track system as well as their inter relationship are shown in Fig. 2.62. The running status of the adjacent vehicles is likely to be different when the train traction/braking control is implemented, or when the train passes through curves and ramps. In these conditions, various postures of the coupler and draft gear systems, such as a large tilt angle of coupler and/or a mismatch of coupling heights, would inevitably emerge, which will cause detrimental dynamic inter-vehicle interactions. The generated in-train forces could be transmitted to the wheelsets through the bogie suspension systems, which is likely to affect the wheel–rail contact relations and the vibrations of track structures. In reverse, the track vibrations will have an effect on the vehicle dynamic responses, and finally influence the working conditions of the coupler and the draft gear packages. Therefore, the coupler and draft gear subsystem, the train subsystem and the track subsystem are closely interrelated.

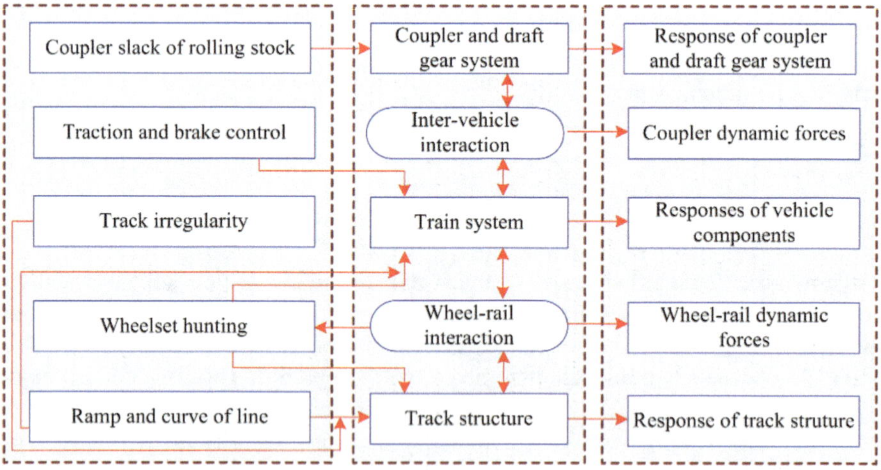

Fig. 2.62 Basic principle of dynamic interaction between train and track [87]

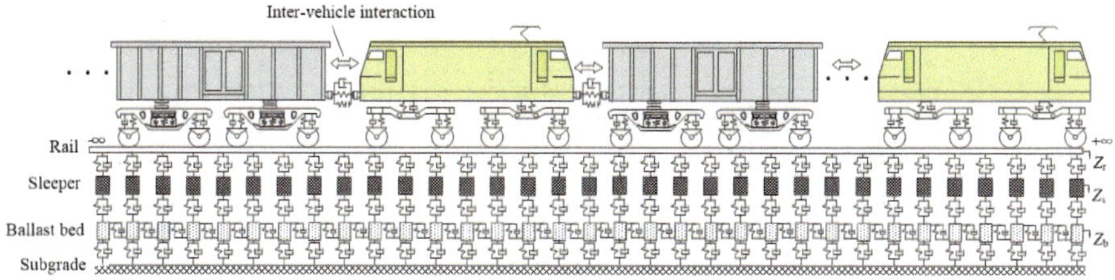

Fig. 2.63 Heavy-haul train–track spatially coupled dynamics model (side view)

2.4.2 *Train–Track Spatially Coupled Dynamics Model*

Based on the basic principle of the train–track interactions introduced in Sect. 2.4.1, the train–track coupled dynamics model is established. The structure characteristics of the train and the track need to be recognized first so as to simulate the dynamic behaviors. The basic components of a certain type of railway vehicles are usually definite. For example, a locomotive is generally composed of the car body, bogie frames, traction motors, wheelsets and suspension systems, while, a freight wagon usually consists of the car body, side frames, bolsters, wheelsets and suspension systems. For the ballasted track structure which is most widely used in the heavy-haul railways, it also has the normative form. Figure 2.63 shows a schematic diagram of the typical heavy-haul train–track coupled dynamics model, in which the locomotives are distributed in different positions of the train (with the Distributed Power mode), and the track is the commonly used ballasted track.

The relevant dynamics models of the locomotive, the freight wagon, and the ballasted track could be referenced to Sect. 2.3 or be found in some literatures [10–13, 112]. Specially, the forces such as the traction force, the braking force, the coupler force and the running resistance should be also considered in the vehicle model when the longitudinal motions are concerned. Furthermore, the possible large creepage between the wheel and the rail needs to be taken into account. Their detailed calculation methods are given in the following subsections.

1. *Calculation of wheel–rail forces*

Large creepage may appear at the wheel–rail contact interface under the train driving or braking conditions. Knothe et al. [113] pointed out that the variable relationship between the wheel–rail tangential force and the creepage could be depicted by introducing the dynamic wheel–rail friction coefficient into the Vermeulen–Johnson model or the Shen–Hedrick–Elkins model, so as to calculate the wheel–rail forces accurately in the case of large creepage. This method is also adopted in our dynamics modeling. Figure 2.64 shows the flow chart for the general calculation of the wheel–rail forces where the large creepage is considered.

To reflect the large creep speed and consider the effect of the rail vibrations, the creepage calculation formulas are given as [87]

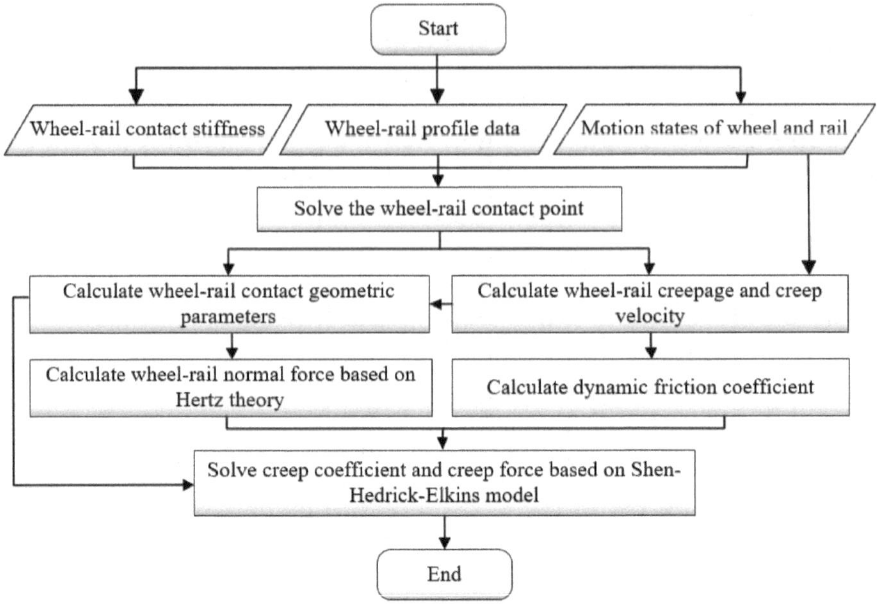

Fig. 2.64 Flowchart for calculation of wheel–rail forces

$$\begin{cases} \xi_x = \frac{V_1 - V_{r1}}{V_1} \\ \xi_y = \frac{V_2 - V_{r2} - V_{Ry}}{V_1} \\ \xi_z = \frac{\Omega_3 - \Omega_{r3}}{V_1} \end{cases} \tag{2.367}$$

where the symbols V_1, V_2 and Ω_3 are successively the instantaneous running speed, the lateral speed, and the spin speed at the wheel mass center; while, V_{r1}, V_{r2}, and Ω_{r3} are the peripheral speed, lateral speed, and spin speed at the wheel contact point; and V_{Ry} represents the lateral speed induced by the rail vibration and track irregularity variation.

According to the field test results of the wheel–rail friction coefficient under the braking conditions, the wheel–rail friction coefficient has a descending trend as the increasing of the creep speed. The empirical formula of the variable friction coefficient is expressed as [114]

$$f_{\text{kin}} = \frac{f_{\text{stat}}}{1 + 0.23|v_s|} \tag{2.368}$$

where, the symbols, f_{stat} and f_{kin}, represent the static and the dynamic friction coefficients, respectively; v_s denotes the wheel–rail creep velocity; and f_{stat} is set to be 0.45 and 0.25 for the dry and the wet rail surface conditions, respectively. Then,

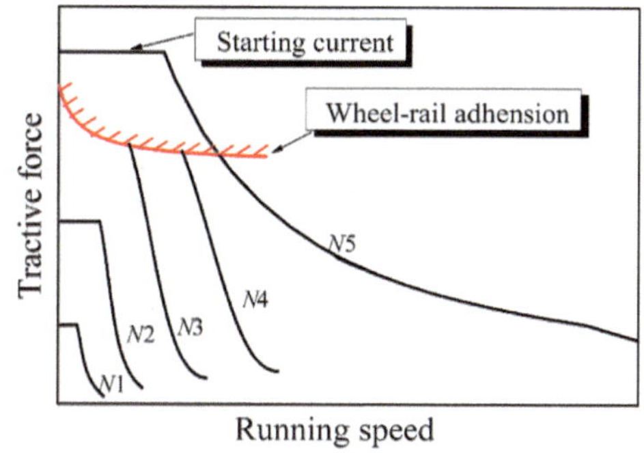

Fig. 2.65 Traction characteristic curves of a locomotive [87]

the wheel–rail creep forces are calculated by introducing the dynamic friction coefficient into the Shen–Hedrick–Elkins model [110].

2. *Calculation of train driving and braking forces*

The train tractive force is an external force which can be adjusted as required in practical train operations. For the locomotive with the property of step speed regulation, its traction characteristics are usually represented by a number of traction force curves which are related to the driver controlling handle positions, as shown in Fig. 2.65. It can be seen that the traction force can be determined by the locomotive running speed and the handle position. It should be noted that the maximum driving force is limited by the starting current and the wheel–rail adhesions during the startup process.

While for the locomotive with the stepless speed regulation, the traction force is dependent on the running speed and the given percentage, which can be calculated as

$$F_{\mathrm{t}} = f(v) \times n\% \tag{2.369}$$

where, $f(v)$ denotes the maximum traction force, $n\%$ is the percentage of control.

In the braking process, the electric braking and the brake shoe braking are the two major types used in the heavy-haul train. For the electric braking which is only applied in the locomotive, it can be regarded as the inverse process of the traction. Thus, the braking force is also limited by the motor power and the wheel–rail adhesion. In the case of the step speed regulation, the electric braking force can be also calculated by interpolating in the electric braking force curves according to the locomotive running speed and the braking handle position. However, if the locomotive has the property of stepless speed regulation, the braking force is dependent on the running speed and the given percentage of control, which is very similar to Eq. (2.369).

However, for the brake shoe braking approach, the action force of the brake shoe braking is from the compressed air (see Fig. 2.66), which distributes in both the

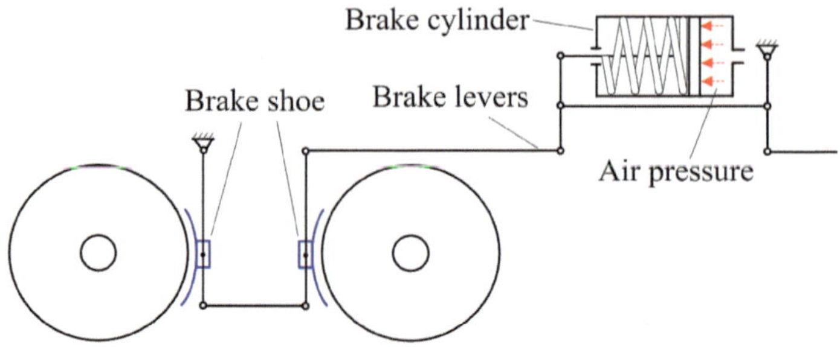

Fig. 2.66 Brake shoe system [87]

locomotives and the freight wagons. In this case, the braking force can be calculated by the product of the brake shoe pressure and the friction coefficient of the contact interface. The brake shoe pressure K is calculated as [115]

$$K = \frac{\pi d_z^2 \cdot p_z \cdot \eta_z \cdot \gamma_z \cdot n_z}{4n_k \times 10^6} (\text{kN}) \tag{2.370}$$

where d_z is the diameter of the brake cylinder (unit: mm); p_z is the air pressure in the brake cylinder (unit: kPa); η_z is the computational transmission efficiency of foundation brake gear; γ_z is the braking leverage; and n_z and n_k are the numbers of cylinder and brake shoes, respectively. The air pressures in the brake cylinders of the wagons and the locomotives could be obtained from the train pneumatic braking tests. And the friction coefficient of brake shoe could be obtained from the experiments or the relevant railway occupation standards.

3. *Dynamic model of coupler and draft gear system*

For the heavy-haul locomotives and freight wagons in China, the nonrigid automatic couplers with a self-centering ability are commonly used. There is usually a free clearance between two couplers connected with each other, which permits them to move relatively in the vertical direction. And the coupler could also sway within a small angle range relative to its draft key in the horizontal direction.

In the longitudinal train dynamics, a pair of connected couplers are usually considered as an entirety without mass. The typical simplified mathematic model [116] is shown in Fig. 2.67. The symbols K_{buf} and C_{buf} represent the stiffness and the damping of the draft gear, respectively. And K_s denotes the structural stiffness of the car body. K_{buf} and C_{buf} are not two constants, but have the nonlinear characteristics. For convenience in simulation, the stiffness and the damping of the draft gear are usually described by the hysteresis curves. It should be noted that the inter-vehicle interactions in the 3D train model will be affected comprehensively by the coupler forces in the longitudinal, the lateral, and the vertical directions, which is different from the coupler forces in the traditional train longitudinal dynamic models. For the typical coupler and draft gear package used in China, the coupler forces can be calculated as below.

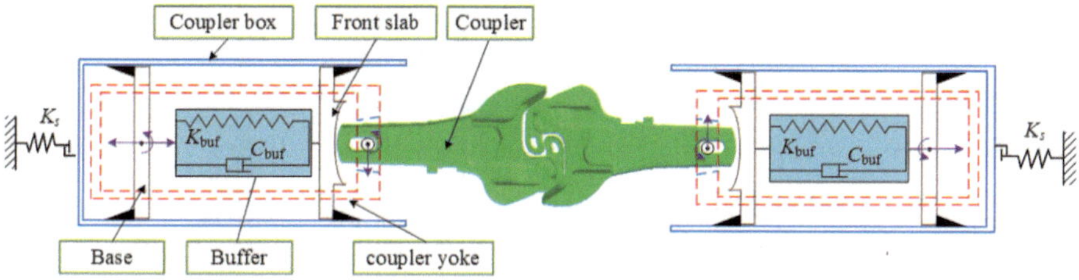

Fig. 2.67 Dynamic model of heavy-haul coupler and draft gear system

Fig. 2.68 Hysteresis characteristic of the draft gear [87]

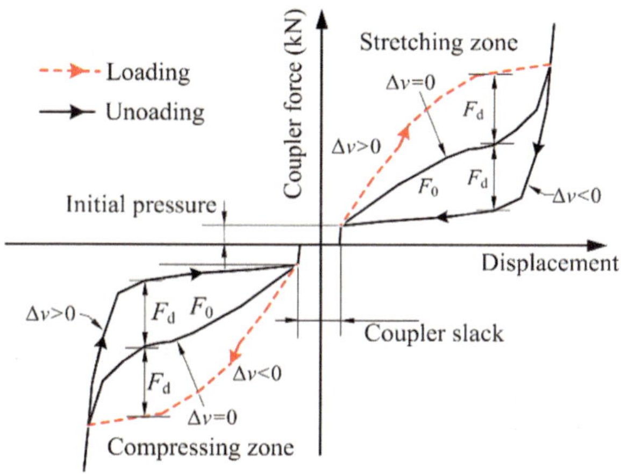

(1) Coupler longitudinal force

The hysteresis characteristic of the draft gear is represented as the misaligned loading and unloading curves, as shown in Fig. 2.68.

The loading or the unloading working status in the draft gear is usually determined by both the relative displacement and the relative speed between the adjacent vehicles. There exists a sudden jump for the coupler force when the draft gear switches between the loading and the unloading conditions, which means that the force variation in the draft gear is discontinuous. The problem of discontinuity point should be solved to guarantee the continuity of the differential equations and the dynamical balance of the system. The handling method of the dry friction damping is used here. The formula for the calculation of the coupler longitudinal force F_{cx} is given as

$$F_{\mathrm{cx}} = \begin{cases} (|F_0| + |F_\mathrm{d}| \cdot \operatorname{sign}(\Delta v)) \cdot \operatorname{sign}(\Delta x) & |\Delta v| > v_\mathrm{f} \\ \left(|F_0| + |F_\mathrm{d}| \cdot \operatorname{sign}(\Delta v)\frac{|\Delta v|}{v_\mathrm{f}}\right) \cdot \operatorname{sign}(\Delta x) & 0 \le |\Delta v| \le v_\mathrm{f} \end{cases} \tag{2.371}$$

where v_f is the switching speed between the loading and the unloading conditions; and F_0 and F_d represent, respectively, the spring force and the damping force of the coupler and draft gear system.

(2) Coupler lateral force

The coupler lateral force includes two parts, namely, the lateral component of the coupler force and the lateral force induced by the coupler restoring moment. The values of these two forces are related to the magnitude of the coupler swing angle in the horizontal plane. In the simulation, the connected two couplers are regarded as a rigid-straight bar without relative rotations. The swing angle of the coupler relative to the car body is defined as positive when it rotates around its draft key clockwise from the top view, otherwise it is negative.

When the train negotiates a curved track, the coupler swing angle will be affected by the line alignment. The coordinates of the track centerline are shown in Fig. 2.69. For the *i*th vehicle, the positions of the front and rear center pins are determined by the coordinates (x_{ti1}, y_{ti1}) and (x_{ti2}, y_{ti2}), respectively. Then, the coordinate values of the car body mass center are obtained as

$$\begin{cases} x_{ci} = (x_{ti1} + x_{ti2})/2 \\ y_{ci} = (y_{ti1} + y_{ti2})/2 \end{cases} \tag{2.372}$$

When the centerline of the draft keys in the car body has displacement D and the yaw angle ψ_c, the coordinates of the front and the rear draft keys in the absolute coordinate system are expressed as

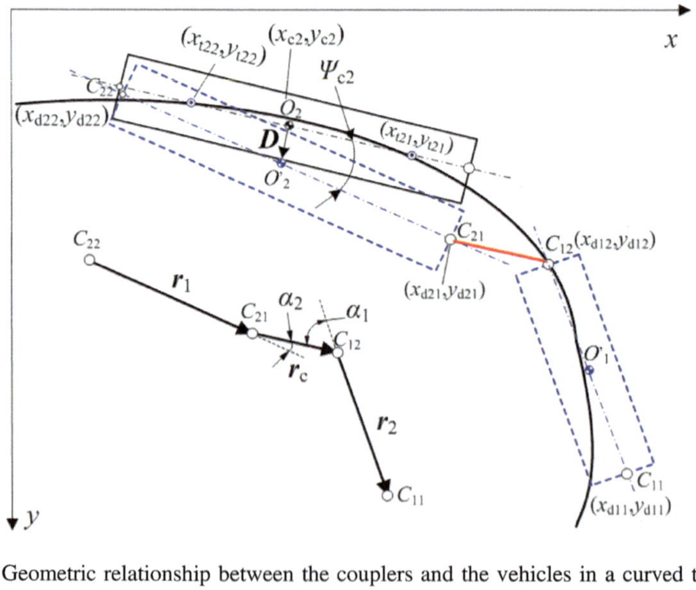

Fig. 2.69 Geometric relationship between the couplers and the vehicles in a curved track [87]

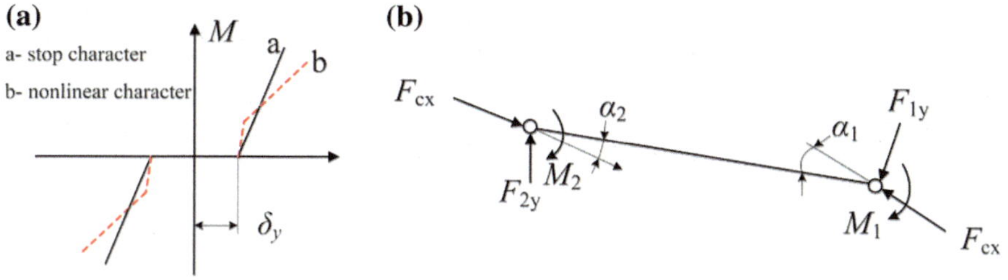

Fig. 2.70 Calculation of lateral coupler force [87]: **a** coupler restoring torque and **b** coupler force analysis

$$\begin{bmatrix} x_{di1} \\ y_{di1} \end{bmatrix} = \begin{bmatrix} x_{ci} \\ y_{ci} \end{bmatrix} + D \cdot \begin{bmatrix} -\sin\psi_c \\ \cos\psi_c \end{bmatrix} + \begin{bmatrix} \cos\psi_c & \sin\psi_c \\ \cos\psi_c & \cos\psi_c \end{bmatrix} \begin{bmatrix} L \\ 0 \end{bmatrix} \tag{2.373}$$

where $D = y_{ci}\text{-}\phi_{ci}h_{cgi}$. L equals l_{cg} and $-l_{cg}$ for the front and rear draft keys, respectively. The coordinates of the draft keys C_{11}, C_{12}, C_{21} and C_{22} of the adjacent two vehicles can be finally calculated by Eq. (2.373). Connecting these four points, three two-dimensional vectors $\boldsymbol{r}_1$, $\boldsymbol{r}_2$, and $\boldsymbol{r}_c$ are obtained. Then the intersection angles of these vectors can be obtained accordingly. The swing angles of the couplers relative to the centerlines of the front and the rear vehicles are calculated as

$$\begin{cases} \alpha_1 = \arccos\left[\frac{\boldsymbol{r}_1 \cdot \boldsymbol{r}_c}{|\boldsymbol{r}_1||\boldsymbol{r}_c|}\right] \cdot \text{sign}[k(\boldsymbol{r}_c) - k(\boldsymbol{r}_1)] \\ \alpha_2 = \arccos\left[\frac{\boldsymbol{r}_2 \cdot \boldsymbol{r}_c}{|\boldsymbol{r}_2||\boldsymbol{r}_c|}\right] \cdot \text{sign}[k(\boldsymbol{r}_c) - k(\boldsymbol{r}_2)] \end{cases} \tag{2.374}$$

where $k(r)$ is the corresponding gradient of the vector $\boldsymbol{r}$ in the global reference system.

Actually, the coupler swing angle could not increase continuously, and its amplitude is limited by the structures of the coupler and the draft gear system. Once the coupler angle exceeds the free swing limit (δ_y), a restoring torque will be generated to resist its swing motion. The restoring torques can be classified into two main forms according to their action principles (see Fig. 2.70a). One is called the rigid stop characteristic with the restoring torque changing linearly with the variation of the swing angle. The other one is called the nonlinear impedance characteristic, which indicates that the restoring torque depends on the compression amount and the load of the draft gear. More specifically, when the coupler swing angle exceeds the free limit value, the coupler needs to first overcome the resistant torque caused by the initial pressure of the draft gear, and then it can compress the draft gear.

The coupler lateral forces acting on the draft keys can be calculated according to the force equilibrium conditions (see Fig. 2.70b):

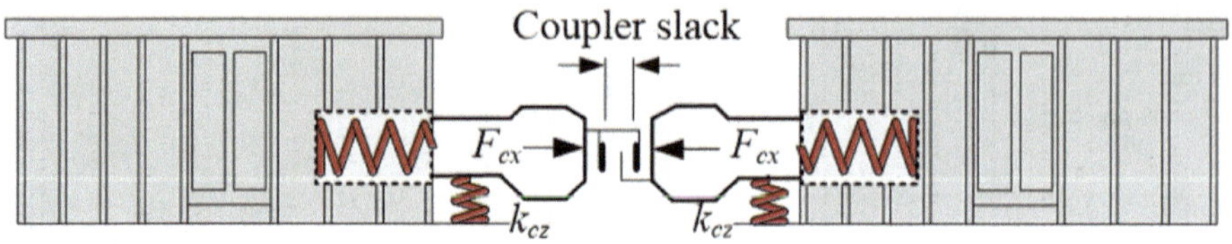

Fig. 2.71 Calculation model of coupler vertical force

$$\begin{cases} F_{1y} = -\frac{F_{cx}L_{cp}\sin\alpha_1 + M_1 + M_2}{L_{cp}\cos(\alpha_1)} \\ F_{2y} = -\frac{F_{cx}L_{cp}\sin\alpha_2 + M_1 + M_2}{L_{cp}\cos(\alpha_2)} \end{cases} \quad (2.375)$$

where L_{cp} is the length of the two connecting couplers, and M_1 and M_2 represent the coupler restoring moments.

(3) Coupler vertical force

The coupler vertical force is mainly caused by the friction force occurring at the coupler contact interface, which can be simplified as a nonlinear spring–friction element with the coupler clearance [116]. Figure 2.71 shows the schematic diagram of the calculation model for the coupler vertical force. The vibrations of the vehicle system are the main reasons that induce the vertical motion of the coupler head. And the vertical and pitching motions of the car body have the primary influence.

The relative displacement and speed between the adjacent vehicles in the vertical direction are calculated as

$$\begin{cases} \Delta z = z_{c1} + l_{cg1}\beta_{c1} - \left(z_{c2} - l_{cg2}\beta_{c2}\right) \\ \Delta\dot{z} = \dot{z}_{c1} + l_{cg1}\dot{\beta}_{c1} - \left(\dot{z}_{c2} - l_{cg2}\dot{\beta}_{c2}\right) \end{cases} \quad (2.376)$$

When the longitudinal relative displacement is smaller than the coupler slack δ_{fc}, the friction force in the vertical direction is $F_{cz} = 0$. Otherwise, two different cases need to be clarified as follows:

(1) If $|F_{cz}| \leq |\mu_0 F_{cx}|$, then

$$F_{cz} = k_z \Delta z \quad (2.377)$$

(2) If $|F_{cz}| > |\mu_0 F_{cx}|$, then

$$\begin{cases} F_{cz} = \left|\frac{\Delta\dot{z}}{v_r}\right| \cdot |F_{cx}| \cdot \mu \cdot \mathrm{sign}(\Delta\dot{z}) & |\Delta\dot{z}| \leq |v_r| \\ F_{cz} = |F_{cx}| \cdot \mu_d \cdot \mathrm{sign}(\Delta\dot{z}) & |\Delta\dot{z}| > |v_r| \end{cases} \quad (2.378)$$

where v_r is the switching speed, and μ_0 and μ_d are the static and dynamic friction coefficients, respectively.

References

1. Prud'homme A, Janin G. The stability of tracks laid with long welded rails, part I and II. Bull Int Railw Congr Assoc. 1969;88:460–80.
2. Lyon D. The calculation of track forces due to dipped rail joints, wheel flats and rail welds. The Second ORE Colloquium on Technical Computer Programs, May 1972.
3. Jenkins HH, et al. The effect of track and vehicle parameters on wheel/rail vertical dynamic forces. Railw Eng J. 1974;3(1):2–16.
4. Newton SG, Clark RA. An investigation into the dynamic effects on the track of wheel flats on railway vehicles. J Mech Eng Sci. 1979;21(4):287–97.
5. Clark RA, Dean PA, Elkins JA, Newton SG. An investigation into the dynamic effects of railway vehicle running on corrugated rails. J Mech Eng Sci. 1982;24(2):65–76.
6. Sato Y. Abnormal wheel load of test train. Perm Way (Tokyo). 1973;14:1–8.
7. Ahlbeck DR, Meacham HC, Prause RH. The development of analytical models for railroad track dynamics. In: Kerr AD, editor. Railroad track mechanics & technology. Pergamon Press; 1978.
8. Zhai WM. Vertical vehicle-track coupled dynamics. Ph.D. thesis. Chengdu, China: Southwest Jiaotong University; 1991 (in Chinese).
9. Zhai WM. The vertical model of vehicle-track system and its coupling dynamics. J China Railw Soc. 1992;14(3):10–21 (in Chinese).
10. Zhai WM, Sun X. A detailed model for investigating vertical interaction between railway vehicle and track. Veh Syst Dyn. 1994;23(Suppl.):603–15.
11. Zhai WM, Cai CB, Guo SZ. Coupling model of vertical and lateral vehicle/track interactions. Veh Syst Dyn. 1996;26(1):61–79.
12. Zhai WM. Vehicle-track coupled dynamics. 1st ed. Beijing: China Railway Publishing House; 1997 (in Chinese).
13. Zhai WM, Cai CB. Theory and application of locomotive-track coupled dynamics. China Railw Sci. 1996;17(2):58–73 (in Chinese).
14. Zhai WM, Han WJ, Cai CB, Wang QC. Dynamic properties of high-speed railway slab tracks. J China Railw Soc. 1999;21(6):65–9 (in Chinese).
15. Knothe K, Grassie SL. Modeling of railway track and vehicle/track interaction at high frequencies. Veh Syst Dyn. 1993;22(3/4):209–62.
16. Popp K, Kruse H, Kaiser I. Vehicle-track dynamics in the mid-frequency range. Veh Syst Dyn. 1999;31(5–6):423–64.
17. Dahlberg T. Vertical dynamic train–track interaction—verifying a theoretical model by full-scale experiments. Veh Syst Dyn. 1995;24(Suppl.):45–57.
18. Nielsen JC, Igeland A. Vertical dynamic interaction between train and track influence of wheel and track imperfections. J Sound Vib. 1995;187(5):825–39.
19. Li JD, Zeng QY. Dynamic analysis of train-tangent-track space-coupling time-varying system. J China Railw Soc. 1997;19(1):101–7.
20. Oscarsson J, Dahlberg T. Dynamic train/track/ballast interaction—computer models and full-scale experiments. Veh Syst Dyn. 1998;28(Suppl.):73–84.
21. Auersch L. Vehicle-track-interaction and soil dynamics. Veh Syst Dyn. 1998;28 (Suppl.):553–8.
22. Wang QC, Cai CB, Luo Q, Cai Y. Allowable values of track deflection angles on high speed railway bridge-subgrade transition sections. J China Railw Soc. 1998;20(3):109–13 (in Chinese).
23. Lei XY, Mao LJ. Dynamic response analyses of vehicle and track coupled system on track transition of conventional high speed railway. J Sound Vib. 2004;271(3–5):1133–46.
24. Wang P. Research on wheel/rail system dynamics on turnout. Ph.D. thesis. Southwest Jiaotong University of China: Chengdu; 1997 (in Chinese).
25. Andersson C. Dahlberg T Wheel/rail impacts at a railway turnout crossing. Proc Inst Mech Eng Part F: J Rail Rapid Transit. 1998;212(2):123–34.

26. Gurule S, Wilson N. Simulation of wheel/rail interaction in turnouts and special track work. Veh Syst Dyn. 1999;33(Suppl.):143–54.
27. Drozdziel J, Sowinski B, Groll W. The effect of railway vehicle-track system geometric deviation on its dynamics in the turnout zone. Veh Syst Dyn. 1999;33(Suppl.):641–52.
28. Ren ZS, Sun S, Zhai W. Study on lateral dynamic characteristics of vehicle/turnout system. Veh Syst Dyn. 2005;43(4):285–303.
29. Cai Z, Raymond GP. Theoretical model for dynamic wheel/rail and track interaction. In: Proceedings of 10th international wheelset congress, Sydney, Australia, Sept. 1992. p. 127–31.
30. Nielsen JCO. Train/track interaction: Coupling of moving and stationary dynamic systems. Ph.D. dissertation. Götebory, Sweden: Chalmers University of Technology; 1993.
31. Diana G, Cheli F, Bruni S, Collina A. Interaction between railroad superstructure and railway vehicles. Veh Syst Dyn 1994;23(Suppl.):75–86.
32. Ripke B, Knothe K. Simulation of high frequency vehicle-track interactions. Veh Syst Dyn. 1995;24(Suppl.):72–85.
33. Frohling RD. Low frequency dynamic vehicle-track interaction: modelling and simulation. Veh Syst Dyn. 1998;28(Suppl.):30–46.
34. Andersson C, Oscarsson J, Nielsen J. Dynamic train/track interaction including state-dependent track properties and flexible vehicle components. Veh Syst Dyn. 1999;33 (Suppl.):47–58.
35. Sun YQ, Dhanasekar M. A dynamic model for the vertical interaction of the rail track and wagon system. Int J Solids Struct. 2002;39:1337–59.
36. Zhai WM, Wang KY, Cai CB. Fundamentals of vehicle–track coupled dynamics. Veh Syst Dyn. 2009;47(11):1349–76.
37. Wu TX, Thompson DJ. On the impact noise generation due to a wheel passing over rail joints. J Sound Vib. 2003;267(3):485–96.
38. Lundqvist A, Dahlberg T. Dynamic train/track interaction including model for track settlement evolvement. Veh Syst Dyn. 2004;41(Suppl.):667–76.
39. Yang YB, Wu YS. Transmission of vibrations from high speed trains through viaducts and foundations to the ground. Jouranl Chin Inst Eng. 2005;28(2):251–66.
40. Lundqvist A, Dahlberg T. Load impact on railway track due to unsupported sleepers. Proc Inst Mech Eng Part F: J Rail Rapid Transit. 2005;219(2):67–77.
41. Kabo E, Nielsen JCO, Ekberg A. Prediction of dynamic train–track interaction and subsequent material deterioration in the presence of insulated rail joints. Veh Syst Dyn. 2006;44(Suppl.):718–29.
42. Cai W, Wen Z, Jin X, Zhai W. Dynamic stress analysis of rail joint with height difference defect using finite element method. Eng Fail Anal. 2007;14(8):1488–99.
43. Wen ZF, Jin XS, Xiao XB, et al. Effect of a scratch on curved rail on initiation and evolution of plastic deformation induced rail corrugation. Int J Solids Struct. 2008;45:2077–96.
44. Li Z, Zhao X, Esveld C, et al. An investigation into the causes of squats—correlation analysis and numerical modeling. Wear. 2008;265(9–10):1349–55.
45. Uzzal RUA, Ahmed AKW, Rakheja S. Analysis of pitch plane railway vehicle–track interactions due to single and multiple wheel flats. Proc Inst Mech Eng Part F: J Rail Rapid Transit. 2009;223:375–90.
46. Kouroussis G, Verlinden O, Conti C. On the interest of integrating vehicle dynamics for the ground propagation of vibrations: the case of urban railway traffic. Veh Syst Dyn. 2010;48(12):1553–71.
47. Zhai W, He Z, Song X. Prediction of high-speed train induced ground vibration based on train-track-ground system model. Earthq Eng Eng Vib. 2010;9(4):545–54.
48. Liu X, Zhai W. Analysis of vertical dynamic wheel/rail interaction caused by polygonal wheels on high-speed trains. Wear. 2014;314(1–2):282–90.
49. Kouroussis G, Connolly DP, Verlinden O. Railway-induced ground vibrations—a review of vehicle effects. Int J Rail Transp. 2014;2(2):69–110.

50. Wang K, Liu P, Zhai W, et al. Wheel/rail dynamic interaction due to excitation of rail corrugation in high-speed railway. Sci China Technol Sci. 2015;58(2):226–35.
51. Zhu S, Cai C, Spanos PD. A nonlinear and fractional derivative viscoelastic model for rail pads in the dynamic analysis of coupled vehicle–slab track systems. J Sound Vib. 2015;335:304–20.
52. Wang K, Zhai W, Lv K, et al. Numerical investigation on wheel-rail dynamic vibration excited by rail spalling in high-speed railway. Shock Vib. 2016;9108780.
53. Zhu S, Cai C, Zhai W. Interface damage assessment of railway slab track based on reliability techniques and vehicle-track interactions. J Transp Eng. 2016;142(10):04016041.
54. Gao J, Zhai W, Guo Y. Wheel–rail dynamic interaction due to rail weld irregularity in high-speed railways. Proc Inst Mech Eng Part F: J Rail Rapid Transit. 2018;232(1):249–61.
55. Zhou S, Zhang X, Di H, He C. Metro train–track–tunnel–soil vertical dynamic interactions —semi-analytical approach. Veh Syst Dyn. 2019;56(12):1945–68.
56. Guo Y, Zhai W. Long-term prediction of track geometry degradation in high-speed vehicle–ballastless track system due to differential subgrade settlement. Soil Dyn Earthq Eng. 2018;113:1–11.
57. Sun Y, Guo Y, Chen Z, et al. Effect of differential ballast settlement on dynamic response of vehicle track coupled systems. Int J Struct Stab Dyn. 2018;18(7):1850091-1-29.
58. Popp K, Knothe K, Pöpper C. System dynamics and long-term behaviour of railway vehicles, track and subgrade: report on the DFG priority programme in Germany and subsequent research. Veh Syst Dyn. 2005;43(6–7):485–538.
59. Xu ZS, Zhai WM. Mechanism of wheel/rail noise for rail transit. Noise Vib Control. 2006;1:52–4 (in Chinese).
60. Zhai WM, Wang KY. Lateral interactions of trains and tracks on small-radius curves: simulation and experiment. Veh Syst Dyn. 2006;44(Suppl.):520–30.
61. Sun YQ, Dhanasekar M, Roach D. A 3D model for lateral and vertical dynamics of wagon-track system. Proc Inst Mech Eng Part F: J Rail Rapid Transit. 2003;217:31–45.
62. Alfi S, Bruni S. Mathematical modelling of train–turnout interaction. Veh Syst Dyn. 2009;47(5):551–74.
63. Ren ZS. Wheel/rail multi-point contacts and vehicle–turnout system dynamic interactions. Beijing: Science Press; 2014 (in Chinese).
64. Yang XW. Theorical analysis and control studies in wheel/rail noise of high speed railway. Ph.D thesis. Chengdu, China: Southwest Jiaotong University; 2010 (in Chinese).
65. Zhai WM, Wang KY. Lateral hunting stability of railway vehicles running on elastic track structures. J Comput Nonlinear Dyn ASME. 2010;5(4):041009-1-9.
66. Torstensson PT, Nielsen JCO. Simulation of dynamic vehicle–track interaction on small radius curves. Veh Syst Dyn. 2011;49(11):1711–32.
67. Kaiser I. Refining the modelling of vehicle–track interaction. Veh Syst Dyn. 2012;50 (Suppl.):229–43.
68. Gialleonardo ED, Braghin F, Bruni S. The influence of track modelling options on the simulation of rail vehicle dynamics. J Sound Vib. 2012;331:4246–58.
69. Zhai W, Xia H, Cai C, et al. High-speed train–track–bridge dynamic interactions—Part I: Theoretical model and numerical simulation. Int J Rail Transp. 2013;1(1–2):3–24.
70. Zhai W, Wang S, Zhang N, et al. High-speed train–track–bridge dynamic interactions–part II: experimental validation and engineering application. Int J Rail Transp. 2013;1(1–2): 25–41.
71. Uzzal RUA, Ahmed AKW, Bhat RB. Modelling, validation and analysis of a three-dimensional railway vehicle–track system model with linear and nonlinear track properties in the presence of wheel flats. Veh Syst Dyn. 2013;51(11):1695–721.
72. Escalona JL, Sugiyama H, Shabana AA. Modelling of structural flexiblity in multibody railroad vehicle systems. Veh Syst Dyn. 2013;51(7):1027–58.
73. Arvidsson T, Karoumi R. Train–bridge interaction—a review and discussion of key model parameters. Int J Rail Transp. 2014;2(3):147–86.

74. Zhu SY, Fu Q, Cai CB, et al. Damage evolution and dynamic response of cement asphalt mortar layer of slab track under vehicle dynamic load. Sci China Technol Sci. 2014;57 (10):1883–94.
75. Zhang X, Zhao C, Zhai W. Dynamic behavior analysis of high-speed railway ballast under moving vehicle loads using discrete element method. Int J Geomech. 2016;17(7):04016157.
76. Askarinejad H, Dhanasekar M. A Multi-body dynamic model for analysis of localized track responses in vicinity of rail discontinuities. Int J Struct Stab Dyn. 2016;16(09):1550058.
77. Xu L, Zhai W. A new model for temporal–spatial stochastic analysis of vehicle–track coupled systems. Veh Syst Dyn. 2017;55(3):427–48.
78. Zhang X, Zhao CF, Zhai WM, et al. Discrete element simulation and its validation on vibration and deformation of railway ballast. Rock Soil Mech. 2017;38(5):1481–8.
79. Ling L, Zhang Q, Xiao X, et al. Integration of car-body flexibility into train–track coupling system dynamics analysis. Veh Syst Dyn. 2018;56(4):485–505.
80. Zhang X, Zhai W, Chen Z, Yang J. Characteristic and mechanism of structural acoustic radiation for box girder bridge in urban rail transit. Sci Total Environ. 2018;627:1303–14.
81. Chen Z, Zhai W, Wang K. Vibration feature evolution of locomotive with tooth root crack propagation of gear transmission system. Mech Syst Signal Process. 2019;115:29–44.
82. Zhai WM, Wang QC. A study on the analytical models for wheel/rail dynamics. J China Railw Soc. 1994;16(1):64–72 (in Chinese).
83. Satoh Hiroshi. Track mechanics. Beijing: China Railway Publishing House; 1981 (in Chinese).
84. Kisilowski J, Knothe K. Advanced railway vehicle system dynamics. Warsaw: Wydawnictwa Naukowo-Tecniczne; 1991.
85. Zhai WM, Sun X, Zhan FS. Computer simulation of vertical dynamic interactions between track and trains. China Railw Sci. 1993;14(1):42–50 (in Chinese).
86. Cole C, McClanachan M, Spiryagin M, et al. Wagon instability in long trains. Veh Syst Dyn. 2012;50(sup1):303–17.
87. Liu P, Zhai W, Wang K. Establishment and verification of three-dimensional dynamic model for heavy-haul train–track coupled system. Veh Syst Dyn. 2016;54(11):1511–37.
88. Evans J, Berg M. Challenges in simulation of rail vehicle dynamics. Veh Syst Dyn. 2009;47 (8):1023–48.
89. Rücker W. Dynamic interaction of railroad-bed with the subsoil. In: Proceedings of soil dynamics & earthquake engineering conference, Southampton. 1982. p. 435–48.
90. Zhai WM, Wang KY, Lin JH. Modelling and experiment of railway ballast vibrations. J Sound Vib. 2004;270(4–5):673–83.
91. Timoshenko S, Young DH, Weaver W, Jr. Vibration problems in engineering. 4th ed. Wiley; 1974.
92. Cao ZY. Theory of shell vibration. Beijing: China Railway Publishing House; 1989 (in Chinese).
93. Zhang T, Chen Z, Zhai W, et al. Establishment and validation of a locomotive-track coupled spatial dynamics model considering dynamic effect of gear transmission. Mech Syst Signal Process. 2019;119:328–45.
94. Zhang T, Chen Z, Zhai W, et al. Effect of the drive system on locomotive dynamic characteristics using different dynamics models. Sci China Technol Sci. 2019;62:308–20.
95. Chen Z, Zhai W, Wang K. A locomotive–track coupled vertical dynamics model with gear transmissions. Veh Syst Dyn. 2017;55(2):244–67.
96. Chen Z, Zhai W, Wang K. Dynamic investigation of a locomotive with effect of gear transmission under tractive conditions. J Sound Vib. 2017;408:220–33.
97. Chen Z, Zhai W, Wang K. Locomotive dynamic performance under traction/braking conditions considering effect of gear transmissions. Veh Syst Dyn. 2018;56(7):1097–117.
98. Chen Z, Shao Y. Mesh stiffness calculation of a spur gear pair with tooth profile modification and tooth root crack. Mech Mach Theory. 2013;62:63–74.
99. Chen Z, Shao Y. Dynamic simulation of spur gear with tooth root crack propagating along tooth width and crack depth. Eng Fail Anal. 2011;18:2149–64.

100. Chen Z, Zhai W, Shao Y, et al. Analytical model for mesh stiffness calculation of spur gear pair with non-uniformly distributed tooth root crack. Eng Fail Anal. 2016;66:502–14.
101. Chen Z. Study on gear mesh nonlinear excitation modelling and vibration features of planetary gear system. Dissertation of Doctor Degree. Chongqing: Chongqing University. 2013 (in Chinese).
102. Chen Z, Zhang J, Zhai W, Wang Y, et al. Improved analytical methods for calculation of gear tooth fillet-foundation stiffness with tooth root crack. Eng Fail Anal. 2017;82:72–81.
103. Chen Z, Shao Y, Lim TC. Non-linear dynamic simulation of gear response under the idling condition. Int J Automot Technol. 2012;13(4):541–52.
104. Mindlin RD. Influence of rotatory inertia and shear in flexural motion of isotropic, elastic plates. J Appl Mech. 1951;18:31–8.
105. Huang TC. The effect of rotary inertia and of shear deformation on the frequency and normal mode equations of beams with simple end conditions. J Appl Mech. 1964;28:579–84.
106. Cao ZY, Yang ST. Theory of thick plate dynamics and its application. Beijing: Science Press; 1983 (in Chinese).
107. Wang KW. The track of wheel contact points and calculation of wheel/rail geometric contact parameters. J Southwest Jiaotong Univ. 1984;1:89–99 (in Chinese).
108. Kalker JJ. On the rolling contact of two elastic bodies in the presence of dry friction. Ph.D. dissertation. Delft, The Netherlands: Delft University of Technology; 1967.
109. Vermeulen JK, Johnson KL. Contact of non-spherical bodies transmitting tangential forces. J Appl Mech. 1964;31:338–40.
110. Shen ZY, Hedrick JK, Elkins JA. A comparison of alternative creep force models for rail vehicle dynamic analysis. In: Proceedings of 8th IAVSD symposium. Cambridge: MIT; 1983. p. 591–605.
111. Sun X. A direct method to determine the wheel/rail contact ellipse. J Southwest Jiaotong Univ. 1985;4:8–21 (in Chinese).
112. Zhai WM. Vehicle–track coupled dynamics. 4th ed. Beijing: Science Press; 2015 (in Chinese).
113. Knothe K, Wille R, Zastrau BW. Advanced contact mechanics-Road and Rail. Veh Syst Dyn. 2001;(4/5):379–407.
114. Bochet B. Nouvelles recherché experimentales sur le Frottement De Glissement. Ann Min. 1961;38:27–120.
115. TB/T 1407-1998. Regulations on railway train traction calculation. Beijing: Ministry of Railways of the People's Republic of China; 1999 (in Chinese).
116. Garg VK, Dukkipati RV. Dynamics of railway vehicle system. Toronto: Academic Press; 1984.

Chapter 3
Excitation Models of Vehicle–Track Coupled System

Abstract Wheel–rail system excitation is the root cause of vibrations of vehicle–track coupled systems. It is necessary to reveal the pattern characteristics, model description and input method of the wheel–rail system excitation. In general, the wheel–rail system excitations can be divided into deterministic excitations and nondeterministic excitations. Nondeterministic excitations mainly refer to track random irregularity. Deterministic excitations are induced by some specific factors of vehicle and track systems. The factors from vehicles are relatively simple, mainly including wheel flats, out-of-round wheels, and eccentric wheels, etc. The factors from tracks are more complicated, not only because of track geometry state, such as rail dipped joints, rail dislocation joints, track geometry irregularities, and rail corrugation, etc., but also due to track structure defects, such as sleeper voids (unsupported sleepers), sudden change of subgrade stiffness, etc. According to their characteristics, the author divides the wheel–rail system excitations into four categories, they are the impact excitation, harmonic excitation, dynamic stiffness excitation, and random excitation. This chapter presents the excitation input method and modeling of these four types of excitations.

3.1 Excitation Input Method

Generally, the wheel–rail system excitations can be input into vehicle–track coupled dynamics model by three methods, namely the fixed-point method, moving-vehicle method, and tracking-window method. In the analysis of vehicle–track coupled dynamics, the three methods could be selected according to practical situation and requirement.

W. Zhai, *Vehicle–Track Coupled Dynamics*,
https://doi.org/10.1007/978-981-32-9283-3_3

3.1.1 Fixed-Point Method

The fixed-point method assumes that the vehicle does not move on the track, while the geometry irregularities of the wheel and rail surfaces move at the train speed in the direction opposite to the train's traveling direction [1]. The irregularities are input to the vehicle–track coupled system through each wheel–rail contact point. This input method is also called the moving-irregularity method, as shown in Fig. 3.1.

When employing the fixed-point method, the shortest track model can be adopted in the calculation (by just having the effective calculation length described in Sect. 4.4). Therefore, this method has the prominent advantages of high calculation speed and efficiency. However, its shortcomings are also obvious. Instead of considering the longitudinal movement of the vehicle along the track, this method considers the wheel–rail excitations moving opposite to the train's traveling speed, which is approximate for the simulation and cannot reflect the dynamic interaction between discrete supporting sleepers and wheelsets that runs along the tracks. When track structures have defects and uneven supporting stiffness, the fixed-point method cannot simulate the complicated dynamic effects of the structures during vehicles' passing.

Then, for what kind of problems should we use the fixed-point method? It is particularly suitable for analyzing the vehicle–track coupled dynamics problem when the dynamic parameters of the track structure is uniform along the longitudinal direction, which usually corresponds to some common problems in the most general cases, such as the riding comfort, lateral stability, and curve negotiation when vehicles move along elastic track structures, as well as the dynamic parameters optimization of the vehicle and track systems. Therefore, the fixed-point method is still a simple and useful method that is widely used by railway engineering dynamics researchers all over the world.

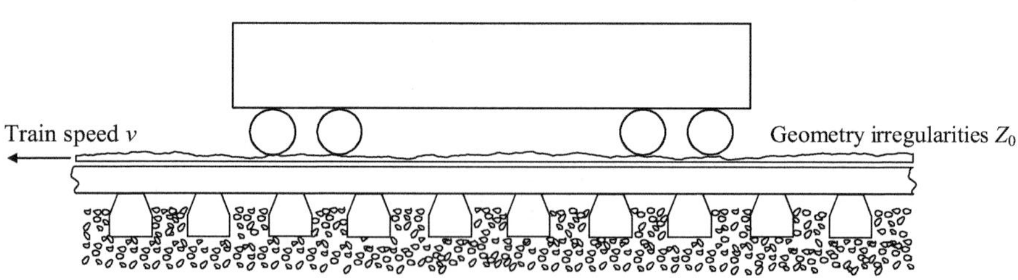

Fig. 3.1 Input of wheel–rail system excitation: fixed-point method

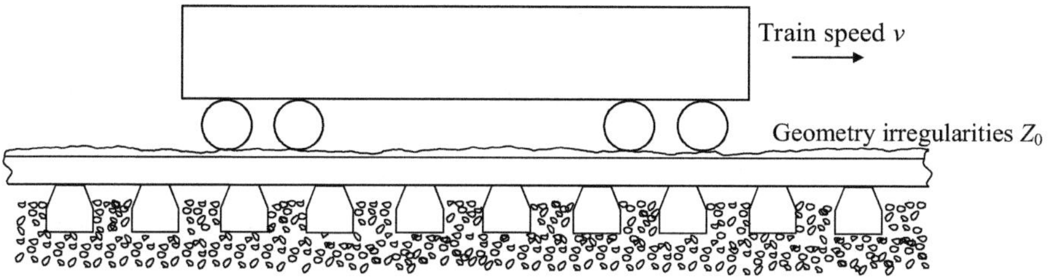

Fig. 3.2 Excitation input of wheel–rail system: moving-vehicle method

3.1.2 Moving-Vehicle Method

In the moving-vehicle method, the wheel–rail system excitation is input into the vehicle–track coupled system while the vehicle moves along the track, as shown in Fig. 3.2. Obviously, it is the most realistic input method.

However, the moving-vehicle method still has its inconvenience. Due to the fact that this method fully considers the spatial position of the vehicle that moves along the track, a sufficiently long track is required in the model. This would lead to a leap increase of model degrees of freedom and computational efforts.

Then, in what case should the moving-vehicle method be employed? The author holds that it is necessary to apply this method when track defects and infrastructure uneven stiffness (namely, dynamic stiffness excitation) are involved. These include vehicle–track interaction at places with failed fasteners, unsupported sleepers, turnouts, or transition zones (subgrade–bridge transition, ballast track–ballastless track transition). In these cases, the track concerned should not be long, so that it is quite convenient and efficient to adopt this input method.

Under the excitations by the rail joints, wheel flats, rail corrugations, and other local track irregularities, the required effective calculation length of track models could be limited, thus it is suitable to adopt the moving-vehicle method for these cases. However, if the variation of infrastructure stiffness is considered, the required calculation length of track models might be very large, and a large number of rail modes will be required in the moving-vehicle method, so that it is quite difficult to efficiently perform a vehicle–track coupled dynamics simulation. Fortunately, the tracking-window method described below can be employed for this case.

3.1.3 Tracking-Window Method

The tracking-window method is an approximation of the moving-vehicle method. It is assumed that the influence of the wheel–rail forces on the vibration of track structure diminishes with the increasing distance to the location where the forces act, and the areas of the track with very small dynamic response can be cut-off in the track modeling. Therefore, only a finite length of track structure under and around the vehicle should be considered in the simulation of the dynamic behavior of the vehicle–track coupled system. In the vehicle–track model using the tracking-window method, a tracking calculation window moving with the vehicle is set for the vehicle–track interaction, where only a finite length l_{TW} of the track structure under and around the vehicle is considered. It can reflect the dynamic behavior of the track structure under a running vehicle through calculation performed only in the moving window.

For the tracking-window method, two key points should be noted: (1) The vibration of track structure within the tracking window is taken into consideration and the vibration of track structure outside the tracking window is assumed to be 0; (2) The vibration characteristic of the track structure possesses an "inheritance relationship" at the instant of the tracking window moving forward. To efficiently reflect the "inheritance relationship" of track vibration in the tracking-window method, the calculation framework is shown in Fig. 3.3, where a unit slab ballastless track is taken as an example. The calculation steps are illustrated as follows:

(1) Input vehicle parameters, track parameters and other parameters required in the vehicle–track interaction simulation, including the length l_{TW} and the moving distance l_m of the tracking window;
(2) Set the number of sub-windows $N_w = 1$ and set the initial position of the vehicle on the track in the first sub-window, where the initial position of the jth wheelset is represented as x_{wj0};

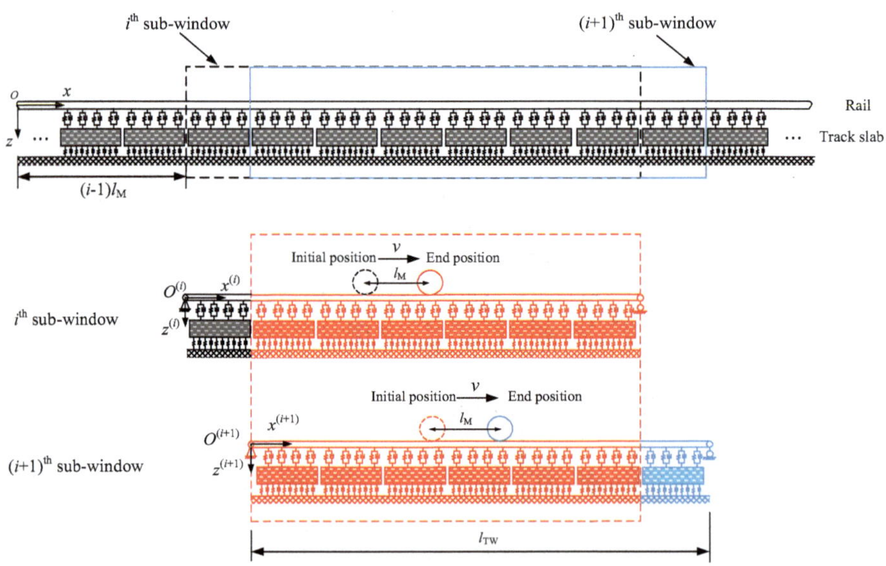

Fig. 3.3 Schematic of the tracking-window method

(3) The vehicle moves forward from the initial position, and the dynamics of the vehicle–track coupled system is calculated based on the traditional moving-vehicle method, in which the position of each wheelset on the track should be obtained. It can be calculated that the moving distance of the vehicle at the *k*th integral step in the sub-window is $x_{mk} = x_{m(k-1)} + v\Delta t$, and the position of the *j*th wheelset on the track can be expressed as $x_{wjk} = x_{wj0} + x_{mk}$;
(4) If $x_{m(k-1)} < l_m$ and $x_{mk} \geq l_m$, that is, when the moving distance of the vehicle at the *k*th integral step is greater than or equal to the distance l_m, the tracking window moves forward a distance of l_m. Meanwhile it satisfies:
   (i) The track vibration in the overlap part of the two adjacent windows must be the same (marked by red in Fig. 3.3), i.e., the "inheritance relationship" is satisfied. At the same time, the track vibration in the nonoverlapping part (marked by blue in Fig. 3.3) is assumed to be 0;
   (ii) As the window moves, the moving distance of the vehicle in the updated sub-window is $x_{mk} = x_{mk} - l_m$;
(5) Set the sub-window number $N_w = N_w + 1$, and then go to Step (3). Continue the cyclic calculation until the calculation time or the calculation distance satisfies the termination condition.

For ballasted tracks on embankment that including rails, fastenings, sleepers, ballast, and subgrade, the length l_{TW} and the moving distance l_m of the window are determined by the sleeper number N_s covered in the tracking window, which is obtained by $l_{TW} = N_s \times l_s$ (l_s is the sleeper spacing). The moving distance l_m is advised to be a sleeper spacing. For ballastless slab tracks on embankment that includes rails, fastenings, slabs, and subgrade, the length l_{TW} and the moving distance l_m of the window are determined by the slab number N_{sb} covered in the tracking window, which is obtained by $l_{TW} = N_{sb} \times l_{sb}$ (l_{sb} is the slab length). The moving distance l_m is advised to be the length of a unit slab. For the ballasted and ballastless tracks on multi-span simply supported bridges, the length l_{TW} and moving distance l_m of the window are determined by the number of bridge span N_b covered in the tracking window, which is obtained by $l_{TW} = N_b \times l_b$ (l_b is the length of a bridge span). The moving distance l_m is advised to be the length of a bridge span. When the calculation length of the tracking window is selected, the vehicle should be set at the center of the tracking window if possible, and the distance from the vehicle to the two ends of the tracking window should be sufficient to ensure that the vibration of the track components outside the calculation window is negligible.

It is noted that the tracking-window method is able to carry out fast calculation of a vehicle moving forward on an infinite track with intricate structures, including the discrete sleepers, slabs, simply supported bridges, transition zones between bridge and embankment, track substructure defects, etc. It overcomes the time-consuming problem of the moving-vehicle method. Another advantage of the tracking-window method is that it can deal with all kinds of track structures neatly such as rail–slab–bridge/embankment structure, rail–sleeper–ballast–bridge/

embankment structure. Therefore, the tracking-window method possesses both the high computational efficiency of the fixed-point method and the comprehensive simulation capability of the moving-vehicle method.

3.2 Impact Excitation Models

When a wheel passes a dipped joint, dislocation joint, joint gap, or surface spalling of rails, the abrupt change of the instantaneous rotation center of the wheel would lead to a vertical impact velocity to the track, which disappears instantly when the wheel runs away from these locations, resulting in a sudden impact and vibrations of the wheel–rail system. Similarly, the same impact vibrations will appear when a wheel with flats moves along a rail. These excitations are defined as impact excitations; they are often input into the wheel–rail system as impact velocities, or as impact displacements.

3.2.1 *Impact Model of Wheel Flat*

During train operation, wheel local scratch and spalling (shown in Fig. 3.4) may occur due to various reasons (braking, wheel spin and slip). These phenomena are collectively known as wheel flats. Wheel flats will induce special dynamic effect during rolling, and the function mechanisms of new and old wheel flats are quite different. An ideal new flat is similar to the chord of a wheel circle, as shown in Fig. 3.4, while an old flat is the wear result of a new flat. An old wheel flat can be

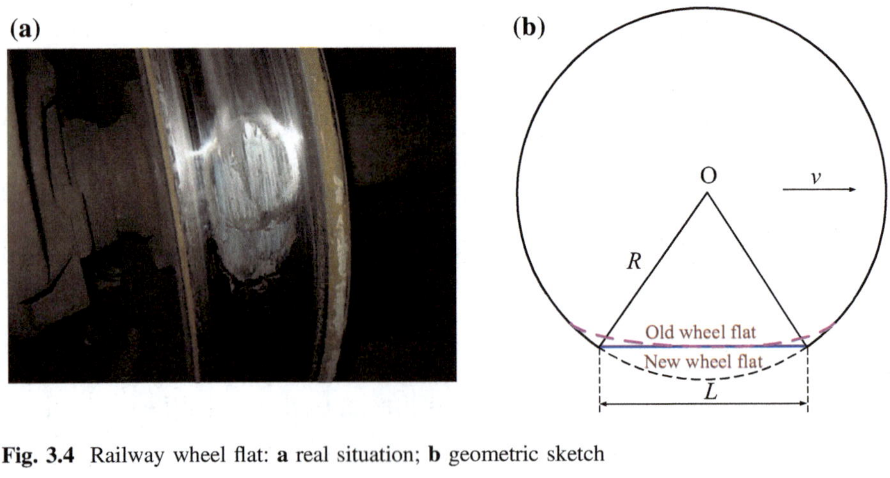

Fig. 3.4 Railway wheel flat: **a** real situation; **b** geometric sketch

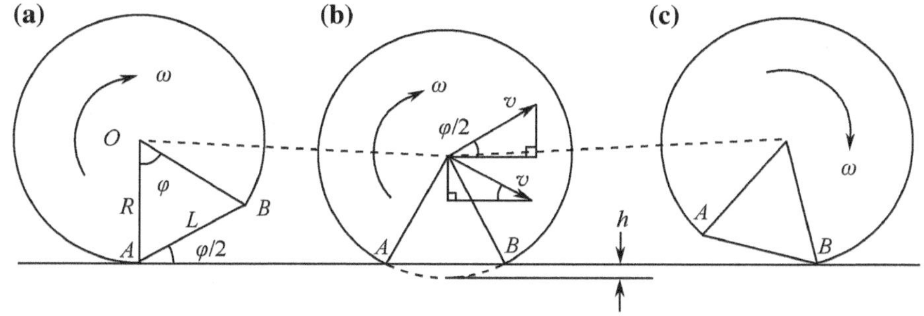

Fig. 3.5 Motion of the wheel with a flat at a low speed

simply described by a cosine curve (as shown in Sects. 3.3.1). Here, the impact mechanism of new wheel scratches will be mainly introduced [2].

1. *Impact mechanism of wheel flat and its critical impact velocity*

As shown in Fig. 3.5, when a wheel rolls at the start point A of a wheel flat at a low speed, the wheel will rotate around point A until the entire flat surface impacts the rail surface, and then the wheel rotates immediately around point B, further exerting a dynamic load on the track until the wheel restores to its normal rolling state. When the wheel rolls at point A at a high speed, it will leave the rail surface and rotate through the air as it moves forward with an inertial motion, and then fall down and make contact with the rail surface at point B, resulting in an impact force on the track, as shown in Fig. 3.6.

Clearly, as the rolling speed increases, the impact characteristics of wheel flats will inevitably have a sudden change at a critical running velocity. When the wheel rolls from location (a) to location (b) shown in Fig. 3.5 at an angular velocity of $\omega = v/R$ (v is the running speed, and R is the wheel radius), the elapsed time t_1 equals to the time t_2 needed for the wheel to fall the distance h. Then the critical state will occur, that is

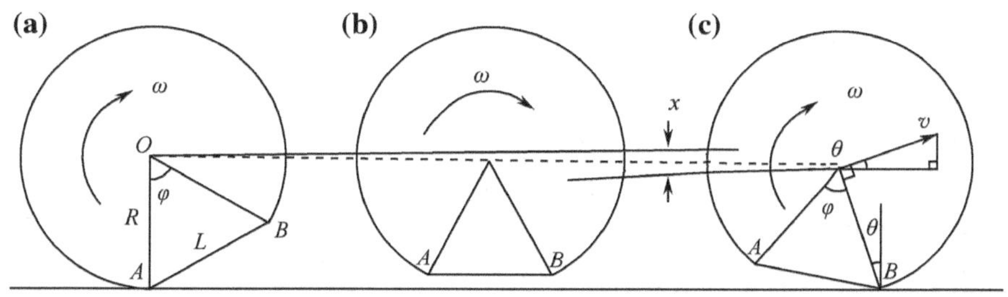

Fig. 3.6 Motion of the wheel with a flat at a high speed

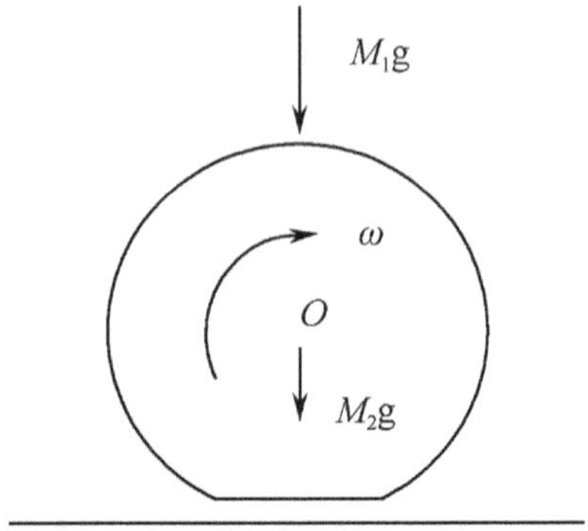

Fig. 3.7 Mechanical analysis of the uncontacted wheel

$$t_1 = \frac{\varphi/2}{\omega} = \frac{R\varphi}{2v_{\mathrm{cr\,0}}} \tag{3.1}$$

$$t_2 = \sqrt{\frac{2h}{\mu}} \tag{3.2}$$

where μ is the falling acceleration of the wheel. According to Newton's second law,

$$\mu = \frac{M_1 + M_2}{M_2} g \tag{3.3}$$

where M_1 and M_2 are the sprung mass and unsprung mass of the primary suspension, respectively; $M_1 g$ is the car body load on the wheel applied through the axle box, as shown in Fig. 3.7.

Setting $t_1 = t_2$, and substituting

$$h = R\left(1 - \cos\frac{\varphi}{2}\right) \approx \frac{1}{8} R\varphi^2$$

into Eq. (3.2) yields the critical running velocity of the wheel flat

$$v_{\mathrm{cr0}} = \sqrt{\mu R} \tag{3.4}$$

2. *Impact velocity equation for low speed* ($v \leq v_{\mathrm{cr0}}$)

At a low speed, the wheel first rotates around point A (Fig. 3.5a) until the whole flat surface AB gets contact with the rail surface (Fig. 3.5b), then it immediately rotates around the point B (Fig. 3.5c), leading to a sudden change of the velocity direction of the wheel center. The wheel impact velocity onto the track consists of two parts, one is the vertical velocity component of the impact velocity induced by the wheel that rotates around point A and hits the rail, written as

$$v_{01} = v \sin\frac{\varphi}{2} = \frac{L}{2R} v \tag{3.5}$$

where L is the length of the flat. The other part is the instantaneous impact component in the opposite direction of the wheel vertical velocity, which is produced at the moment when the rail hinders the wheel rotating around point B, expressed as

$$v_{02} = \gamma \frac{L}{2R} v \tag{3.6}$$

where γ is the coefficient for the transformation from rotational inertia to reciprocating inertia [3].

Therefore, the impact velocity can be obtained as

$$v_0 = (1 + \gamma) \frac{L}{2R} v \tag{3.7}$$

Obviously, the impact velocity at low speed is proportional to the flat length L and the running speed v, and is inversely proportional to the wheel radius R.

3. *Impact velocity equation for high speed* ($v > v_{cr0}$)

For $v > v_{cr0}$, the impact velocity also consists of two parts, which are the falling speed from the air to the rail surface

$$v_{01} = \mu t \tag{3.8}$$

and the vertical component of the wheel center velocity due to the rotation

$$v_{02} = \gamma v \sin\theta \approx \gamma v \theta \tag{3.9}$$

In Fig. 3.6, the elapsed time for the angular $\varphi - \theta$ of the wheel rotating from the location (a) to the location (b) can be given as

$$t = \frac{\varphi - \theta}{\omega} = \frac{(\varphi - \theta)R}{v} \tag{3.10}$$

The falling height of the wheel center is

$$x = \frac{1}{2}\mu t^2 = \frac{\mu(\varphi - \theta)^2 R^2}{2v^2} \tag{3.11}$$

By substituting Eq. (3.11) into the geometric constraint relation,

$$\cos\theta = (R - x)/R \tag{3.12}$$

one can obtain

$$\theta = \frac{\varphi\sqrt{\mu R}}{v + \sqrt{\mu R}} \tag{3.13}$$

and

$$\varphi \approx L/R \tag{3.14}$$

Therefore,

$$\theta = \frac{L}{v + \sqrt{\mu R}} \cdot \sqrt{\frac{\mu}{R}} \tag{3.15}$$

Substituting Eqs. (3.14) and (3.15) into Eq. (3.10) leads to

$$t = \frac{L}{v + \sqrt{\mu R}} \tag{3.16}$$

and thus

$$v_{01} = \frac{\mu L}{v + \sqrt{\mu R}} \tag{3.17}$$

$$v_{02} = \frac{\gamma v L}{v + \sqrt{\mu R}} \cdot \sqrt{\frac{\mu}{R}} \tag{3.18}$$

Therefore, the impact velocity can be written as

$$v_0 = \frac{L}{v + \sqrt{\mu R}}\left(\mu + \gamma v\sqrt{\frac{\mu}{R}}\right) \tag{3.19}$$

As can be seen form Eq. (3.19), the impact velocity at a high speed is also proportional to the wheel flat length L, while it slightly decreases with an increase of the velocity, and finally, it goes to a constant value

$$\bar{v}_0 = \lim_{v\to\infty} v_0 = \gamma L\sqrt{\frac{\mu}{R}} \tag{3.20}$$

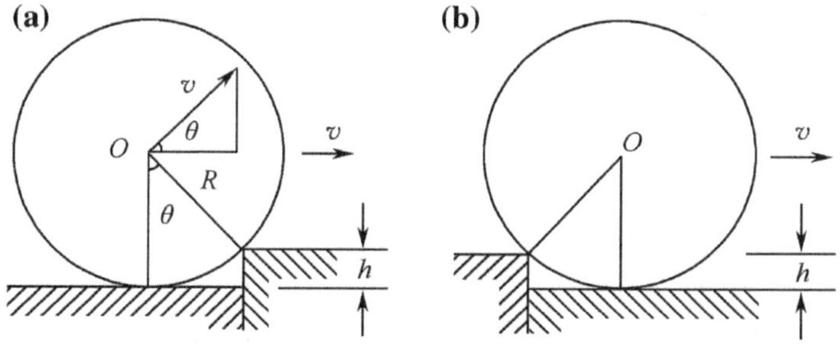

Fig. 3.8 Rail dislocation joint: **a** forward and **b** backward

3.2.2 Model of Rail Dislocation Joint

A rail dislocation joint refers to an abnormal joint which has height difference between the adjacent rail surfaces of a rail joint. According to the vehicle running direction, the rail dislocation joints can be divided into forward and backward dislocation joints. As shown in Fig. 3.8a, a backward dislocation joint corresponds to the case that the wheel runs from a lower rail surface to a higher one, while a forward dislocation joint is presented in Fig. 3.8b, where h is the height difference.

It is known from Fig. 3.8a that the impact velocity of the backward dislocation joint can be expressed as

$$v_0 = v\sin\theta = v\frac{\sqrt{R^2 - (R-h)^2}}{R} \approx v\sqrt{\frac{2h}{R}} \tag{3.21}$$

Clearly, v_0 increases as the vehicle speed increases.

However, for the forward dislocation joint, the case will be different. As shown in Fig. 3.9, when the wheel passes through a forward dislocation joint at a low speed, the wheel–rail contact point will change from point A to point B, and the wheel center velocity will have a sudden change, resulting in a vertical component Δv, as shown in Fig. 3.9a. The impact velocity of which is the same as (3.21). When the wheel passes through the forward dislocation joint at a high speed, as shown in Fig. 3.9b, wheel–rail separation will occur. The wheel is horizontally thrown out from point A at the higher rail and falls freely onto the lower rail under the gravity force, impacting at point D. Obviously, the wheel vertical impact velocity, in this case, will be a function of the height difference h only; it does not vary with vehicle speed.

$$v_0 = \sqrt{2\mu h} \tag{3.22}$$

Setting the critical impact velocity as v_{cr0}, a wheel with an initial horizontal speed of v_{cr0} is supposed to fall down at the point B, then

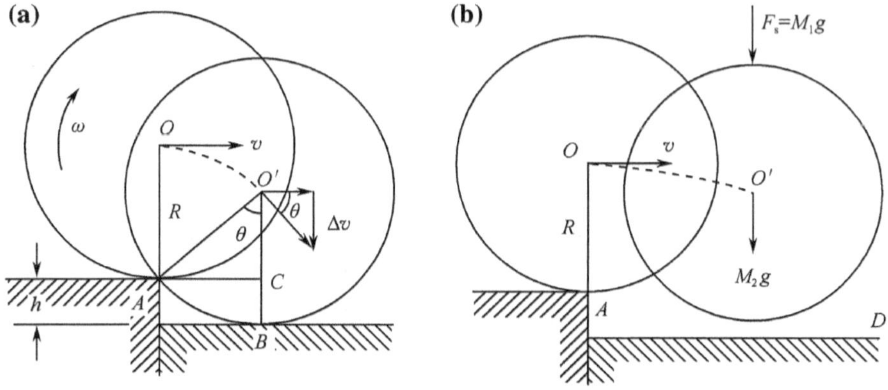

Fig. 3.9 Schematic of wheel passing through a forward dislocation joint at **a** low speed and **b** high speed

$$v_{\mathrm{cr}\,0}t_0 = |\mathrm{AC}| = \sqrt{2Rh} \tag{3.23}$$

where t_0 is the time for the wheel to fall a height h at an acceleration of μ

$$t_0 = \sqrt{\frac{2h}{\mu}} \tag{3.24}$$

By substituting Eq. (3.24) into Eq. (3.23), the critical impact velocity is

$$v_{\mathrm{cr}\,0} = \sqrt{\mu R}$$

which is the same as that of the wheel flat as shown in Eq. (3.4).

It is noted that the above analysis is based on the assumption of rigid rail. In fact, the rail is elastically supported, so that the impact velocity is smaller than that of rigid support, according to Ref. [4]. Therefore, Eq. (3.21) should be modified as

$$v_0 = \frac{m_{\mathrm{eq}}}{m_{\mathrm{w}}} v \sqrt{\frac{2h}{R}} \tag{3.25}$$

Equation (3.22) is revised as

$$v_0 = \frac{m_{\mathrm{eq}}}{m_{\mathrm{w}}} \sqrt{2\mu h} \tag{3.26}$$

where m_{w} is the wheel mass and m_{eq} is the equivalent impact mass of the track, written as

$$m_{\mathrm{eq}} = m\left[\Gamma\left(\frac{3}{4}\right)\Gamma\left(\frac{5}{4}\right)\sqrt{2}\right]^{4/3}\left(\frac{4EI}{K_{\mathrm{H}}}\right)^{1/3} \tag{3.27}$$

where K_{H} is Hertz linear contact stiffness at wheel–rail interface and m is the rail (including the sleepers) mass per unit length.

For a general wheel–rail contact condition

$$m_{\mathrm{eq}} \approx 0.4m \tag{3.28}$$

3.2.3 *Model of Dipped Rail Joint*

Dipped rail joints are the most common impact excitation sources for jointed track lines, which present themselves as a dip at the rail joints, as shown in Fig. 3.10. The impact velocity of a dipped rail joint is usually expressed by the product of the joint angles α_1, α_2 and the vehicle speed v.

$$v_0 = 2\alpha v = (\alpha_1 + \alpha_2)v \tag{3.29}$$

where 2α is the total angle of the dipped joint (see Fig. 3.10).

3.2.4 *Impact Model of Turnout*

When a vehicle passes a turnout in the through route, the vehicle–turnout interaction mainly presents vertical impact and vibrations at the crossing frog, while lateral interaction between the vehicle and turnout becomes dominant when the vehicle passes via the divergent route.

1. *Vertical impact model of turnout frog*

The vertical impact at a turnout crossing mainly occurs at the fixed frog. As shown in Fig. 3.11, there is a harmful gap between the point rail and the wing rail, leading

Fig. 3.10 Dipped rail joint

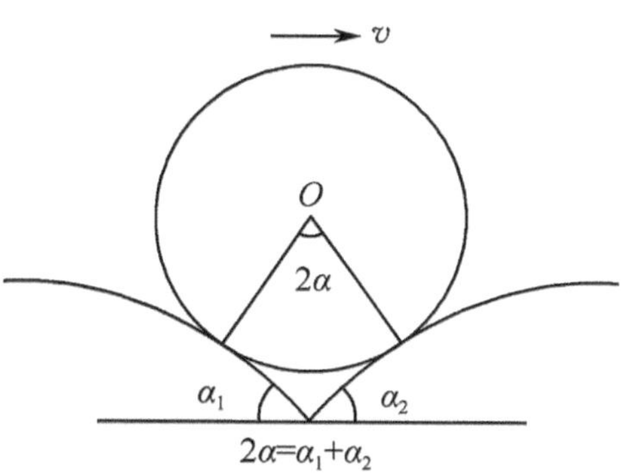

Fig. 3.11 Turnout with fixed frog

to a discontinuous rolling path of the wheel and inducing the wheel–frog impact. When a wheel rolls from the wing rail to the point rail, the wheel rolling radius becomes increasingly smaller as the wheel moves away from the wing rail, causing a descent of the wheel gravity center. To avoid collision between the wheel and the point rail, the rail surface height at this location decreases dramatically, and then increases gradually to the normal height. In this way, the wheel gradually rolls back to the original height when it is completely on the point rail.

The traveling path of the wheel center on the frogs was investigated by measuring the wear of the wheel tread and frog in the former Soviet Union. China Academy of Railway Sciences also measured the vertical irregularity of the Hadfield steel railway frogs and obtained a similar traveling path of the wheel center. This irregularity can be described with the schematic curve as shown in Fig. 3.12, where the first half part is sinusoidal, and the latter part is triangular. The mathematical expression can be written as

$$Z_0 = \begin{cases} h_0 \sin(2\pi x/L_0) & (0 \le x \le x_1) \\ h_d(x - x_1)/(x_2 - x_1) & (x_1 < x \le x_2) \\ h_d(x_3 - x)/(x_3 - x_2) & (x_2 < x \le x_3) \\ 0 & (x > x_3) \end{cases} \tag{3.30}$$

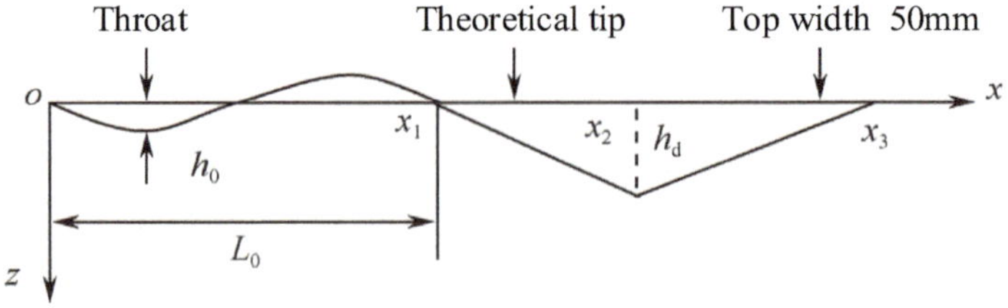

Fig. 3.12 Model of vertical irregularity of a fixed frog

Table 3.1 Relationship between vertical irregularity parameters and wear level of fixed frog

Wear level	Slight wear	Medium wear	Severe wear
Wear value/mm	2–4	4–6	>6
h_0/mm	0.7	0.8	0.9
h_d/mm	3.7	4.3	7.1

where the irregularity amplitude h_0 and h_d are related to the wear level of frogs. Table 3.1 lists the values from measurement [5].

For the No. 12 Hadfield steel frogs of CN60 rail commonly used in Chinese railways, $L_0 = 0.87$ m, $x_1 = 0.87$ m, $x_2 = 1.276$ m, $x_3 = 1.816$ m. The irregularity function can be simplified as

$$Z_0 = \begin{cases} h_0 \sin(70.222x) & (0 \leq x \leq 00.87) \\ 20.463 h_d (x - 00.87) & (00.87 < x \leq 10.276) \\ 10.852 h_d (10.816 - x) & (10.276 < x \leq 10.816) \\ 0 & (x > 10.816) \end{cases} \quad (3.31)$$

2. *Lateral impact model of turnout switch rail*

When a vehicle enters the divergent route of a turnout in the facing direction, the wheel will impact the switch rail as shown in Figs. 3.13 and 3.14. The induced lateral impact force is dependent upon the impact angle β between the wheel and the switch rail, passing speed v, and the types of the vehicle and switch rail. Large lateral impact force can cause severe wear and damage of turnout switch and vehicle bogie. Further, the wheel can climb up on the switch rail, which could potentially lead to derailment. Obviously, changing the straight switch rail (Fig. 3.13) to a curved switch rail (Fig. 3.14) can reduce the impact angle, thus decrease the lateral wheel–rail impact when a vehicle passes through the divergent route.

When a vehicle passes through the divergent route of a turnout in the facing direction, the lateral impact velocity from the wheel to the switch rail can be given as

$$v_{0L} = v \sin \beta \approx v\beta \quad (3.32)$$

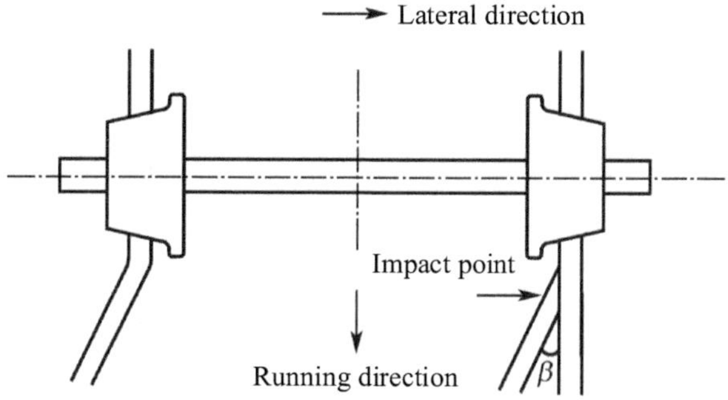

Fig. 3.13 Impact at straight switch rail

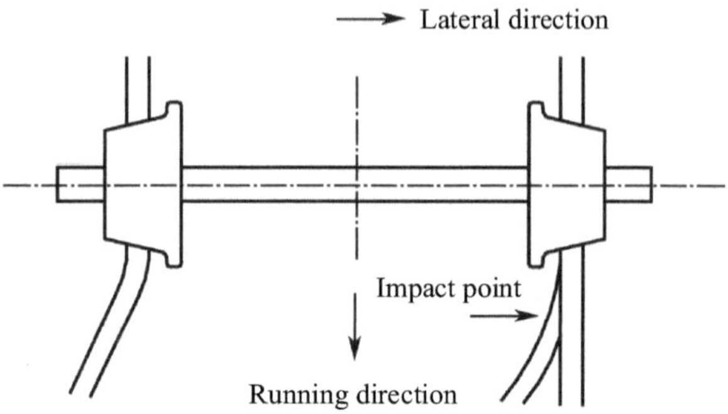

Fig. 3.14 Impact at curved switch rail

For the straight switch rail, β is the angle between the straight switch rail and the stock rail. For the curved switch rail, β is determined by the equation

$$\beta = \sqrt{2\delta/R} \tag{3.33}$$

where δ is the gap between the outside wheel flange and the switch rail; according to the limit size method or probability theory, δ is usually set to be 47 mm; R is the radius of the turnout.

Therefore, when a vehicle passes through the divergent route, the excitation between the wheelset and switch rail can be simply applied to the lateral velocity of the wheelset by an instantaneous impact velocity v_{0L}. Indeed, the lateral wheel–rail interaction is very complicated for a vehicle passing through the divergent route and the excitation description for the entire turnout still needs further investigation.

3.2.5 *Other Impulsive Excitation Models*

Welds of continuous welded rails can induce the same impact effect as that of the backward dislocation joints due to the convex weld surface (Fig. 3.15) caused by poor welding process. The impact velocity can be found with Eq. (3.25).

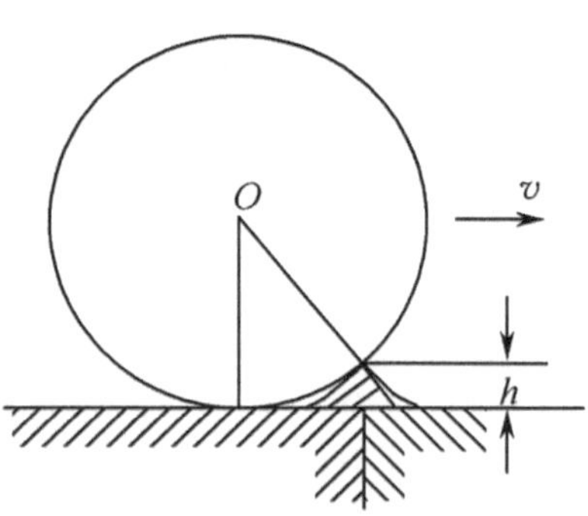

Fig. 3.15 Irregularity model for convex rail joint

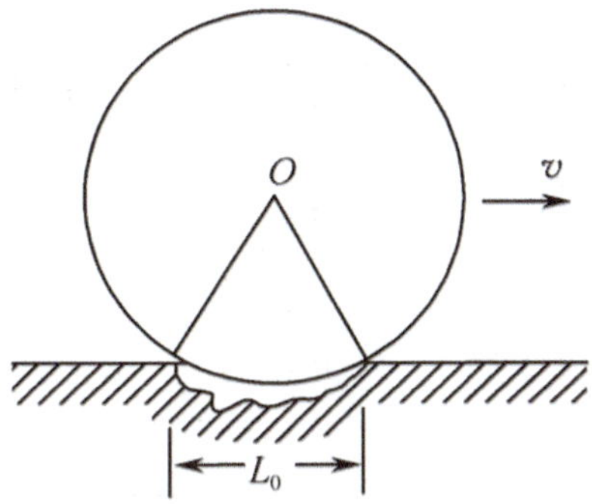

Fig. 3.16 Model for rail surface spalling

Fig. 3.17 Wide rail joint gap: **a** photograph, and **b** model

The excitation models of rail surface spalling (Fig. 3.16) and wide rail joint gaps (Fig. 3.17) are similar to that of wheel flat by simply replacing the L in Eqs. (3.7) and (3.19) with the spalling length L_0 or the rail gap width H.

3.3 Harmonic Excitation Models

In many cases, track irregularities can be described by a single harmonic wave or multiple harmonic waves. For example, squats occur at poorly welded rail joints under repeated wheel loads; they belong to the single harmonic excitation. For another example, the corrugation that widely exists all over the world present undulated wave on rail surface (shown in Fig. 3.18); this is a typical continuous harmonic excitation. In addition, periodic harmonic excitation of wheel–rail system can also be induced by the deviation of wheel center from the geometric center of gravity. It is reasonable to employ sine (cosine) functions to describe all these excitations.

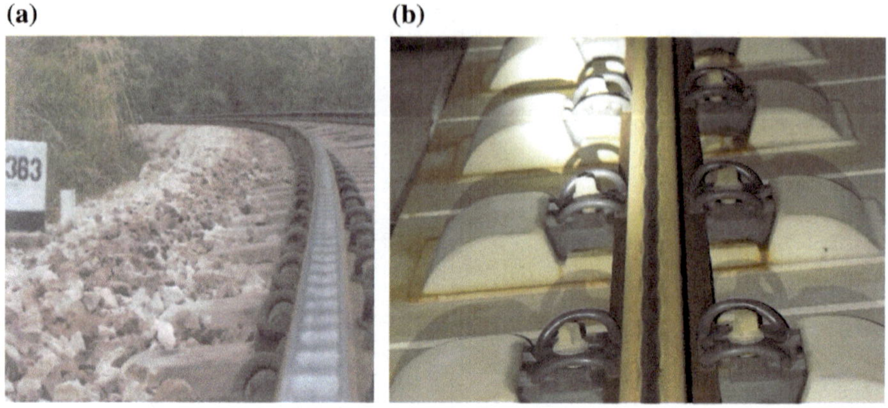

Fig. 3.18 Rail corrugation on **a** ordinary railways and **b** high-speed railways

3.3.1 Displacement Input Model of Harmonic Excitation

1. *Displacement input function of harmonic excitation*

For the single harmonic excitation as shown in Fig. 3.19, the cosine function can be simply employed to describe the rail surface irregularity, written as

$$Z_0(t) = \frac{1}{2}a(1 - \cos \omega t) \quad \left(0 \leq t \leq \frac{L}{v}\right) \tag{3.34}$$

where

$$\omega = \frac{2\pi v}{L} \tag{3.35}$$

where L is the irregularity wavelength, and a is the irregularity wave depth.

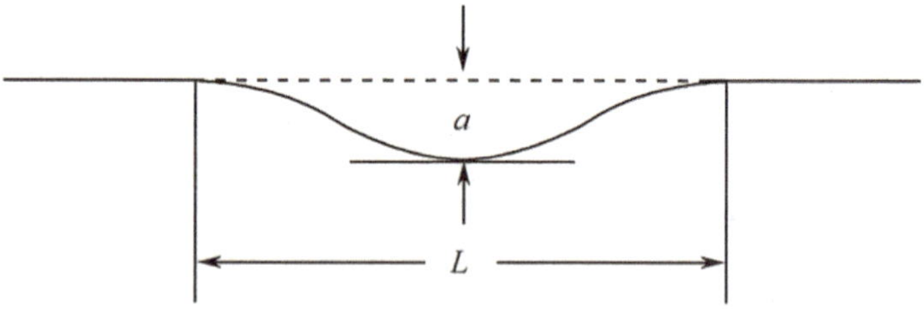

Fig. 3.19 Excitation with a single harmonic wave

For the excitation with multiple waves, the variable t in the displacement input function (3.34) should satisfy

$$0 \leq t \leq \frac{nL}{v} \tag{3.36}$$

where n is the number of excitation waves.

2. *Common railway harmonic displacement excitations*

Rail squats and dipped rail surface are common displacement excitations with a single harmonic wave. For rail squats, generally, the wavelength L = 200–500 mm, and the wave depth a = 0.5–1.0 mm, while for the dipped rail surface, the wave length and depth have a wide variation range.

Vertical irregularities of turnout switch rail (Fig. 3.20) and movable-point frog (Fig. 3.21) are typical single harmonic excitation; they exist at the transition section between the stock rail and point rail (movable-point rail). Here the turnout switch area is taken for explanation. When a wheel rolls from the stock rail to the point rail, the wheel starts to descend slightly due to the surface of the switch rail being lower than that of the stock rail. When the wheel leaves the stock rail and comes to the

Fig. 3.20 Turnout switch rail

Fig. 3.21 Turnout movable-point frog

point rail, it rolls at a smaller rolling radius. When the wheel totally rolls on the point rail, the wheel gradually rises due to the heights of the point rail gradually ascending to the same level as the stock rail. As a result, the irregularity in the vertical plane will be presented as shown in Fig. 3.22. For the Chinese No. 12 turnout with 60AT rail, the irregularity wavelength at the point rail $L \approx 2$ m, and the wave depth a can be determined according to the wear level, which are 2–4 mm for slight wear, 4–6 mm for medium wear, and above 6 mm for severe wear. The vertical irregularity for the movable-point frog is similar to that of the turnout switch area, the only difference is that the change rate of the irregularity at the point rail is larger than that of the switch rail, namely, its wavelength is smaller. For the Chinese No. 12 movable-point frog of CN60 rail, $L \approx 1$ m.

Wheel out-of-roundness is also a typical harmonic irregularity, which includes the wheel tread local defect and wheel polygon. For wheel local defect due to eccentric tread wear (Fig. 3.23), the harmonic irregularity function as illustrated in Fig. 3.19 can be adopted. Considering the fact that the wheel local defect is a periodic irregularity at wheel–rail interface, it can be expressed as

$$Z_0(t) = \begin{cases} \frac{1}{2}a\left[1 - \cos\left(\frac{2\pi v}{L} \times \mathrm{mod}\left(t, \frac{2\pi R}{v}\right)\right)\right] & \left(\mathrm{mod}\left(t, \frac{2\pi R}{v}\right) \le \frac{L}{v}\right) \\ 0 & \left(\mathrm{mod}\left(t, \frac{2\pi R}{v}\right) > \frac{L}{v}\right) \end{cases} \tag{3.37}$$

where mod is the remainder function, R is the wheel radius, and L and a are the wavelength and wave depth of the wheel local defect, respectively. Generally, L = 250–800 mm, a = 0.4–3.5 mm.

Wheel polygon refers to periodic radial deviation formed by wheel nonuniform wear. Figure 3.24 shows a real wheel polygon of a Chinese high-speed train, and Fig. 3.25 shows the field measured wheel radial profile of a Chinese high-speed train, which indicates that this wheel has obvious polygonal features. Similar to the

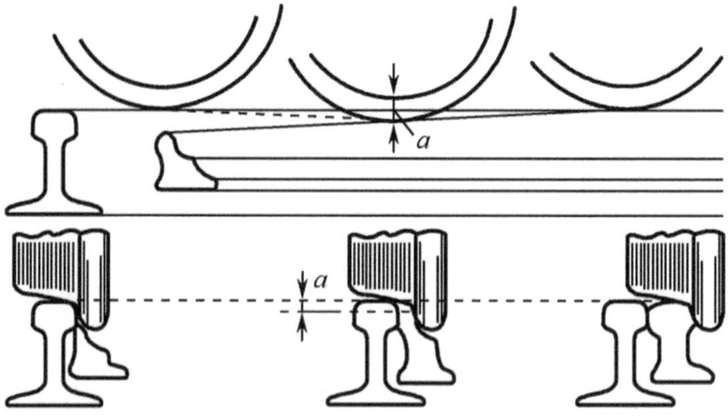

Fig. 3.22 Schematic diagram of a wheel passing a turnout switch

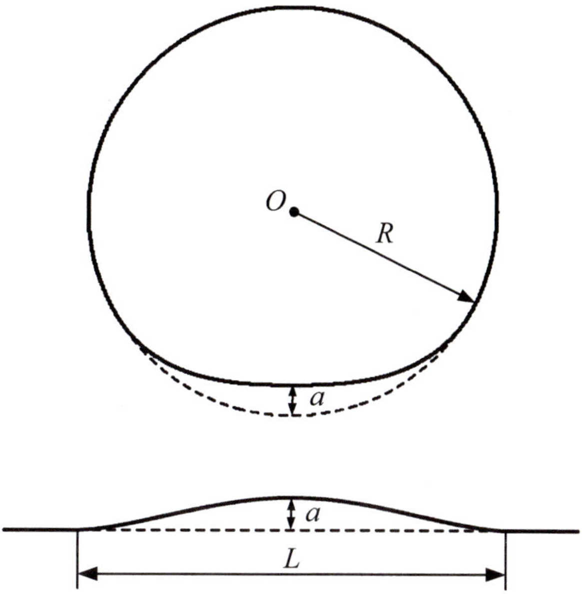

Fig. 3.23 Out-of-round wheel and its expansion

Fig. 3.24 Wheel polygon

wheel local defect, wheel polygon is also a periodic irregularity along the wheel circumference, which can be represented by the following equation using Fourier series:

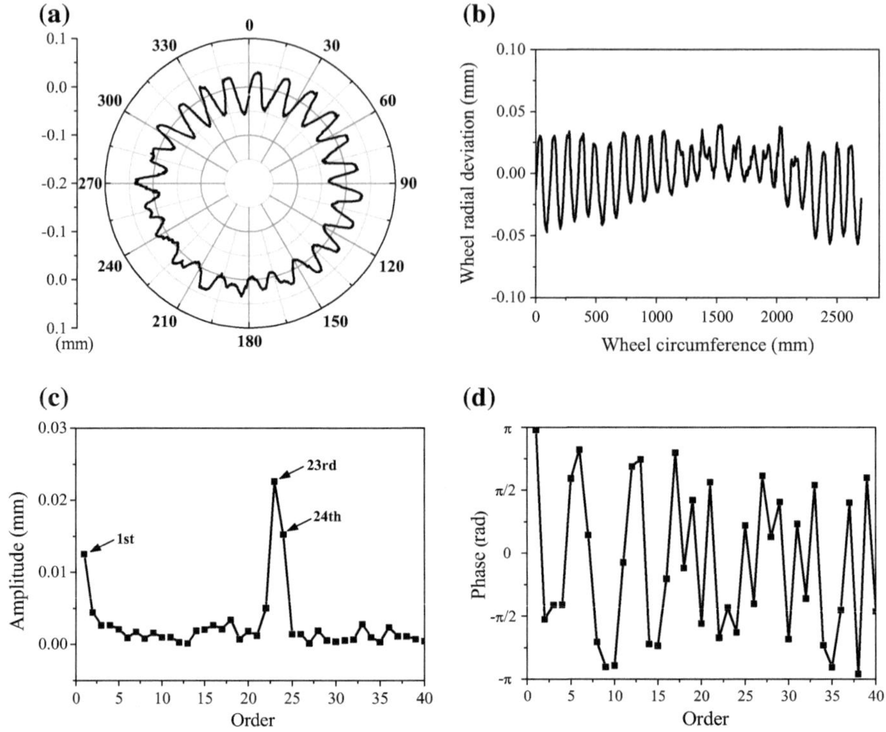

Fig. 3.25 Field measured wheel polygon. **a** Polar diagram of wheel radial deviation; **b** expansion of wheel radial deviation along its circumference; **c** amplitude of each order of wheel polygon; **d** phase of each order of wheel polygon

$$Z_0(t) = \sum_{i=0}^{\infty} A_i \sin\left[i\left(\frac{v}{R}\right)t + \varphi_i\right] \tag{3.38}$$

where i is the order of the wheel polygon, A_i is the amplitude of the ith harmonic wave, φ_i is the corresponding phase, A_i and φ_i can be obtained by performing discrete Fourier transform of the measured radial deviation of the wheel along its circumference. In the equation, A_0 is the overall deviation of the test data from the nominal wheel that has no radial deviation; it is negligible for the vehicle–track system dynamics. Figure 3.25 shows the amplitude and phase distributions as functions of the order of wheel polygon presented in Fig. 3.24. As can be seen, the 1st-, 23rd-, and 24th-order harmonic waves are the most obvious components; they are the high-order wheel polygons and have some eccentric wheel wear. The amplitudes of the other harmonic waves are quite small. The phase distributes randomly versus the order.

In fact, most wheel polygons present the main harmonic components within the 40th order; components with higher order usually have a small amplitude.

Moreover, due to the wheel–rail contact filtering effect, the high order components have a negligible effect on the vehicle–track coupled dynamics. Considering the first N-order dominant components and ignoring the component A_0, Eq. (3.38) can be further written as

$$Z_0(t) = \sum_{i=1}^{N} A_i \sin\left[i\left(\frac{v}{R}\right)t + \varphi_i\right] \tag{3.39}$$

There is a saddle-shaped irregular wear at rail joints in Chinese traditional railways, as shown in Fig. 3.26. They commonly appear in the transition section of 300 mm between the hardened zone and non-hardened zone of rail end. The rail surface has severe depressions of a few millimeter deep. This irregularity can be described by two single waves separated by a nondepressed rail surface of length L_B, expressed as

$$Z_0(t) = \begin{cases} \frac{1}{2}a(1-\cos\omega t) & \left(0 \leq t \leq \frac{L}{v}\right) \\ 0 & \left(\frac{L}{v} < t \leq \frac{L+L_B}{v}\right) \\ \frac{1}{2}a\left[1-\cos\omega\left(t-\frac{L+L_B}{v}\right)\right] & \left(\frac{L+L_B}{v} < t \leq \frac{2L+L_B}{v}\right) \end{cases} \tag{3.40}$$

According to the characteristics of rail corrugations in terms of wavelength and wave depth, short-wave rail corrugations usually appear in straight railway lines in plain regions. An example is illustrated with the K137–K161 section of Chinese Shanghai–Hangzhou railway line, where the wavelength is found to be 170–260 mm and the wave depth is 0.4–1.5 mm. At small radius curves in long stretches of slope in mountain areas, corrugations with a longer wavelength and a larger wave depth have a high occurrence rate. For example, the author made an investigation at the Jinzhai–Tongqingshu section of Qian–Gui railway, where corrugations are found at many places, with wavelength over 300 mm and wave depth around 1 mm. For another example, the corrugations at small radius curves have the wavelength of 300–600 mm and the wave depth of around 0.5–2.5 mm in the Fengzhou section of Baoji–Chengdu railway, the K379+550 section of Chengdu–Chongqing railway and

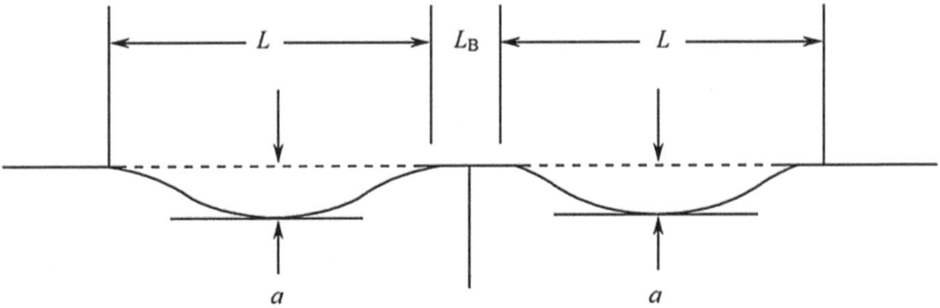

Fig. 3.26 Saddle-shaped irregularity at rail joints

the Shijiazhuang–Yangquan section of Shijiazhuang–Taiyuan railway. In recent years, rail corrugation is also increasingly becoming a problem in Chinese high-speed railways. For instance, typical corrugations with short waves (as shown in Fig. 3.18b) were observed with the wavelength $L \approx 0.14$ m and the wave depth $a \approx 0.1$ mm in some curved sections of the high-speed ballastless tracks. Although the wave depth is quite small, it could induce very intensive wheel–rail interactions under the train speed of 300–350 km/h; thus, much attention should be paid to this issue.

3. *Particularity of extremely short-wave harmonic irregularity of rail surface*

It is worth pointing out that when the wavelength of a harmonic irregularity, such as abrasions in rail surfaces, is extremely short, impact phenomenon similar to that induced by new wheel flats will occur. In this case, the irregularity should be considered as an impulsive excitation by using impact velocity input, rather than harmonic displacement input.

As shown in Fig. 3.27, when the irregularity depth $a > h$, the wheel cannot contact the rail surface in the irregularity area. This will induce impact vibrations at the starting and ending points of the irregularity. If, according to the geometry relationship in Fig. 3.27, the irregularity wavelength L and wave depth a meet the relationship

$$L < 2\sqrt{2Ra - a^2} \tag{3.41}$$

then, this harmonic irregularity would become an impulsive excitation in reality.

Take the interaction between common wagons (R = 420 mm) and tracks as an example, Table 3.2 lists the critical wavelengths for common irregularity depths. Further, Fig. 3.28 shows the corresponding harmonic/impulsive excitation areas.

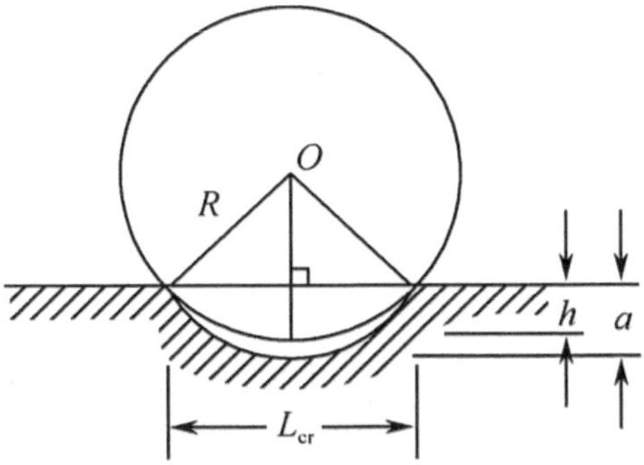

Fig. 3.27 Sketch for a wheel passing an extremely short-wave harmonic irregularity

Table 3.2 Critical wavelength L_{cr} of impulsive excitation induced by harmonic irregularity

Depth a/mm	0.5	1.0	1.5	2.0	2.5	3.0	4.0	5.0
L_{cr}/mm	40.98	57.93	70.93	81.88	91.52	100.22	115.65	129.23

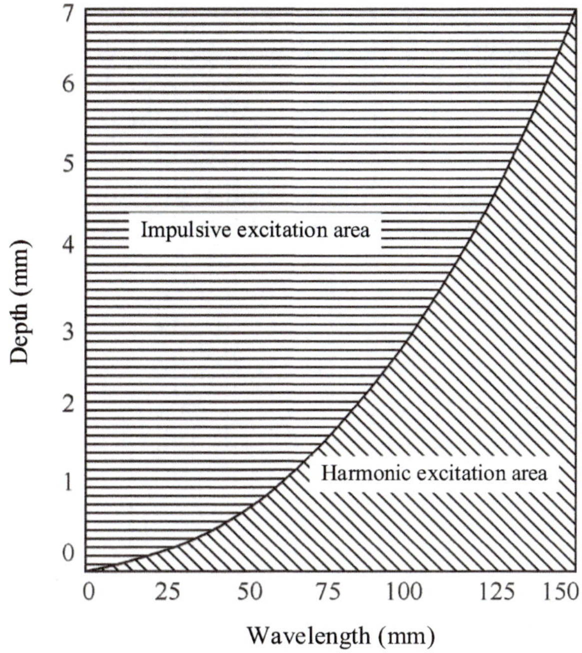

Fig. 3.28 Impulsive excitation area of extremely short-wave harmonic irregularity

3.3.2 Input Method of Common Track Irregularities

Generally, track geometry irregularities refer to the deviations of actual rail geometry from its ideal state. As shown in Fig. 3.29, common track irregularities mainly consist of alignment, gauge, height, and cross-level irregularities.

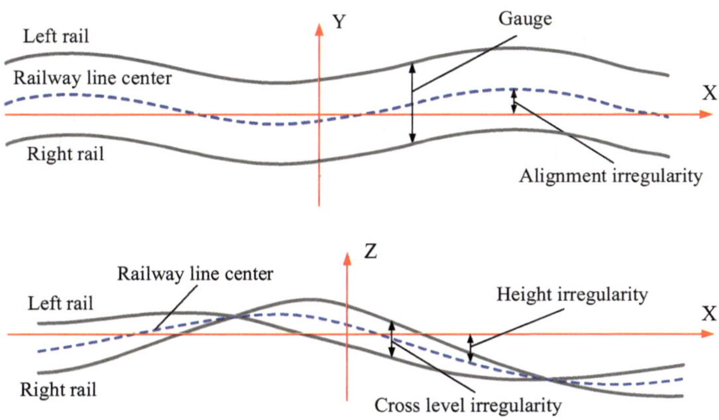

Fig. 3.29 Common track irregularities

The alignment irregularity is the lateral deviation of the railway center line due to lateral deviations of left and right rails, expressed as

$$y_t = \frac{1}{2}(y_l + y_r) \tag{3.42}$$

where y_l and y_r are the horizontal coordinates of the left and right rails, respectively.

The gauge irregularity is the gauge variation due to lateral deviations of the left and right rails, written as

$$g_t = y_l - y_r - g_0 \tag{3.43}$$

where g_0 is the nominal gauge.

The height irregularity is the vertical deviation of the railway center line due to vertical deviations of the left and right rail surfaces, given as

$$z_t = \frac{1}{2}(z_l + z_r) \tag{3.44}$$

where Z_l and Z_r are the vertical coordinates of the left and right rails, respectively.

The cross-level irregularity is the height difference between the left and right rails due to rail vertical deviation, defined as

$$\Delta z_t = z_l - z_r \tag{3.45}$$

In addition, there are two particular kinds of track irregularities, namely the torsion irregularity and the superposed irregularity. The torsion irregularity refers to the torsion of left and right rail surfaces with respect to the track plane, namely, firstly the left rail surface is higher than the right rail surface, and then inverse state appears, commonly known as the track triangle twist, and vice versa. The superposed irregularity refers to the case that the vertical and lateral irregularities simultaneously appear at the same location of railway lines.

The above track geometry irregularities can be input as system excitation by using displacement function. For the common detection method of the irregularities given in the code of Chinese railway maintenance and repair, the geometry deviation over a certain chord length is generally adopted to measure the irregularities. Therefore, the input of different track irregularities (Fig. 3.30a–f) can be realized by simply applying the single-wave cosine irregularity (as shown in Eq. 3.34) with the same or opposite directions, being in phase or out of phase, to one or two of the rails. In the figure, L is the wavelength and A is the wave depth.

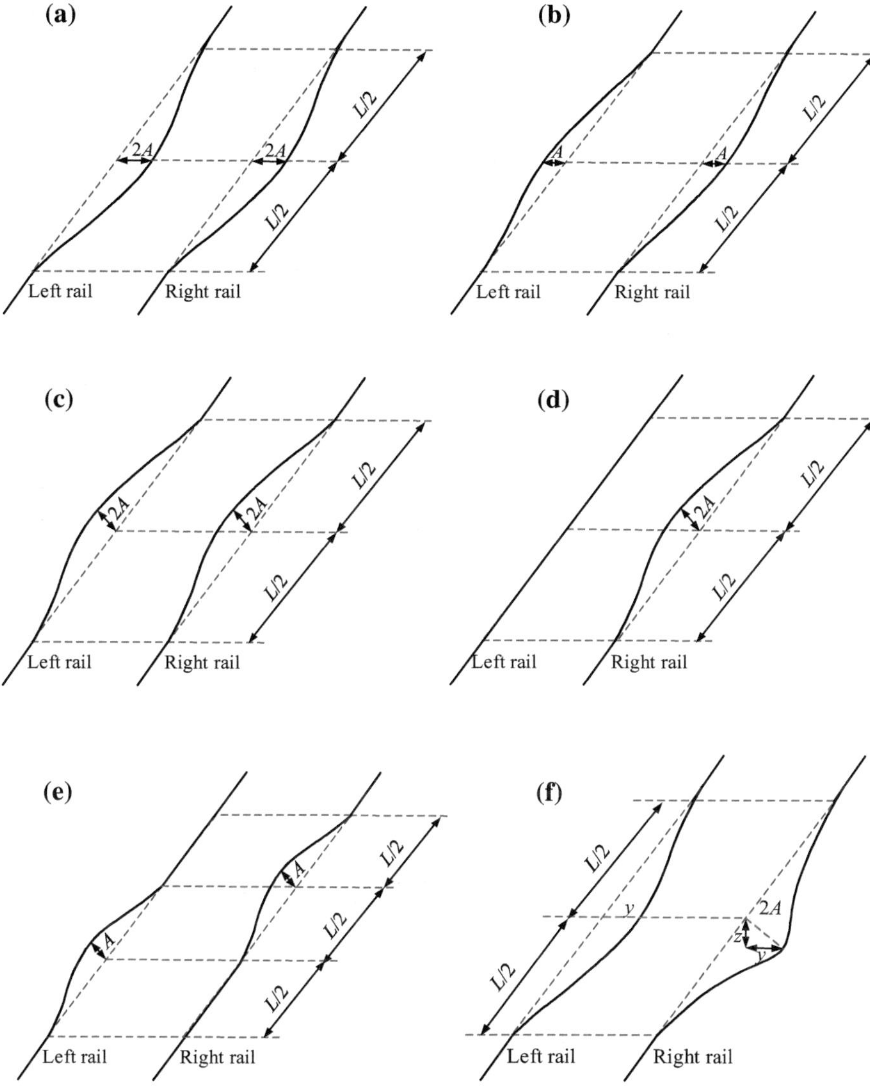

Fig. 3.30 Input models of track irregularities. **a** Alignment irregularity; **b** gauge irregularity; **c** height irregularity; **d** cross-level irregularity; **e** torsion irregularity; **f** superposed irregularity

3.3.3 Input Function of Periodic Harmonic Force

If a wheel center does not coincide with its geometry center with an eccentricity, as shown in Fig. 3.31, an unbalanced inertia force with a constant value pointing outwards would be produced, expressed as

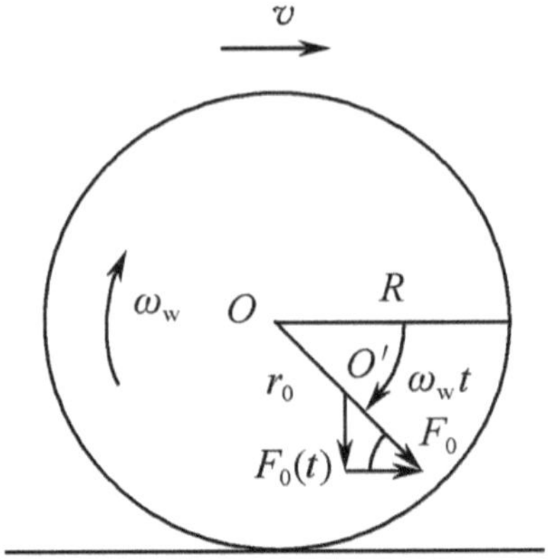

Fig. 3.31 Eccentric wheel

$$F_0 = M_w \omega_w^2 r_0 = \frac{M_0}{g} \omega_w^2 \tag{3.46}$$

where M_w is the mass of the wheel, ω_w is the angular velocity of wheel, r_0 is the eccentricity, and M_0 is the static moment of eccentricity.

$$\omega_w = \frac{v}{R} \tag{3.47}$$

$$M_0 = M_w g r_0 \tag{3.48}$$

The vertical component of the inertia force is the periodic harmonic force due to the wheel eccentricity

$$F_0(t) = M_w \left(\frac{v}{R}\right)^2 r_0 \sin\left(\frac{v}{R} t\right) \tag{3.49}$$

or

$$F_0(t) = \frac{M_0}{g} \left(\frac{v}{R}\right)^2 \sin\left(\frac{v}{R} t\right) \tag{3.50}$$

3.4 Excitation Model of Track Dynamic Stiffness Irregularity

Dynamic stiffness excitation refers to the phenomenon of longitudinal uneven elasticity of a track due to infrastructure condition variations or defects. The elasticity (stiffness) irregularity might not show any visible abnormalities, but track deformation and impact that differ from those at normal track would occur when a vehicle passes over a stiffness irregularity. Vibrations due to the stiffness irregularity will over time lead to local permanent deformation of tracks, worsening track geometry irregularity, and in return, intensifying wheel–rail interactions. Generally, the cause of track stiffness irregularity mainly includes loose or disabled rail fasteners, voided sleepers, ballast hardening or loosening, and transition zones (between subgrade and bridge, subgrade and culvert, subgrade and tunnel, and ballast track and ballastless track), etc.

3.4.1 Stiffness Irregularity at Track Transition Sections

When the supporting conditions of railway infrastructure change, it will cause longitudinal uneven stiffness of the track, inducing the track stiffness irregularity. For example, dynamic stiffness irregularities appear at the transition zones between subgrade and bridge abutment, subgrade and culvert, subgrade and tunnel, the beginning and the end of a turnout, and ballast track and ballastless track, etc. Among them, the transition zone between subgrade and bridge is the most typical; it is further explained as follows.

On the one hand, the big difference between subgrade and bridge stiffness would induce an abrupt change of track supporting stiffness. On the other hand, the settlements of subgrade and bridge are different near the transition point, causing a bending deformation at rail surface. When a train passes this transition section, it will inevitably intensify the interactions between the train and the track, affect the track structural stability, and may even threaten the operational safety of the train. This phenomenon will be more severe with the increase in train speed and the operation of heavy-haul trains. As a result, frequent maintenance and repair works have to be carried out to ensure the smoothness requirement of tracks. Inspection of some sections of the Beijing–Jiujiang railway shows that there are obvious track deformations at the ends of many bridges. This deformation develops rapidly in the initial operation period and some can reach up to 20 mm within about 2 months. A large number of surveys and analyses show that the degradation at the subgrade–bridge transition is quite extensive and severe in Chinese railway lines, and repeated maintenance and repair cause serious ballast capsules at the subgrade next to bridge abutments, which extend longitudinally up to 10–30 m.

A typical example is the severe deterioration due to dynamic interactions at the transition between the Strandmoelle bridge and subgrade (shown in Fig. 3.32) in Danish State Railway (Danske Statsbaner, DSB), where the author together with professor Hans True from Technical University of Denmark carried out field inspection and investigation in August 1998 [6]. The configuration of the track structures was complicated, as shown in Fig. 3.33. The rail was UIC60, and the rail pad thickness was 5 mm. Concrete half-sleepers were employed in the normal tracks on subgrade, and ballastless tracks with concrete long-sleepers were used on the bridge deck. Under-sleeper rubber pads of 10 mm thick were applied to both ends of the long-sleepers to fix them by glue on the bridge abutments. In addition,

Fig. 3.32 DSB Strandmoelle bridge–subgrade transition (bridge abutment is on the right)

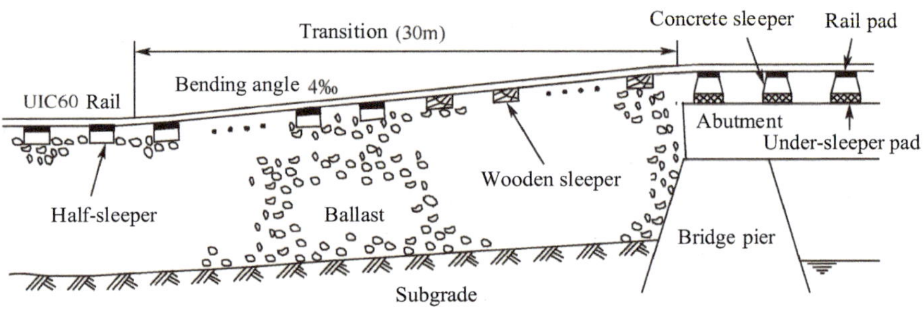

Fig. 3.33 Track structure at DSB Strandmoelle bridge–subgrade transition

23 wooden sleepers were used in the subgrade transition zone. As it can be seen that from the subgrade to the bridge abutment, not only the infrastructure supporting stiffness changed significantly, but also the sleeper supporting conditions were quite different, leading to a large stiffness variation of the tracks at the subgrade–bridge transition. In this case, differential settlement in the transition zone was quite obvious, presenting a rail bending section with a length of 30 m and a bending angle of 4‰. This substantially increased the maintenance and repair work of the track structure, which adversely made the ballast looser and caused larger track deformation. This typical dynamic stiffness irregularity of track may not be neglected in the vehicle–track coupled dynamics analysis, and the discrete supported track model has to be employed to describe in details the track stiffness parameters at the transition.

To avoid sudden stiffness change, a certain length of transition zone can usually be set between the subgrade and bridge abutment, see Fig. 3.34. It is able to render

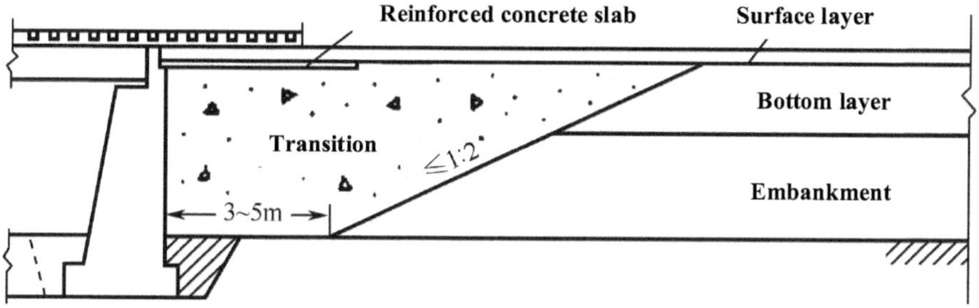

Fig. 3.34 Example of railway subgrade–bridge transition

Fig. 3.35 Subgrade stiffness irregularity model at track transition

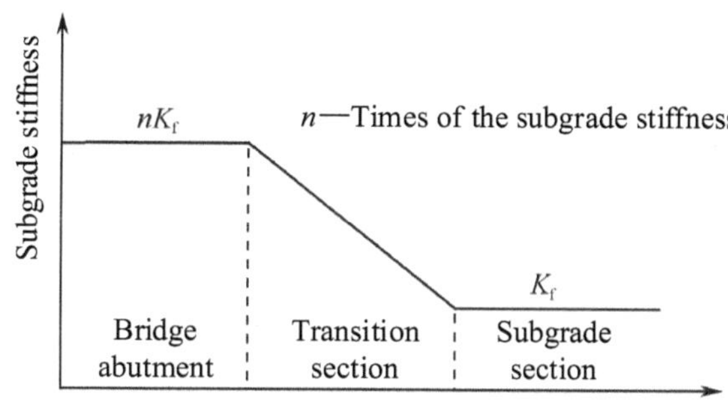

a gradual change of the stiffness in a certain range, and to reduce the variation rate of height difference of rail surface (bending angle of rail surface). As a result, it alleviates vibrations of trains and tracks and reduces track dynamic interactions, ensuring the operation safety and stability of trains. The variation of the subgrade stiffness in the transition zone can be simply illustrated with the piece-wise linear curve shown in Fig. 3.35, where the bridge abutment stiffness is assumed to be n times of the subgrade stiffness. This case can be extended by analogy to the transition zones of subgrade–culvert, subgrade–tunnel, etc.

3.4.2 *Track Stiffness Irregularity at Turnout Section*

The special configuration of turnouts determines the fact that track stiffness at turnouts is larger than that in normal tracks. Figure 3.36 shows the elastic track deformation at a turnout in the exit section of Linying station on Zhengzhou–Wuchang railway line measured by an inspection vehicle for track deformation [7]. Being the inverse of the stiffness, the deformation indirectly reflects the track stiffness variation. As can be seen, the track elastic deflection in the turnout section (especially at the frog and the switch rail) is one third smaller than that in the normal track, and it changes gradually at both ends of the turnout. The reasons that

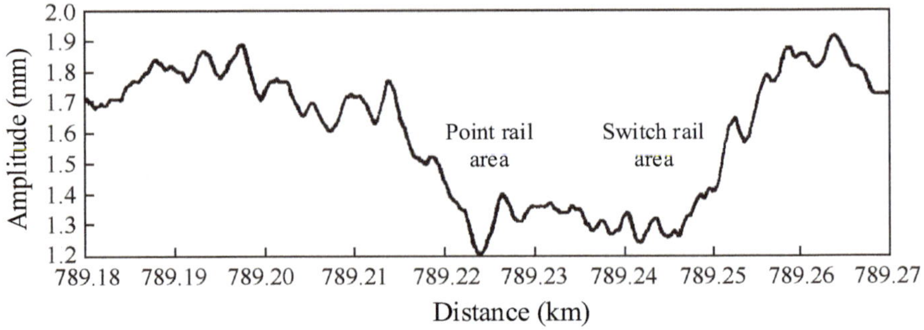

Fig. 3.36 Rail deformation (corresponding to track stiffness irregularity) at a turnout section

cause the larger and varied stiffness at the turnout mainly include different length of turnout sleepers that leads to different supporting areas of sleepers, more rails and complicated cross section of the railheads that cause a significant change of the rail bending stiffness, uneven tamping of ballast bed, etc.

3.4.3 Modeling of Rail Infrastructure Defects

Generally, when the stiffness or damping of rail infrastructure changes unevenly in the longitudinal direction, excitations can be input by simply setting in the model the values for each supporting stiffness and damping. For the cases of rail infrastructure defects, special methods can be adopted to input different kinds of dynamic stiffness irregularity as follows.

1. *Disabled rail fastener*

When a rail fastener is fractured, loose or fully disabled (shown in Fig. 3.37), one can set $K_{pi} = C_{pi} = 0$ in the model for this supporting point.

Fig. 3.37 Fracture of rail fastener

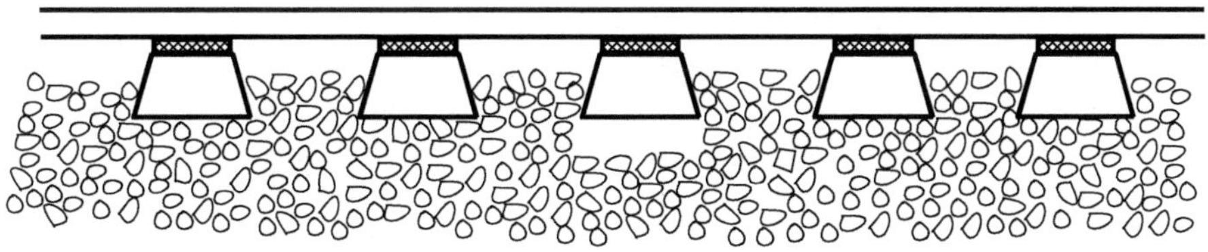

Fig. 3.38 Schematic of voided sleeper

2. *Voided sleeper*

Subgrade settlement, track deflection, and tamping could cause a local hidden pit of ballast bed, inducing a void under sleeper, as shown in Fig. 3.38. If the sleeper void δ_{gap} is large, the ballast bed at this location will completely lose its function. Correspondingly, $K_{bi} = C_{bi} = 0$ can be used in the model. If δ_{gap} is not large so that the ballast bed still has partial bearing capacity, namely the sleeper still makes contact with the ballast bed and interaction forces are induced, the following equation can be employed instead of normal interaction force between sleeper and ballast bed.

$$F_{bs}(t) = \begin{cases} K_b\left[Z_s(t) - Z_b(t) - \delta_{gap}\right] & (Z_s - Z_b - \delta_{gap} \geq 0) \\ 0 & (Z_s - Z_b - \delta_{gap} < 0) \end{cases} \tag{3.51}$$

where K_b is the ballast stiffness; Z_s and Z_b are the dynamic displacements of the sleeper and the ballast, respectively.

3. *Hardening or loosening ballast bed*

During ballast bed laying and maintenance and due to natural conditions during operation, etc., ballast hardening or loosening could occur at certain sections, which leads to significant variations of stiffness and damping of ballast bed. The stiffness and damping at the corresponding supporting point can be expressed as

$$\begin{cases} K'_{bi} = \eta_k K_{bi} \\ C'_{bi} = \eta_c C_{bi} \end{cases} \tag{3.52}$$

where η_k and η_c are coefficients of variations in the stiffness and damping of ballast bed, respectively. For different cases, η_k and η_c can be 0.1–10.

3.5 Excitation Model of Random Track Irregularity

The system excitations introduced in Sects. 3.2–3.4 belong to the category of typical deterministic excitations. Another category is the nondeterministic and is usually called the random track irregularity. The geometrical state of real-life railways always shows clear randomness, which is caused by many factors, including initial rail bending; wear and damage of rails; nonuniform sleeper spacing; nonuniform gradation, strength, loosening, contamination, and hardening of ballast bed; subgrade stiffness variation and differential settlement, and so on. Through the combined action of all these factors, the randomness of the track irregularity is formed. Under the excitation of the random track irregularity, the vehicle–track coupled system will vibrate in a stochastic way, which affects the ride comfort of passengers and the stability of cargo in one way, and the fatigue breakage and serviceability of the rolling stock structural components in another way. At the same time, the fatigue damage of track structures and the accumulation of track deformation are also influenced, which in turn aggravate the deterioration of track geometrical state.

An actual track irregularity is a superposition of random harmonic waves of different wavelengths, phases, and amplitudes; it is a complicated random process depending on the location along the track. In general, Power Spectral Density (PSD) is the most important and most commonly used statistical function for the representation of random track irregularity which is usually considered as a stationary stochastic process, hence, the so-called track irregularity PSD. The PSD charts are usually adopted in engineering to show the relationship between spectral densities and the corresponding frequencies. The PSD charts of track irregularities are continuously varied curves with the spectrum density being the ordinate and the frequency or wavelength the abscissa, in which the relationship between irregularity amplitude and frequency is clearly shown.

The statistical characteristics of random track irregularities can only be obtained by field measurement. The test work begins in 1964 by British Railway who was one of the earliest to study the random track irregularity problem. At present, many countries including USA, Germany, UK, Japan, Russia, India, and Czech Republic have measured their own track irregularity PSDs and determined the related functions. Much research work has also been done in China. In 1982, various kinds of methods for track irregularity measurement are discussed by Luo [8] who worked for China Academy of Railway Sciences (CARS). In his article, the track irregularity data were measured by the so-called "Inertia Reference Method" based on a track inspection car. These data were analyzed and processed, and several typical PSDs of the track irregularity samples were illustrated. In 1985, the research group on random vibration of Changsha Railway Institute [9] divided the track irregularity into elastic irregularity and geometric irregularity, different types of track irregularity PSDs were obtained by three successive field measurements on the Beijing–Guangzhou railway line, and analytical expressions of Chinese mainline track irregularity PSDs were found. It should be realized that due to the shortage of samples (only several hundred meters and dozens of kilometers of measurements from Changsha Railway Institute and from CARS, respectively), the resolution of the track irregularity PSDs obtained by the earlier research of the two institutions was not high, so they could hardly represent the statistical characteristics of track irregularities of the Chinese railways.

In view of this, intensive research was carried out by CARS in the late 1990s on Chinese railway track irregularities. Data were acquired on about 40,000 km of the railway main lines all over the country, mainly by train-borne measurement with the track inspection cars and partially by on-track measurement. The data were filtrated, classified and analyzed statistically; the track irregularity PSDs (including height, cross-level, and alignment, as well as some long wavelength track irregularities) of Chinese railway main lines (heavy-haul lines, speedup lines, high-speed test lines, with various track structures, super-large bridges, etc.) were proposed [10].

The large-scale launch and operation of the Chinese new high-speed railway lines in recent years have put the measurement and study of high-speed railway track irregularity PSDs high on the agenda. To meet the need for research and maintenance of the Chinese high-speed railway, CARS together with Southwest Jiaotong University (the research group of the author), based on the statistical analysis of high-speed railway ballastless track irregularity data measured by the high-speed track inspection cars from Beijing–Tianjin, Wuhan–Guangzhou, Zhengzhou–Xi'an, Shanghai–Hangzhou, Shanghai–Nanjing, and Beijing–Shanghai high-speed railways, put forward the first Chinese standard on high-speed railway ballastless track irregularity PSDs in October 2014 [11]. Then in June 2016, CARS et al., based on the statistical analysis of high-speed railway ballasted track irregularity data measured from Hangzhou–Shenzhen, Nanning–Guangzhou, Nanchang–Fuzhou, Hengyang–Liuzhou, Jinan–Qingdao, and Hefei–Wuhan high-speed railways, published the PSDs of ballasted track irregularities of Chinese high-speed railways [12].

Typical track irregularity PSDs are listed below, and the features are clarified through analysis and comparison, so as for the readers to conveniently choose the appropriate random track irregularity excitation model for the vehicle–track coupled dynamics analysis.

3.5.1 Track Irregularity PSDs of United States of America

Based on massive measurement data, track irregularity PSDs were obtained by Federal Railway Administration (FRA) of America. The PSDs were fitted by even rational functions containing cut-off frequencies and roughness constants, with wavelength ranging from 1.524 to 304.8 m. Six track classes were considered [13].

(1) Height

$$S_{\mathrm{v}}(\phi) = \frac{A_{\mathrm{v}}\phi_{\mathrm{v2}}^2\left(\phi^2 + \phi_{\mathrm{v1}}^2\right)}{\phi^4\left(\phi^2 + \phi_{\mathrm{v2}}^2\right)} \tag{3.53}$$

(2) Alignment

$$S_{a}(\phi) = \frac{A_{a}\phi_{a2}^{2}\left(\phi^{2} + \phi_{a1}^{2}\right)}{\phi^{4}\left(\phi^{2} + \phi_{a2}^{2}\right)} \tag{3.54}$$

(3) Cross-level

$$S_{c}(\phi) = \frac{A_{c}\phi_{c2}^{2}}{\left(\phi^{2} + \phi_{c1}^{2}\right)\left(\phi^{2} + \phi_{c2}^{2}\right)} \tag{3.55}$$

(4) Gauge

$$S_{g}(\phi) = \frac{A_{g}\phi_{g2}^{2}}{\left(\phi^{2} + \phi_{g1}^{2}\right)\left(\phi^{2} + \phi_{g2}^{2}\right)} \tag{3.56}$$

where $S(\phi)$ is the track irregularity PSD [$m^2/(1/m)$]; ϕ is the spatial frequency of track irregularity (1/m); A is the roughness constants (m); ϕ_1, ϕ_2 are the cut-off frequencies (1/m).

The roughness constants and cut-off frequencies of the six track classes are listed in Table 3.3. In this table, the allowable maximum traffic speeds for the different track classes are also shown according to the FRA safety standard.

Table 3.3 Parameter values of FRA track irregularity PSDs

Types	Parameter	Track class					
		First	Second	Third	Fourth	Fifth	Sixth
Height	$A_v/(10^{-7}$ m)	16.7217	9.5250	5.2917	2.9633	1.6722	0.9525
	$\phi_{v1}/(1/m)$	0.0233	0.0233	0.0233	0.0233	0.0233	0.0233
	$\phi_{v2}/(1/m)$	0.1312	0.1312	0.1312	0.1312	0.1312	0.1312
Alignment	$A_a/(10^{-7}$ m)	10.5833	5.9267	3.3867	1.8838	1.0583	0.5927
	$\phi_{a1}/(1/m)$	0.0328	0.0328	0.0328	0.0328	0.0328	0.0328
	$\phi_{a2}/(1/m)$	0.1837	0.1837	0.1837	0.1837	0.1837	0.1837
Cross-level	$A_c/(10^{-7}$ m)	4.8683	3.3867	2.3283	1.5663	1.0583	0.7197
	$\phi_{c1}/(1/m)$	0.0233	0.0233	0.0233	0.0233	0.0233	0.0233
	$\phi_{c2}/(1/m)$	0.1312	0.1312	0.1312	0.1312	0.1312	0.1312
Gauge	$A_g/(10^{-7}$ m)	10.5833	5.9267	3.3867	1.8838	1.0583	0.5927
	$\phi_{g1}/(1/m)$	0.0292	0.0292	0.0292	0.0292	0.0292	0.0292
	$\phi_{g2}/(1/m)$	0.2329	0.2329	0.2329	0.2329	0.2329	0.2329
Allowable maximum speeds/(km/h)	Freight car	16	40	64	96	128	176
	Passenger car	24	48	96	128	144	176

3.5.2 Track Irregularity PSDs of Germany

German railway divides the track irregularity PSDs into two categories, namely "low disturbance" and "high disturbance", which are described as unified form in the following contents [14]:

(1) Height

$$S_{\mathrm{v}}(\Omega) = \frac{A_{\mathrm{v}}\Omega_{\mathrm{c}}^2}{(\Omega^2 + \Omega_{\mathrm{r}}^2)(\Omega^2 + \Omega_{\mathrm{c}}^2)} \tag{3.57}$$

(2) Alignment

$$S_{\mathrm{a}}(\Omega) = \frac{A_{\mathrm{a}}\Omega_{\mathrm{c}}^2}{(\Omega^2 + \Omega_{\mathrm{r}}^2)(\Omega^2 + \Omega_{\mathrm{c}}^2)} \tag{3.58}$$

(3) Cross-level

$$S_{\mathrm{c}}(\Omega) = \frac{A_{\mathrm{v}} \cdot b^{-2} \cdot \Omega_{\mathrm{c}}^2 \cdot \Omega^2}{(\Omega^2 + \Omega_{\mathrm{r}}^2)(\Omega^2 + \Omega_{\mathrm{c}}^2)(\Omega^2 + \Omega_{\mathrm{s}}^2)} \tag{3.59}$$

where the units of height and alignment PSDs are m^2/(rad/m); As the cross-level irregularity was measured by the inclination angle, the unit for $S_{\mathrm{c}}(\Omega)$ becomes 1/(rad/m); Ω is the spatial frequency of track irregularity (rad/m); Ω_{c}, Ω_{r} and Ω_{s} are the cut-off frequencies (rad/m); A_{v} and A_{a} are the roughness constants ($\mathrm{m}^2 \cdot \mathrm{rad/m}$); b is half the distance between two the rolling circles of a wheelset (m), normally it is set as 0.75 m.

(4) Gauge

No equation was given in Ref. [14] for the gauge PSD; it just stipulated that the range of gauge variation be between −3 and 3 mm. In general, the PSD expressions for gauge and cross-level irregularities are in similar form, so the equation of the gauge PSD can be expressed as

$$S_{\mathrm{g}}(\Omega) = \frac{A_{\mathrm{g}}\Omega_{\mathrm{c}}^2\Omega^2}{(\Omega^2 + \Omega_{\mathrm{r}}^2)(\Omega^2 + \Omega_{\mathrm{c}}^2)(\Omega^2 + \Omega_{\mathrm{s}}^2)} \quad \mathrm{m}^2/(\mathrm{rad/m}) \tag{3.60}$$

The roughness coefficients and cut-off frequencies are shown in Table 3.4, in which A_{g} are the reference values calculated under the assumption that the gauge variation is between −3 and 3 mm.

"Low disturbance" is suitable for German high-speed railways whose operational speeds are 250 km/h and above, while "high disturbance" is suitable for German ordinary railways whose speeds are below 250 km/h.

Table 3.4 Roughness coefficients and cut-off frequencies for German track irregularity PSDs

Track class	Ω_c (rad/m)	Ω_r (rad/m)	Ω_s (rad/m)	A_a ($m^2 \cdot$ rad/m)	A_v ($m^2 \cdot$ rad/m)	A_g ($m^2 \cdot$ rad/m)
Low disturbance	0.8246	0.0206	0.4380	2.119×10^{-7}	4.032×10^{-7}	5.32×10^{-8}
High disturbance	0.8246	0.0206	0.4380	6.125×10^{-7}	1.08×10^{-6}	1.032×10^{-7}

3.5.3 *Track Irregularity PSDs of China*

To date, there is still no complete system of track irregularity PSD standards, which can represent the various track geometry states in China. But as mentioned above, many studies have been carried out by the institutions on the track irregularity PSDs; the PSD formula of various track types have been given according to the measurement results. These are introduced as follows:

1. *Track irregularity PSD of Chinese conventional main lines*

This track irregularity PSD reflects the track geometry state of the Chinese existing conventional main lines after the speedup renovation. A same analytic expression is adopted for track height, cross-level, and alignment PSDs, but with different coefficient values [10].

$$S(f) = \frac{A(f^2 + Bf + C)}{f^4 + Df^3 + Ef^2 + Ff + G} \tag{3.61}$$

where the unit of $S(f)$ is $mm^2/(1/m)$; f is the spatial frequency(1/m), A, B, C, D, E, F, G are the characteristic parameters of the corresponding track irregularity PSDs, their values vary with different track classes and irregularity types.

The values of the track irregularity characteristic parameters for three conventional main lines after raising train speed, i.e., the Beijing–Shanghai, Beijing–Guangzhou, and Beijing–Harbin conventional lines, are given in Table 3.5. They are suitable for conventional railway lines with maximum operational speedup to 160 km/h.

2. *Track irregularity PSDs of Chinese high-speed railways*

The ballastless track is widely used in the Chinese high-speed railways with the operational speed between 300 and 350 km/h, and the ballasted track is frequently

Table 3.5 Characteristic parameters of the track irregularity PSDs for Chinese existing speedup main lines

Parameter	*A*	*B*	*C*	*D*	*E*	*F*	*G*
Height (left)	1.1029	−1.4709	0.5941	0.8480	3.8016	−0.2500	0.0112
Height (right)	0.8581	−1.4607	0.5848	0.0407	2.8428	−0.1989	0.0094
Alignment (left)	0.2244	−1.5746	0.6683	−2.1466	1.7665	−0.1506	0.0052
Alignment (right)	0.3743	−1.5894	0.7265	0.4353	0.9101	−0.0270	0.0031
Cross-level	0.1214	−2.1603	2.0214	4.5089	2.2227	−0.0396	0.0073

employed for the high-speed railway with operational speed between 200 and 250 km/h. The track irregularity PSDs of the high-speed railways were obtained based on the field measurement data of typical ballastless and ballasted tracks. Piece-wise fitting by the power function was adopted for the PSDs and each wavelength segment of the PSD curves had the same expression as follows [11, 12]:

$$S(f) = \frac{A}{f^n} \tag{3.62}$$

where the unit of $S(f)$ is $mm^2/(1/m)$; f is the spatial frequency(1/m), A and n are the fitting coefficients. The fitting coefficients of the mean ballastless and ballasted track irregularity PSDs for the Chinese high-speed railways are given in Table 3.6 and Table 3.7, respectively. There are four segments for the fitting coefficients, their corresponding cut-off spatial frequencies and wavelengths are shown in Tables 3.8 and 3.9. These parameters are suitable for ballastless track with operational speed between 300 and 350 km/h and ballasted track with operational speed between 200 and 250 km/h. Note that the ballastless track irregularity PSD is applicable to the spatial frequency range of 0.005–0.5 m^{-1} and the corresponding wavelength range of 2–200 m; and the ballasted track irregularity PSD is applicable to the spatial frequency range of 0.01–0.5 m^{-1} and the corresponding wavelength range of 2–100 m.

Previous study has shown that track irregularity PSDs estimated from a large amount of irregularity data approximately obey the χ^2 distribution with two degrees of freedom. The values given in Tables 3.6 and 3.7 correspond to the so-called mean PSDs. For different track geometrical states, the percentile PSDs can be estimated according to the mean track irregularity PSDs of the high-speed railways.

$$S_\alpha(f) = C \cdot S(f) \tag{3.63}$$

where α is the percentile and C is the transformation coefficient as shown in Table 3.10.

3. *Short wavelength track irregularity PSD*

The wavelength of the above track irregularity PSDs ranges normally from several meters to dozens of meters, applicable to the low-frequency random vibration analysis of vehicle, track and/or bridge systems. However, it could not be applicable for analysis of the high-frequency wheel–rail interaction and track structure random vibration, as the main vibration frequencies of the unsprung mass and the track structure can be several hundred Hertz and up to several thousand Hertz, which are excited by short wavelength wheel and rail irregularities. For this reason, a short wavelength track height irregularity was in-track measured on the Shijiazhuang–Taiyuan railway line [15]. By regression analysis, the short-wavelength vertical irregularity PSD of the track with the Chinese 50 kg/m rail was approximated as

Table 3.6 Fitting coefficients for the mean ballastless track irregularity PSDs of the Chinese high-speed railways

Type of irregularity	Segment 1		Segment 2		Segment 3		Segment 4	
	A	n	A	n	A	n	A	n
Height	1.0544×10^{-5}	3.3891	3.5588×10^{-3}	1.9271	1.9784×10^{-2}	1.3643	3.9488×10^{-4}	3.4516
Alignment	3.9513×10^{-3}	1.8670	1.1047×10^{-2}	1.5354	7.5633×10^{-4}	2.8171	–	–
Cross-level	3.6148×10^{-3}	1.7278	4.3685×10^{-2}	1.0461	4.5867×10^{-3}	2.0939	–	–
Gauge	5.4978×10^{-2}	0.8282	5.0701×10^{-3}	1.9037	1.8778×10^{-4}	4.5948	–	–

Table 3.7 Fitting coefficients for the mean ballasted track irregularity PSDs of the Chinese high-speed railways

Type of irregularity	Segment 1		Segment 2		Segment 3		Segment 4	
	A	*n*	*A*	*n*	*A*	*n*	*A*	*n*
Height	8.1981×10^{-2}	1.6284	4.6880×10^{-5}	3.4194	3.8215×10^{-2}	1.2306	2.4609×10^{-4}	4.1663
Alignment	2.5934×10^{-4}	2.6607	1.3245×10^{-1}	0.8041	7.1498×10^{-4}	3.2979	–	–
Cross-level	9.9714×10^{-3}	1.6103	5.5743×10^{-2}	1.1336	1.6833×10^{-3}	3.1315	–	–
Gauge	1.3635×10^{-1}	0.6903	2.1384×10^{-2}	1.5653	8.1641×10^{-4}	4.1305	–	–

Table 3.8 Spatial frequencies and corresponding wavelengths at the segment points for the ballastless track irregularity PSDs of the Chinese high-speed railways

Type of irregularity	Point separating first and second segments		Point separating second and third segments		Point separating thirrd and fourth segments	
	Spatial frequency (1/m)	Spatial wavelength m	Spatial frequency (1/m)	Spatial wavelength m	Spatial frequency (1/m)	Spatial wavelength m
Height	0.0187	53.5	0.0474	21.1	0.1533	6.5
Alignment	0.0450	22.2	0.1234	8.1	–	–
Cross-level	0.0258	38.8	0.1163	8.6	–	–
Gauge	0.1090	9.2	0.2938	3.4	–	–

Table 3.9 Spatial frequencies and corresponding wavelengths at the segment points for the ballasted track irregularity PSDs of the Chinese high-speed railways

Type of irregularity	Point separating first and second segments		Point separating second and third segments		Point separating thirrd and fourth segments	
	Spatial frequency (1/m)	Spatial wavelength m	Spatial frequency (1/m)	Spatial wavelength m	Spatial frequency (1/m)	Spatial wavelength m
Height	0.0155	64.5	0.0468	21.4	0.1793	5.6
Alignment	0.0348	28.7	0.1232	8.1	–	–
Cross-level	0.0270	37	0.1735	5.8	–	–
Gauge	0.1204	8.3	0.2800	3.6	–	–

Table 3.10 Transformation coefficients between the mean and percentile track irregularity PSDs of the Chinese high-speed railways

Transformation coefficient	Percentile (%)										
	10.0	20.0	25.0	30.0	50.0	60.0	63.2	70.0	75.0	80.0	90.0
C	0.105	0.223	0.288	0.357	0.693	0.916	1.000	1.204	1.386	1.609	2.303

$$S(f) = 0.036f^{-3.15} \tag{3.64}$$

where the unit of $S(f)$ is $\text{mm}^2/(1/\text{m})$ and f is the spatial frequency (1/m). Its applicable wavelength range is 0.01–1 m.

To evaluate high-frequency vibrations caused by short-wavelength rail surface irregularity in high-speed railways, the Train and Track Research Institute at Southwest Jiaotong University (the author's research group) carried out field measurements in Tianjing–Qinhuangdao high-speed railway line using the "muller-BBM" rail surface roughness gauge. The test section was located at the China Railway Track System (CRTS) II ballastless track on a bridge, as shown in Fig. 3.39. The result is shown in Fig. 3.40, which could only be regarded as a sample of the short-wavelength rail vertical irregularity for Chinese high-speed railway.

Fig. 3.39 Field measurements of short wavelength rail height irregularity

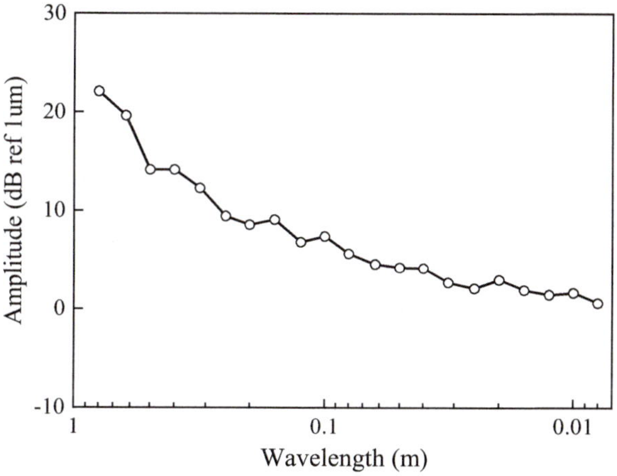

Fig. 3.40 One-third octave band spectrum of short wavelength rail vertical irregularity for Chinese high-speed railway ballastless track

3.5.4 Comparison of Typical Track Irregularity PSDs

Two kinds of Chinese track irregularity PSDs are introduced in the previous section based on the measured track geometric parameters, namely the track irregularity PSDs of conventional railway speedup lines and of the high-speed railway lines. However, it is not clear what their essential features are and what the differences are compared to the typical track irregularity PSDs? It is necessary to answer these questions to facilitate the rational selection of the random track irregularity disturbances for the dynamics analysis. In Ref. [16], the differences of vehicle dynamic performances under the excitation of Chinese conventional main lines and typical foreign track irregularity PSDs were compared through dynamics simulation. Here, from the viewpoint of track irregularity PSD characteristics itself, the differences between the Chinese track irregularity PSDs and the American, German track irregularity PSDs are analyzed and compared.

1. *Comparison of track irregularity PSDs for conventional railway lines*

The track irregularity PSDs under the ordinary operation speed are compared first. Takes the Chinese existing speedup main lines (i.e., three conventional main lines Beijing–Shanghai, Beijing–Guangzhou and Beijing–Harbin) as the examples of the conventional railway lines, their track irregularity PSDs (see Eq. (3.61) and Table 3.5) are applicable to the speed of 160 km/h and below. The track irregularity PSDs of American FRA 5th (applicable to the speed below 144 km/h) and FRA 6th (applicable to the speed below 176 km/h) track classes are selected for comparison.

Figure 3.41 shows the comparison of the alignment and height irregularities for the aforementioned three track categories within the wavelength range of 1–30 m, respectively.

It can be seen from Fig. 3.41a that the alignment PSD of the Chinese conventional railway lines is generally higher than those of the FRA 5th and 6th class tracks, indicating that the alignment geometrical state of the Chinese conventional railway lines is worse. Specific comparison results are as follows: Compared with the FRA 6th class track PSD, the Chinese conventional railway lines' alignment PSD is apparently bigger under wavelength of 25 m, and is slightly smaller only for wavelength above 25 m; Compared with the FRA 5th track class PSD, the Chinese conventional railway lines' alignment PSD is relatively bigger below wavelength of 20 m, and is smaller above 20 m.

It can be seen from Fig. 3.41b that the height irregularity PSD of the Chinese conventional railway lines is generally higher than those of the FRA 5th and 6th class tracks, the specific comparison results are similar to those of the alignment PSDs.

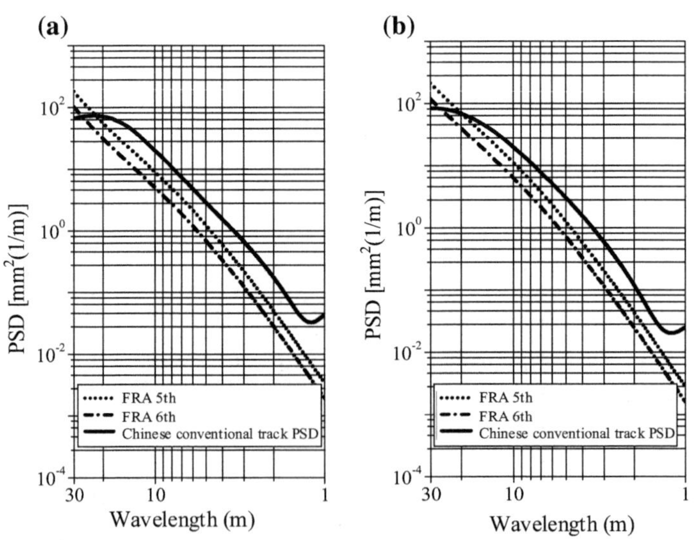

Fig. 3.41 Comparison of the PSDs for conventional railways: **a** alignment and **b** height

2. *Comparison of track irregularity PSDs for high-speed railways*

With regards to the track irregularity PSDs under high-speed operation conditions, the most typical standard should be the German track irregularity PSDs, including low disturbance PSD (suitable for speed 250 km/h and above) and high disturbance PSD (suitable for speed below 250 km/h). In China, the standards for the ballastless and ballasted track irregularity PSDs of the high-speed railways have been published (see Eq. (3.62), Tables 3.6, 3.7, 3.8 and 3.9). They are compared as follows.

The alignment and height irregularities of the aforementioned four categories of high-speed railway track irregularity PSDs are compared in Fig. 3.42, where the wavelength range is from 1 to 200 m. It is worth noting that for the Chinese ballastless track irregularity PSDs, the effective wavelength range of the PSDs is between 2 and 200 m and they are suitable for ballastless track with operational speed from 300 to 350 km/h. For the Chinese ballasted track irregularity PSDs, the effective wavelength range is between 2 and 100 m; they are suitable for ballasted track with operational speed from 200 to 250 km/h.

According to Fig. 3.42, it can be concluded as follows:

For the alignment irregularity (Fig. 3.42a), the ballastless track irregularity PSD of the Chinese high-speed railways is generally better than the German low disturbance track irregularity PSD in the whole wavelength range of 2–200 m, and much better than the German high disturbance PSD, especially for wavelength longer than several meters. In the wavelength range of 2–100 m, the Chinese ballasted track PSD is slightly poorer than the Chinese ballastless track PSD, but it is still generally better than the German low and high disturbance PSDs. As longer wavelength alignment irregularity has a critical effect on the ride comfort of high-speed trains, it can be deduced that the lateral ride comfort will be better under the excitation of the Chinese ballastless or ballasted track irregularities.

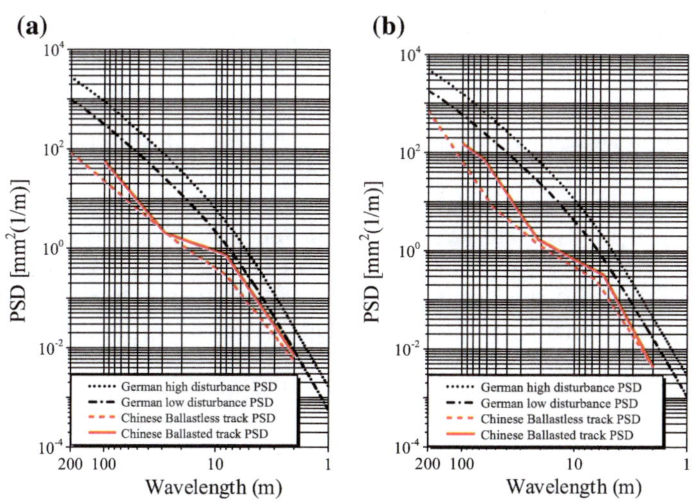

Fig. 3.42 Comparison of the PSDs for high-speed railways: **a** alignment and **b** height

For the height irregularity (Fig. 3.42b), the Chinese ballastless track irregularity PSD is generally better than both the German high and low disturbance PSDs in the whole wavelength range of 2–200 m, especially for the wavelength from 10 to 100 m. The Chinese ballasted track PSD is slightly poorer than the Chinese ballastless track PSD, but it is obviously better than German low and high disturbance PSDs in the wavelength range of 2–100 m, especially for the wavelengths from 10 to 60 m. Similarly, it can be deduced that the vertical passenger ride comfort should be good under the excitation of the Chinese ballastless or ballasted track irregularities.

3. *Conclusions*

The geometrical state of the Chinese conventional railway lines is generally more irregular than the FRA 5th and 6th class tracks. The amplitudes are obviously higher, especially when the wavelengths are shorter than 20 m. But for the wavelength above 20 m, the PSDs of the Chinese conventional main lines are better than that of the FRA 5th class track, and are better than that of the FRA 6th class track when the wavelength is above 25 m.

The ballastless and ballasted track irregularity PSDs of the Chinese high-speed railways are obvious smoother than the German low disturbance track irregularity PSD in the wavelength range of 2–200 m and 2–100 m, respectively, which is more noticeable for the longer wavelength irregularity PSDs above 10 m. The comparison shows that the Chinese high-speed tracks have rather excellent geometrical smoothness.

3.5.5 Numerical Simulation Method for Random Track Irregularity Time-Domain Samples Transformed from Track Irregularity PSDs

It can be seen from above that the power spectral density function is usually employed for representing the random track irregularity. However, for the nonlinear vehicle–track coupled dynamics model established in Chap. 2, time-domain input is generally adopted as system excitation for convenience of numerical simulation. Therefore, it is necessary to convert the random track irregularity PSD functions into spatial irregularity samples varying with distance (time-domain samples can be obtained accordingly). For this purpose, proper time–frequency conversion method should be used, and the accuracy of the method is crucial for the true representation of an actual track spatial geometry state.

At present, there are several existing methods to realize time–frequency conversion, such as the quadratic-filtering method [17], trigonometric-series method [18], white-noise filtering method, and so on. These methods have various problems when they are applied to the numerical simulation of random track irregularities. For example, suitable filters have to be designed for different track irregularity types in the case of the quadratic-filtering method, which thus lacks versatility. Here a new algorithm based on frequency-domain PSD equivalence is introduced [19], in which the spectral amplitude and random phase are first obtained according to the random track irregularity PSDs and then the track irregularity time-domain samples are computed through Inverse Fast Fourier Transform (IFFT).

According to Ref. [19], the PSD $S_{xx}(k)$ has a definite relationship with the signal spectrum at the discrete sampling points as

$$\begin{aligned} S_{xx}(k) &= \left\{\frac{1}{N}\sum_{s=0}^{N-1} x_s \exp\left[i\left(k\frac{2\pi}{N}\right)s\right]\right\}\left\{\frac{1}{N}\sum_{j=0}^{N-1} x_j \exp\left[-i\left(k\frac{2\pi}{N}\right)j\right]\right\} \\ &= \frac{1}{N^2}|\mathrm{DFT}[x_s]|^2 = \frac{1}{N^2}[X^*(k)X(k)] \end{aligned} \tag{3.65}$$

where $X(k)$ is the spectrum of the time series $\{x_s\}$ ($s = 0, 1, \ldots, N-1$), $k = 0, 1, \ldots, (N-1)$.

The PSDs of railway track irregularities are unilateral spectra, so a unilateral spectrum is first converted into the bilateral spectrum $S_x(f)$. Assuming $\lambda_{\min}$ to be the shortest wavelength of a track irregularity, $\lambda_{\max}$ the longest wavelength, $v_{\max}$ the maximum vehicle speed, then $f_{\max} = v_{\max}/\lambda_{\min}$ represents the maximum time frequency, $f_{\min} = v_{\max}/\lambda_{\max}$ stands for the minimum time frequency. Since the main eigenfrequencies of rolling stocks are generally around 1 Hz, so $f_{\min} < 1$ Hz should be ensured. According to the sampling theorem, the sampling period $\Delta T \leq 1/(2f_{\max})$. Assuming T_s to be the total simulation time, then $T_s/\Delta T$ describes the number of time-domain sampling points. It is generally necessary to add zeros, in the end, to ensure that the number of sampling points is an integer power of 2, namely N_r. The PSD estimated by the periodogram method is periodic and symmetric, therefore an even symmetric sequence $S_x(f = k\Delta f)$, $k = 0, 1, \ldots, N_r - 1$, $\Delta f = 1/(N_r\Delta T)$ with $N_r/2$ as the center is formed, as shown in Fig. 3.43.

From Eq. (3.65), the spectrum modulus of the time-domain sequence is

$$\begin{aligned} |X(k)| &= |\mathrm{DFT}[x_s]| = \sqrt{N_r^2 \times S_k(k)} = N_r\sqrt{S_k(k)} \\ &= N_r\sqrt{S_x(k\Delta f)\Delta f} \quad (k = 0, 1, \ldots, N_r - 1) \end{aligned} \tag{3.66}$$

Fig. 3.43 Sampling of the periodogram PSD

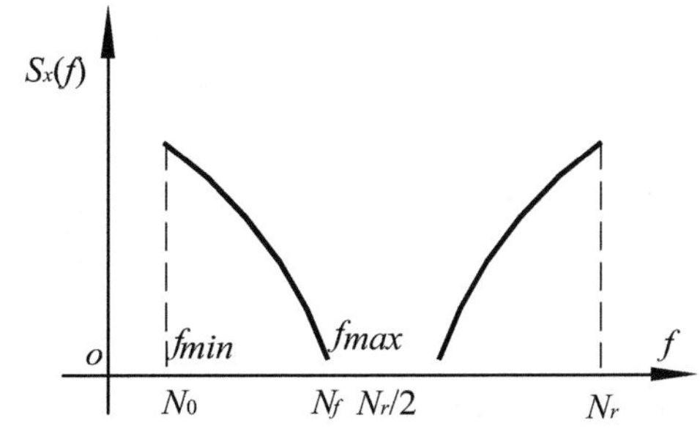

As the time sequence $X(k)$ is a random process, the spectrum phase will incorporate randomness. Suppose ξ_n is the independent phase sequence with zero mean value and $|\xi_n| = 1$. Because the Fourier transform of a real sequence will be a complex sequence (the real part is even symmetry, the imaginary part is odd symmetry), ξ_n should be complex. Therefore,

$$\xi_n = \cos\phi_n + i\sin\phi_n = \exp(i\phi_n) \tag{3.67}$$

where ϕ_n distributes uniformly between 0 and 2π.

Since the real and imaginary parts of $X(k)$ are even and odd symmetric about $N_r/2$, respectively, only the spectrum of 0–$N_r/2$ are required. According to Eqs. (3.66) and (3.67),

$$X(k) = \xi_n|X(k)| = N_r\xi_n\sqrt{S_n(k\Delta f)\Delta f} \quad (k = 0, 1, \ldots, N_r/2) \tag{3.68}$$

Obviously, it is easy to get $X(k)$ from the symmetric condition, where $k = 0, 1, \ldots, N_r - 1$. Then the simulated time-domain samples of the track irregularity can be obtained by IFFT of the complex sequence $X(k)$.

An example is given here to illustrate the conversion precision of the proposed method. Take the PSD of the FRA 6th class track as an example, see Sect. 3.5.1 for its equation and parameter values. Figure 3.44a shows the simulated time-domain sequence of track irregularity at the speed of 100 km/h. Figure 3.44b shows the simulated PSD by calculating the periodogram of the simulation sample in Fig. 3.44a and it is completely in agreement with the original analytical PSD. It can be seen that the simulation precision of this method is quite high and the calculation speed is also fast. This method has also been applied to generate the excitation input for the train–track–bridge system dynamics, see Ref. [20] for more information.

As a specific application of the above numerical simulation method (based on the algorithm of frequency-domain PSD equivalence), and also to further understand the geometrical irregularity states of the ballastless and ballasted tracks of the Chinese high-speed railways, numerical conversions of the ballastless and ballasted

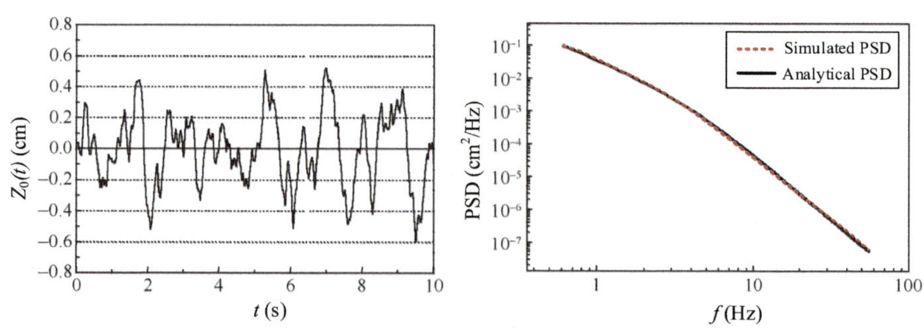

Fig. 3.44 An example on the time/frequency conversion of random track irregularity based on the frequency-domain PSD equivalence. **a** Simulation results of random track irregularity time series; **b** comparison of simulated and analytical PSD

track irregularity PSDs of the Chinese high-speed railways (as shown in Eq. (3.62)) are carried out, respectively, below. The track irregularity samples as functions of distance are obtained, as shown in Figs. 3.45 and 3.46, where the wavelength range of the ballastless track irregularities is from 2 to 200 m, and the wavelength range

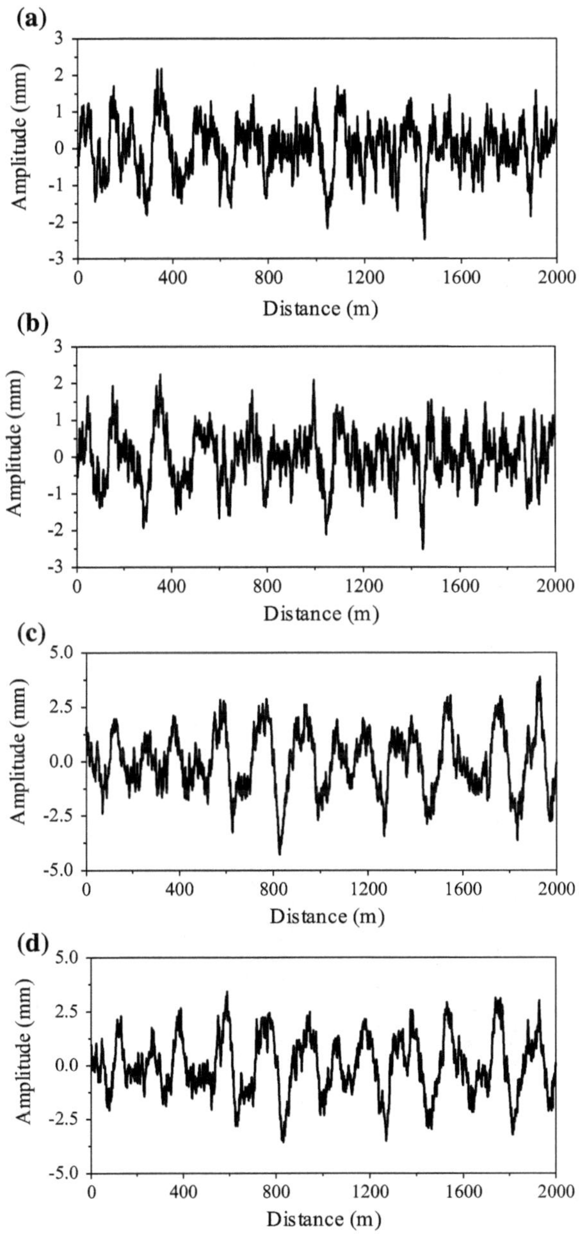

Fig. 3.45 Numerical simulation of random track irregularity samples of ballastless track of Chinese high-speed railway. **a** Alignment irregularity (left rail); **b** alignment irregularity (right rail); **c** height irregularity (left rail); **d** height irregularity (right rail)

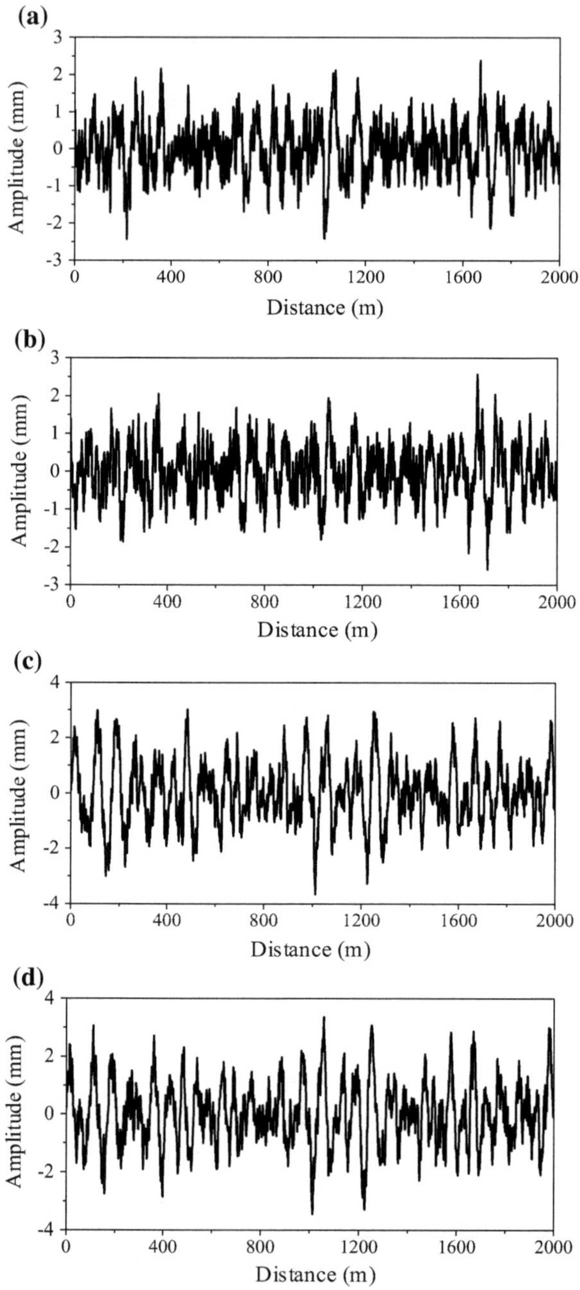

Fig. 3.46 Numerical simulation of random track irregularity samples of ballasted track of Chinese high-speed railway. **a** Alignment irregularity (left rail); **b** alignment irregularity (right rail); **c** height irregularity (left rail); **d** height irregularity (right rail)

of ballasted track irregularities is from 2 to 100 m. It can be seen in Fig. 3.45 that for the ballastless tracks of the Chinese high-speed railways, the amplitude of the height irregularity changes within the range of −5 to 5 mm and the amplitude of the alignment irregularity varies between −3 and 3 mm, showing very good track geometrical condition. For the ballasted tracks shown in Fig. 3.46, the amplitude of the height irregularity changes within the range of −4 to 4 mm and the amplitude of the alignment varies between −3 and 3 mm, also showing pretty good track geometrical condition.

References

1. Knothe K, Grassie SL. Modeling of railway track and vehicle/track interaction at high frequencies. Veh Syst Dyn. 1993;22(3/4):209–62.
2. Zhai WM. The dynamic effect of flat scar on railway wheels. Roll Stock. 1994;7:1–5 (in Chinese).
3. Satoh Y. Dynamic effect of a flat wheel on track deformation. Bull Int Railw Congr Assoc. 1965;42(8/9):547–53.
4. Ver IL, Ventre CS, Myles MM. Wheel/rail noise—Part III: Impact noise generation by wheel and rail discontinuities. J Sound Vib. 1976;46(3):395–417.
5. Zhao X. A finite element computational method for dynamic analysis of railway train frog system and its application. Master thesis. Beijing: China Academy of Railway Sciences; 1988. (in Chinese).
6. Zhai WM, True H. Vehicle–track dynamics on a ramp and on the bridge: simulation and measurements. Veh Syst Dyn. 1999;33(Suppl):604–15.
7. China Academy of Railway Sciences, Zhengzhou Railway Bureau. Research report on the comprehensive operation test of the high-speed train with maximum speed of 240 km/h on Zhengzhou–Wuhan railway. TY-1345. Beijing: China Academy of Railway Sciences; 1998 (in Chinese).
8. Luo L. Track random excitation functions. China Railw Sci. 1982;13(1):74–110 (in Chinese).
9. Research Group on Random Vibration of Changsha Railway Institute. Research on random excitation functions of rolling stock/track system. J Chang Railw Inst. 1985;(2):1–36 (in Chinese).
10. Railway Engineering Research Institute of China Academy of Railway Sciences. Study on the track irregularity power spectrum density of Chinese main lines, TY-1215. Beijing: China Academy of Railway Sciences; 1999 (in Chinese).
11. Kang X, Zhai WM, Liu XB, et al. PSD of ballastless track irregularities of high-speed railway. Standard of National Railway Administration of People's Republic of China: TB/T 3352-2014. Beijing: China Railway Publishing House; 2014 (in Chinese).
12. Li GQ, Gao L, Zhai WM, et al. PSD of ballast track irregularities of high-speed railway. Standard of China Railway Corporation: Q/CR 508-2016. Beijing: China Railway Publishing House; 2016 (in Chinese).
13. Garg VK, Dukkipati RV. Dynamics of railway vehicle systems. Ontario: Academic Press Canada; 1984.
14. Munich Research Center of German Federal Railway. ICE technology assignment for inter-city express train; 1993.
15. Wang L. Random vibration theory of rail/track structure and its application in the rail/track vibration isolation. Ph.D. thesis. Beijing: China Academy of Railway Sciences; 1988 (in Chinese).

16. Chen G, Zhai WM, Zuo HF. Comparing track irregularities PSD of Chinese main lines with foreign typical lines by numerical simulation computation. J China Railw Soc. 2001;23(3):82–7 (in Chinese).
17. Ontes RK, Enochson L. Digital time series analysis. New York: Wiley; 1972.
18. Katsu H. Random vibration analysis. Beijing: Seismological Press; 1977.
19. Chen G, Zhai WM. Numerical simulation of the stochastic process of railway track irregularities. J Southwest Jiaotong Univ. 1999;34(2):138–42 (in Chinese).
20. Zhai WM, Xia H. Train–track–bridge dynamic interaction: theory and engineering application. Beijing: Science Press; 2011 (in Chinese).

Chapter 4
Numerical Method and Computer Simulation for Analysis of Vehicle–Track Coupled Dynamics

Abstract As can be seen from Chap. 2, the vehicle–track coupled system belongs to a large-scale dynamic system including strong nonlinearities. It is impossible to theoretically solve dynamic response for such a complicated system. Time-stepping integration provides the best way for the numerical solution of the equations of motion of the vehicle–track coupled dynamics system. This chapter discusses the application of time integration methods to the analysis of vehicle–track coupled dynamics, focusing on the application of a new simple fast time integration method (Zhai in Int J Numer Meth Eng. 39(24):4199–214, 1996 [1]), and introduces associated computer simulation programs.

4.1 Time Integration Methods for Solving Large-Scale Dynamic Problems

Step-by-step time integration methods are widely used to solve the equations of multi-degree of freedom systems in mechanical and structural dynamics, especially in nonlinear system dynamics. For large-scale dynamic problems, as is frequently the case in modern dynamic analysis in practical engineering problems, the calculation efficiency is always a crucial concern, especially in the beginning of the study of the vehicle–track coupled dynamics. At that moment, the computer technology was not advanced. Therefore, it is necessary to develop more efficient time integration algorithms for large-scale complex system analysis.

Basically, there are two general classes of algorithms for dynamic problems: implicit and explicit. When implicit algorithms, such as the Newmark-β method [2] and the Wilson-θ method [3], are applied to large-scale dynamic problems, the computational time and the cost increase dramatically with the degrees of freedom of system because large-scale simultaneous algebraic equations must be solved in each time step although large time steps are permitted due to their good numerical stabilities. On the contrary, explicit schemes tend to be inexpensive. Hoff and Taylor [4] have pointed out that if lumped mass and damping matrices are used, an explicit scheme probably consists of pure vector operations. This is very convenient

W. Zhai, *Vehicle–Track Coupled Dynamics*,
https://doi.org/10.1007/978-981-32-9283-3_4

for computers with vector processors and the disadvantage of the explicit schemes in the aspect of conditional stability can be effectively alleviated through a vectorized implementation. Therefore, these explicit algorithms become more competitive on large-scale problems compared to the more stable implicit algorithms.

As we know, advances have been made to improve explicit methods. A state of the art in high-order explicit schemes for use in structural dynamic applications has been summarized in [4] by Hoff and Taylor in 1990. As indicated by these authors, it seems that the second-order accurate central difference method still remains the most popular explicit algorithm although several new developed explicit methods have their own advantages. These newly developed methods usually have lower stability than the central difference method, and some of them need more than one function evaluation (calculation of the internal forces), that is the most expensive part in nonlinear problems which are discretized by finite elements, finite differences or boundary elements.

When applied to nonlinear system dynamics, the central difference method is still supposed to solve a set of linear algebraic equations in each step unless the problem is an ideal one which satisfied the following conditions: (a) the mass matrix is diagonal, and (b) the damping matrix can be neglected or is proportional to the mass matrix. However, condition (b) is difficult to meet for practical engineering problems, while condition (a) can usually be observed. For the problem of vehicle–track coupled dynamics, the damping is not only non-negligible but also complicated and generally, the damping matrix is not proportional to the mass matrix, while the mass matrix is usually diagonal or is easily diagonalized. Thus, there is no obvious advantage for the central difference method to solve the vehicle–track coupled dynamics problem because a set of large-scale linear algebraic equations has to be solved in each time step.

Fortunately, a new simple explicit method was developed by the author in the early 1990s [1], originally aiming to fast solve long train longitudinal dynamics problem at that time. This new explicit method has at least the same stability limit as the central difference method, and needs only simple vector operations in each time step as long as the mass matrix of the solved system is diagonal, no matter what form the damping matrix of the system is. The computational efficiency is greatly enhanced. It has been found that it is fast, convenient, and economical to use this explicit method to analyze large-scale nonlinear dynamic problems in engineering. Therefore, it is suggested to employ this new simple fast explicit method to numerically solve the vehicle–track coupled dynamics response.

It is worth mentioning that this explicit method has been positively reviewed and widely adopted as so-called “Zhai method” or “Zhai algorithm” in international journals on engineering computation method, mechanics, vibration and structural engineering during the past decade [5–18]. Representatives are as follows: Rio on Advances in Engineering Software [5], Rezaiee–Pajand on Engineering Computations [6], Zhang on Journal of Sound and Vibration [7], Chen on ASME Journal of Vibration and Acoustics [8] and on Journal of Sound and Vibration [9], Zhou on Vehicle System Dynamics [10], Banimahd on Proceedings of the

Institution of Civil Engineers-Transport [11], Zhou on Acta Mechanica Sinica [12], and Wang on ASME Journal of Vibration and Acoustics [13].

In the following section, the new explicit scheme and its stability, accuracy, and numerical dissipation are presented in detail.

4.2 New Simple Fast Explicit Time Integration Method: Zhai Method

From Chap. 2, the equations of vehicle–track coupled dynamics can be expressed as a unified matrix form in Eq. (4.1a, 4.1b), which is a set of second-order ordinary differential equations and generally the matrix equation of multibody system dynamics or structural dynamics is the same as this.

$$\mathbf{M}\ddot{\mathbf{X}} + \mathbf{C}\dot{\mathbf{X}} + \mathbf{K}\mathbf{X} = \mathbf{F} \tag{4.1a}$$

or

$$\mathbf{M}\mathbf{A} + \mathbf{C}\mathbf{V} + \mathbf{K}\mathbf{X} = \mathbf{F} \tag{4.1b}$$

where $\mathbf{M}$, $\mathbf{C}$, and $\mathbf{K}$ are the mass, damping, and stiffness matrices, respectively; $\mathbf{F}$ is the vector of applied loads of the system (a given function of time, $\mathbf{F} = \mathbf{F}(t)$); $\mathbf{X}$, $\mathbf{V}$ and $\mathbf{A}$ (i.e., $\mathbf{X}$, $\dot{\mathbf{X}}$, and $\ddot{\mathbf{X}}$) are the vectors of displacements, velocities, and accelerations, respectively. The initial-value problem consists of finding a function $\mathbf{X} = \mathbf{X}(t)$, which satisfies Eq. (4.1a, 4.1b), and the initial conditions

$$\begin{cases} \mathbf{X}(0) = \mathbf{X_0} \\ \mathbf{V}(0) = \mathbf{V_0} \end{cases} \tag{4.2}$$

Here $\mathbf{X_0}$ and $\mathbf{V_0}$ are given vectors of initial displacements and velocities, respectively.

4.2.1 *Integration Scheme of Zhai Method*

Inspired by the well-known Newmark-β implicit method, the new explicit integration scheme for approximate solutions of Eqs. (4.1a, 4.1b) and (4.2) is constructed as follows:

$$\begin{cases} \mathbf{X_{n+1}} = \mathbf{X_n} + \mathbf{V_n}\Delta t + (\frac{1}{2} + \psi)\mathbf{A_n}\Delta t^2 - \psi\mathbf{A_{n-1}}\Delta t^2 \\ \mathbf{V_{n+1}} = \mathbf{V_n} + (1 + \varphi)\mathbf{A_n}\Delta t - \varphi\mathbf{A_{n-1}}\Delta t \end{cases} \tag{4.3}$$

where $\mathbf{X_n}$, $\mathbf{V_n}$, and $\mathbf{A_n}$ are the approximations to $\mathbf{X}(t = n\Delta t)$, $\mathbf{V}(t = n\Delta t)$ and $\mathbf{A}(t = n\Delta t)$, respectively; Δt is the time step, and ψ and φ are free parameters that control the stability and numerical dissipation of the algorithm.

Substituting Eq. (4.3) into Eq. (4.1a, 4.1b) at time step $t = (n + 1)\Delta t$

$$\mathbf{MA_{n+1}} + \mathbf{CV_{n+1}} + \mathbf{KX_{n+1}} = \mathbf{F_{n+1}} \tag{4.4}$$

and rearranging the terms yield

$$\mathbf{A_{n+1}} = \mathbf{M}^{-1}\tilde{\mathbf{F}}_{\mathbf{n+1}} \tag{4.5}$$

where

$$\begin{aligned}\tilde{\mathbf{F}}_{\mathbf{n+1}} = \mathbf{F_{n+1}} - \mathbf{KX_n} - (\mathbf{C} + \mathbf{K}\Delta t)\mathbf{V_n} - \left[(1+\varphi)\mathbf{C} + \left(\tfrac{1}{2} + \psi\right)\mathbf{K}\Delta t\right]\mathbf{A_n}\Delta t \\ + (\varphi\mathbf{C} + \psi\mathbf{K}\Delta t)\mathbf{A_{n-1}}\Delta t\end{aligned} \tag{4.6}$$

in which $\mathbf{F_{n+1}} = \mathbf{F}[t = (n+1)\Delta t]$.

To start the integration procedure, one can easily let $\varphi = \psi = 0$ at the first time step and use the initial conditions Eq. (4.2) as well as

$$\mathbf{A_0} = \mathbf{M}^{-1}(\mathbf{F_0} - \mathbf{CV_0} - \mathbf{KX_0}) \tag{4.7}$$

Therefore, the scheme is self-starting.

If the mass matrix is diagonal, as is the case in vehicle–track coupled dynamics, the new integration algorithm needs not to solve any equations at each time step.

4.2.2 *Stability of Zhai Method*

To investigate the stability of Zhai method, we need only consider the linear homogeneous form of Eq. (4.4) without damping for the single-degree of freedom case as

$$a_{n+1} + \omega^2 x_{n+1} = 0 \tag{4.8}$$

where

$$\omega = \sqrt{k/m} \tag{4.9}$$

The difference form of Eq. (4.8) is

$$x_{n+2} + \left[\left(\frac{1}{2}+\psi\right)\Omega^2 - 2\right]x_{n+1} + \left[\left(\frac{1}{2}+\varphi-2\psi\right)\Omega^2 + 1\right]x_n + (\psi-\varphi)\Omega^2 x_{n-1} = 0 \tag{4.10}$$

where

$$\Omega = \omega\Delta t \tag{4.11}$$

The eigenvalue equation of Eq. (4.10) takes the following form:

$$\lambda^3 + \left[\left(\frac{1}{2}+\psi\right)\Omega^2 - 2\right]\lambda^2 + \left[\left(\frac{1}{2}+\varphi-2\psi\right)\Omega^2 + 1\right]\lambda + (\psi-\varphi)\Omega^2 = 0 \tag{4.12}$$

The requirement for stability is $|\lambda| \le 1$. By use of the transformation

$$\lambda = \frac{1+Z}{1-Z} \tag{4.13}$$

Equation (4.12) becomes

$$[4+2(\varphi-2\psi)\Omega^2]Z^3 + [4+(4\psi-4\varphi-1)\Omega^2]Z^2 + 2\varphi\Omega^2 Z + \Omega^2 = 0 \tag{4.14}$$

and then the requirement for stability is simply that $R_e(Z) \le 0$. According to the Routh–Hurwitz criterion, stable steps can be derived, which have been given in Table 4.1 in detail.

It can be seen from Table 4.1 that the range of stable steps is very wide. When $\varphi = \psi = 1/2$, the stability limit is $\Delta t = 2/\omega$, which is the same as that of the central difference method. And if $\varphi = 1/8$ and $\psi = 1/4$, the stable step range is $\frac{2}{\omega}\sqrt{\frac{6}{5}} \le \Delta t \le \frac{2}{\omega}\sqrt{\frac{4}{3}}$, in which the maximum stable step is larger than $2/\omega$.

Table 4.1 Conditions of stability of the Zhai method

φ	ψ	Δt
$\varphi > \frac{1}{2}$	$\psi < \varphi$	$\Delta t \le \frac{1}{\omega}\sqrt{\frac{2\varphi-1}{(\varphi-\psi)(2\varphi+1)}}$
$\varphi > \frac{1}{2}$	$\psi \ge \varphi$	$\Delta t \le \frac{1}{\omega}\sqrt{\frac{2}{2\psi-\varphi}}$
$\varphi = \frac{1}{2}$	$\psi \ge \frac{1}{2}$	$\Delta t \le \frac{2}{\omega}\sqrt{\frac{1}{4\psi-1}}$
$0 < \varphi < \frac{1}{2}$	$\varphi < \psi < \varphi + \frac{1}{4}$	$\frac{1}{\omega}\sqrt{\frac{1-2\varphi}{(\psi-\varphi)(2\varphi+1)}} \le \Delta t \le \mathrm{Min}\left\{\frac{1}{\omega}\sqrt{\frac{2}{2\psi-\varphi}}, \frac{2}{\omega}\sqrt{\frac{1}{4(\varphi-\psi)+1}}\right\}$
$0 < \varphi < \frac{1}{2}$	$\psi \ge \varphi + \frac{1}{4}$	$\frac{1}{\omega}\sqrt{\frac{1-2\varphi}{(\psi-\varphi)(2\varphi+1)}} \le \Delta t \le \frac{1}{\omega}\sqrt{\frac{2}{2\psi-\varphi}}$

4.2.3 Accuracy of Zhai Method

Applying Taylor formula to $\mathbf{A_{n-1}}$, we have

$$\mathbf{A_{n-1}} = \mathbf{A_n} - \dot{\mathbf{A}}_\mathbf{n}\Delta t + \frac{1}{2}\ddot{\mathbf{A}}_\mathbf{n}\Delta t^2 - \cdots\cdots \tag{4.15}$$

where a supposed dot denotes a time derivative. Substituting Eq. (4.15) into Eq. (4.3) yields

$$\begin{cases} \mathbf{X_{n+1}} = \mathbf{X_n} + \mathbf{V_n}\Delta t + \frac{1}{2}\mathbf{A_n}\Delta t^2 + \psi\dot{\mathbf{A}}_\mathbf{n}\Delta t^3 - \frac{1}{2}\psi\ddot{\mathbf{A}}_\mathbf{n}\Delta t^4 + \mathbf{O}(\Delta t^5) \\ \mathbf{V_{n+1}} = \mathbf{V_n} + \mathbf{A_n}\Delta t + \varphi\dot{\mathbf{A}}_\mathbf{n}\Delta t^2 - \frac{1}{2}\varphi\ddot{\mathbf{A}}_\mathbf{n}\Delta t^3 + \mathbf{O}(\Delta t^4) \end{cases} \tag{4.16}$$

Local truncation errors can then be written as

$$\begin{cases} \mathbf{E}(\mathrm{X}) = \left(\frac{1}{6} - \psi\right)\dot{\mathbf{A}}_\mathbf{n}\Delta t^3 + \left(\frac{1}{24} + \frac{1}{2}\psi\right)\ddot{\mathbf{A}}_\mathbf{n}\Delta t^4 + \mathbf{O}(\Delta t^5) \\ \mathbf{E}(\mathrm{V}) = \left(\frac{1}{2} - \varphi\right)\dot{\mathbf{A}}_\mathbf{n}\Delta t^2 + \left(\frac{1}{6} + \frac{1}{2}\varphi\right)\ddot{\mathbf{A}}_\mathbf{n}\Delta t^3 + \mathbf{O}(\Delta t^4) \end{cases} \tag{4.17}$$

If $\psi = 1/6$ the order of accuracy of $\mathbf{E}(\mathrm{X})$ is $\mathbf{O}(\Delta t^4)$, and if $\varphi = 1/2$ the order of accuracy of $\mathbf{E}(\mathrm{V})$ will be $\mathbf{O}(\Delta t^3)$. Otherwise, the orders of accuracy decrease to $\mathbf{O}(\Delta t^3)$ and $\mathbf{O}(\Delta t^2)$, respectively. Obviously, the Zhai method has the same order of accuracy as that of the implicit Newmark-β method.

4.2.4 Numerical Dissipation and Dispersion

In order to show numerical dissipation and dispersion of the Zhai algorithm, the analytical procedure used by Hilber et al. [19] is adopted here. Rewriting Eq. (4.10) as follows:

$$x_{n+2} - 2P_1x_{n+1} + P_2x_n - P_3x_{n-1} = 0 \tag{4.18}$$

where

$$\begin{cases} P_1 = 1 - \frac{1}{2}\left(\psi + \frac{1}{2}\right)\Omega^2 \\ P_2 = 1 + \left(\frac{1}{2} + \varphi - 2\psi\right)\Omega^2 \\ P_3 = (\varphi - \psi)\Omega^2 \end{cases} \tag{4.19}$$

The general solution of Eq. (4.18) is

$$x_n = \sum_{i=1}^{3} c_i \lambda_i^n \tag{4.20}$$

where λ_i (i = 1, 2, 3) is the root of eigenvalue Eq. (4.12) and C_i is the constant defined by the initial data. If $\varphi = \psi$, there is a zero root in Eq. (4.12) which is the so-called spurious root, λ_3. And the other roots

$$\lambda_{1,2} = P_1 \pm \sqrt{P_1^2 - P_2} \tag{4.21}$$

when

$$\Omega < \frac{4}{2\varphi + 1} \tag{4.22}$$

λ_1 and λ_2 become two complex conjugate eigenvalues of Eq. (4.12), called principal roots, which satisfy $|\lambda_{1,2}| \leq 1$. Let's write $\lambda_{1,2}$ as

$$\lambda_{1,2} = P \pm Qi = \exp[\bar{\Omega}(-\bar{\zeta} \pm i)] \tag{4.23}$$

where

$$\begin{cases} P = P_1 \\ Q = \sqrt{P_2 - P_1^2} \\ \bar{\Omega} = \tan^{-1}(|Q/P|) \\ \bar{\zeta} = -\ln(P^2 + Q^2)/(2\bar{\Omega}) \end{cases} \tag{4.24}$$

In this case, the solution of Eq. (4.18) could be written in the form

$$x_n = \exp(-\bar{\zeta}\bar{\omega}t_n)[c_1 \cos(\bar{\omega}t_n) + c_2 \sin(\bar{\omega}t_n)] \tag{4.25}$$

where

$$\begin{cases} \bar{\omega} = \bar{\Omega}/\Delta t \\ t_n = n\Delta t \end{cases} \tag{4.26}$$

As measures of the numerical dissipation and dispersion, we consider the amplitude decay function *AD* and the relative period error *TD*, respectively:

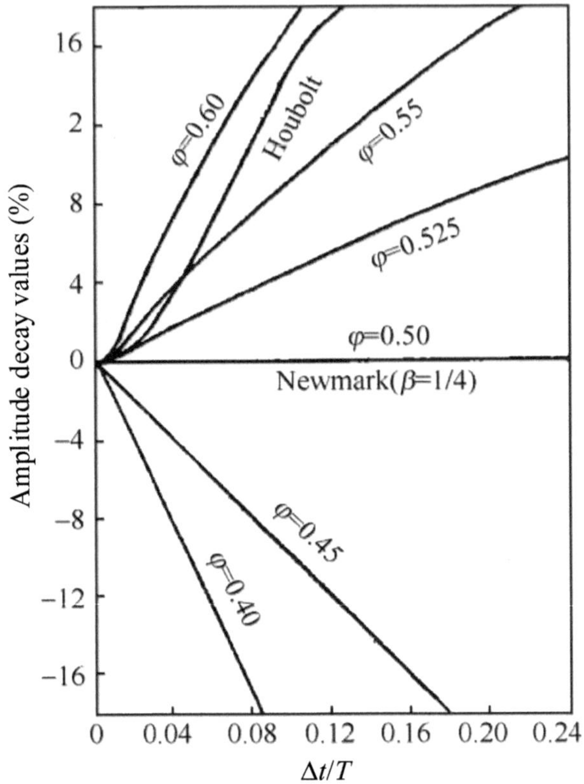

Fig. 4.1 Algorithmic damping characteristics of the Zhai method: amplitude decay function for various φ (=ψ) (Reprinted from Ref. [1], Copyright 1996, with permission from John Wiley & Sons.)

$$\begin{cases} AD = 1 - \exp(2\pi\bar{\zeta}) \\ TD = (\bar{T} - T)/T = \Omega/\bar{\Omega} - 1 \end{cases} \tag{4.27}$$

where $T = 2\pi/\omega$ and $\bar{T} = 2\pi/\bar{\omega}$.

In Fig. 4.1, algorithmic amplitude decay values of the Zhai method for various φ (=ψ) are shown. The continuous control of numerical dissipation in this explicit method is evident. When both φ and ψ are larger than 1/2 or one of them is larger than 1/2 and the other equals to 1/2, there will be algorithmic damping to decrease amplitudes. And when $\varphi = \psi = 1/2$, there is no numerical dissipation.

In Fig. 4.2, the relative period errors of the Zhai method are plotted versus $\Delta t/T$ for various cases. The relative period errors of the Houbolt and Newmark-β ($\beta = 1/4$) methods are also depicted in Fig. 4.2 for comparison. The period obtained with the Zhai method is less than the actual period, whereas the calculated periods with the Houbolt, Newmark, and Park methods are longer than the actual period. It is clear from Fig. 4.2 that the relative period error of the Zhai method is much less than those of the other methods. For example, when $\varphi = \psi = 1/2$ the period error is only about half of the error of the trapezoidal rule (i.e., the Newmark-β method with $\beta = 1/4$).

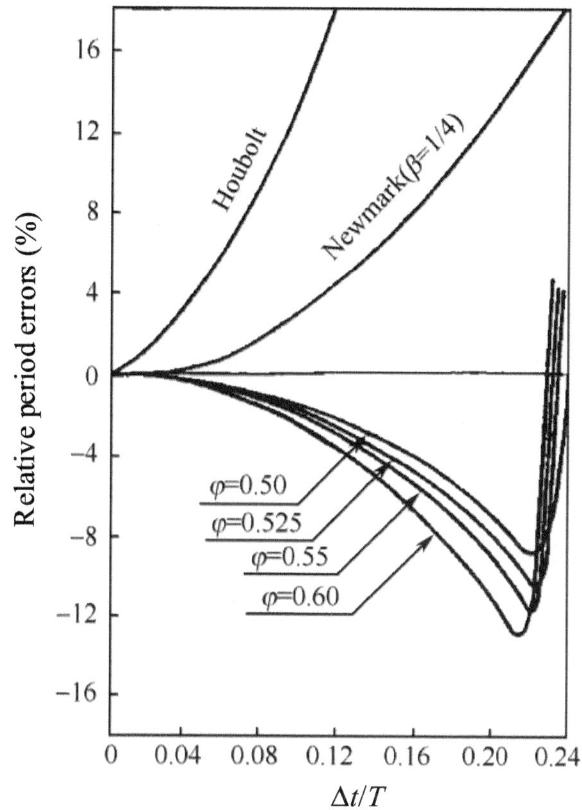

Fig. 4.2 Relative period errors of Zhai method ($\varphi = \psi$), Houbolt and Newmark methods (Reprinted from Ref. [1], Copyright 1996, with permission from John Wiley & Sons.)

4.2.5 Numerical Examples for Verification

In order to examine the actual integration accuracy and stability of the Zhai method, three numerical examples including a linear, a piecewise linear, and a nonlinear problem are employed and analyzed in this section.

1. *Example I*

In the linear case, we consider the following initial-value problem:

$$\begin{cases} \ddot{x} + x = 0 \\ \dot{x}(0) = 0 \\ x(0) = 1 \end{cases} \tag{4.28}$$

When the time step is sufficiently small, e.g., $\Delta t = 0.05$ s, the numerical result with the Zhai method is almost the same as the exact solution. Comparison of the Zhai method with the trapezoidal rule and with the Houbolt method is shown in Fig. 4.3, with a time step of $\Delta t = 0.5$ s.

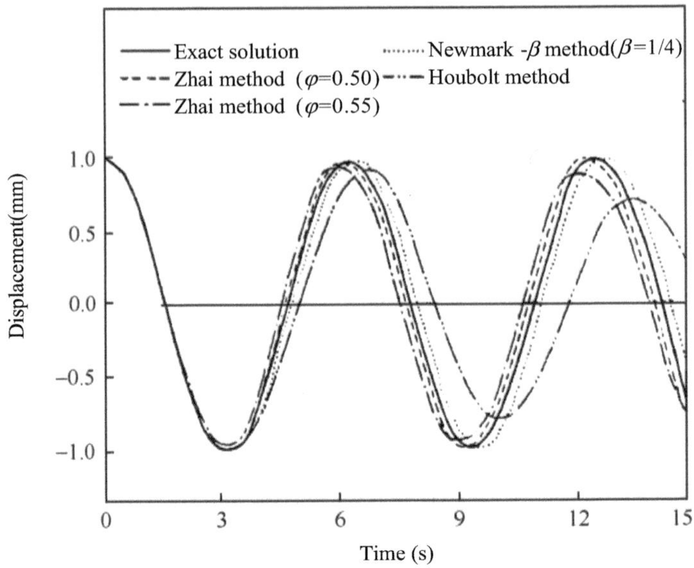

Fig. 4.3 Some integrated results with time step $\Delta t = 0.5$ s for linear case (Reprinted from Ref. [1], Copyright 1996, with permission from John Wiley & Sons.)

To examine the stability, we consider the explicit scheme with the same value, 0.5, both for φ and ψ. According to Table 4.1, the critical stable step is $\Delta t_{cr} = 2/\omega = 2$ s for this example, which is well proved in Fig. 4.4. A little increment of Δt from 2.0 s to 2.01 s causes a divergent result.

Fig. 4.4 Examination of stability of the new explicit method with a critical step $\Delta t_{cr} = 2$ s when $\varphi = \psi = 1/2$ (Reprinted from Ref. [1], Copyright 1996, with permission from John Wiley & Sons.)

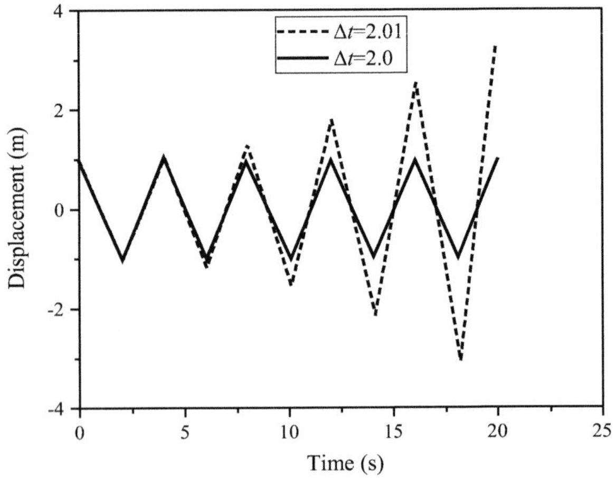

2. *Example II*

Consider the piecewise linear system such as

$$\begin{cases} \ddot{x} + kx = 0 \\ \dot{x}(0) = 10 \\ x(0) = 0 \end{cases} \tag{4.29}$$

where the variation of parameter k is given in Fig. 4.5. Numerical results obtained by the Zhai method are analyzed and compared for different values of φ and ψ. It is shown that if $\varphi = \gamma = 1/2$, the best accuracy and stability of the new explicit scheme is obtained. Therefore, the optimum integration parameters of the Zhai method ($\varphi = \psi = 1/2$) will be adopted in the following calculation including the simulation of vehicle–track coupled dynamics.

Table 4.2 lists the comparison of the Zhai method ($\varphi = \psi = 1/2$) with the exact solution for two kinds of time steps: $\Delta t_1 = 0.005$ s and $\Delta t_2 = 0.05$ s. It can be seen from Table 4.2 that when the time step is set to 0.005 s, the integration results of the Zhai method are almost the same as the exact solution. When the time step increases to 0.05 s, the maximum error is only 2.1%.

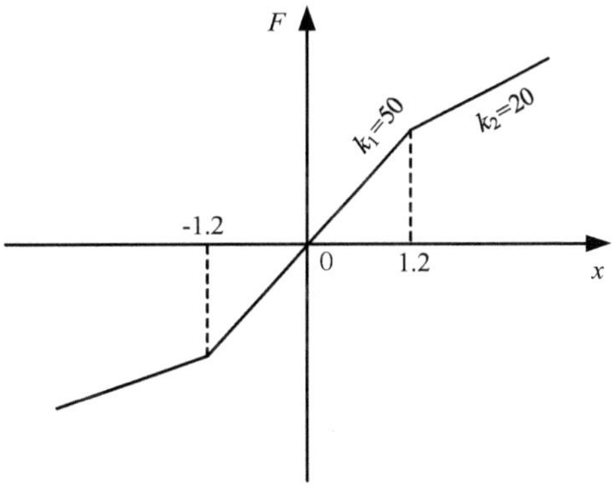

Fig. 4.5 Piecewise linear–spring characteristic used in Example II (Reprinted from Ref. [1], Copyright 1996, with permission from John Wiley & Sons.)

Table 4.2 Comparison of integration results and the exact solution for the piecewise linear system

$\frac{t}{\Delta t_1}$	Exact solution	$\Delta t_1 = 0.005$ s	$\Delta t_2 = 0.05$ s
10	0.48965	0.48975	0.50000
20	0.91873	0.91891	0.93750
30	1.23416	1.23439	1.25781
40	1.40091	1.40115	1.42523
50	1.40829	1.40852	1.43139
60	1.25591	1.25612	1.27598
Error	–	0.0204%	2.1%

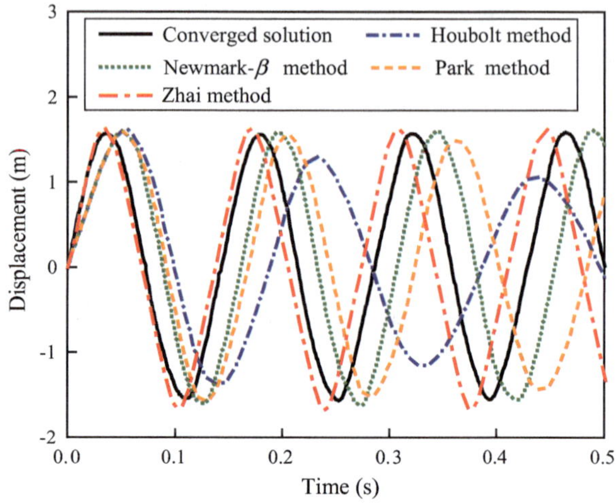

Fig. 4.6 Comparison of responses obtained with Zhai method, Newmark-β method (β = 1/4), Houbolt method, and Park method

3. *Example III*

Consider a nonlinear system with cubically hardening spring quoted in Ref. [20]

$$\begin{cases} \ddot{x} + 100x + 1000x^3 = 0 \\ \dot{x}(0) = 60 \\ x(0) = 0 \end{cases} \tag{4.30}$$

The same time step, Δt = 0.015 s, employed by Park in Ref. [20] is adopted here. Numerical response processes with the Zhai method, the Newmark-β method (β = 1/4), the Houbolt method, and the Park method are plotted in Fig. 4.6. It can be seen that the response by the Zhai method traces the actual behavior more closely than that by the Newmark method, and much more closely than those by the Park method and by the Houbolt method, see Ref. [1].

4.3 Application of Zhai Method to Analysis of Vehicle–Track Coupled Dynamics

Up till now, the Zhai method has been widely applied to analysis of railway dynamics problems, especially for vehicle–track system dynamics [7, 11, 16, 17, 21–24], for train–track–bridge dynamic interaction [25–30], and even for train–track–tunnel–soil dynamic interactions [10]. Nowadays, it becomes the most common method for analyzing the vehicle–track coupled dynamics problems.

4.3.1 Numerical Integration Procedure

It is quite convenient to employ the Zhai method to numerically analyze the vehicle–track coupled dynamics problem. For each time step, the integration method is, respectively, applied to the vehicle subsystem and the track subsystem to explicitly calculate the responses of displacements and velocities. The nonlinear wheel–rail contact forces can then be determined based on the calculated displacements and velocities of wheels and rails. With these known results, the accelerations of the vehicle and the track subsystems are finally calculated from the equations of motion of each subsystem.

Let's rewrite Eq. (4.1a, 4.1b) for the vehicle subsystem and for the track subsystem, respectively,

$$\mathbf{M}_V\mathbf{A}_V + \mathbf{C}_V\mathbf{V}_V + \mathbf{K}_V\mathbf{X}_V = \mathbf{F}_V \tag{4.31}$$

$$\mathbf{M}_T\mathbf{A}_T + \mathbf{C}_T\mathbf{V}_T + \mathbf{K}_T\mathbf{X}_T = \mathbf{F}_T \tag{4.32}$$

where $\mathbf{M}_V$ is the mass matrix of the vehicle; $\mathbf{C}_V$ and $\mathbf{K}_V$ are the damping and the stiffness matrices which can depend on the current state of the vehicle subsystem to describe nonlinearities within the suspension; $\mathbf{X}_V$, $\mathbf{V}_V$, and $\mathbf{A}_V$ are the vectors of displacements, velocities, and accelerations of the vehicle subsystem, respectively; $\mathbf{F}_V$ is the vehicle subsystem load vector representing the nonlinear wheel–rail contact forces determined by the wheel–rail coupling model and the external forces including gravitational forces and forces resulting from the centripetal acceleration when the vehicle is running through a curve; $\mathbf{M}_T$ is the mass matrix of the track structure; $\mathbf{C}_T$ and $\mathbf{K}_T$ are the damping and the stiffness matrices of the track subsystem; $\mathbf{X}_T$, $\mathbf{V}_T$, and $\mathbf{A}_T$ are the vectors of displacements, velocities, and accelerations of the track subsystem; and $\mathbf{F}_T$ is the load vector of the track subsystem representing the nonlinear wheel–rail forces.

The main solution procedure for the vehicle–track coupled dynamics consists of the following six steps:

Step 1 Calculate the displacement $\mathbf{X}_{V,n+1}$ and the velocity $\mathbf{V}_{V,n+1}$ of the vehicle subsystem at the time $(n + 1)\Delta t$ by using Eq. (4.3) based on the states of the system at time $n\Delta t$ and time $(n - 1)\Delta t$;
Step 2 Calculate the displacement $\mathbf{X}_{T,n+1}$ and the velocity $\mathbf{V}_{T,n+1}$ of the track subsystem at the time $(n + 1)\Delta t$ by using Eq. (4.3) based on the states of the system at time $n\Delta t$ and time $(n - 1)\Delta t$;
Step 3 Estimate the wheel–rail normal contact forces and creep forces with the wheel–rail coupling model based on the calculated displacements and velocities of wheels and rails from steps 1 and 2;
Step 4 Calculate the load vectors $\mathbf{F}_{V,n+1}$ and $\mathbf{F}_{T,n+1}$ at the time $(n + 1)\Delta t$ from the wheel–rail contact forces and from the external forces due to the curvature, superelevation, etc., for the current position of each body of the vehicle;

Step 5 Compute the acceleration $\mathbf{A}_{\mathrm{V},n+1}$ of the vehicle subsystem from Eq. (4.31) at time $(n + 1)\Delta t$:

$$\mathbf{A}_{\mathrm{V},n+1} = \mathbf{M}_{\mathrm{V}}^{-1}[\mathbf{F}_{\mathrm{V},n+1} - \mathbf{C}_{\mathrm{V}}\mathbf{V}_{\mathrm{V},n+1} - \mathbf{K}_{\mathrm{V}}\mathbf{X}_{\mathrm{V},n+1}] \tag{4.33}$$

Step 6 Compute the acceleration $\mathbf{A}_{\mathrm{T},n+1}$ of the track subsystem from Eq. (4.32) at time $(n + 1)\Delta t$:

$$\mathbf{A}_{\mathrm{T},n+1} = \mathbf{M}_{\mathrm{T}}^{-1}[\mathbf{F}_{\mathrm{T},n+1} - \mathbf{C}_{\mathrm{T}}\mathbf{V}_{\mathrm{T},n+1} - \mathbf{K}_{\mathrm{T}}\mathbf{X}_{\mathrm{T},n+1}] \tag{4.34}$$

Because the mass matrices $\mathbf{M}_{\mathrm{V}}$ of the vehicle subsystem and $\mathbf{M}_{\mathrm{T}}$ of the track subsystems are diagonal matrices, no algebraic equations have to be solved to obtain the inverse mass matrices $\mathbf{M}_{\mathrm{V}}^{-1}$ and $\mathbf{M}_{\mathrm{T}}^{-1}$ required in Eqs. (4.33) and (4.34). Thus, computational efficiency is greatly enhanced.

4.3.2 Determination of Time Step of Zhai Method

One of the most important issues for the Zhai method to be applied to analyzing the vehicle–track coupled dynamics problem is the determination of the integration time step. How to obtain the critical stable time step of the Zhai method for such a complicated engineering dynamics system with strong nonlinearities? How to determine a rational time step in order to ensure the numerical solution with sufficiently high accuracy? The author believes that the numerical trial method is able to provide the possibility to deal with these problems.

A meticulous numerical trial has been carried out for the Zhai method to solving dynamic responses of the vehicle–track coupled system. The variation of system dynamic responses with the change of integration time step was carefully investigated. Each key dynamics indices of the whole coupled system were observed. Here is just shown some results of an example taken for the vertical vehicle–track coupled dynamic system in the case of a freight car passing over a dipped rail joint on a traditional ballasted track. Figure 4.7 shows the variations of the calculated vertical wheel–rail forces with the time step Δt, in which P_1 represents the high-frequency wheel–rail impact force and P_2 denotes the low-frequency wheel–rail force. Figure 4.8 shows the variations of the calculated wheel–rail system accelerations with the time step Δt, where a_{w} is the wheelset vertical vibration acceleration as representative of the vehicle subsystem and a_{b} is the ballast vertical vibration acceleration as representative of the track subsystem. It can be seen from both figures that the critical stable time step of the Zhai method is

Fig. 4.7 Stability of the Zhai method for vertical vehicle–track coupled dynamics: variation of the calculated wheel–rail forces P_1 and P_2 with the time step Δt

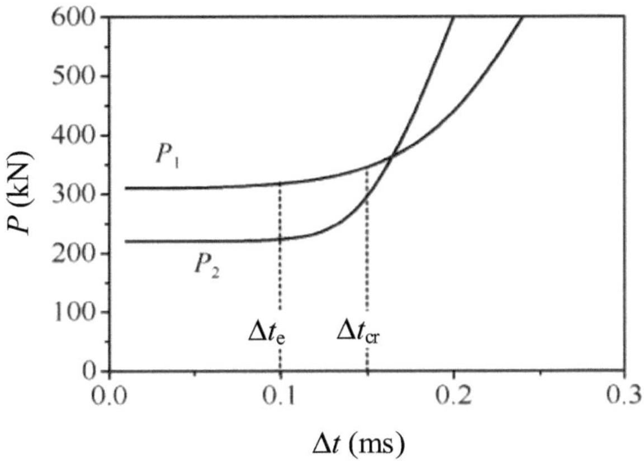

Fig. 4.8 Stability of the Zhai method for vertical vehicle–track coupled dynamics: variation of the calculated wheel–rail system accelerations a_w and a_b with the time step Δt

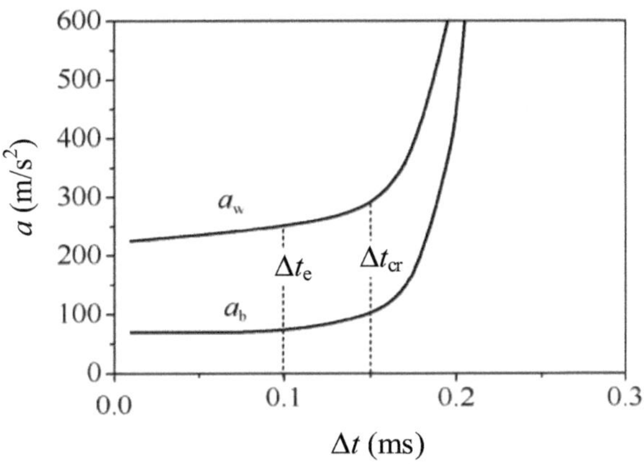

$\Delta t_{cr} = 1.5 \times 10^{-4}$ s when used for solution of the present vertical vehicle–track dynamics problem. In order to get a high calculation accuracy, however, the actually adopted effective time step Δt_e is suggested to be smaller than the critical time step Δt_{cr}, for example, $\Delta t_e = 1.0 \times 10^{-4}$ s.

As to the lateral dynamics problem of the vehicle–track coupled system, numerical trial results indicate that the effective time step Δt_e could be 1.0×10^{-4} s for the passenger coach and track system and 5.0×10^{-5} s for the freight wagon and track system. Although the time step has to be very small in order to reflect high-frequency wheel–rail contact vibration, the computational efficiency is quite high due to the needless of solving a large-scale set of algebraic equations at each time step. A common micro-computer is enough to implement the simulation of the vehicle–track coupled dynamics.

4.4 On Some Key Issues in Solving Process of Vehicle–Track Coupled Dynamics

4.4.1 Determination of Calculated Length of Track and Mode Number of Rail

There are two undetermined issues arising from the established vehicle–track coupled dynamics models in Chap. 2. One is the calculated length of track, l, in the actual simulation. The other is the mode number of rail, NM, adopted in the calculation. Here, the numerical trial method will be used again to determine these two values.

Taking the vertical vehicle–track coupled dynamics model as an example, the numerical trial is carried out for the track length under the same case as used in above Sect. 4.3.2, i.e., a freight car passing over a dipped rail joint on a traditional ballasted track. Figures 4.9 and 4.10 give the numerical trial results of the influence

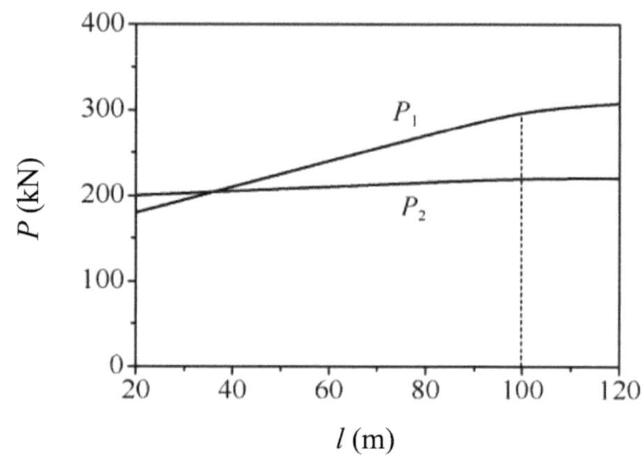

Fig. 4.9 Determination of the calculated length of track for vertical vehicle–track coupled dynamics model: influence of the length on wheel–rail forces P_1 and P_2

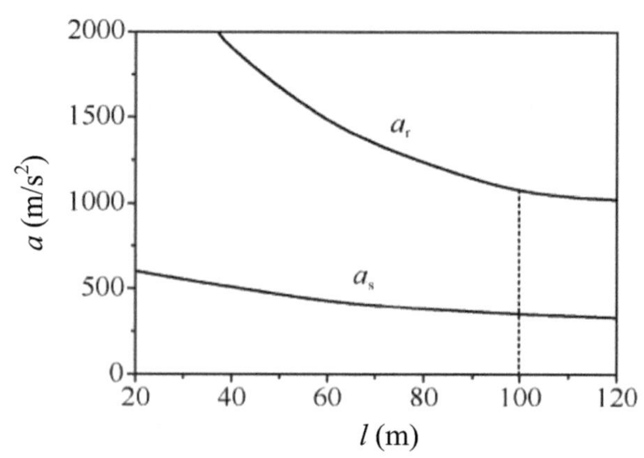

Fig. 4.10 Determination of the calculated length of track for vertical vehicle–track coupled dynamics model: influence of the length on track accelerations a_r and a_s

of the length on the vertical wheel–rail forces and on track accelerations, where a_r and a_s are the vertical accelerations of rail and sleeper. It is observed from these two figures that the track length has obvious influence on calculated dynamic responses, however, little influence can be found if the calculated track length is larger than 100 m. Therefore, it is enough for the calculated track length to be set as 100 m in the simulation of vertical vehicle–track interaction. For the lateral dynamics problem of the vehicle–track coupled system, numerical trial reveals the similar results of the vertical problem. Thus a conclusion can be made: the track length could be set as 100 m for simulation of general vehicle–track coupled dynamics problems due to local short-wavelength track irregularities such as rail joints, void sleeper, and defects on the wheel and rail surfaces. In these cases, the moving distance of a vehicle needed in calculation could be short and the excitation can be input with the moving-vehicle method as introduced in Sect. 3.1.2 of Chap. 3. 100 m is also long enough for the calculated track length if the excitation is input by the fixed-point method as shown in Sect. 3.1.1 of Chap. 3. For long-distance track simulation with the moving-vehicle method, we can adopt the tracking window method to deal with the infinite length of the track. More detail can be referred to Sect. 3.1.3 in Chap. 3.

The mode number of the rail is important to describe the rail vibration behavior. Different rail mode numbers reflect different frequency components of the vibration. The higher the rail mode number selected, the higher the rail vibration frequency included. In order to felicitously simulate the vehicle–track system dynamics behavior, especially the high-frequency wheel–rail contact vibration, the maximum mode number of rail should be selected sufficiently high. Generally, it is required that the highest frequency reflecting by the maximum mode number of rail should be at least twice as high as the concerned rail frequency in simulation. Rational maximum mode number can also be determined by use of the numerical trial method. Our investigation indicates that there is a matching relation between the rational rail mode number and the calculated length of track

$$NM = N_l \tag{4.35}$$

where NM is the maximum mode number of rail and N_l is the number of rail fastening supporting points within the calculated track length. If the calculated track length increases, the rail mode number should increase accordingly.

4.4.2 *Solving Technique for the Train–Track Coupled Dynamics*

In the train–track spatially coupled dynamics model which has been introduced in Sect. 2.4 of Chap. 2, there exists a huge number of DOFs. To achieve the fast solution on the dynamic responses of this large-scale system, a fast explicit numerical integration algorithm, namely the Zhai method [1], is adopted. However, it is still quite time-consuming if each vehicle is regarded as a detailed three-dimensional model (3D model) and the track elasticity is also considered simultaneously. It is challenging for a common computer to accomplish massive calculations. Therefore, an efficient solving technique is necessary to be developed on the premise of satisfying the precision requirements. For increasing the solving speed, the dynamic model can be simplified to a certain degree according to the analysis purpose. The new idea for solving this problem is proposed from the following two simplifications: (1) if the train longitudinal dynamics is more concerned, each vehicle can be simplified as the single-mass model; and (2) if the dynamic performance of the vehicle subjected to the largest coupler force is focused, the focused vehicles can be modeled in detail while others are simulated by the mass block. The following contents will make a detail introduction to the second case.

1. *Solving procedure*

For the long train–track coupled dynamics system, the solving flow is illustrated in Fig. 4.11. The basic processes are given as follows:

(1) Carry out the analysis of the longitudinal train dynamics, and every vehicle is simplified as the one-dimensional mass point.
(2) Obtain the distributions of the longitudinal coupler forces in the whole train, then find out the occurring position of the largest coupler force. And the vehicles near the maximum coupler force are of being focused thoroughly.
(3) Convert the single-mass models of vehicles near the largest coupler force to the three-dimensional vehicle–track coupled models (see Fig. 4.12).
(4) Recalculate the train dynamic response of the mixed model, and comprehensively evaluate the indices of the wheel–rail forces, derailment coefficient, track displacements, coupler swing angle, and so on.

By applying this solving procedure, the computation efforts for solving such a huge system can be greatly reduced, and the analysis accuracy can also meet requirements.

An important issue for the accurate calculation using the developed mixed dynamics model is to determine the number N of the three-dimensional vehicle models in the concerned position of the train. When the train negotiates a curved track or runs on an irregular track, there are longitudinal, lateral and vertical dynamic forces between the two adjacent vehicles. However, the spatial motions of the 3D vehicles will be constrained by the vehicles modeled as the one-dimensional single mass. Therefore, an inevitable problem needs to be answered. That is, for the concerned position, how many vehicles represented by the 3D model are required to reduce the negative effects of the single-mass models on the dynamic responses of the focused vehicles.

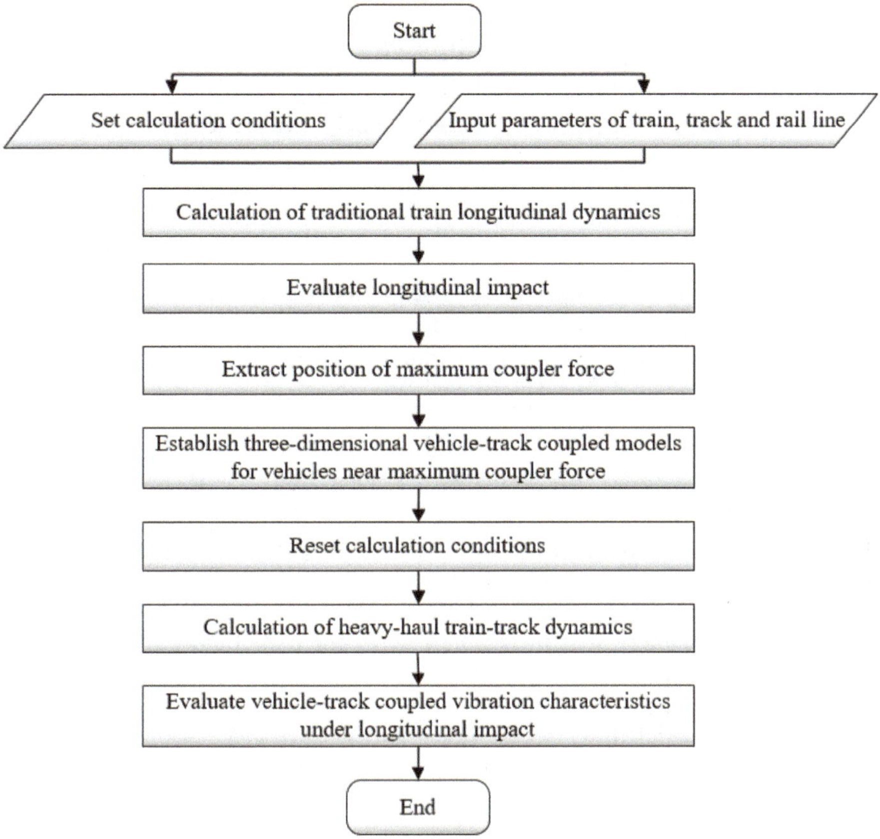

Fig. 4.11 Flow chart for calculation of train–track coupled dynamics model

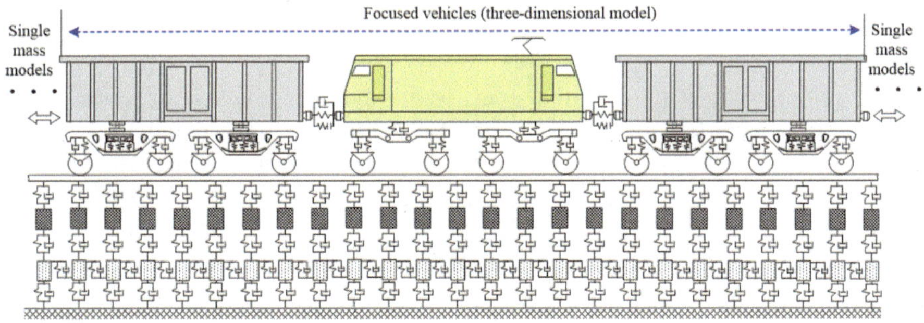

Fig. 4.12 Mixed dynamics model of a heavy-haul train

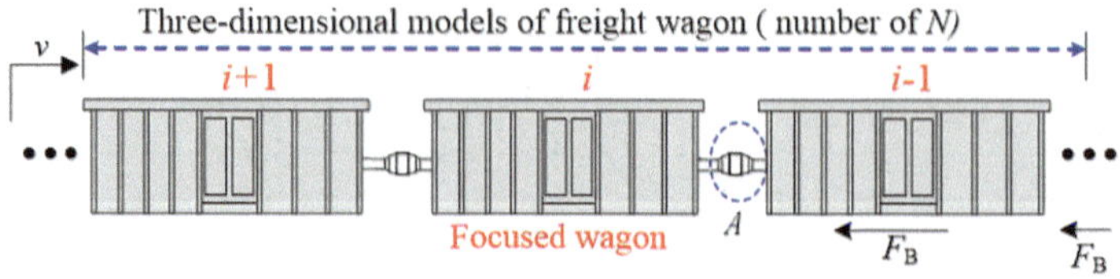

Fig. 4.13 Short train model composed only by freight wagons

In order to clarify this question, the simulation experiments for two short trains with the freight wagon marshaling mode and the locomotive–wagon marshaling mode are carried out to determine the minimum value of the number N required.

2. *Determination of the required number of the vehicles modeled as a three-dimensional model*

Two cases of a heavy-haul train formation are considered: (1) the short train is only composed of the freight wagons, as shown in Fig. 4.13. The vehicles at both the train ends are modeled as the single-mass model, while other wagons are modeled as the 3D model; (2) for the combined train, the locomotives are usually distributed at the head, the middle and the end positions. The dynamic behaviors of the locomotive may be more sensitive to the coupler force than the wagons due to its softer horizontal suspension stiffness. For these reasons, once the focused vehicle in the train is a locomotive, all of the centralized locomotives at this position should be considered as the detailed 3D model. Then the key question arises that whether it is necessary to use the 3D model to simulate the neighboring freight wagons of the locomotives. Figure 4.14 displays the dynamic model of the marshaling train used for this case.

In the numerical experiments, the heavy-haul locomotive HXD_2 and freight wagon C80, which are widely used in China are applied for the analysis under the braking conditions. The dynamics indices of the focused freight wagon are extracted to illustrate the effect of the number of the vehicles represented by the 3D model. For example, the lateral wheelset forces for the two cases are shown in Fig. 4.15.

The principle of using the 3D model for the dynamic analysis of a heavy-haul train can be concluded based on the results of numerical experiments as: (1) if the maximum coupler force occurs in the freight wagon which is far away from the

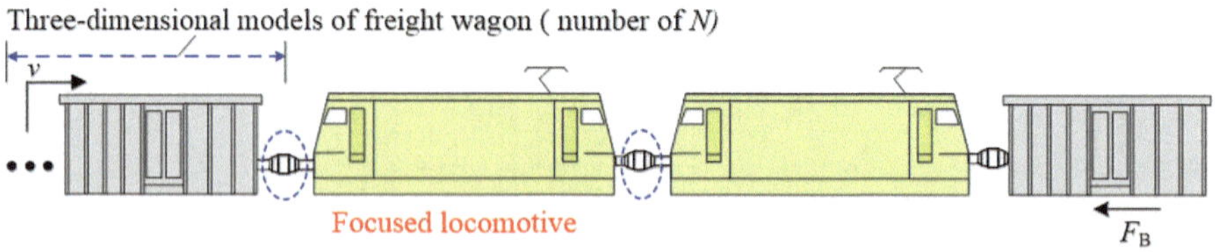

Fig. 4.14 Dynamic model of a train composed by locomotives and freight wagons

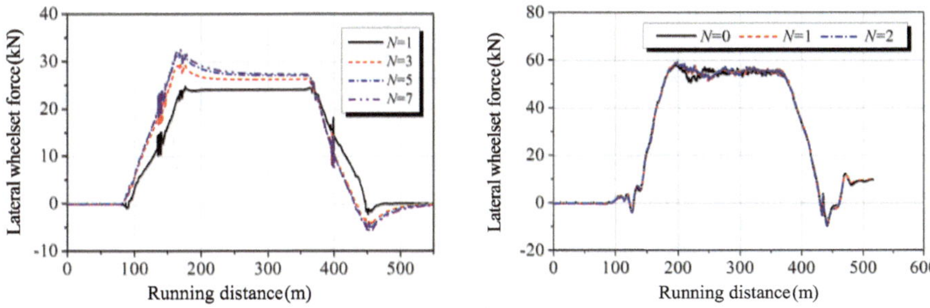

Fig. 4.15 Wheelset lateral force of the first axle in the focused vehicle with different number of 3D model: **a** case 1, and **b** case 2

locomotives, at least three wagons around the maximum coupler force should be modeled as the 3D models; (2) if the maximum coupler force appears in the locomotives, the centralized distributing locomotives and adjacent two freight wagons in this area ought to be simulated simultaneously by the 3D models.

4.5 Computer Simulation of Vehicle–Track Coupled Dynamics

Computer simulation is a key technique to obtain the detailed responses of such a large dynamic system. On the basis of the established dynamics models in Chap. 2 and the excitation models in Chap. 3, three computer simulation programs were developed by use of the above fast integration method to analyze the vehicle–track coupled dynamics.

4.5.1 Vehicle–Track Vertically Coupled Dynamics Simulation

The first simulation software is called VICT, which was developed by the author in early 1990s based on the vehicle–track vertically coupled dynamics model and used for analyzing the vertical dynamic interactions between railway vehicles and tracks, especially for evaluating dynamic effects of vehicles on track structures. Figure 4.16 gives the flowchart of the VICT simulation system.

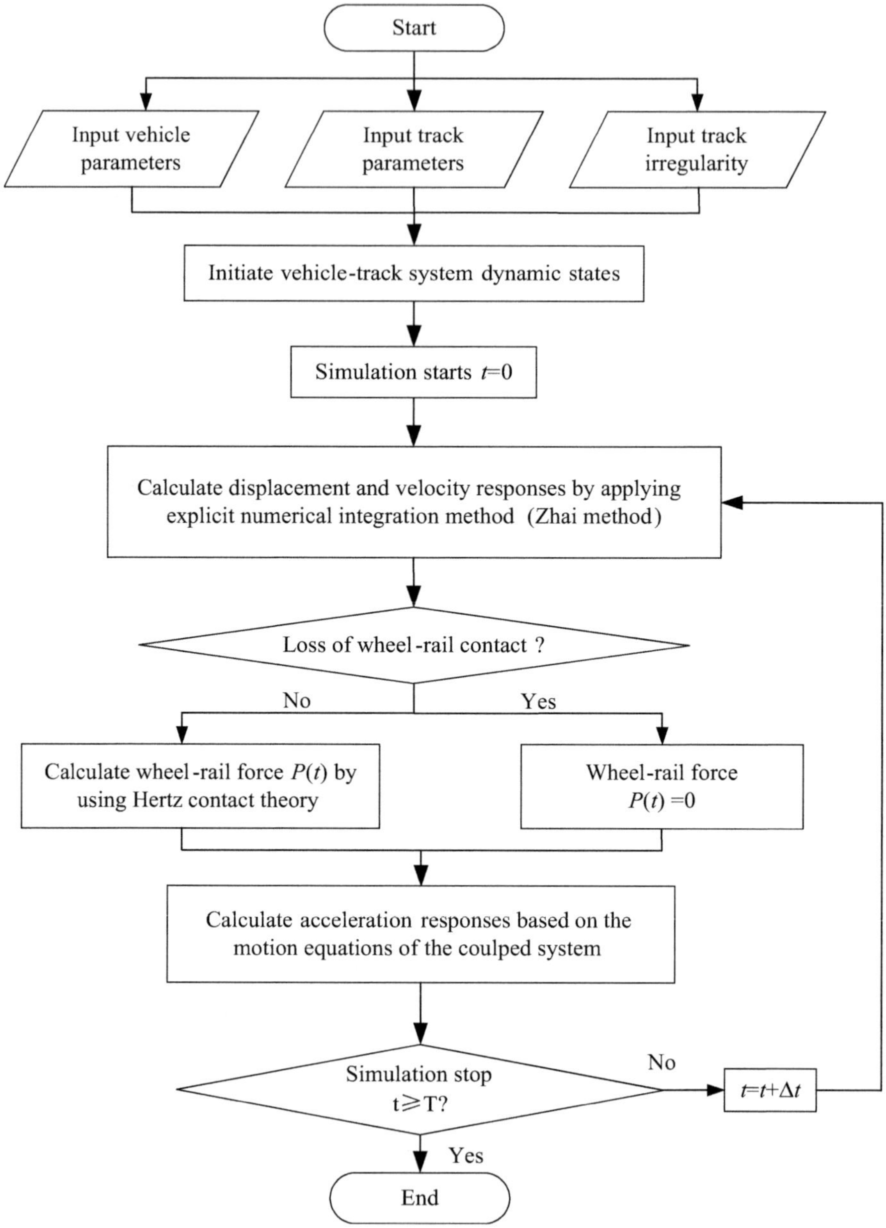

Fig. 4.16 Flowchart of VICT simulation system

4.5.2 Vehicle–Track Spatially Coupled Dynamics Simulation

Another simulation program, named as TTISIM, was developed by the author in early 2000s at Train & Track Research Institute of Southwest Jiaotong University [31]. It is based on the vehicle–track spatially coupled dynamics model involving the new dynamic wheel–rail coupling model. Figure 4.17 shows the module structure of the TTISIM simulation program. The flow chart of the complete simulation process is illustrated in Fig. 4.18.

Unlike the VICT, which is mainly used to analyze vertical dynamic effects of vehicles on tracks, TTISIM can be used to investigate the dynamic behavior of vehicles running on flexible track structures including the lateral hunting stability, the ride comfort, and the curving performance, etc. The major difference between TTISIM and other commercial vehicle system dynamics software might be that the track structure components are modeled in detail in TTISIM. Actually measured profiles of wheels and rails can be directly inputted to TTISIM as initial data. TTISIM can produce complete responses of the vehicle–track coupled system including the vertical and lateral wheel–rail forces, the rail pad forces, the vertical and lateral accelerations of car body, bogies, and wheelsets, the vertical and lateral displacements and accelerations of rails, sleepers, ballasts, etc.

4.5.3 Train–Track Spatially Coupled Dynamics Simulation

Furthermore, the third computer simulation software is developed based on the train–track spatially coupled dynamics model, which is able to analyze three-dimensional train–track system dynamics problems, with special emphasis on the longitudinal dynamic interaction between vehicles.

The modular simulation idea is adopted here for the simulation of the train–track spatially coupled dynamics system, which is illustrated in Fig. 4.19 [32]. The simulation system consists of four submodules: the train control submodule, the train submodule, the wheel–rail contact submodule, and the track submodule. Like in VICT and TTISIM software, the train subsystem and the track subsystem are coupled through the wheel–rail nonlinear contact relationship. It is worthy of noting in this simulation that the interactions between the vehicles in the train formation are included through the couplers, and the train control module providing the control strategies (e.g. determination of driving torque, running resistance, traction and braking characteristics) is considered for the train operation under different railway line conditions.

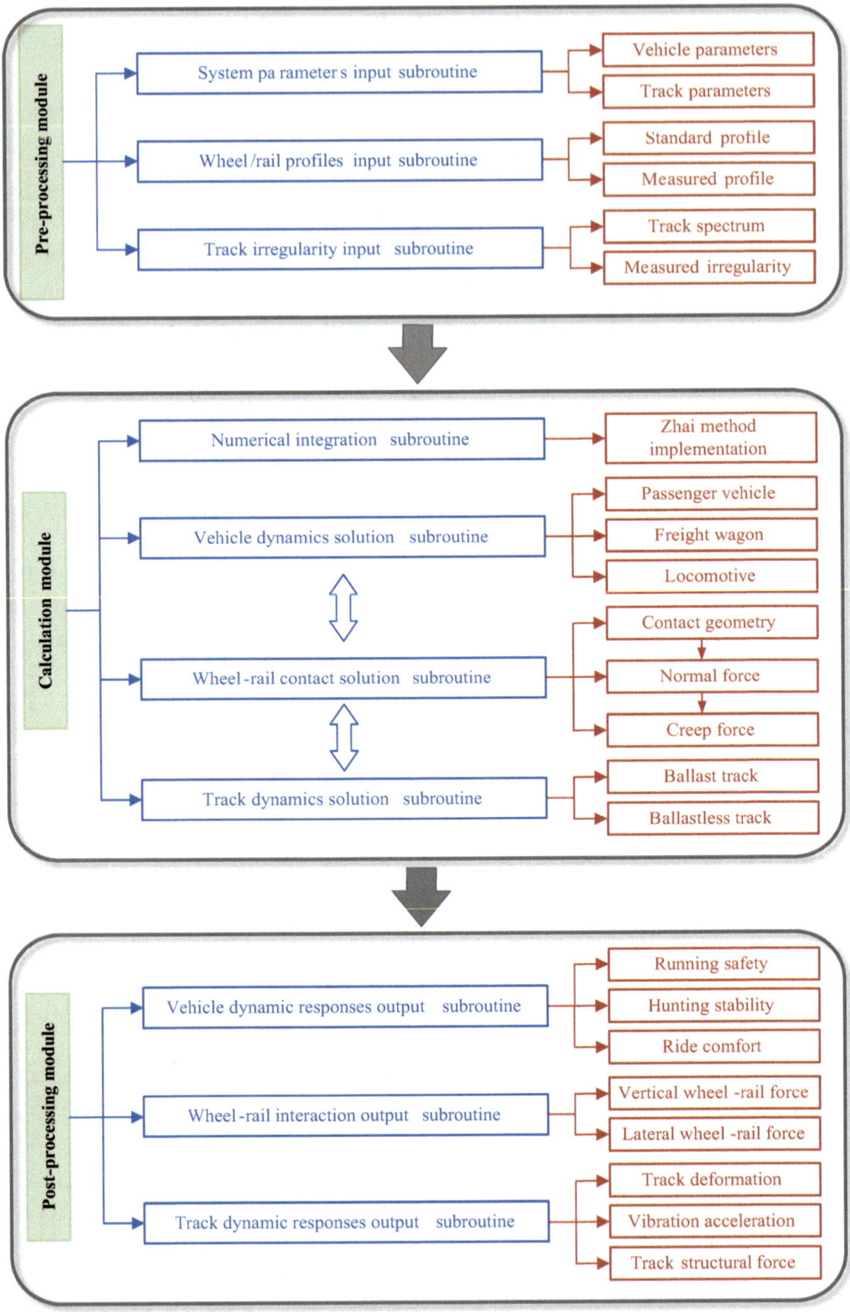

Fig. 4.17 Module structure of TTISIM simulation program

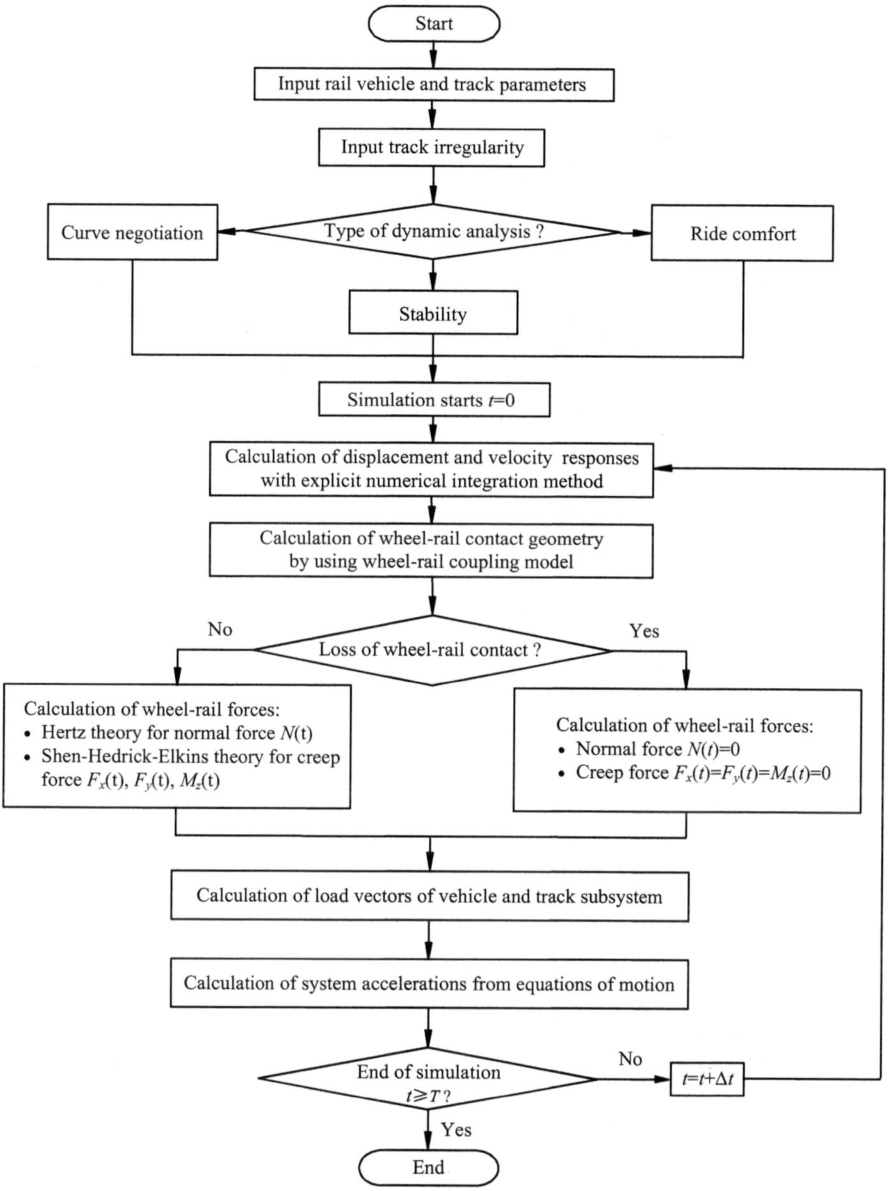

Fig. 4.18 Flow chart of TTISIM simulation system (Reprinted from Ref. [31], Copyright 2009, with permission from Taylor & Francis.)

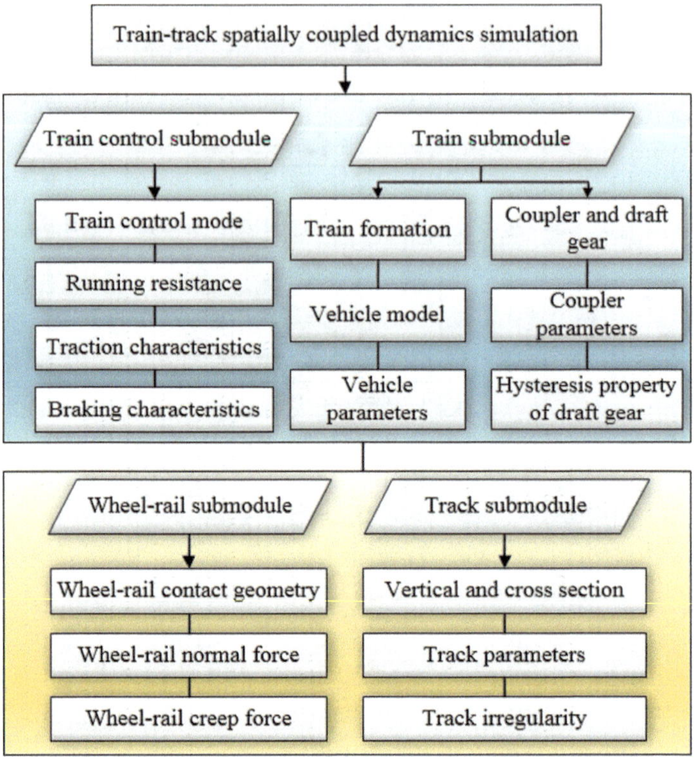

Fig. 4.19 Modular simulation of the train–track spatially coupled dynamics

References

1. Zhai WM. Two simple fast integration methods for large-scale dynamic problems in engineering. Int J Numer Meth Eng. 1996;39(24):4199–214.
2. Newmark NM. A method of computation for structural dynamics. J Eng Mech Div ASCE. 1959;85(2):67–94.
3. Wilson EL, Farhoomand I, Bathe KJ. Nonlinear dynamic analysis of complex structure. Earthquake Eng Struct Dyn. 1973;1:241–52.
4. Hoff C, Taylor RL. Higher derivative explicit one step methods for nonlinear dynamic problems. Part I: Design and theory. Int J Numer Methods Eng. 1990;29:275–90.
5. Rio G, Soive A, Grolleau V. Comparative study of numerical explicit time integration algorithms. Adv Eng Softw. 2005;36(4):252–65.
6. Rezaiee-Pajand M, Alamatian J. Numerical time integration for dynamic analysis using a new higher-order predictor-corrector method. Eng Comput. 2008;25(6):541–68.
7. Zhang J, Gao Q, Tan SJ, Zhong WX. A precise integration method for solving coupled vehicle–track dynamics with nonlinear wheel–rail contact. J Sound Vib. 2012;331:4763–73.
8. Chen G. A new rotor-ball bearing-stator coupling dynamics model for whole aero-engine vibration. J Vib Acoust ASME. 2009;131(4):061009-1 ~ 9.
9. Chen G. Simulation of casing vibration resulting from blade–casing rubbing and its verifications. J Sound Vib. 2016;361:190–209.

10. Zhou S, Zhang X, Di H, He C. Metro train–track–tunnel–soil vertical dynamic interactions—semi-analytical approach. Veh Syst Dyn. 2019;56(12):1945–68.
11. Banimahd M, Woodward PK, Kennedy J, Medero GM. Behaviour of train–track interaction in stiffness transitions. Proc Inst Civ Eng Transport. 2012;165:205–14.
12. Zhou J, Zhou Y. A new simple method of implicit time integration for dynamic problems of engineering structures. Acta Mech Sin. 2007;23:91–9.
13. Wang HF, Chen G, Song PP. Simulation analysis of casing vibration response and its verification under blade–casing rubbing fault. J Vib Acoust ASME. 2016;138:031004-1 ~ 14.
14. Rezaiee-Pajand M, Alamatian J. Implicit higher-order accuracy method for numerical integration in dynamic analysis. J Struct Eng ASCE. 2008;36(6):973–85.
15. Rezaiee-Pajand M, Alamatian J. Nonlinear dynamic analysis by dynamic relaxation method. Struct Eng Mech. 2008;28(5):549–70.
16. Chen Y, Zhang B, Zhang N, Zheng M. A condensation method for the dynamic analysis of vertical vehicle–track interaction considering vehicle flexibility. J Vib Acoust ASME. 2015;137:041010-1 ~ 8.
17. Wang W, Zhang Y, Ouyang H. An iterative method for solving the dynamic response of railway vehicle vehicle–track coupled systems based on prediction of wheel-rail forces. Eng Struct. 2017;151:297–311.
18. Yuan P, Li D, Cai CS, Xu G. An efficient decoupling dynamic algorithm for coupled multi-spring-systems. Comput Struct. 2018;209:44–56.
19. Hilber HM, Hughes TJR, Taylor RL. Improved numerical dissipation for time integration algorithms in structural dynamics. Earthq Eng Struct Dyn. 1977;5:283–92.
20. Park KC. An improved stiffly stable method for direct integration of nonlinear structural dynamic equations. J Appl Mech ASME. 1975;42:464–70.
21. Ling L, Han J, Xiao XB, Jin XS. Dynamic behavior of an embedded rail track coupled with a tram vehicle. J Vib Control. 2017;23(14):2355–72.
22. Varandas JN, Paixão A, Fortunato E. A study on the dynamic train–track interaction over cut-fill transitions on buried culverts. Comput Struct. 2017;189:49–61.
23. Guo Y, Zhai WM. Long-term prediction of track geometry degradation in high-speed vehicle–ballastless track system due to differential subgrade settlement. Soil Dyn Earthq Eng. 2018;113:1–11.
24. Sun Y, Guo Y, Chen ZG, Zhai WM. Effect of differential ballast settlement on dynamic response of vehicle–track coupled systems. Int J Struct Stab Dyn. 2018;18(7):1850091-1 ~ 29.
25. Arvidsson T, Karoumi R. Train–bridge interaction—a review and discussion of key model parameters. Int J Rail Transp. 2014;2(3):147–86.
26. Zhai WM, Xia H, Cai CB, Gao MM, Li XZ, Guo XR, Zhang N, Wang KY. High-speed train–track–bridge dynamic interactions—Part I: Theoretical model and numerical simulation. Int J Rail Transp. 2013;1(1–2):3–24.
27. Li XZ, Liu QM, Pei SL, Song LZ, Zhang X. Structure-borne noise of railway composite bridge: Numerical simulation and experimental validation. J Sound Vib. 2015;353:378–94.
28. Zhang X, Zhai W, Chen Z, et al. Characteristic and mechanism of structural acoustic radiation for box girder bridge in urban rail transit. Sci Total Environ. 2018;627:1303–14.
29. Zhang X, Li XZ, Li XD, Liu QM, Zhang Z. Train-induced vibration and noise radiation of a prestressed concrete box-girder. Noise Control Eng J. 2013;61(4):425–35.
30. Chen ZW, Zhai WM, Yin Q. Analysis of structural stresses of tracks and vehicle dynamic responses in train–track–bridge system with pier settlement. Proc Inst Mech Eng Part F: J Rail Rapid Transit. 2018;232(2):421–34.
31. Zhai WM, Wang KY, Cai CB. Fundamentals of vehicle–track coupled dynamics. Veh Syst Dyn. 2009;47(11):1349–76.
32. Liu PF, Zhai WM, Wang KY. Establishment and verification of three-dimensional dynamic model for heavy-haul train–track coupled system. Veh Syst Dyn. 2016;54(11):1511–37.

Chapter 5
Field Test on Vehicle–Track Coupled System Dynamics

Abstract Field test and theoretical analysis are the two fundamental investigation approaches in the field of vehicle–track coupled dynamics. Field test is the essential step in validating the dynamics model and the simulation system. The validated dynamics model and simulation system can then be used to optimize the dynamic performance of the system, leading to a shortened designing phase and a reduced need for the costly field test. Furthermore, field test is also the last step in examining the reliability of the vehicle–track system design. Section 5.1 introduces the field test methods including field test methods of vehicle dynamics and track dynamics. Typical tests that were carried out by the author and his team are then presented, which includes dynamic performance tests for typical high-speed passenger train and freight train (Sect. 5.2), as well as vehicle–track dynamic interaction test for a high-speed train on slab track and track dynamics tests for a heavy-haul train passing a rail joint and for an ordinary train negotiating a small-radius curve (Sect. 5.3).

5.1 Field Test Methods of Vehicle–Track Coupled System Dynamics

On-track dynamics test of vehicle–track coupled systems mainly include vehicle dynamics test and track dynamics test, with emphasis on the wheel–rail dynamic interaction test. Simultaneous field dynamics tests for vehicle and track are quite essential for systematic assessment of the dynamic performance of a vehicle running on a track. As the vehicle–track coupled effects increase with the vehicles developing toward high-speed and heavy-haul, vehicle–track coupled dynamics test has become increasingly important. This section will focus on field test methods of vehicle–track coupled system dynamics based on vehicle dynamics test and track dynamics test that is widely carried out and well defined by a series of regulations and guidelines.

W. Zhai, *Vehicle–Track Coupled Dynamics*,
https://doi.org/10.1007/978-981-32-9283-3_5

5.1.1 Field Test Methods of Vehicle Dynamics

Field tests of vehicle dynamics covers the areas of running stability, running safety and ride comfort.

For low-speed operation tests in China, the test methods should follow standards GB5599-85 'Railway vehicles—Specification for evaluation the dynamic performance and accreditation test' [1] and TB/T2360-93 'Identification method and evaluation standard for dynamic performance test of railway locomotives' [2]. For high-speed train tests, the test methods should follow "Code for testing of high-speed electric multiple unit on completion of construction" [3] and TB10761-2013 "Technical regulations for dynamic acceptance for high-speed railways construction" [4]. The test methods for low and high-speed vehicles are generally similar, although the acceptance criteria are different.

1. *Hunting stability*

Bogie lateral accelerations are normally measured to assess the vehicle running stability. A 0.5–10 Hz bandpass filter will be applied to the measured acceleration signal. The vehicle is classified as unstable if the acceleration magnitude reaches or exceeds the threshold (normally 8–10 m/s^2) for 6 consecutive peaks, otherwise, it is classified as stable.

2. *Running safety*

Vehicle running safety can be assessed using the derailment coefficient, the wheel unloading rate, and the rollover factor. These parameters can all be calculated using measurements from instrumented wheelsets. Instrumented wheelsets can (continuously or discretely) measure wheel vertical force P and lateral force Q, hence the derailment coefficient can be computed as Q/P and the wheel unloading rate is worked out as

$$\Delta P/P_0 = (P_0 - P)/P_0 \tag{5.1}$$

where ΔP is the wheel load variation relative to P_0 and P_0 is the static wheel load.

The rollover factor is used to evaluate the possibility of rollover of a vehicle under the action of crosswind, lateral vibrations and centrifugal forces. It is assessed with the wheel load variation for each wheel on one side of the vehicle defined as

$$D = P_\mathrm{d}/P_0 \tag{5.2}$$

where P_d is the dynamic load of the wheel. The rollover safety can only be assessed with the rollover factors for all wheels on one side of the vehicle. If the rollover factors for all wheels on one side of the vehicle reach or exceed the threshold (normally 0.8), the vehicle is assumed dangerous (more details can be found in [1]).

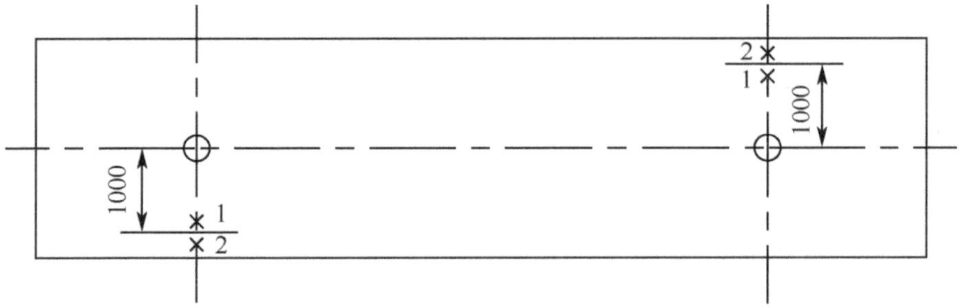

1-Vertical acceleration sensor, 2- Lateral acceleration sensor

Fig. 5.1 Sensors arrangement for passenger car body acceleration measurement (unit: mm)

3. *Ride comfort*

Vehicle ride comfort is assessed by the ride comfort indices, including vertical and lateral ride comfort indices, which are calculated from measured vertical and lateral accelerations of the car body.

Low-frequency accelerometers are used to measure the vertical and lateral car body accelerations. For passenger vehicles, the sensors for measuring the vertical and lateral car body accelerations should be mounted on the floor level and at the position 1000 mm away from the center pivots, as shown in Fig. 5.1. For freight vehicles, the sensors should be mounted on the floor level (in case of an empty car) or under-frame level (in case of a loaded car) and within 1000 mm away from the center pivot as shown in Fig. 5.2. For locomotives, the sensors should be mounted on both ends of the traction beam along the centreline of the under-frame. For driver cab, the sensors should be mounted at the center of the floor in the cabs.

5.1.2 Field Test Methods of Track Dynamics

Field tests of track dynamics require the measurements of wheel–rail dynamic forces, track, and substructure displacements and track structure vibrations.

The measurement of dynamic effects of vehicles on the track must be conducted onsite, and assessed by processing the measurements taken when the vehicles are passing the instrumented track section. Many test techniques have been well developed in China and abroad through continuous improvements [5–7].

1. *Wheel–rail force*

Wheel–rail force is the most important indicator for evaluating vehicle–track interaction. Low-frequency vertical and lateral wheel–rail forces can be measured using strain gauges or 90° strain gauge rosettes.

To measure vertical wheel–rail force, strain gauges should be placed on both sides of the rail web, along the neutral axis, and 110 mm away from the mid-point between two sleepers. The strain gauge orientation and the track longitudinal

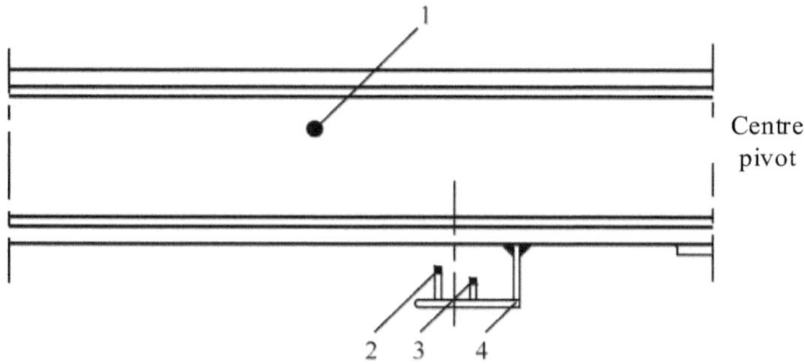

1-Center sill, 2- Vertical acceleration sensor, 3- Lateral acceleration sensor, 4-installed steel plate

Fig. 5.2 Sensors arrangement for freight car body acceleration measurement

direction should form an angle of 45°. The mounting positions and the electrical bridge diagram are shown in Fig. 5.3.

To measure lateral wheel–rail force, strain gauges should be placed on the top surface of the rail foot, 20 mm away from its outer edge, and 110 mm away from the mid-point between two sleepers. The strain gauge orientation and the track longitudinal direction should form an angle of 45°. The mounting positions and the electrical bridge diagram are shown in Fig. 5.4.

2. *Track structural deflection*

Track structural deflection is also a key indicator for assessing vehicle–track interaction, which includes rail vertical and lateral displacements, gauge widening and sleeper (or slab) vertical and lateral displacements.

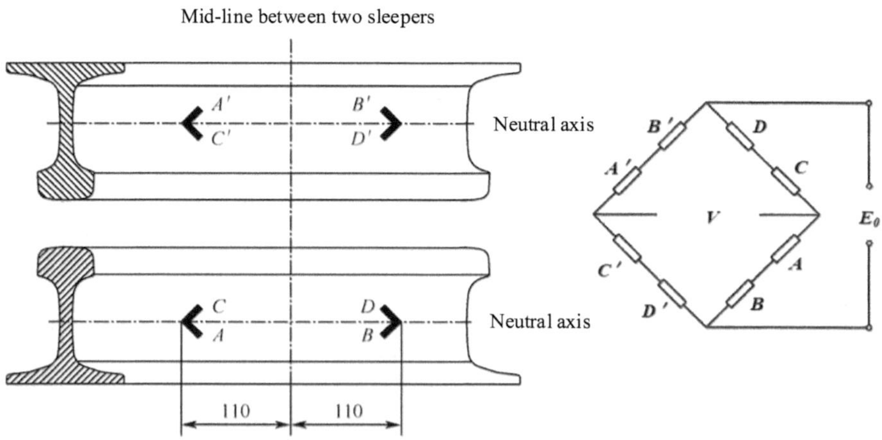

Fig. 5.3 Mounting positions and the electrical bridge diagram for measuring vertical wheel–rail force (unit: mm)

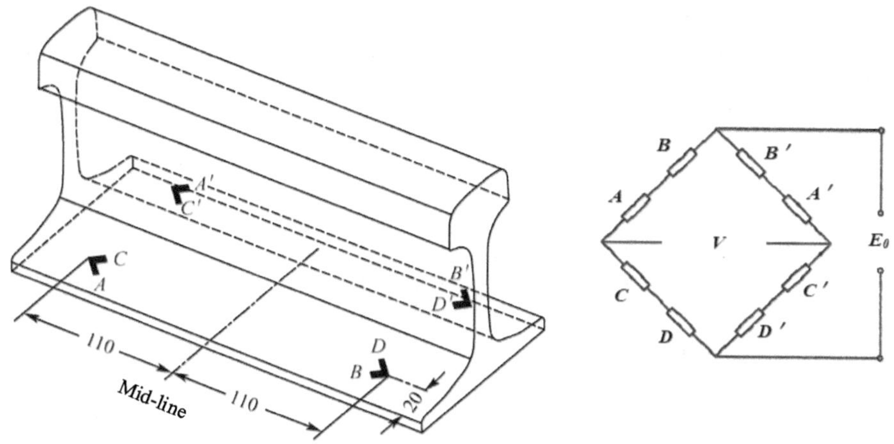

Fig. 5.4 Mounting positions and the electrical bridge diagram for measuring lateral wheel–rail force (unit: mm)

Leaf spring displacement sensor can be used to measure the dynamic displacements of track components conveniently. The sensor should be made to suit the test site conditions (space-wise) for easy installation. A nonuniform cross-section cantilever with uniform strength can be made from spring steel as shown in Fig. 5.5. Two strain gauges mounted at the same position on each side of the cantilever and form an electrical bridge as shown in Fig. 5.5, where R_0 is a standard resistance of 120 Ω in normal cases. The dynamic movement of the free end of the cantilever (in contact with the measured component) can be calculated from the deflection of the cantilever. This sensor can be calibrated using feeler gauge at the test site.

It is essential to ensure that the measuring datum is securely fixed when measuring track displacement. Displacement piles (steel rods) can be used as the

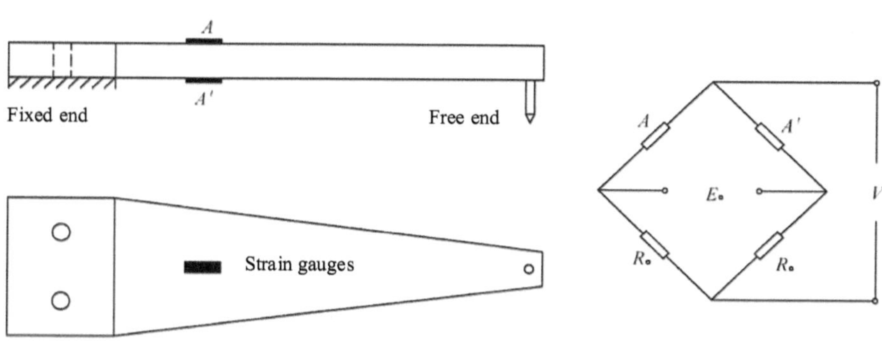

Fig. 5.5 Leaf spring displacement sensor with nonuniform cross section and the corresponding electrical bridge diagram of strain gauges

Table 5.1 Recommended accelerometer specifications for test of track structural vibration

Test object	Rail	Sleeper	Slab	Ballast
Working frequency (Hz)	0–5000	0–1000	0–2000	0–500
Measuring range (*g*)	1000	100	100	20

absolute datum for the ballasted track system. The displacement sensors can be fixed on to the pile with various suitable ways to measure the track vertical and lateral displacements (see details in Sect. 5.3). For slab track system or ballasted track sections on bridges or in tunnels, the displacement sensors can be fixed accordingly to measure the relative displacements between rail and slab, and between slab and bridge beam.

3. *Track structural vibration*

Track structural vibration is resulted from vehicle–track interaction, which can be measured using piezoelectric accelerometers. The recommended accelerometer specifications for rail, sleeper (or slab) and ballast can be found in Table 5.1.

For ballasted track systems, the accelerometers could be mounted at the locations marked in Fig. 5.6. The accelerometer mounted on rail at point A (20 mm away from the edge of rail foot) measures rail vibration; the accelerometer mounted on sleeper at point B (250 mm away from the center line of the rail) measures sleeper vibration; the accelerometer embedded in ballast at point C (150 mm under the bottom of the sleeper along the rail center line) measures ballast acceleration

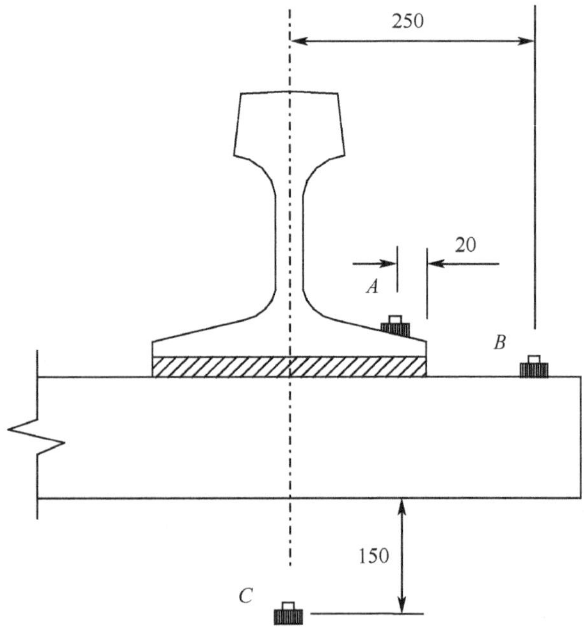

Fig. 5.6 Accelerometers layout for track structures (unit: mm)

Fig. 5.7 Steel protective box for ballast accelerometer (Reprinted from Ref. [8], Copyright 2003, with permission from Elsevier.)

It is worth noting that measuring ballast acceleration is particularly challenging because the accelerometer must be embedded in the ballast [8]. A steel protective box is designed to protect the direct contact between the accelerometer and the ballast granules (Fig. 5.7), preventing the accelerometer being damaged. The bottom of the box is a thin steel plate, which ensures the vertical positioning of the accelerometer. The size of the box should be similar to the size of the ballast granule in order to keep the original ballast configuration (see Fig. 5.7), which is important for measurements to obtain the actual ballast vibration level. The test can only proceed when the ballast is well settled after placing the box under the ballast. Our experiments have proved that this method is effective.

For slab track system, the main difference in the measurement method is that the accelerometer should be placed on top of the slab, and 450 mm outside of the track center line.

5.2 Typical Dynamics Tests of Vehicles Running on Tracks

5.2.1 Dynamic Test for a Typical High-Speed Train on Slab Track

Long-term monitoring for a CRH high-speed train was carried out by State Key Laboratory of Traction Power, Southwest Jiaotong University from August 2010, to get a better understanding of vehicle vibration in operation and to provide fundamental knowledge for high-speed train running safety. This section presents measurement results of the vibration accelerations of the CRH high-speed train running at 350 km/h on a slab track in Ref. [7].

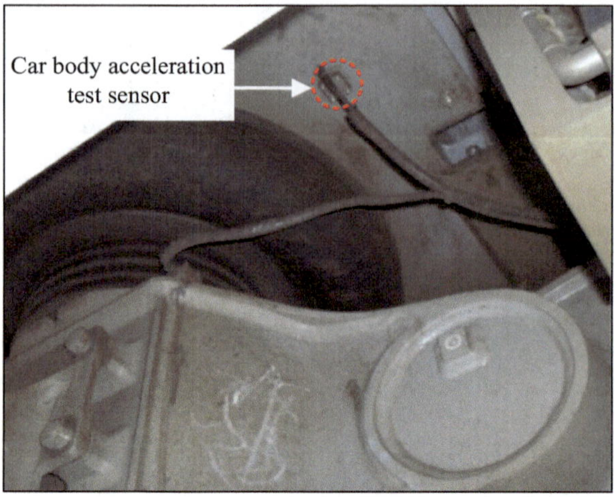

Fig. 5.8 Accelerometer layout for car body acceleration of high-speed train [7]

The first to fourth cars of this high-speed train are tested. Figures 5.8 and 5.9 show the onsite photos of different sensors installed on the corresponding components. Car body acceleration sensor is installed on the lower surface of car body floor (Fig. 5.8), and the bogie acceleration sensor is fixed on the top surface of bogie frame near the primary coil spring (Fig. 5.9), the axle-box acceleration sensor is installed on the upper side of the axle-box end cover (Fig. 5.9). In order to obtain the vehicle vibration characteristics in a wide frequency range, the sampling frequency for the axle-box acceleration sensor is set to be 5000 Hz while others are 2000 Hz.

1. *Vibration characteristics of the high-speed vehicle system*

The vibration of the main components of the high-speed train can be measured during field tests, which can then be used to assess the stability and ride

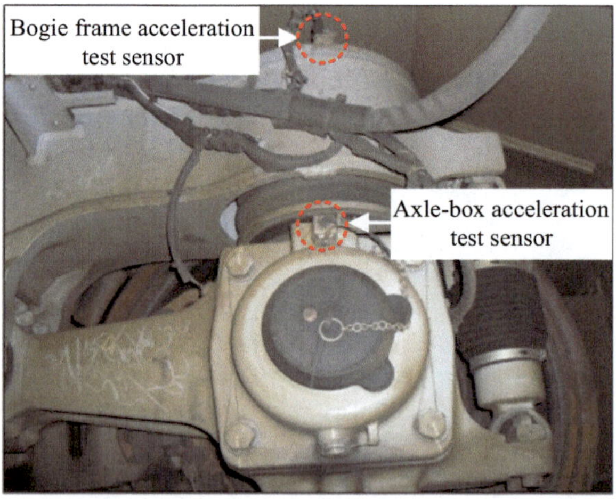

Fig. 5.9 Accelerometer layout for bogie and axle-box accelerations of high-speed train [7]

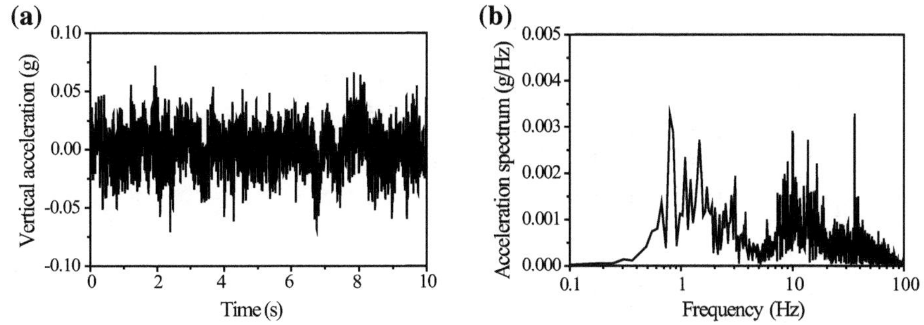

Fig. 5.10 Test results of car body vertical acceleration at train speed of 350 km/h: **a** time history and **b** frequency spectrum

performance of the vehicle. The vibration measurements of the car body, bogie, and axle box of a car of the high-speed train running at 350 km/h will be discussed in this section.

Figures 5.10 and 5.11 show the test results of the vertical and the lateral acceleration responses of the car body, respectively. Both the results in time and frequency domains are displayed in the figures. As shown in Fig. 5.10a, the peak value of the car body vertical acceleration is less than 0.075 g (g = 9.8 m/s^2). It is indicated in Fig. 5.10b that there are three distinct dominant frequencies in the vertical vibration of the car body. The first main frequency is near 1 Hz, which represents the natural vibration frequency of the car body's vertical suspension. The second main frequency is about 10 Hz, close to the first-order natural vibration frequency of the car body vertical bending. The third main frequency is 34 Hz, which reflects the forced vibration induced by the excitations from the wheel, such as the first-order out-of-round of the wheel, wheelset dynamical unbalance, local defects in wheel profile and other possible factors. Figure 5.11 shows that the lateral vibration acceleration of the car body is usually less than 0.05 g. There also

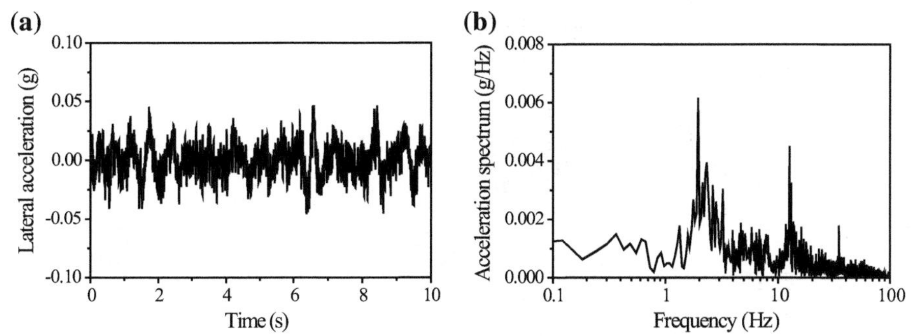

Fig. 5.11 Test results of car body lateral acceleration at train speed of 350 km/h: **a** time history and **b** frequency spectrum

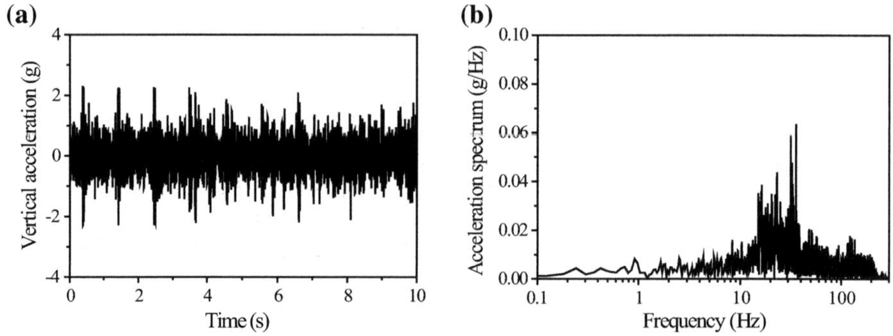

Fig. 5.12 Test results of bogie frame vertical acceleration at train speed of 350 km/h: **a** time history and **b** frequency spectrum

exist three distinct dominant frequencies. The first main frequency is 1.9 Hz, which is determined by the natural vibration frequency of the suspension system. The second one is 13 Hz, corresponding to the natural frequency of the car body torsional vibration. The third one is 34 Hz, which indicates that the forced vibration induced by wheel perimeter also affects the car body's lateral vibration. In short, the car body works at a low vibration level when the train runs at high speed. Both its vertical and lateral accelerations are far less than the limit of 2.5 m/s^2 defined by the "Code for Testing of High-speed Electric Multiple Unit on Completion of Construction" [3]. It indicates that the high-speed train can be in service with excellent ride comfort. A significant reason is the high geometry quality of the ballastless track in Chinese high-speed railway lines.

Figures 5.12 and 5.13 are the measured results of the vertical and the lateral vibration accelerations for the bogie frame, respectively. It can be found in Fig. 5.12 that the bogie frame vibrates more violently than the car body in the vertical direction. Its vertical vibration acceleration varies within the range of

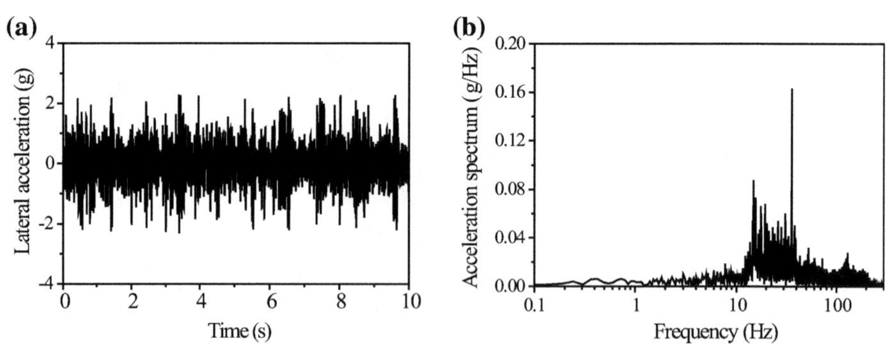

Fig. 5.13 Test results of bogie frame lateral acceleration at train speed of 350 km/h: **a** time history and **b** frequency spectrum

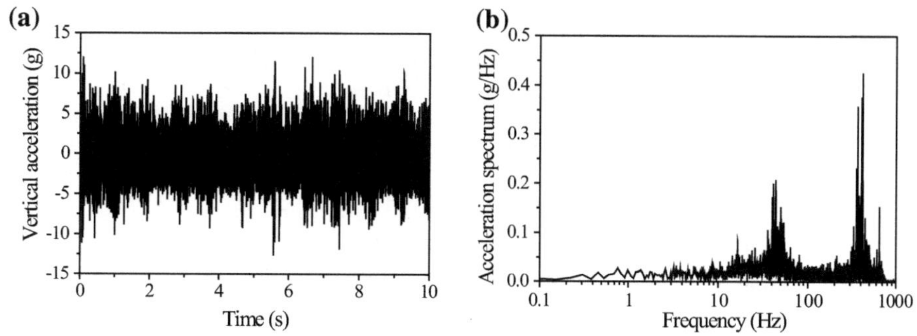

Fig. 5.14 Test results of axle-box vertical acceleration at train speed of 350 km/h: **a** time history and **b** frequency spectrum

±2.5 *g*. In the frequency domain, the vibration energy distributes mainly in the range of 15–35 Hz which contains the low-order elastic modal frequencies of the bogie frame. Especially, the forced vibration induced by the wheel perimeter dominates a large component at 34 Hz. Figure 5.13 indicates that the maximum value of the lateral acceleration in the bogie frame is also less than 2.5 *g*, and its main frequencies are similar to those of the vertical vibration.

The time history of the axle-box vertical acceleration and its corresponding frequency spectrum are shown in Fig. 5.14, and those of the axle-box lateral acceleration are displayed in Fig. 5.15. It can be seen from Fig. 5.14 that during full-speed operation the axle-box vibrates strongly and the peak value of its vertical acceleration is approximately 13 *g*. Due to the large contact stiffness between the wheel and the rail, the high-frequency contact vibration in wheel–rail interface excited by the track irregularity is liable to be transmitted to the axle-box. The vertical vibration distributes in a wide frequency range below 700 Hz, which covers two distinct dominant frequency ranges. The first one is the frequency range of

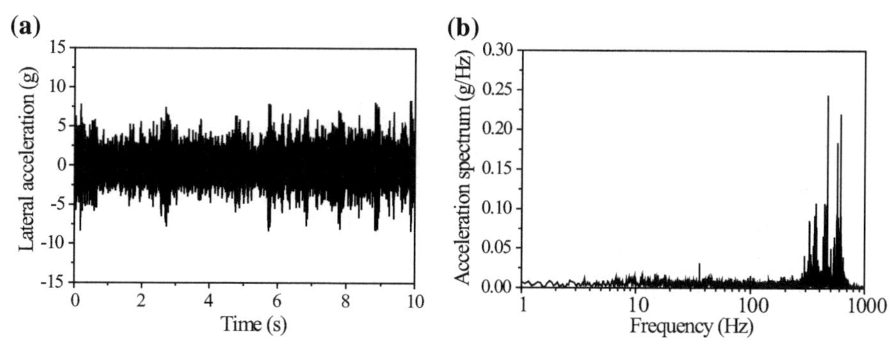

Fig. 5.15 Test results of axle-box lateral acceleration at train speed of 350 km/h: **a** time history and **b** frequency spectrum

30–50 Hz, which is mainly related to the vibrations induced by wheel perimeter and the elastic vibration of the bogie frame. The second one is the high-frequency range of 350–500 Hz, which reflects the high-frequency Hertzian contact vibration occurring in the wheel–rail interface, as well as the elastic vibration of the wheelset. As shown in Fig. 5.15, the axle-box lateral vibration is weaker than its vertical vibration (Fig. 5.14). The lateral vibration energy mainly distributes in the frequency range of 290–650 Hz. The vibration acceleration below 290 Hz is much lower, and there is a very small peak value at 34 Hz, which indicates that the periodic excitation caused by wheel perimeter also affects the axle-box lateral vibration.

The measured results indicate that the vibration was significantly dissipated when passed through the primary and secondary suspensions to the bogie and to car body. The axle-box acceleration is reduced to 1/3–1/5 of its original value when transferred to the bogie; which is further reduced by an order of magnitude when transferred to the car body. The high-frequency components above 50 Hz are eliminated by the primary and secondary suspension systems.

It is worth noting that the measured results discussed above include the elastic modes of the car body, bogie, and wheelset. However, such elastic modes are not included in rigid body dynamics simulations, whereas such features should be modeled in elastic body dynamics simulations.

2. *Evaluation of high-speed vehicle running stability*

The lateral stability of a high-speed vehicle running on a track can be evaluated using the tested response of lateral vibration acceleration of the bogie frame as mentioned above. The basic principle is, by monitoring the vibration acceleration responses of the bogie frame, to evaluate whether a continuous lateral oscillation of the bogie can decay or not. The assessment of vehicle lateral stability was not defined in China in the early days [1, 2], however, the evaluation method of vehicle running stability is now explicitly given in Chinese Code [3, 4]. A bandpass filter between 0.5 and 10 Hz needs to be applied to the measured bogie lateral acceleration, and the vehicle running stability could be evaluated through counting if the acceleration peaks reach or exceed 8–10 m/s^2 for 6 consecutive times. If it is the case, the vehicle is regarded as losing its stability.

The above method is applied to the bogie acceleration measurement discussed in the above section. The filtered signal from the time history of the bogie frame lateral acceleration in Fig. 5.13a is plotted in Fig. 5.16, where the peak is much lower than 8 m/s^2, proving that the high-speed train has outstanding stability.

The above measurements and assessment indicate that the bogie and car body vibrations are isolated and dissipated by the primary and secondary suspension systems to a very low level. Hence, the high-speed train has remarkable ride comfort and stability performance running at 350 km/h on the slab track.

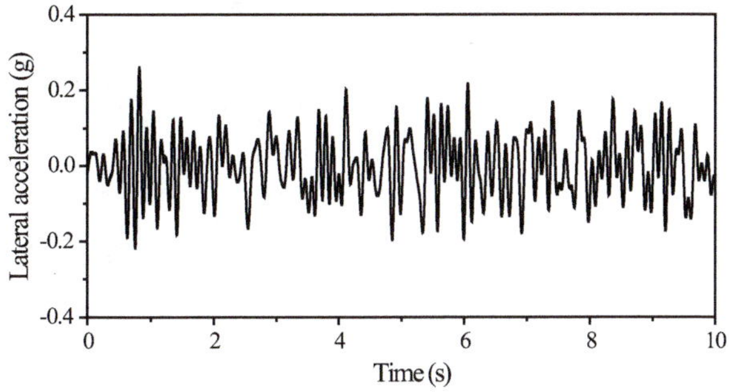

Fig. 5.16 Lateral acceleration time history of bogie frame after 0.5–10 Hz bandpass filtering

5.2.2 Dynamic Test for a Typical Freight Vehicle on Ballasted Track

This section will discuss a field test regarding a less usual freight vehicle behavior. A freight vehicle dynamic test was carried out by the author and his team on the Shuohuang railway line in June 2004, which captured the lateral hunting movement of a freight vehicle with three-piece bogies (Z8A) [9].

The measured freight wagon was in empty condition, and well maintained. 66 C_{64} wagons were towed by two SS_{4B} locomotives, and the measurements were taken on the 45th wagon (number 12591).

The lateral acceleration of the car body, bogie side frame, and wheelset, as well as the lateral displacement between the car body and side frame were measured. The car body accelerometer was mounted on the wagon floor (empty car); the side frame accelerometer was mounted on a rigid plate on the right-hand side of the front bogie (Fig. 5.17); the wheelset accelerometer was mounted on the right-hand side axle box of the first wheelset (Fig. 5.17).

Fig. 5.17 Accelerometer layout for side frame and wheelset accelerations

Fig. 5.18 Measurement for lateral displacement between the car body and side frame

The lateral displacement between the car body and side frame was measured using the string potentiometer. One end of the string was fixed on the car body of the wagon, and the other end on a steel frame fixed perpendicular to the side frame, ensuring the string horizontal, as shown in Fig. 5.18.

The lateral accelerations of the car body, side frame, and axle box, and the lateral displacement between the car body and side frame were measured at speeds of 60, 70, and 75 km/h. The test results show that the C_{64} empty wagon experienced severe hunting movement at 75 km/h, which was demonstrated by the extraordinarily large periodic acceleration wave (Fig. 5.19). The maximum lateral acceleration of the car body reached at 0.77 *g*, which greatly exceeded its safety value, 0.5 *g*, allowed for the freight car used in Chinese Railways. The side frame also experienced severe lateral vibration (Fig. 5.20), and peaked at 10.81 *g*. The lateral displacement between the wagon and side frame was also significantly increased compared to normal working conditions, and the vibration magnitude changed periodically as shown in Fig. 5.21.

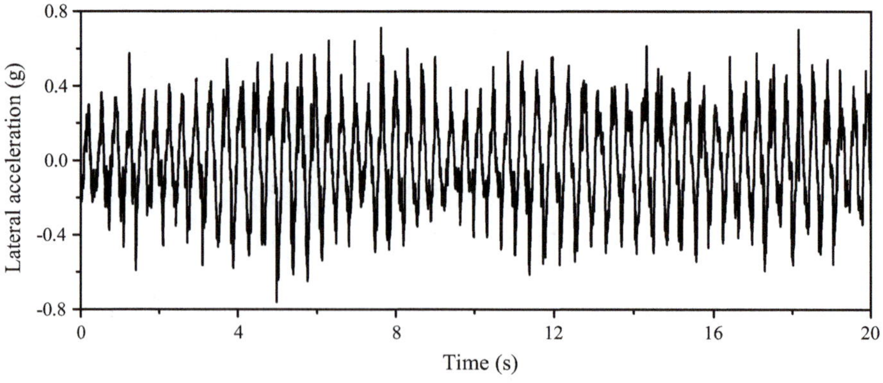

Fig. 5.19 Lateral acceleration of wagon (at the center plate) measured at 75 km/h

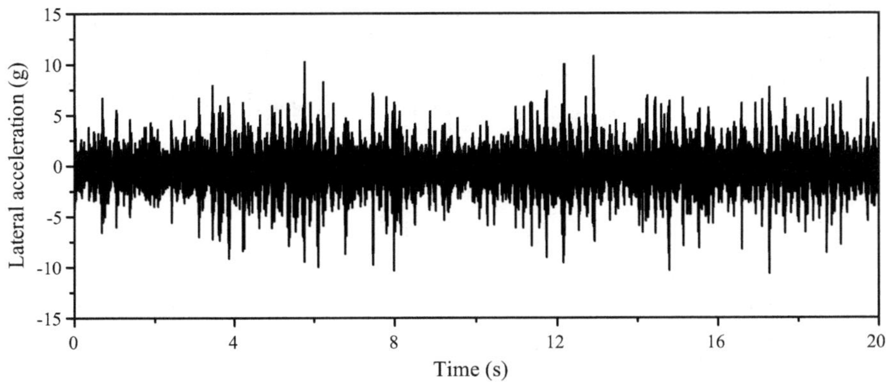

Fig. 5.20 Lateral acceleration of side frame measured at 75 km/h

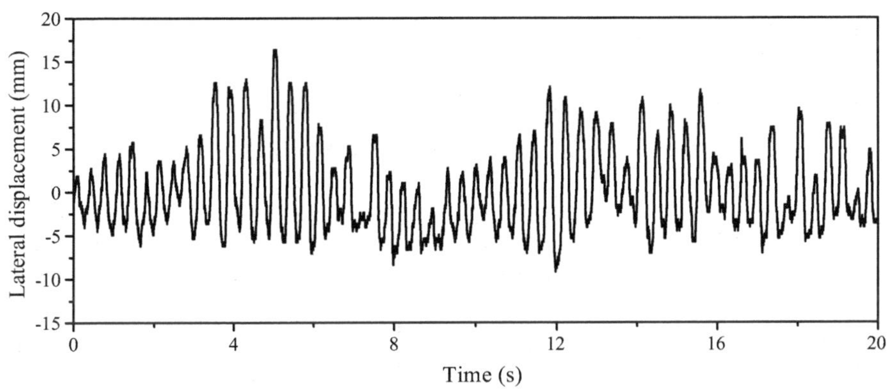

Fig. 5.21 Lateral displacement between the car body and side frame measured at 75 km/h

It can be concluded from the experimental results that the lateral dynamical system is very close to the unstable state at the speed of 75 km/h. Figure 5.22 depicts the measured frequency spectra of the car body lateral accelerations, in which the hunting frequency of the car body is 2.69 Hz.

The results above were in good agreement with those obtained by the vehicle–track coupled dynamics analysis. In our analysis, the theoretical critical speed of such vehicle under empty condition is 78 km/h [9]. Therefore, running the C_{64} empty wagon at a speed of 75 km/h is at the stability margin and deemed unsafe.

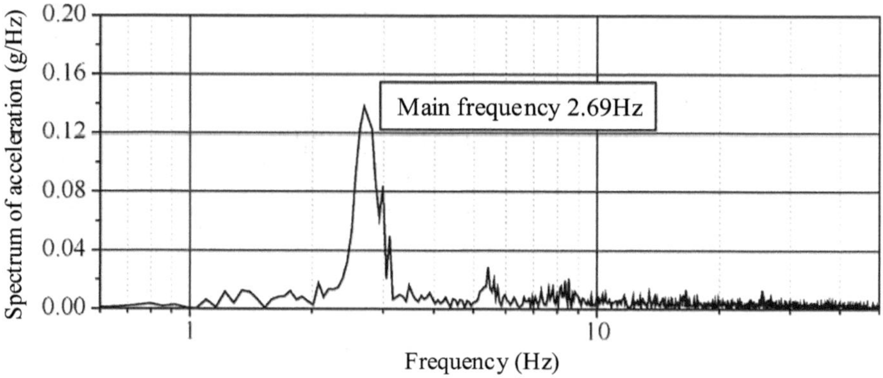

Fig. 5.22 Frequency content of wagon lateral acceleration (at the center plate) measured at 75 km/h

5.3 Typical Vehicle–Track Dynamic Interaction Tests

5.3.1 Wheel–Rail Interaction Test with a High-Speed Train on Qinshen Passenger Dedicated Line

Qinshen Passenger Dedicated Line is the first passenger dedicated line built in China from Qinhuangdao to Shenyang, with a designed speed of 200 km/h, including a 66.8 km high-speed test section with a designed speed of 300 km/h.

A series of high-speed train tests were organized by the Ministry of Railways at the end of 2002, and set the train speed record at that moment in China, reaching 321.5 km/h. The author led the wheel–rail dynamic interaction tests on the slab track system on Shuanghe Bridge, and on the ballasted track system on Xing-Yan Bridge. This section will discuss the dynamics test on the slab track system as an example.

This test is to assess the risk of high-speed wheel–rail interaction and the adequacy of the slab track system design. The vertical and lateral wheel–rail forces, the rail pad force, vertical displacement between the rail and slab, lateral rail displacement and gauge widening, as well as the vertical accelerations of the rail and slab were measured in the test. The layout of the sensors is shown in Fig. 5.23, and Figs. 5.24, 5.25, 5.26, and 5.27 are the photos taken at the test site.

The "China Star" high-speed test train, which is comprised of 2 motor cars and 4 trailer cars, was used in this test. The wheel–rail dynamic interaction indices were measured when the test train passed the slab track section on Shuanghe Bridge at 160, 180, 200, 220, and 225 km/h, respectively, as shown in Fig. 5.28. As examples, the measured results at the speed of 200 km/h are shown in Figs. 5.29, 5.30, 5.31, 5.32, 5.33, 5.34, 5.35, and 5.36. The characteristics of dynamic impacts on the track components can be clearly seen in these graphs, where the motor car had a much larger impact on the track than the trailer cars due to its higher axle

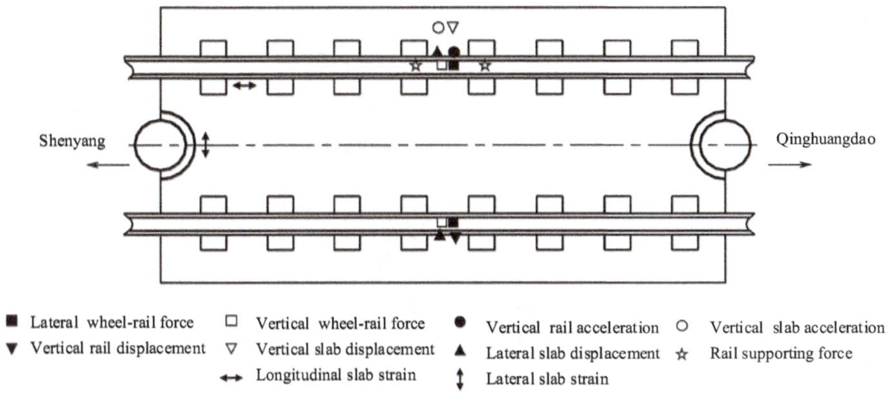

Fig. 5.23 Sensor layout of main test section (at the middle of the second span of the bridge)

Fig. 5.24 Measurement of vertical and lateral wheel–rail forces

Fig. 5.25 Instrumented load bearing plate for rail pad force measurement

Fig. 5.26 Displacement test device for slab track on the bridge

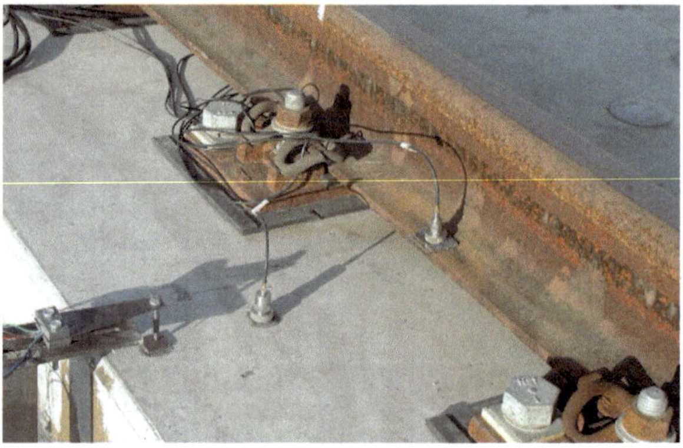

Fig. 5.27 Accelerometer layout for measuring slab track accelerations

Fig. 5.28 The "China Star" high-speed test train was passing the slab track section on Shuanghe Bridge

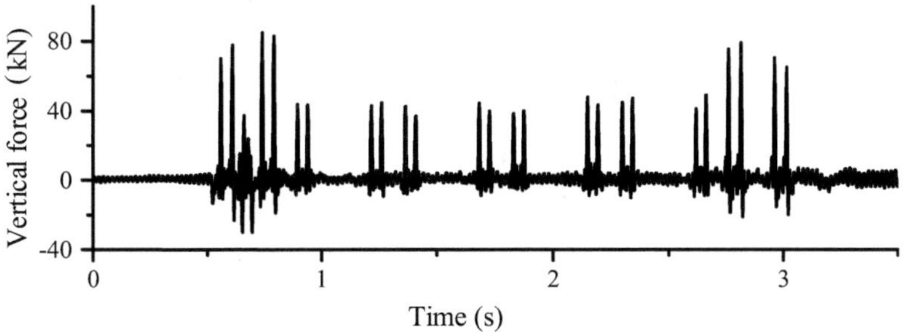

Fig. 5.29 Measured vertical wheel–rail force when the high-speed train passed at 200 km/h

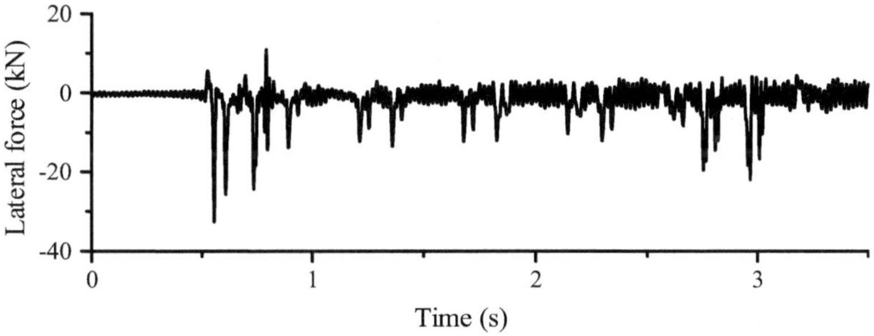

Fig. 5.30 Measured lateral wheel–rail force when the high-speed train passed at 200 km/h

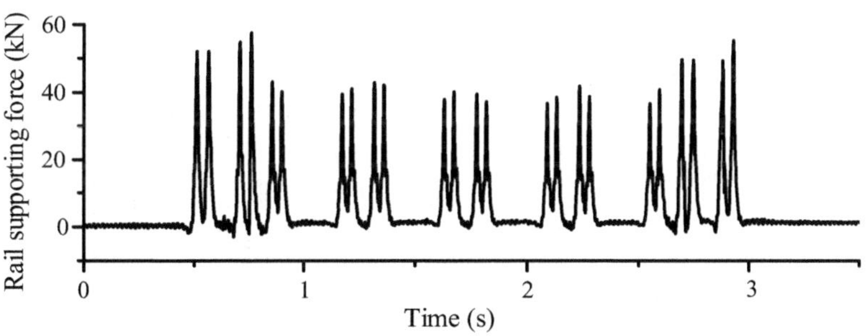

Fig. 5.31 Measured rail supporting force when the high-speed train passed at 200 km/h

load. The measured results proved that all the measured values were significantly lower than the safety thresholds. The vertical dynamic wheel–rail force peaked at 111.34 kN, which is much lower than the threshold of 300 kN set in 'On-Bridge Ballasted Track Design Specification for Qinshen Passenger Dedicated Line'. The lateral wheel–rail forces were generally lower than 30 kN, in compliance with the

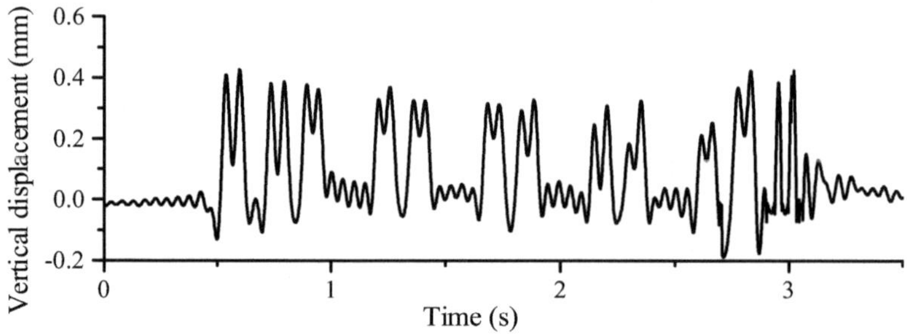

Fig. 5.32 Measured vertical rail displacement when the high-speed train passed at 200 km/h

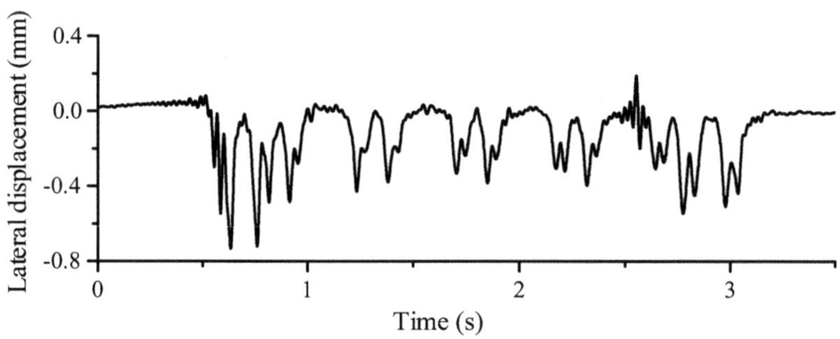

Fig. 5.33 Measured lateral rail displacement when the high-speed train passed at 200 km/h

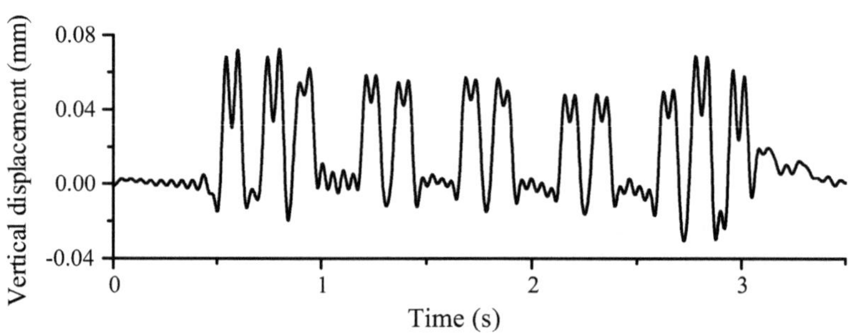

Fig. 5.34 Measured vertical slab displacement when the high-speed train passed at 200 km/h

thresholds (78 kN for the power car, 51.16 kN for the trailer car). The derailment coefficient was 0.27 and the wheel unloading rate was 0.29, which were much lower than their safety thresholds.

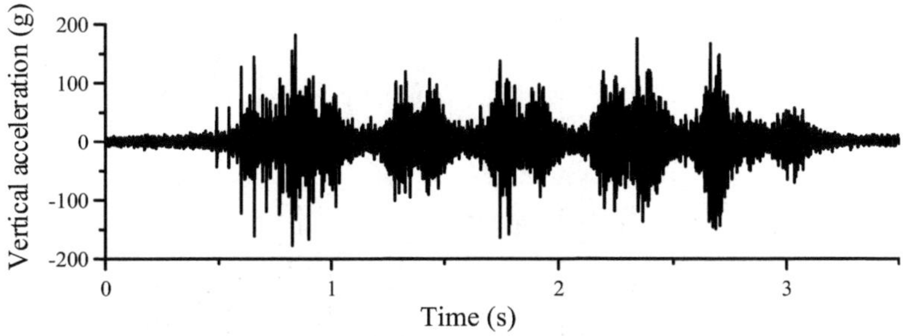

Fig. 5.35 Measured vertical rail acceleration when the high-speed train passed at 200 km/h

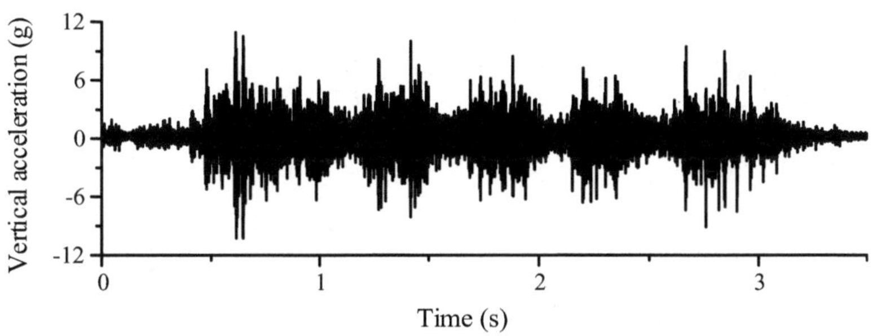

Fig. 5.36 Measured vertical slab acceleration when the high-speed train passed at 200 km/h

5.3.2 Track Dynamics Test with a 10,000-Tonne Heavy-Haul Train on Daqin Line

The Datong–Qinhuangdao Line (Daqin Line for short) is the first heavy-haul coal-transportation line in China. Two 10,000 t heavy-haul train tests were taken during the early 1990s to improve the coal-carrying capacity. This section will introduce the test carried out in October 1993 on the Upper Line at K318+70, and the photo in Fig. 5.37 was taken during the test. The test aimed to provide fundamental data to study the dynamic impact of the 10,000 t train on the track structure.

The test focused on the rail, sleeper and ballast vibrations when the heavy-haul train passed a rail joint with a 0.5 mm height difference. The test vehicle in the 10,000 t heavy-haul train is the Chinese freight wagon C_{61} equipped with 3-piece bogies (type Z8A), and the axle load was 21 t. Accelerometers were arranged as in Fig. 5.6 in Sect. 5.1.2. Figures 5.38, 5.39, and 5.40 plot the vibration acceleration responses of the rail, sleeper, and ballast when the wagon passed the rail joint with a speed of 52 km/h.

Our measured results indicated that the dynamic impact of the heavy-haul freight vehicle was not significantly increased, comparing to normal vehicles. The

Fig. 5.37 Test site of wheel–rail dynamic interaction of a 10,000 t heavy-haul train in Daqin Line

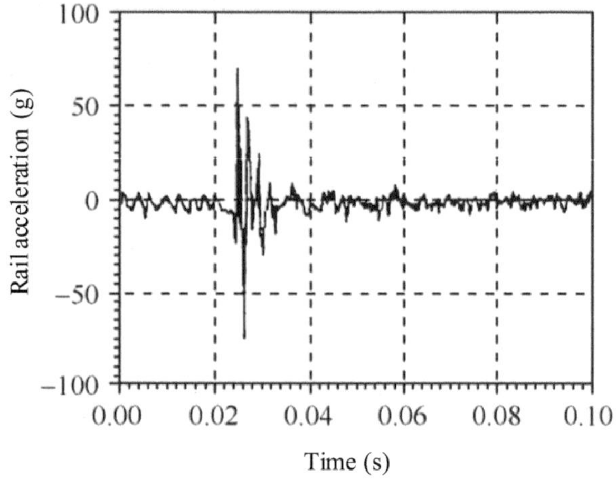

Fig. 5.38 Measured rail acceleration when a 10,000 t heavy-haul train passed the rail joint at 52 km/h

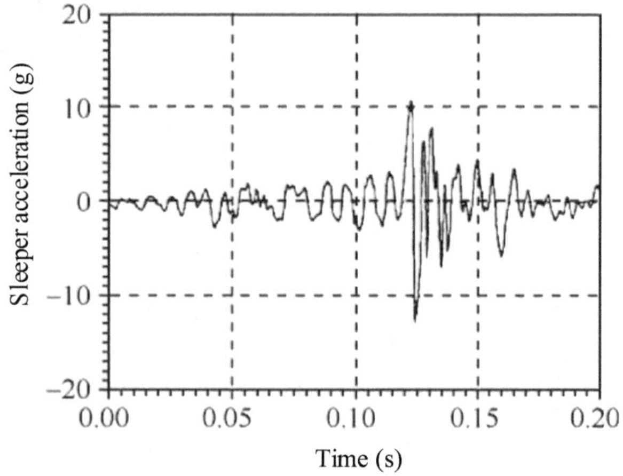

Fig. 5.39 Measured sleeper acceleration when a 10,000 t heavy-haul train passed the rail joint at 52 km/h

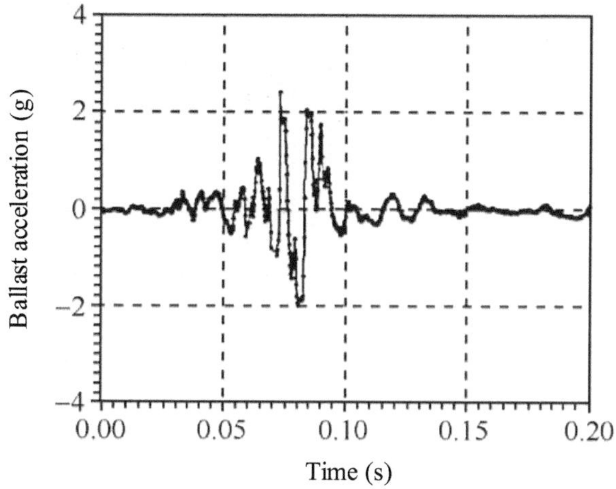

Fig. 5.40 Measured ballast acceleration when a 10,000 t heavy-haul train passed the rail joint at 52 km/h

acceleration peaks for the rail, sleeper, and ballast were 76.12 *g*, 13.16 *g*, and 2.64 *g*, respectively, because the wagon used for both heavy-haul and typical freight trains were the same, despite the heavy-haul train had much more wagons. The track structure will experience much more cycles of the dynamic impact due to the extra wagons, therefore the accumulated damage was much higher, requiring more frequent maintenance on the track.

5.3.3 Wheel–Rail Interaction Test on a Small-Radius Curve in Mountain Area Railway

Railway lines in mountainous areas have a much larger proportion of small-radius curves, which is a key factor reducing the safety of train operation. The condition of the track structures in these curves was not good before 2003 in China. The 50 kg/m rails and old wooden sleepers were widely used in the curves in mountain area. Large wheel–rail dynamic interaction resulted in serious rail flats, sleeper cracks, and gauge enlargement. The speed of train passing through these kinds of curves was usually below 50 km/h. In order to raise the train speed and enhance the running safety, Chinese Railways tried to strengthen the track structures of small-radius curves at that time, based on some scientific investigation including vehicle–track coupled dynamics analysis and field experiment [6]. The author and his research group carried out a series of field tests focusing on the lateral wheel–rail dynamic interaction at two locations (K444 and K70) on Chengdu–Chongqing Line in southwest China in April 2003. The tests were organized by Chengdu Railway Bureau, and supported by the Chinese Ministry of Railways. The track structure was strengthened on half part of the small-radius curves, by replacing the wood sleepers with concrete ones. The effectiveness of the strengthening with the

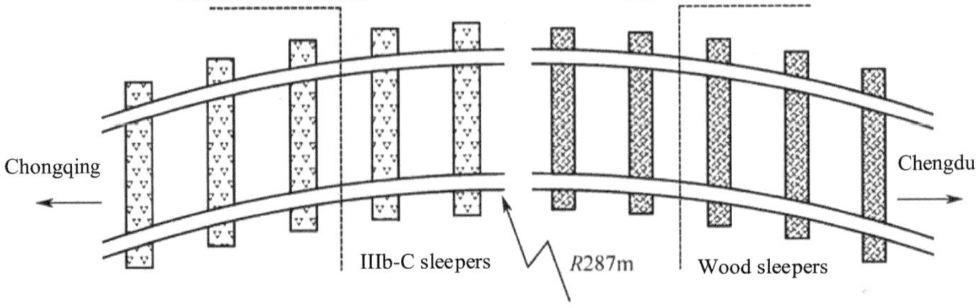

Fig. 5.41 Plan sketch of K444 test site at a small-radius curve on Chengdu–Chongqing line

concrete sleepers was investigated by comparing the wheel–rail interaction indicators for both types of sleepers. This section will present the test results at the site between K444+519.7 and K444+765.55 as an example.

The curve radius for this track section was only 287 m, with 125 mm cant. The curved track section from the middle of the curve towards Chengdu laid the wooden sleepers, while the track section toward Chongqing laid the concrete sleepers for comparison purposes. Figure 5.41 shows two sections selected for the test. The sensors were installed as illustrated in Fig. 5.42. The measured wheel–rail dynamic interaction indexes include the vertical and lateral wheel–rail forces, vertical and lateral rail displacements, vertical and lateral sleeper displacements, vertical rail and sleeper accelerations. Leaf spring displacement sensors were fixed to displacement piles to measure rail and sleeper displacements, as demonstrated in Figs. 5.43 and 5.44, respectively.

Thirty trains passed the two test sections during the test period, including 18 freight trains (C_{62A} wagons hauled by SS_3 locomotives as shown in Fig. 5.45) and 12 passenger trains (YZ_{22} passenger coaches hauled by SS_3 locomotives). The train speed varied between 49.5 and 68.5 km/h.

Figures 5.46, 5.47, and 5.48 show a typical set of measured results, representing a passenger train passing the first test section (with IIIb-C type concrete sleepers) at a speed of 65.5 km/h. The measured results showed that exceptional lateral wheel–rail interaction was observed when the locomotive passed the small-radius curve. The test results for the freight wagon C62A passing through the test sections with different speeds can be found in Ref. [6], and thus not given here.

It can be seen for Fig. 5.46 that the lateral wheel–rail force exceeded 100 kN on the outer rail when the locomotive negotiated the 287 m radius curve (approximately 50 kN for the case of passenger coaches), which resulted in severe rail elastic deformation (Fig. 5.47) and obvious sleeper lateral displacement (Fig. 5.48). Lateral rail elastic deformation could further lead to dynamic gauging widening. Table 5.2 listed the gauge widening induced by the locomotive (SS_3), freight wagon (C_{62A}), and passenger coach (YZ_{22}) at both test sections at speeds between 49.5 and 68.5 km/h. The gauge widening reached the largest under the action of locomotives, and the smallest under the action of passenger coaches. The maximum

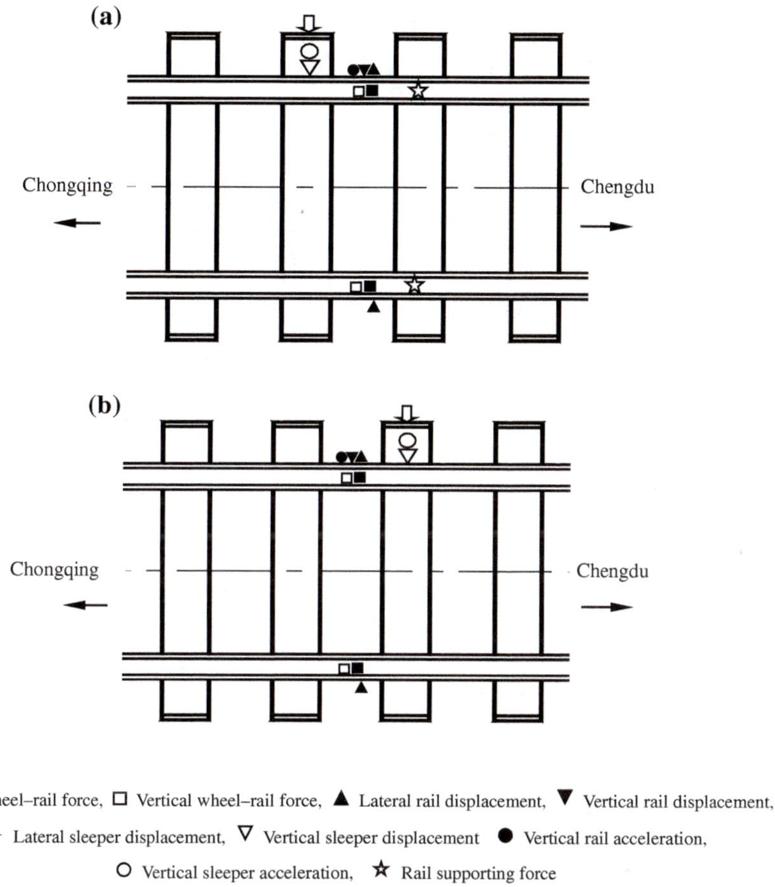

Fig. 5.42 Sensor layout of each test section in K444 test site on Chengdu–Chongqing line: **a** the first test section (with IIIb-C type concrete sleeper) and **b** the second test section (with timber sleeper)

Fig. 5.43 Rail lateral displacement test

Fig. 5.44 Sleeper displacement test

Fig. 5.45 Freight train passed the test sections with a small-radius curve on Chengdu–Chongqing Line

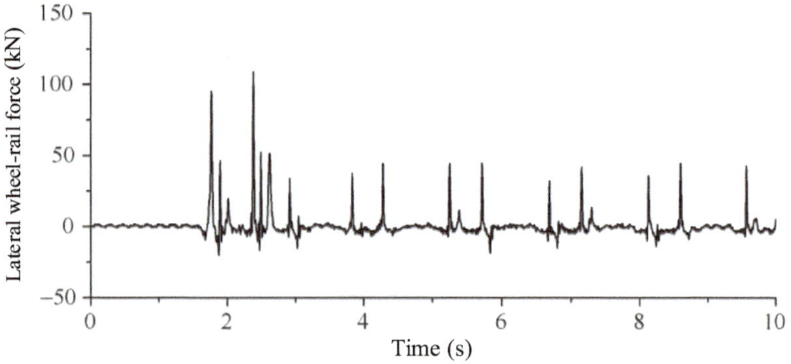

Fig. 5.46 Measured lateral wheel–rail force at the outer rail side

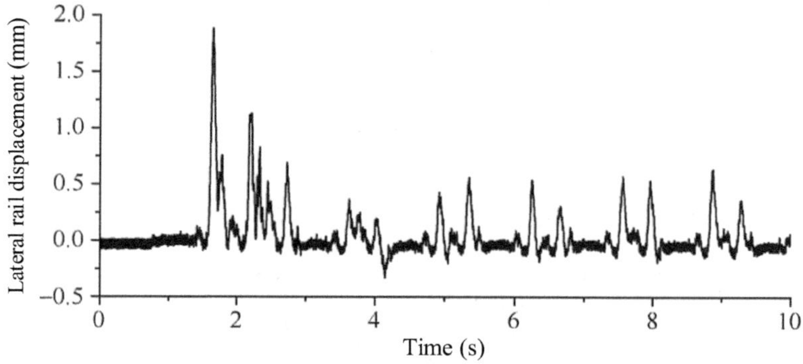

Fig. 5.47 Measured lateral rail displacement at the outer rail side

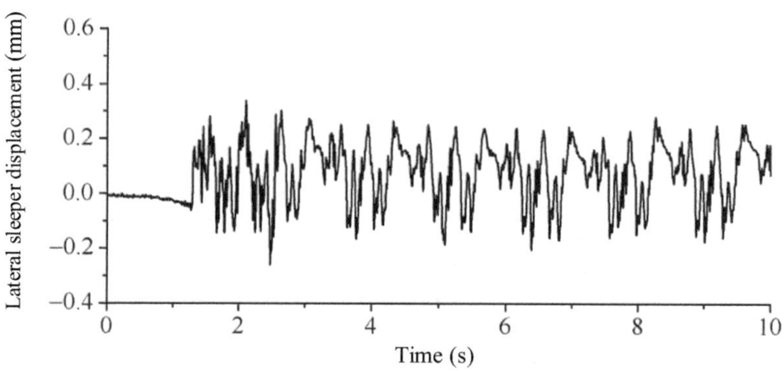

Fig. 5.48 Measured lateral sleeper displacement at the outer rail side

Table 5.2 Measured gauge widening induced by different vehicles at both test sections (unit: mm)

Vehicle type	Locomotive (SS_3)	Freight wagon (C_{62A})	Passenger coach (YZ_{22})
Wooden sleeper track	11.8	6.37	3.84
Concrete sleeper track	3.02	2.15	1.69

gauge widening on the wood sleeper section was 11.8 mm, and 3.02 mm on the concrete sleeper section. The gauge widening can be greatly reduced if the concrete sleepers were adopted. Therefore, the lateral wheel–rail interaction on small-radius curves could be severe, and the technical measures discussed in [6] must be applied to ensure the train running safety.

Nowadays, the wood sleeper tracks on the small-radius curves in the mountainous areas railways have been completely strengthened with the concrete sleepers, and the train speed has been raised accordingly.

References

1. GB5599-85. Railway vehicles—specification for evaluation the dynamic performance and accreditation test. Beijing: China Planning Press; 1985 (in Chinese).
2. TB/T2360-93. Identification method and evaluation standard for dynamic performance test of railway locomotives. Beijing: China Railway Publishing House; 1993 (in Chinese).
3. The Bureau of Transportation. Code for testing of high-speed electric multiple unit on completion of construction. Beijing: Ministry of Railways of the People's Republic of China; 2008 (in Chinese).
4. TB10761-2013. Technical regulations for dynamic acceptance for high-speed railways construction. Beijing: China Railway Publishing House; 2013 (in Chinese).
5. TB/T2489-94. Track side test methods of vertical and lateral wheel-rail forces. Beijing: China Railway Publishing House; 1994 (in Chinese).
6. Zhai WM, Wang KY. Lateral interactions of trains and tracks on small-radius curves: simulation and experiment. Veh Syst Dyn. 2006;44(sup1):520–30.
7. Zhai WM, Liu PF, Lin JH. Experimental investigation on vibration behaviour of a CRH train at speed of 350 km/h. Int J Rail Transp. 2015;3(1):1–16.
8. Zhai WM, Wang KY, Lin JH. Modelling and experiment of railway ballast vibrations. J Sound Vib. 2004;270(4–5):673–83.
9. Zhai WM, Wang KY. Lateral hunting stability of railway vehicles running on elastic track structures. J Comput Nonlinear Dyn. 2010;5(4):041009-1 ~ 9.

Chapter 6
Experimental Validation of Vehicle–Track Coupled Dynamics Models

Abstract Any theoretical model must be validated by experiments so as to confirm its correctness and reliability, especially for the dynamic models related to engineering problems. Only the theoretical model or the analysis software that is validated by field experiments could be applied to practical engineering in order to ensure the safety and reliability of critical engineering. Based on the typical field tests introduced in Chap. 5 and the plenty of wheel–rail system dynamics field tests on Chinese speedup lines and high-speed railway lines, this chapter performs systematic validations of the three theoretical models established in Chap. 2: (i) the vehicle–track vertically coupled dynamics model and the corresponding simulation analysis system, VICT; (ii) the vehicle–track spatially coupled dynamics model and the corresponding simulation analysis system, TTISIM; (iii) the train–track spatially coupled dynamics model.

6.1 Experimental Validation on the Vehicle–Track Vertically Coupled Dynamics Model

In order to check the correctness of the vehicle–track vertically coupled dynamics model and ensure the reliability of the corresponding simulation analysis system, VICT, comparisons between the theoretical analysis results and the substantial field test results under various conditions are performed. These field tests mainly include the following:

(1) Comprehensive dynamics tests of vehicle and track carried out on the ring railway test line in Beijing eastern suburb by China Academy of Railway Sciences in June 1993;
(2) Comprehensive wheel–rail dynamic interaction tests finished by China Academy of Railway Sciences in 1984;
(3) Corrugated rail dynamics tests performed on Fengtai–Shacheng railway line by Southwest Jiaotong University in 1990;

W. Zhai, *Vehicle–Track Coupled Dynamics*,
https://doi.org/10.1007/978-981-32-9283-3_6

(4) 10,000-ton train–track dynamic interaction tests carried out on Datong–Qinhuangdao heavy-haul railway by Southwest Jiaotong University in October 1993;
(5) Dynamics tests with respect to the effect of rail irregularities on the train operation performed on Guangzhou–Shenzhen railway line by China Academy of Railway Sciences from September to November 1994;
(6) Field tests on the vibration reduction of track structure with high elasticity rail pads carried on Chengdu–Kunming railway line by Southwest Jiaotong University in October 2001.

Some major results are selected for analysis in the following subsections.

6.1.1 Comparison of Vehicle Vibrations Between Theoretical and Measured Results

The dynamics test of train operation on the Guangzhou–Shenzhen railway line performed by China Academy of Railway Sciences in 1994 provides results for validating the analysis of car body vertical vibration acceleration. The vehicle vibration acceleration measured under excitations of rail corrugations with a different combination of wavelength λ and depth a, which are shown in Table 6.1.

The corresponding simulations were performed according to the actual test conditions by using the software VICT of the author. The parameters selected for the simulations were those of the speedup vehicle in China and of the continuously welded 60 kg/m rail with heat treatment by quenching over the full length. Besides, the vehicle speed and the rail vertical irregularities were the same as those in the tests. The simulated results are listed in Table 6.1. It can be seen that the maximum values of the simulated and the measured car body vertical vibration accelerations under the excitations of the various rail irregularities coincide with each other very well.

In June 1993, the axle-box vertical accelerations of the main kind of freight cars, namely, the China freight car C_{62A} with an axle load of 21 t and the heavy axle load car (C_{75}) with an axle load of 25 t, were measured on the ring test line by China Academy of Railway Sciences. Here, the impulsive vibrations excited by rail joints

Table 6.1 Comparison between measured and simulated car body vertical accelerations

Test condition	Vertical irregularity	Wavelength λ (m)	10	12	12	24	24
		Wave depth a (mm)	10	9	9	16	20
	Vehicle speed v (km/h)		160	135	150	160	160
Peak value of measured result (g)			0.12	0.06	0.08	0.12	0.13
Peak value of simulated result (g)			0.104	0.078	0.085	0.096	0.120

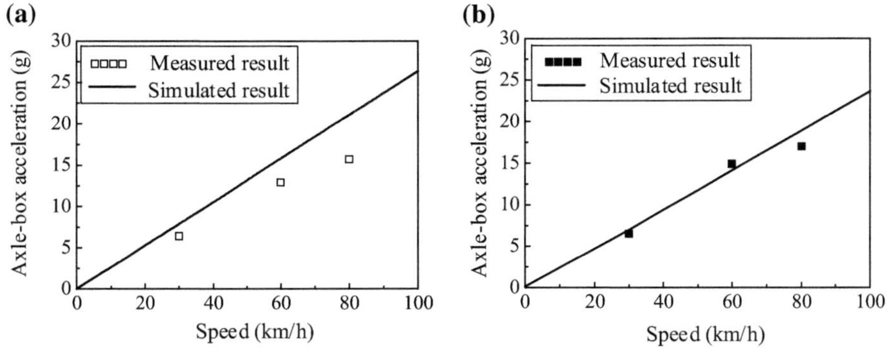

Fig. 6.1 Comparison between simulated and measured vehicle axle-box accelerations of **a** C_{62A} car and **b** C_{75} car

are especially concerned [1]. These test data could be applied to check the correctness of the simulated impulsive accelerations of the vehicle wheelsets. As shown in Fig. 6.1, the simulated axle-box vibration accelerations are compared with the measured results for the C_{62A} and the C_{75} cars, respectively. Here, the vehicles were running under different speeds on 50 kg/m rails through the dipped rail joints with a dipped angle of 0.02 rad.

6.1.2 Comparison Between Theoretical and Measured Vibrations of Track Structure

Acceleration of track components is the main index for track structure vibration. The validation analysis is performed based on the tests carried out on the Datong–Qinhuangdao heavy-haul railway line, focusing on the dynamic effect of 10,000 t trains on the track structure. The tested track conditions consist of 60 kg/m rail of U_{74} type; J-2 concrete sleeper with mass of $M_s = 251$ kg and with 1840 sleepers per kilometer; spring bar fastener of ω type, common rail pads with thickness of 10 mm and with stiffness of $K_p = 7.8 \times 10^7$ N/m; ballast bed with thickness of 450 mm. The tested vehicles are of the C_{61} type for coal transportation with three-piece bogies and with 21 t axle load. These cars have similar parameters with the C_{62A} type cars. In the test, the train was running over a rail joint with a height difference of 0.5 mm in reverse direction at a speed of 52 km/h. The measured and theoretically simulated peak values of the impulsive vibration accelerations are compared for the rail, the sleeper, and the ballast. The measured and simulated results shown in Table 6.2 indicate a good agreement.

In addition, the measured results from Ref. [2] are also referenced here for validation. The simulated results using the vehicle–track vertically coupled dynamics model are compared with the measured results obtained from the

Table 6.2 Measured and simulated impulsive vibration accelerations of track at a rail joint on Datong–Qinchuangdao railway line

Comparison index	Rail acceleration (*g*)	Sleeper acceleration (*g*)	Ballast acceleration (*g*)
Measured peak value	76.12	13.16	2.64
Simulated peak value	79.58	12.62	2.65

regression equation in Ref. [2], as shown in Fig. 6.2. These results are for the common wagons of 21 t axle load running reversely over a rail joint with a height difference of 1 mm. It can be seen that the theoretical and the measured results coincide well with each other in general, especially for the rail acceleration, and then the sleeper acceleration.

Further, the simulated ballast vibration is especially validated due to its particularity. The validation data are mainly from the vibration attenuation test of track structure performed on the Chengdu–Kunming railway line, where the ballast vibration acceleration was measured by the author [3]. The track conditions of the test section are as follows: buffer area of the continuous welded line with 60 kg/m rail; concrete sleeper of 69 type with 1840 sleepers per kilometer; high elastic rail pads; ballast bed with the thickness of 450 mm and common limestone ballast. There is an apparent rail joint depression at the test position, the wavelength of which is measured to be about 80 mm with a wave depth of about 0.4 mm.

Figure 6.3a displays a field measured impact waveform of ballast acceleration when a Chinese main kind of freight wagon (C_{62A} loaded car) was passing over the rail joint at a speed of 60 km/h. Figure 6.3b gives the simulated result using VICT based on the vehicle–track vertically coupled dynamics model. In the simulation, the ballast parameters were derived by using the ballast modified model (see

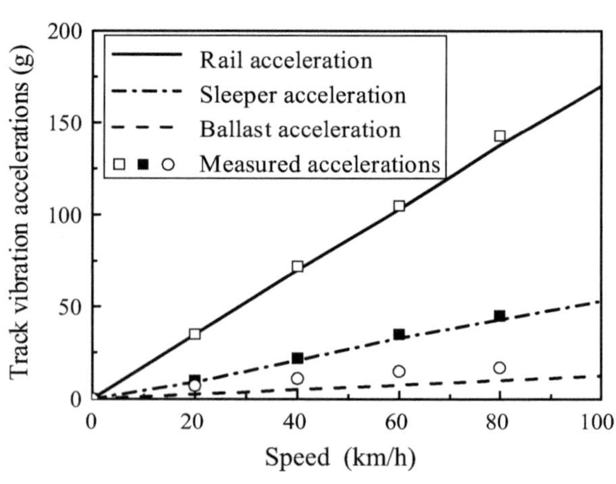

Fig. 6.2 Comparison between simulated and measured track vibration accelerations at a rail joint with a height difference of 1 mm

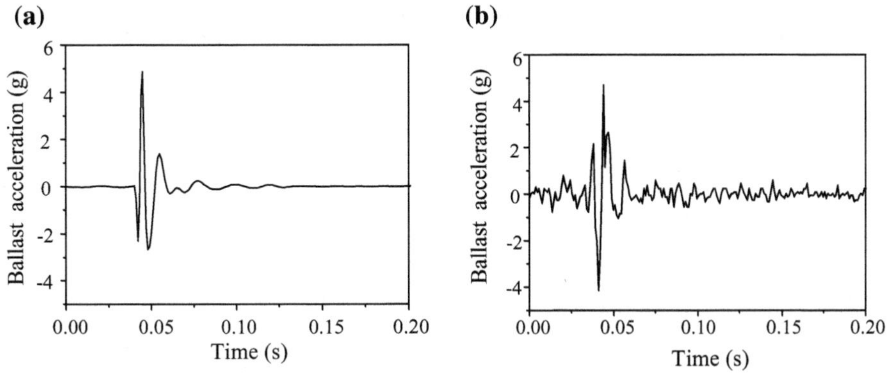

Fig. 6.3 Comparison between **a** simulated and **b** measured time histories of ballast vibration acceleration (Reprinted from Ref. [3], Copyright 2003, with permission from Elsevier.)

Fig. 2.13). According to the measured results of the Chinese railway track parameters, the ballast density is $\rho_b = 1800$ kg/m^3, the ballast elasticity modulus $E_b = 110$ MPa and the subgrade modulus $E_f = 90$ MPa/m (K_{30} value). Thus, it can be derived by Eqs. (2.5)–(2.9) that the mass of ballast is $M_b = 531.4$ kg, the ballast stiffness $K_b = 137.75$ MN/m and the subgrade stiffness $K_f = 77.5$ MN/m. This simulated ballast stiffness is close very much to the stable value (about 140 MN/m) measured on a Chinese conventional railway line by China Academy of Railway Sciences [3]. Consequently, it indicates that the modified model of ballast could reasonably predict the vertical supporting stiffness of granular ballast bed, which is a key parameter for analysis of ballast vibration.

It can be seen from the comparison between Fig. 6.3a, b that, the measured and the simulated results coincide with each other very well for both the waveform and the amplitude; namely, it is 4.69 g for the measured result and 4.97 g for the simulated result. The corresponding spectral density curves of the ballast acceleration are shown in Fig. 6.4, from which the spectral features from the theoretical modeling results agree well with the field measured results except for some minor detailed discrepancies. The main frequency range of the measured ballast acceleration is 70–100 Hz, while that of the theoretical result is 80–110 Hz.

Rail displacement is another index revealing the track dynamics properties. It should be pointed out that the rail displacement in the transition zone between subgrade and bridge is especially remarkable and is used here for the model validation. As introduced in Sect. 3.4.1 of Chap. 3, the dynamic performance degradation problem of the transition zone between the subgrade and the Strandmoelle bridge of the Danish State Railways (DSB) had perplexed the railway maintenance department for a long period. So, the DSB carried out several dynamic tests successively in 1990, 1995, and 1997. Regarding this problem, some collaboration between the Technical University of Denmark and the author was carried out to calculate the rail displacement in the subgrade–bridge transition zone by using the

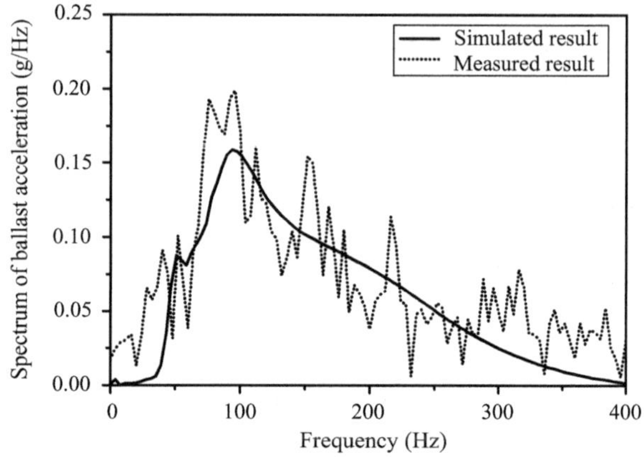

Fig. 6.4 Comparison of ballast vibration acceleration spectrum between field measured and simulated results (Reprinted from Ref. [3], Copyright 2003, with permission from Elsevier.)

Table 6.3 Comparison of measured and simulated rail displacements in the subgrade–bridge transition zone

Position (see Fig. 3.33)	Subgrade zone (half-sleeper zone)	Transition zone (wooden sleeper zone)	Bridge abutment zone (long sleeper zone)
Measured value (mm)	0.7	1.0	0.4
Simulated value (mm)	0.72	1.06	0.41

VICT software; it demonstrated that the simulated and the measured results illustrated in Table 6.3 agree well with each other [4].

6.1.3 Comparison Between Computed and Measured Wheel–Rail Dynamic Forces

Figure 6.5a displays measured time history of the vertical wheel–rail force induced by a wheel flat of a wagon equipped with the Chinese Z8A type of bogie. The vehicle speed through the test section was 27 km/h. The test data were acquired at Luoyang Eastern railway station by using a wheel flat detection device developed by the Research Institute of Zhengzhou Railway Administration. This wheel flat with a length of 52.8 mm and a depth of 1 mm is typical of an old worn one. It is approximated by a cosine function. The simulated results using the VICT software are shown in Fig. 6.5b. Comparison between Fig. 6.5a and b reveals that the theoretical simulation is capable of reproducing the waveform of the actual dynamic interaction force caused by the wheel flat [5].

Wheel–rail interaction forces, P_1 and P_2, are important indexes that reveal the wheel–rail vertical impact action when a vehicle passes over impulsive excitations

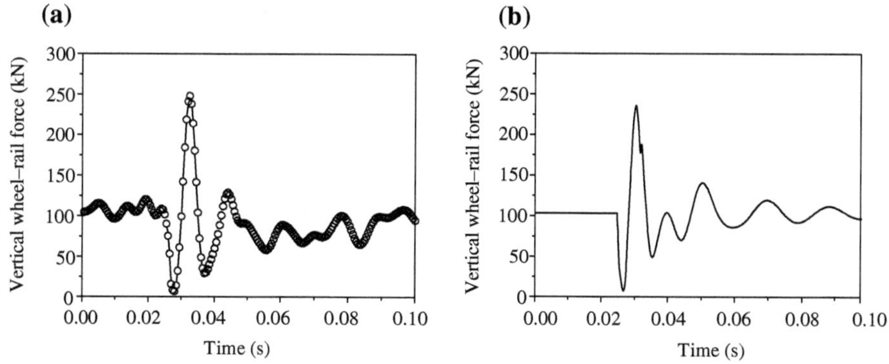

Fig. 6.5 Comparison between **a** measured and **b** simulated vertical wheel–rail forces caused by a typical wheel flat

such as rail joints. Comparison between measured and simulated wheel–rail impact forces is made in Fig. 6.6 when a C_{62A} vehicle was running reversely through a rail joint with a height difference of 1.5 mm. The simulated and the measured P_2 forces are compared in Fig. 6.7 for the C_{75} and C_{62A} cars, respectively, when they passed through a dipped rail joint ($2\alpha = 0.02$ rad) with various speeds [1]. It is seen from the two figures that the simulated values coincide with the measured results quite well in the entire speed range.

The vertical wheel–rail forces excited by rail corrugations are now compared for harmonic excitations. The tests were carried out at the curved section K38+25 (with a radius of 600 m) of the Fengtai–Shacheng railway line, since there existed apparent short pitch rail corrugations with a wavelength of about 200 mm. The maximum values of the vertical wheel–rail forces were measured for rail corrugations with wave depth of 0 mm (after grinding), 0.5 mm, 0.7 mm, and 1.1 mm, respectively. The measured and the simulated results are compared in Fig. 6.8, indicating a good agreement between them.

Fig. 6.6 Comparison between measured and simulated wheel–rail impact forces caused by a rail joint with height difference

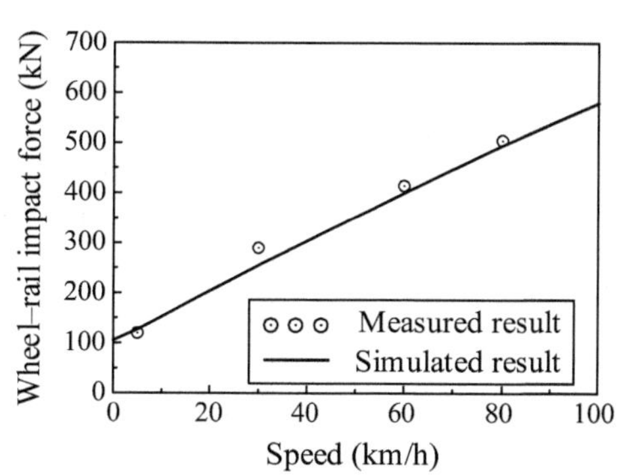

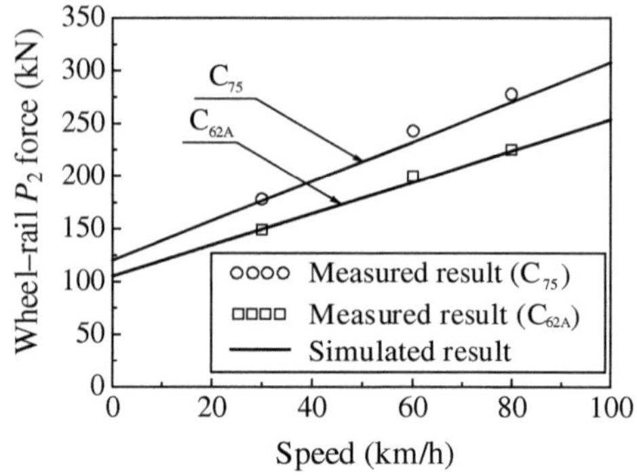

Fig. 6.7 Comparison between measured and simulated wheel–rail P_2 forces caused by a dipped rail joint

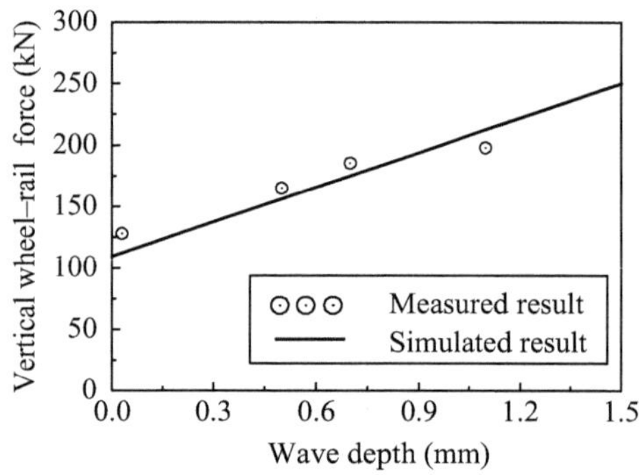

Fig. 6.8 Comparison of measured and simulated vertical wheel–rail forces due to rail corrugation

6.1.4 Conclusions

Based on the aforementioned validation analyses under various test conditions through different dynamics indexes, it can be concluded that the theoretical results simulated by using the vehicle–track vertically coupled dynamics model agree well with the field measured data, whatever for the vehicle vibration, track vibration, and wheel–rail force, or for the response waveform and the amplitude. This was also commented by Knothe and Grassie in an early review paper [6] as "The attraction of such a model is that it offers the possibility of obtaining better correlation between the simulated and measured response."

Consequently, a conclusion can be drawn that the vehicle–track vertically coupled dynamics model is effective and reliable with satisfactory analysis accuracy, and application of the VICT simulation system to analyzing the vertical vehicle–track coupled vibration and the wheel–rail vertical dynamic interaction is feasible.

6.2 Experimental Validation of the Vehicle–Track Spatially Coupled Dynamics Model

In order to validate the vehicle–track spatially coupled dynamics model and the corresponding software TTISIM, especially to validate the dynamic wheel–rail coupling model, a number of experimental studies of a variety of aspects are introduced in this section for comparison purpose. These mainly include the dynamics tests on the speedup lines or the high-speed lines performed by China Academy of Railway Sciences and Southwest Jiaotong University. They are as follows:

(1) Freight train derailment test in the straight section of Beijing ring railway test line in December 1999;
(2) Field test on Beijing–Qinhuangdao railway line after the reconstruction for train operation speeds up to 200 km/h on the first try in December 2000;
(3) First high-speed train running test in China on Qinhuangdao–Shenyang passenger dedicated line in December 2002;
(4) Wheel–rail dynamic interaction test in a curve with a small radius in Chengdu–Chongqing mountain railway line in April 2003.

6.2.1 Experimental Validation by Field Test on Beijing–Qinhuangdao Speedup Line

During the period of December 510, 2000, a field test was performed for train operation speed raised up to 200 km/h from previous 120 km/h on the existing Beijing–Qinhuangdao railway line by Beijing Railway Bureau together with China Academy of Railway Sciences. Indexes of the train operation safety on straight sections, curved sections and turnout zones were measured in the speed range of 160–210 km/h. The measured results of the train passing through a curved section at speed of 160 km/h and results of the train running at a straight section with speed of 200 km/h are adopted to validate the vehicle–track spatially coupled dynamics model and the corresponding software TTISIM.

The test line consisted of 60 kg/m rails, type II concrete sleepers and a common ballast bed used in Chinese conventional railway lines. The measured results in the representative curved section (K121+233 ~ K121+575) are selected for the validation of the vehicle–track lateral dynamic interaction. This curved section had a circular part of 142.25 m with a radius of 1200 m; the length of the transition curve was 100 m and the superelevation was 100 mm. The normal operation speed limit was 120 km/h and the actual speed was, however, up to 160 km/h in this test. Due to the fact that there is no track irregularity data measured on this line, the track irregularity data measured on a similar line, i.e., the Zhengzhou–Wuhan line, is used in the simulation. The test vehicle was a 4-axle double-deck passenger car.

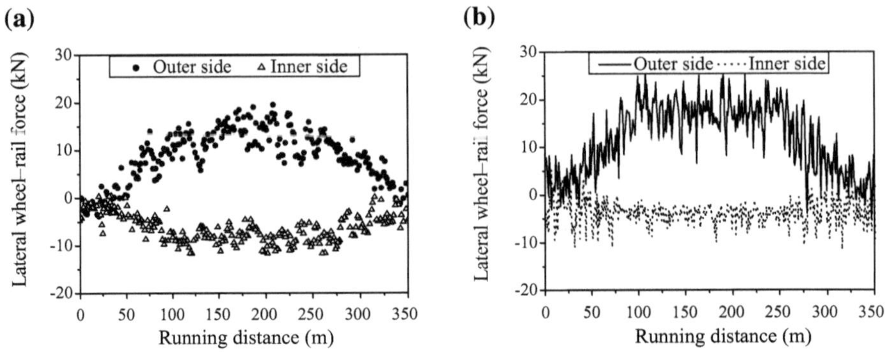

Fig. 6.9 Comparison of **a** measured and **b** simulated lateral wheel–rail forces on the curved section

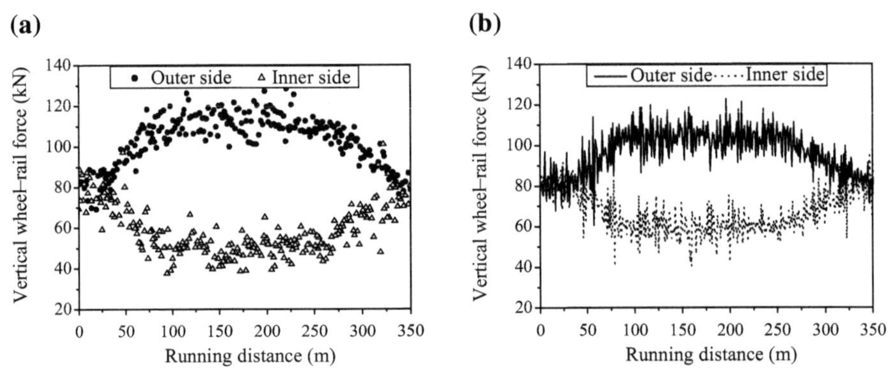

Fig. 6.10 Comparison of **a** measured and **b** simulated vertical wheel–rail forces on the curved section

The measured and the simulated results of wheel–rail interaction force, the car body acceleration and the derailment coefficient for the test train passing over the curved section under the speed of 160 km/h are given in Figs. 6.9, 6.10, 6.11, and 6.12, respectively. In general, the simulated dynamic responses are very similar to the measured results. The results in Fig. 6.9 show that the maximum lateral wheel–rail force is 20 kN for the test and 25 kN for the calculation. The vertical wheel–rail forces shown in Fig. 6.10 also display a good agreement between the measured and the simulated results with maximum values of 128.5 kN for the test and 122.6 kN for the calculation, and the minimum values of 37.7 kN for the test and 40.6 kN for the calculation. For the car body vibration acceleration in Fig. 6.11, the maximum value of the measured result is greater than that of the simulated result for both the lateral and the vertical vibrations, while the trends of the time response curves are the same. Finally, the derailment coefficients computed from the lateral and vertical

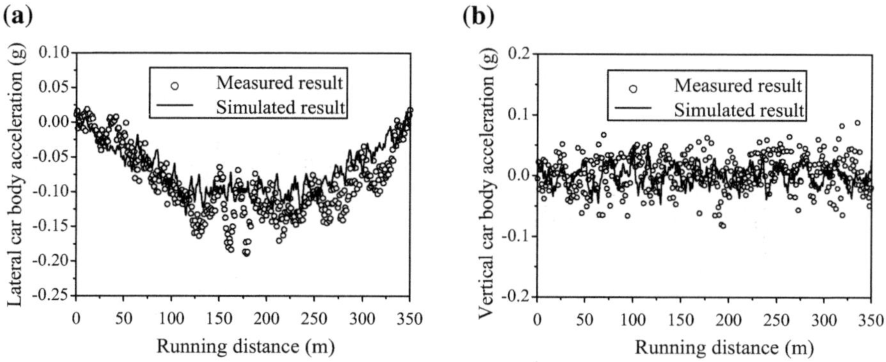

Fig. 6.11 Comparison of measured and simulated car body accelerations on the curved section: **a** lateral acceleration, and **b** vertical acceleration

Fig. 6.12 Comparison of measured and simulated derailment coefficients on the curved section

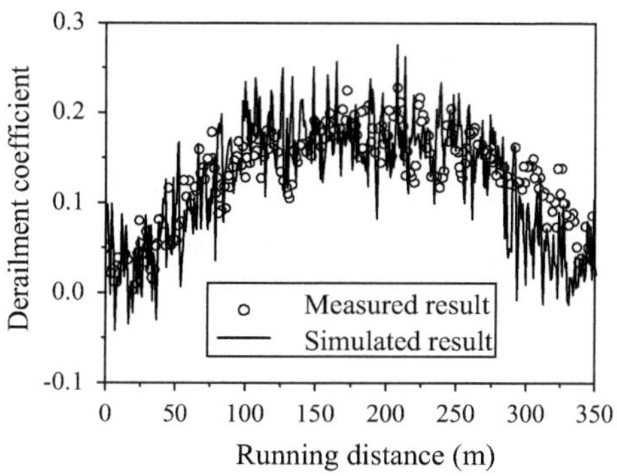

wheel–rail forces, as shown in Fig. 6.12, also coincide with each other due to the good agreement between the measured and the simulated wheel–rail forces.

In addition, the measured results in the straight section (downstream K110 +158 ~ K110+508) of the test line are also employed for validation. The measured and the simulated results of the wheel–rail interaction force, the car body acceleration and the derailment coefficient under the speed of 200 km/h are shown in Figs. 6.13, 6.14, 6.15 and 6.16, respectively. It can be seen that the simulated response waveforms coincide well with the measured results. Variations of the wheel–rail dynamic forces are small when the test vehicle runs on the straight section at a high speed. Both the measured and the simulated lateral wheel–rail forces vary within 15 kN (see Fig. 6.13a), and the vertical wheel–rail forces vary between 55 and 115 kN (see Fig. 6.13b). For the car body vibration, the lateral acceleration responses in Fig. 6.14a are in the range of −0.05 to 0.05 g, while the measured vertical acceleration in Fig. 6.14b is greater than that of the simulated result. A good agreement between the measured and the simulated derailment coefficients can be observed in Fig. 6.15.

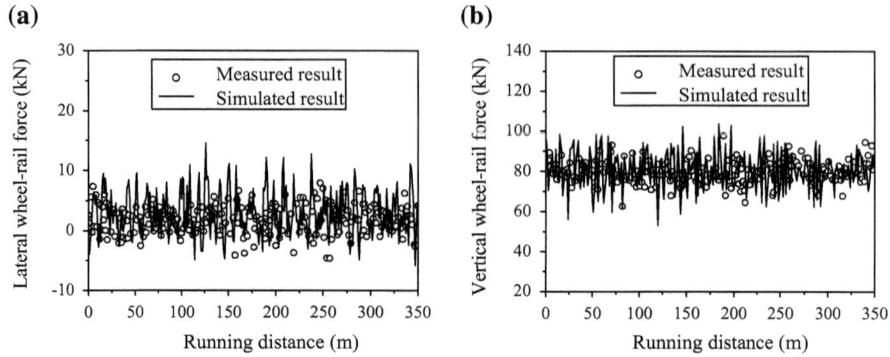

Fig. 6.13 Comparison of measured and simulated wheel–rail forces on the straight section: **a** lateral force, and **b** vertical force

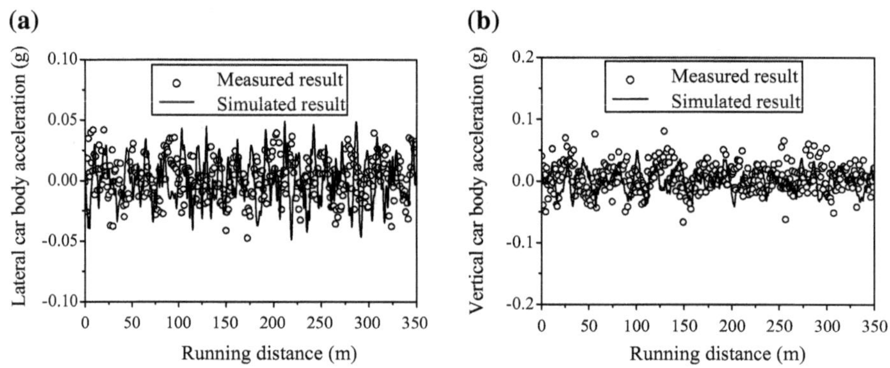

Fig. 6.14 Comparison of measured and simulated car body accelerations on the straight section: **a** lateral acceleration, and **b** vertical acceleration

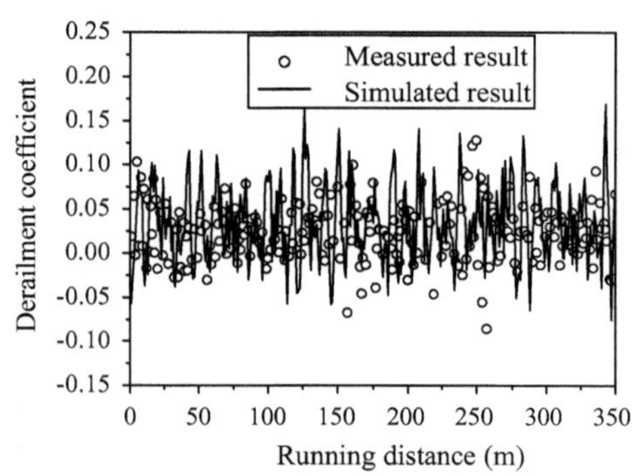

Fig. 6.15 Comparison of measured and simulated derailment coefficients on the straight section

In all, the simulated wheel–rail forces and the car body vibrations agree well with the measured results for both the curved and the straight sections. There exist, of course, inevitably some minor discrepancies between the simulated and the measured results due to the fact that it is impossible for the track irregularities used in the simulation (from Zhengzhou–Wuhan line) to be completely the same as those of the test line (the Beijing–Qinhuangdao line).

6.2.2 *Validation by High-Speed Train Running Test on Qinshen Passenger Dedicated Line*

At the end of 2002, the Chinese first high-speed train running test was performed on the Qinshen (Qinhuangdao–Shenyang) passenger dedicated line, including both a wheel–rail dynamic interaction test (specified in Sect. 5.3.1 of Chap. 5) and a train dynamics test. Key dynamic performance indexes such as the lateral wheel–rail force, the derailment coefficient, and the lateral and vertical car body ride comfort indexes were measured. In this subsection, the dynamic performance indexes measured when the "Chinese Star" high-speed test train was passing through the curved test section near K197+412 are employed to validate the theoretical calculation results. The design parameters of the test track were as follows: curve radius of 4000 m, transition curve length of 390 m, and superelevation of 115 mm. The speeds of the test train passing through the test region were 160 km/h, 180 km/h, 200 km/h, 220 km/h, and 225 km/h, respectively. The actual track irregularities measured on the Qinshen passenger dedicated line are applied as the excitation to the wheel–rail system in the simulation. The simulated and the measured values of the lateral wheel–rail force, lateral wheelset force, derailment coefficient, wheel unloading ratio, and ride comfort index of the high-speed vehicle are compared in Figs. 6.16, 6.17, 6.18, 6.19, and 6.20.

It can be seen by the comparisons in Figs. 6.16, 6.17, 6.18, 6.19, and 6.20 that the theoretical calculation results coincide satisfactorily with the measured results,

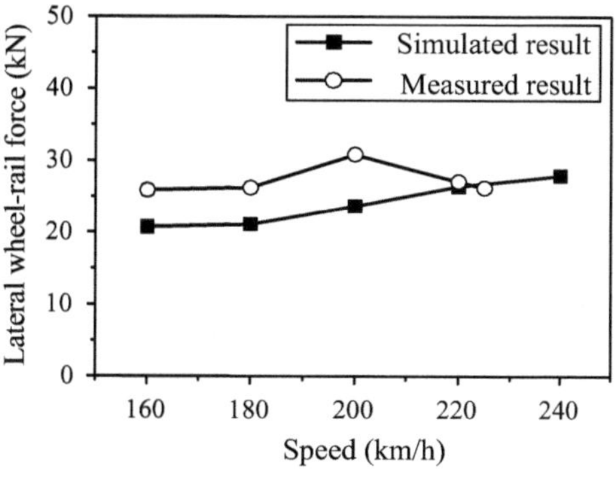

Fig. 6.16 Lateral wheel–rail force versus speed

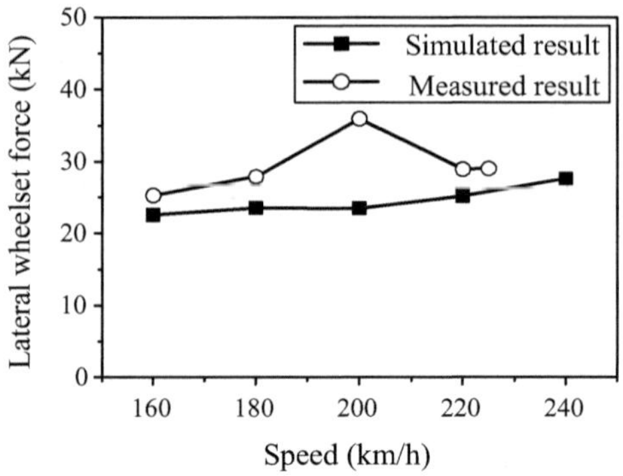

Fig. 6.17 Lateral wheelset force versus speed

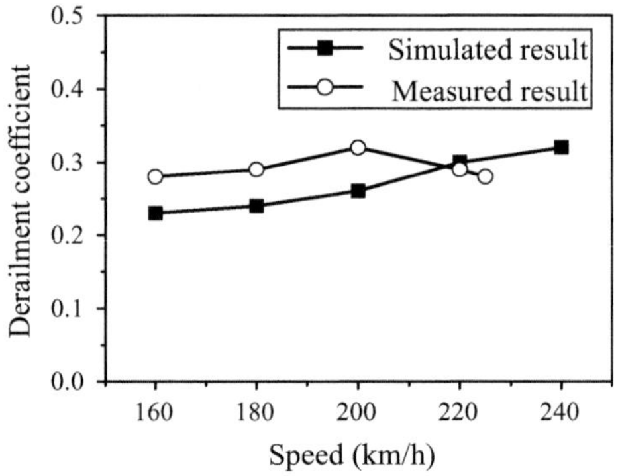

Fig. 6.18 Derailment coefficient versus speed

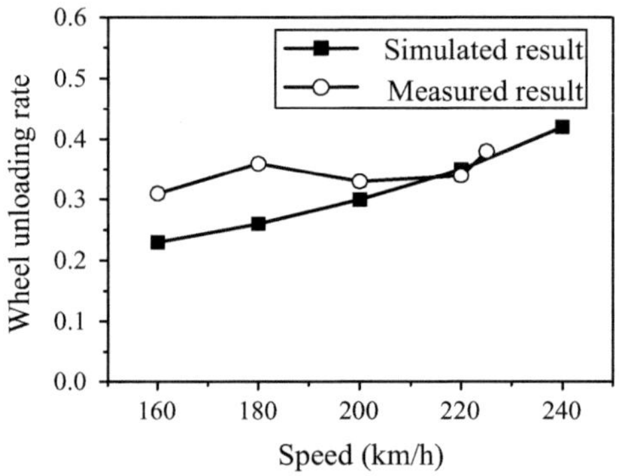

Fig. 6.19 Wheel unloading ratio versus speed

especially for the ride comfort indexes (see Fig. 6.20). The discrepancies between the simulated and the measured results only appear at some places, e.g., at the speed of 200 km/h for the lateral wheelset force and at the speeds of 160 and 180 km/h for the wheel unloading ratio. Consequently, the vehicle–track spatially coupled

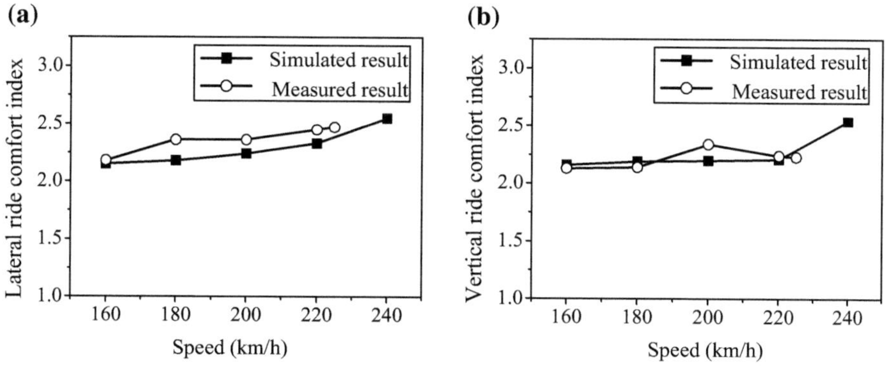

Fig. 6.20 Ride comfort indexes versus speed: **a** lateral, and **b** vertical

dynamics model and the corresponding software implementation could be applied to reliably analyze the wheel–rail dynamic interaction and dynamic behavior of vehicles under high-speed operation.

6.2.3 Validation by Derailment Experiment for Freight Train Running on Straight Line

At the early stage in the implementation of the Chinese railway speedup project (around 1997), a new problem presented, i.e., empty freight wagons derailed on some straight sections of the speedup lines. It threatened train operation safety seriously. Therefore, a freight train derailment experiment was performed in the straight section of the Beijing ring test line by the previous Ministry of Railways together with China Academy of Railway Sciences from September 18, 1999 to January 26, 2000. The dynamic responses of both the vehicle and the track systems were measured for a comprehensive evaluation of system dynamic performance.

The measured results indicated that apparent hunting motions occurred when the empty C_{62} wagons equipped with the Z8A three-piece bogies were running in the straight-line section at the speed of 78 km/h, which was higher than before. The hunting motion could cause a very drastic fluctuation of the lateral wheel–rail force, such as the measurement result shown in Fig. 6.21a with the maximum values beyond 50 kN. The vehicle–track spatially coupled dynamics model was adopted to calculate the dynamic responses under the same conditions as those in the test. The test track conditions included: 60 kg/m rails, concrete sleepers, and a common ballast bed. The track irregularity for such line might be similar to that of FRA 5th track class, and thus this irregularity spectrum was used in the calculation. Time history of the theoretically simulated lateral wheel–rail force is displayed in Fig. 6.21b. Similar hunting motion can be found in the theoretical result with the

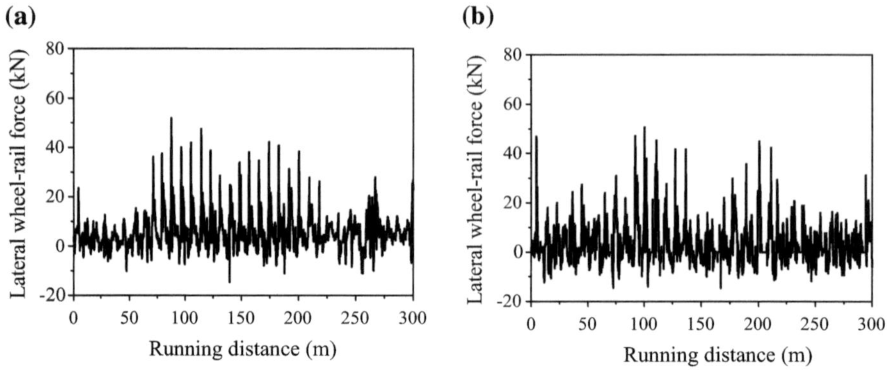

Fig. 6.21 Lateral wheel–rail force of empty C_{62} wagon running at 78 km/h on straight line: **a** measured result, and **b** simulated result

maximum lateral wheel–rail force up to 50 kN, agreeing well with the measured result.

The corresponding vertical wheel–rail force responses from measurement and calculation are shown, respectively, in Fig. 6.22. The measured vertical wheel–rail force varies between 5 and 80 kN, which is close to the simulated result varying in the range of 0–70 kN. In addition, both the measured and the simulated results reveal that serious wheel unloading occurs, which brings about a huge possibility of the derailment, seriously threatening vehicle operation safety.

It can be concluded that the vertical and lateral wheel–rail force responses obtained by the theoretical calculation using the vehicle–track spatially coupled dynamics model agree very well with those from the field tests. It is also inevitable that some discrepancies exist between the theoretical results and the measured ones due to too many complicated external influencing factors, such as the differences in

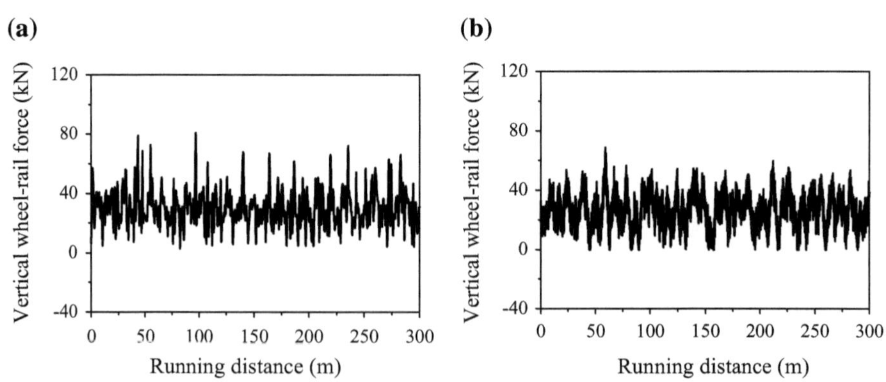

Fig. 6.22 Vertical wheel–rail force of empty C_{62} wagon running at 78 km/h on straight line: **a** measured result, and **b** simulated result

the track irregularities and the wear of the vehicle components, causing the deviations of the actual parameters from the simulated ones. However, both the theoretical results and the measured results were able to reveal the same essence, namely, drastic hunting motions would happen to the empty C_{62} wagons with the Z8A three-piece bogies running on straight lines at the speed of 78 km/h.

6.2.4 Experimental Validation by Wheel–Rail Dynamic Interaction Test on a Small Radius Curve of Mountain Railway

This subsection is on the validation of the simulated dynamic responses of track structure by a field test of wheel–rail dynamic interaction on a small radius curve in a mountain area of the Chengdu–Chongqing line. The test was carried out by the author's team of the Train and Track Research Institute at Southwest Jiaotong University (specified in Sect. 5.3.3 of Chap. 5).

The curved line had a radius of only 287 m, the length of the transition curve was 70 m, the superelevation of the outer rail was 125 mm, and the track gauge was widened by 15 mm. Besides, half of the curved track was equipped with wooden sleepers, while the other half with concrete sleepers of IIIb-C type. Here, the measured results at the concrete sleeper's section are used for the validation analysis.

The rail displacement, the sleeper displacement, and the track structure accelerations were measured and simulated, respectively, when the test train went through the curve at the speed range of 49.5–68.5 km/h. Their maximum values are compared in Tables 6.4, 6.5, and 6.6.

The results listed in Tables 6.4, 6.5 and 6.6 show that the theoretically simulated track dynamic responses under the action of the locomotive, the freight wagon, and the passenger car agree well with the measured results. For example, the lateral rail displacements, as one of the significant indexes of the lateral wheel–rail interaction in small radius curves, are 3.45 mm and 3.37 mm for the field test and for the theoretical calculation, respectively; the sleeper vertical displacements are 1.83 mm for test and 2.06 mm for calculation. Besides, the simulated accelerations of the rail and the

Table 6.4 Maximum values of measured and simulated rail displacements (unit: mm)

Vehicle type	Vertical displacement		Lateral displacement		Dynamic gauge widening	
	Measured	Simulated	Measured	Simulated	Measured	Simulated
Locomotive	2.08	2.55	3.45	3.37	4.98	4.10
Freight wagon	1.72	1.82	3.04	2.31	2.78	3.38
Passenger car	1.63	1.46	1.99	1.59	1.92	1.95

Table 6.5 Maximum values of measured and simulated sleeper displacements (unit: mm)

Vehicle type	Vertical displacement		Lateral displacement	
	Measured	Simulated	Measured	Simulated
Locomotive	1.83	2.06	0.49	0.68
Freight wagon	1.41	1.40	0.38	0.36
Passenger car	1.29	1.23	0.24	0.33

Table 6.6 Maximum values of measured and simulated track vertical accelerations (unit: *g*)

Vehicle type	Rail		Sleeper	
	Measured	Simulated	Measured	Simulated
Locomotive	58.37	56.83	4.12	3.82
Freight wagon	42.51	37.34	2.98	3.21
Passenger car	29.96	18.64	2.04	2.99

sleeper agree also well with the measured results. In general, the vehicle–track spatially coupled dynamics model is capable of analyzing the track structure vibration performance (including the lateral dynamic performance) in various operation conditions with good accuracy. This is because the model adopts the dynamic wheel–rail coupling relationship and fully considers the track structure vibrations.

6.2.5 Conclusions

The aforementioned experimental validation analyses demonstrate that the dynamic performance indexes of both the vehicle and the track simulated by using the vehicle–track spatially coupled dynamics model could agree well with the field test results. The corresponding simulation software, TTISIM, could be applied not only to the analysis of dynamic safety and ride comfort of vehicles running on different track conditions at various speeds, especially to investigating the lateral wheel–rail dynamic interaction performances, but also to dynamic performance evaluation of the track structure under various conditions.

6.3 Experimental Validation of the Train–Track Spatially Coupled Dynamics Model

In this section, the train–track spatially coupled dynamics model are validated by comparing simulated results with those from the field measurements, with special emphasis on the validation of the longitudinal dynamic interaction between vehicles. The detailed validation under several conditions can be found in [7] and are briefly presented in the following.

6.3.1 Validation by Measured Coupler Longitudinal Forces of a Heavy-Haul Combined Train Under Braking Conditions

Operation tests of a 20,000 t heavy-haul combined train with multiple locomotives under braking conditions were performed [8] and the measured results are employed here for validation of our theoretical model. The tests were carried out on the Datong–Qinhuangdao heavy-haul railway line in 2004. The train consisted of 4 locomotives (SS_4) and 204 freight wagons (C_{80}) with the formation of $SS_4 + 51 \times C_{80} + SS_4 + 51 \times C_{80} + SS_4 + 51 \times C_{80} + SS_4 + 51 \times C_{80}$. Field tests in the full service and emergency braking conditions were conducted on a track section with a slope gradient of −12‰. The initial speed at the brake application was 75 km/h.

Figure 6.23 shows both the simulated and the measured results of the maximum coupler forces along the train under two different operation modes: (a) emergency braking and (b) full-service braking. It is clear from Fig. 6.23a that the varying trend of the simulated result broadly coincides with that of the measured result for the emergency braking condition. Similarly, for the full-service braking condition, the simulated coupler force distribution is also generally consistent with the measured result, as shown in Fig. 6.23b.

6.3.2 Validation by Tested Train Dynamic Characteristics Under Electric Braking Conditions

In June 2008, Taiyuan Railway Administration of Chinese Railways organized an operational safety test of a 10,000 t heavy-haul train. The test was carried out in straight sections of the North Tongpu railway and the Datong–Qinhuangdao heavy-haul railway [9]. The purpose of the test was to verify the working stability of the coupler and the draft gear system when they were subjected to a large coupler compression force. By applying electric braking at different levels, the compression force of the coupler system increased gradually. The train formation is shown in Fig. 6.24 and the focus of the measurement was on the front, the middle and the rear couplers. An instrumented wheelset was equipped as the third wheelset of the third locomotive.

Comparisons between the measured and simulated results of the coupler swing angles under the electric braking condition are made and shown in Fig. 6.25 [7]. The measured results from Fig. 6.25a show that the couplers sway near the central

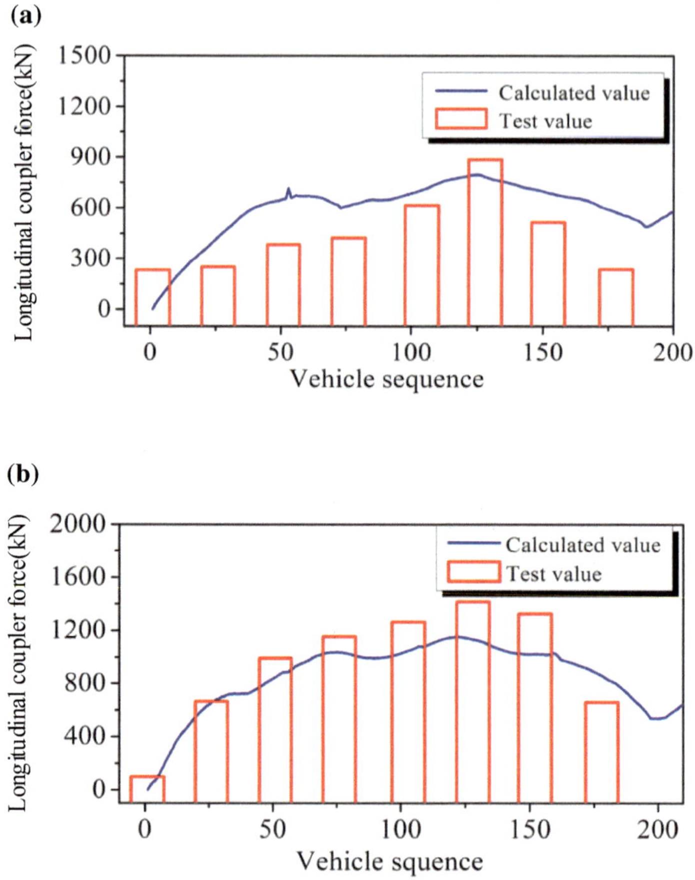

Fig. 6.23 Comparison of measured and simulated longitudinal coupler forces of 20,000 t combined train under braking conditions: **a** emergency braking, and **b** full service braking

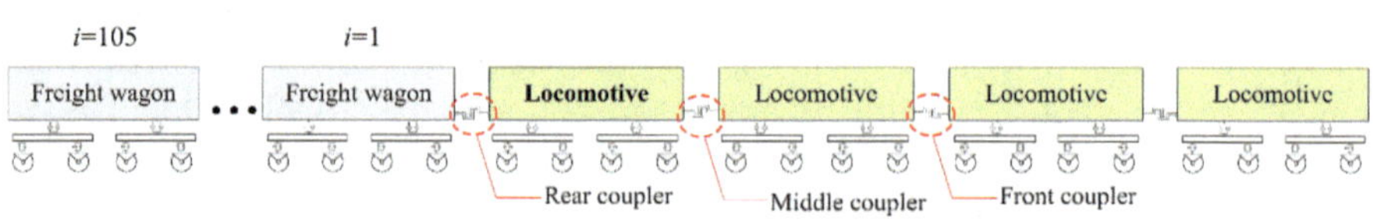

Fig. 6.24 Formation of the 10,000 t heavy-haul train [7]

position under pulling coupler force, and the swing amplitudes decrease with the increase of the pulling force. Conversely, the couplers will tilt to one side as the coupler compression force increases to a certain magnitude. The calculation results in Fig. 6.25b indicate that the variation of the coupler swing angle follows similar trend as the measured results in Fig. 6.25a. It can be found that the swing angles of

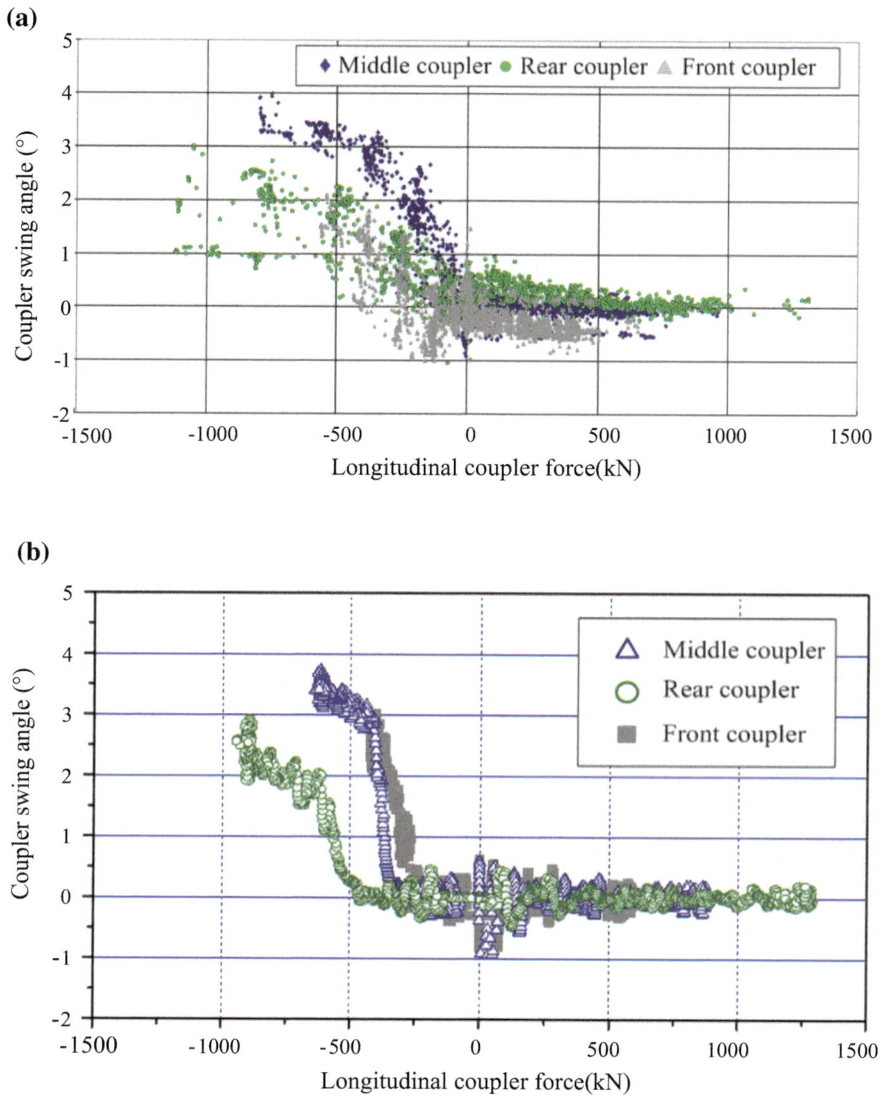

Fig. 6.25 Comparison of **a** measured and **b** simulated coupler swing angles

the middle and the rear couplers are slightly below 4° and 3°, respectively. Overall, the simulated and the measured results coincide well with each other.

The lateral wheelset forces obtained from the field test and the simulation are compared and displayed in Fig. 6.26 [7]. The results reveal that the lateral wheelset forces present statistically a unidirectional increase when the compression coupler force increases. The measured wheelset force is slightly lower than the simulated

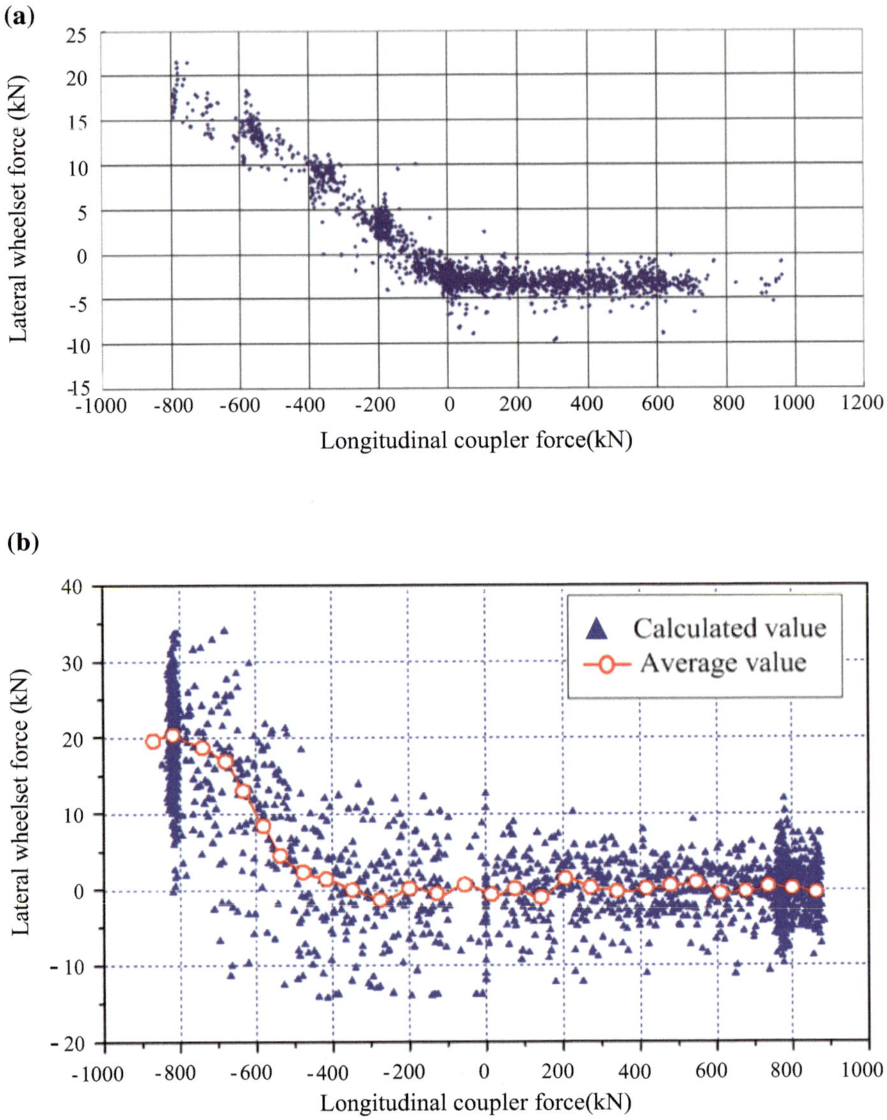

Fig. 6.26 Comparison of **a** measured and **b** simulated lateral wheelset forces of the locomotive under the electric braking condition

value, which is most likely caused by the differences in the data processing method and the track irregularities. It is worth pointing out that the measured data are closed to the average values of the simulated results.

6.3.3 Validation by Measured Results of Heavy-Haul Train Curving Performance

For the validation of wheel–rail dynamic interaction of heavy-haul trains passing on curves, the measured curving performance of a train composed of the C_{70} freight wagons in Ref. [7] is employed here. The simulated results are compared with the measured results under the same operation condition. The radius of curve is 500 m. Due to the wheel–rail wear, the actual profiles of the outer and the inner rails present asymmetry as shown in Fig. 6.27. To reflect the actual track states, the measured rail profiles are used in the simulation. The train runs at a constant speed of 60 km/h.

Comparison between the simulated and measured lateral wheel–rail forces, the vertical wheel–rail forces and the derailment coefficients are displayed in Figs. 6.28, 6.29, and 6.30. It is seen that the variation of each simulated dynamics index agrees well with that of the measured result. The measured maximum values of the lateral wheel–rail force, the vertical wheel–rail force and the derailment coefficient on the outer rail are 65 kN, 127 kN and 0.52, respectively. The correspondingly simulated peak values of these indexes are 59 kN, 142 kN and 0.48, which are close to the measured values. For the wheel–rail dynamic indexes on the inner rail, the measured amplitudes of the lateral wheel–rail force, the vertical wheel–rail force and the derailment coefficient are 34 kN, 133 kN, and 0.33, respectively, while the corresponding simulated values are 32 kN, 133 kN, and 0.29, indicating again a good agreement.

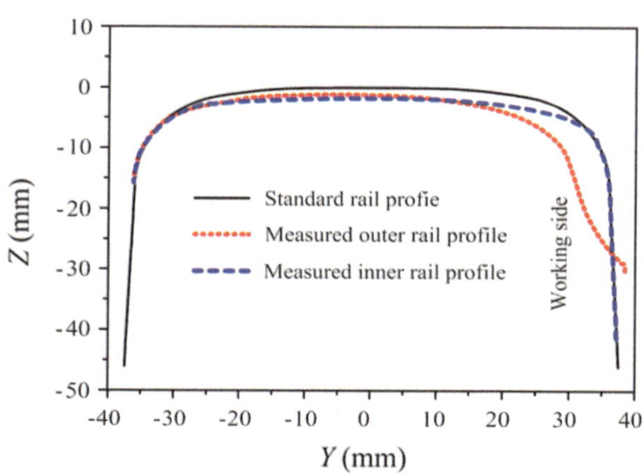

Fig. 6.27 Measured rail profiles on the test curved track

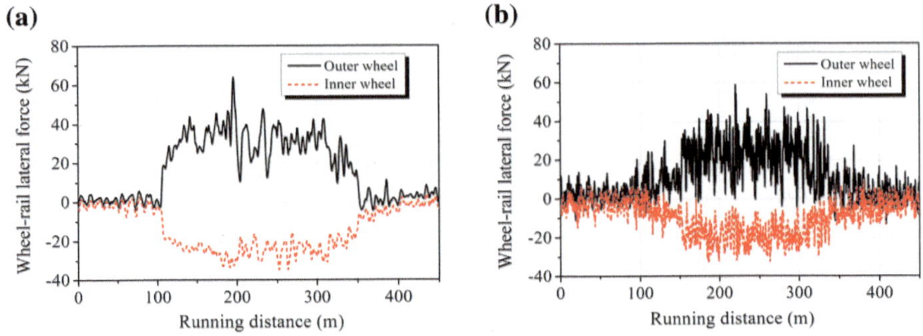

Fig. 6.28 Comparison of **a** measured and **b** simulated lateral wheel–rail forces

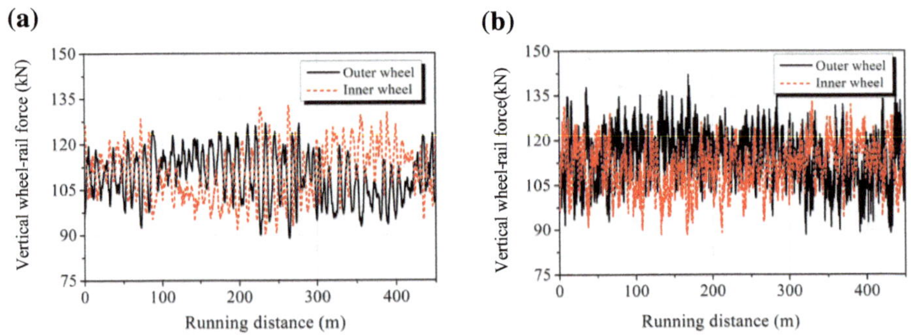

Fig. 6.29 Comparison of **a** measured and **b** simulated vertical wheel–rail forces

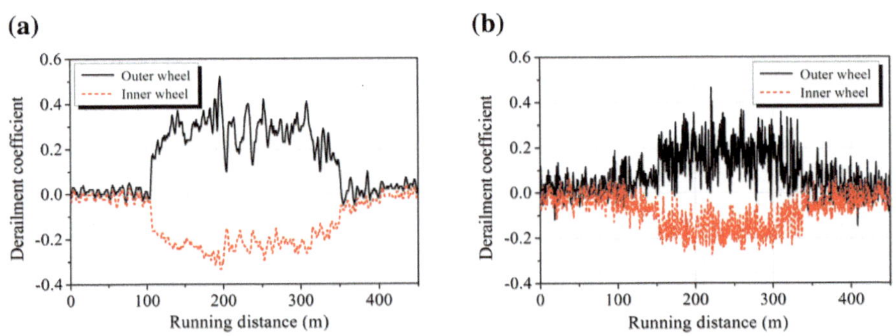

Fig. 6.30 Comparison of **a** measured and **b** simulated derailment coefficients

6.3.4 Conclusions

Consequently, the reliability of the train–track spatially coupled dynamics model has been validated in good consistence between the simulated and the measured results for different operation conditions of different trains. This theoretical dynamics model can then be used in the analysis and evaluation of the dynamic interactions between vehicles and tracks, especially for longitudinal dynamic interaction between vehicles in long heavy-haul trains.

It can be predicted that the experimentally validated vehicle–track coupled dynamics theory and its corresponding simulation software would have a broad application prospect in practical railway engineering.

References

1. Wang PQ. Influence of large axle vehicle on vertical track dynamics. Roll Stock. 1984;11:8–14 (in Chinese).
2. Zeng SG. Dynamic test of heavy-haul track structure. J China Railw Soc. 1988;10(2):66–77 (in Chinese).
3. Zhai WM, Wang KY, Lin JH. Modelling and experiment of railway ballast vibrations. J Sound Vib. 2004;270(4–5):673–83.
4. Zhai WM, True H. Vehicle–track dynamics on a ramp and on the bridge: simulation and measurements. Veh Syst Dyn. 1999;33(Suppl):604–15.
5. Zhai WM, Cai CB, Wang QC, Lu ZW, Wu XS. Dynamic effects of vehicles on tracks in the case of raising train speed. J Rail Rapid Transit. 2001;215(2):125–35.
6. Knothe K, Grassie SL. Modeling of railway track and vehicle/track interaction at high frequencies. Veh Syst Dyn. 1993;22(3/4):209–62.
7. Liu PF, Zhai WM, Wang KY. Establishment and verification of three-dimensional dynamic model for heavy-haul train–track coupled system. Veh Syst Dyn. 2016;54(11):1511–37.
8. Chen L. Introduction to the evaluation of railway wagon performance. Beijing: China Railway Publishing House; 2010 (in Chinese).
9. Yang JJ. On study of the carrying characteristic and structure adaptability of the locomotive coupler–draft gear based on the "1+1" 20kt built-up train mode. Ph.D. thesis. Beijing: Beijing Jiaotong University; 2009 (in Chinese).

Chapter 7
Computational Comparison of Vehicle–Track Coupled Dynamics and Vehicle System Dynamics

Abstract It is of prime importance to ascertain the specific differences in the computational results between the vehicle–track coupled dynamics model (hereinafter referred to as the "coupled model") and the traditional vehicle system dynamics model (hereinafter referred to as the "traditional model") for further understanding the necessity of research on the theory of vehicle–track coupled dynamics. It can also provide the basis for defining the application scopes of both the traditional model and the coupled model. This chapter will analyze and compare the computational results of the two models from three aspects: railway vehicle nonlinear hunting stability, ride comfort, and curving performance.

7.1 Comparison of Computational Results on Vehicle Hunting Stability

7.1.1 Numerical Calculation Method of Vehicle Nonlinear Hunting Stability

1. *Vehicle nonlinear stability and critical hunting speed*

The evaluation index of railway vehicle nonlinear hunting stability is the critical speed. When vehicle running speed is less than the critical speed, the vehicle system stays dynamically stable; once it exceeds the critical speed, the vehicle system will become unstable. It indicates that the critical hunting speed is a state parameter characterizing the vibration state of convergence or instability of the vehicle system.

For a nonlinear vehicle system running on a straight track, the lateral stability characteristic could be described as an "S" shape limit cycle curve, as shown in Fig. 7.1 [1], where the solid line and the dashed line represent the stable and the unstable limit cycles, respectively, and the horizontal axis represents the equilibrium position of the system. When the train speed v is less than v_D, the system vibration will stabilize to the equilibrium position for any external disturbance. When the train speed v is larger than v_B, a stable limit cycle vibration with large amplitude will be observed no matter what amplitude of the external disturbance is.

W. Zhai, *Vehicle–Track Coupled Dynamics*,
https://doi.org/10.1007/978-981-32-9283-3_7

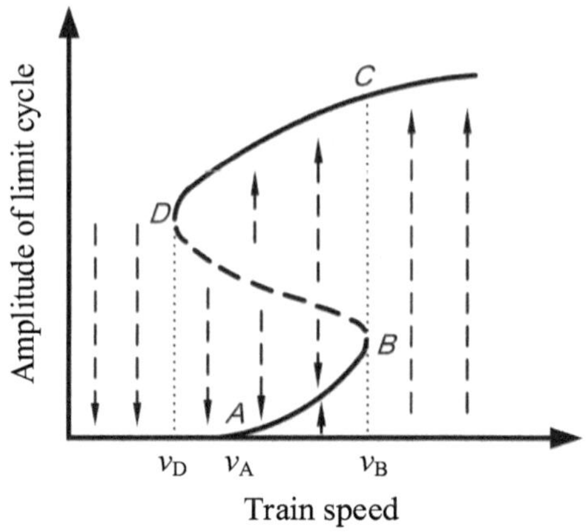

Fig. 7.1 Limit cycle diagram of the nonlinear vehicle system

In the interval $v_D < v < v_B$, the stability of the vehicle system depends on the amplitude of the external disturbance. With the continuous increase of train speed, the vehicle system will firstly exhibit the limit cycle vibration at point *D* when subjected to a large initial disturbance, whereas the limit cycle vibration at point *A* will first occur when subjected to a small initial disturbance. The corresponding speed v_D at point *D* is called the nonlinear critical hunting speed and v_A at point *A* is called the linear critical hunting speed. For the investigation of the stability of railway vehicle system in field practice, the nonlinear critical hunting speed v_D is generally considered.

2. *Numerical method to determine the critical hunting speed*

In consideration of the fact that the vehicle–track coupled dynamics theory deals with a very large-scale dynamic system involving strong nonlinearities, it is impossible to obtain the system stability with any analytical method. Therefore, a direct numerical method based on the characteristic of time histories of vehicle components is proposed to determine the critical hunting speed of a vehicle running on a track. The simple fast time-integration method (i.e., the Zhai method) is adopted to numerically solve the system equations and to obtain time responses of vehicle components due to initial disturbance. The time histories of lateral displacements of vehicle components are monitored as the train speed increases step-by-step from very low value. If the time responses of the lateral displacements induced by large initial disturbance decay with time and regain their equilibrium states at one speed, the vehicle system is stable at this speed. Then another calculation is needed to increase the speed, until the critical situation occurs where at least one of the time histories of the lateral displacements does not decay to its equilibrium state and sustains variation almost with the same amplitude. According to Fig. 7.1, the speed at this critical situation is regarded as the nonlinear critical hunting speed. It should be noted that a little decrease of the speed from the nonlinear critical hunting speed will result in the occurrence of the stable equilibrium state of the system. Therefore very small step of the speed should be used in the calculation when the speed approaches the nonlinear critical hunting speed.

Due to the occurrence of hunting instability, the motion of each rigid body of the vehicle system shows the same form of bifurcation and limit cycle characteristics, which means the Höpf bifurcation and saddle-node bifurcation will simultaneously appear on the motion of a rigid body under the same speed [2]. Therefore, the critical speed of the vehicle system can be determined according to the time history and phase plane graph of a rigid body, for instance, the first wheelset of the vehicle.

It is very important to use a large initial lateral disturbance so as to ensure that the large vibration of the vehicle system can be excited and the nonlinear critical hunting speed is ascertained. In order to inflict a large initial lateral disturbance to a railway vehicle system, it is suggested to use the actual track irregularity spectra of the railway line on which the analyzed vehicle will operate. Once the lateral vibration of the vehicle is completely induced, the track irregularities should be cut off in the calculation so that the running vehicle can vibrate freely without any external excitation [3]. This initial disturbance obtained by this manner is larger enough and closer to the real operating condition than the commonly used disturbance using a small lateral displacement given to the leading wheelset. It is noticeable that the initial condition may also affect the dynamic behavior of the nonlinear vehicle system. So, the length of the irregular track section used in the calculation should be sufficiently long so as to be able to make each component of the vehicle system be fully excited.

3. *Numerical example*

The computation of the critical hunting speed of a typical Chinese high-speed vehicle is illustrated here as an example. The vehicle first runs on an elastic track with random track irregularity (200 m long), and then moves on to the straight track without irregularity. The lateral motion of the first wheelset is investigated. By increasing the vehicle speed continuously, it can be found that the lateral displacement of the wheelset at speed of 361 km/h gradually attenuates to the equilibrium position (zero position) as shown in Fig. 7.2a, while it starts the periodic motion with a constant amplitude as the speed increases to 362 km/h as shown in Fig. 7.2b. In the corresponding phase plane graph (Fig. 7.3), the phase locus of the lateral velocity of the wheelset at speed 361 km/h converges to the singular point of the limit cycle in Fig. 7.3a. While at the vehicle speed of 362 km/h, the wheelset lateral velocity no longer attenuates to zero but tends to achieve a stable limit cycle as shown in Fig. 7.3b. Thus, 362 km/h is the actual critical hunting speed of this vehicle.

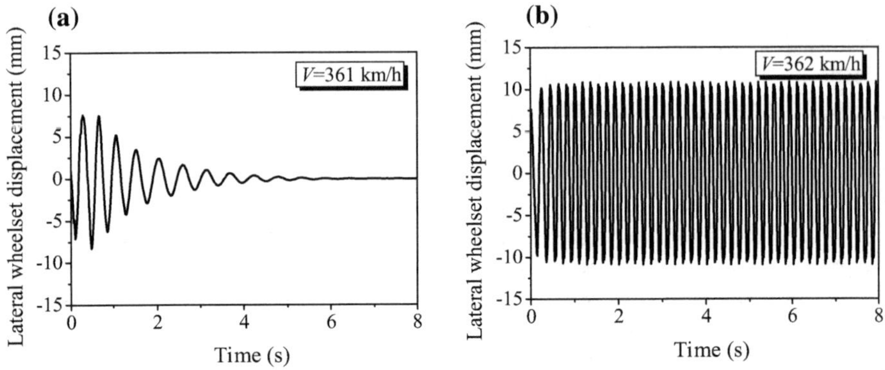

Fig. 7.2 Time history of wheelset lateral vibration near the critical state of vehicle instability: **a** equilibrium solution, and **b** periodic solution

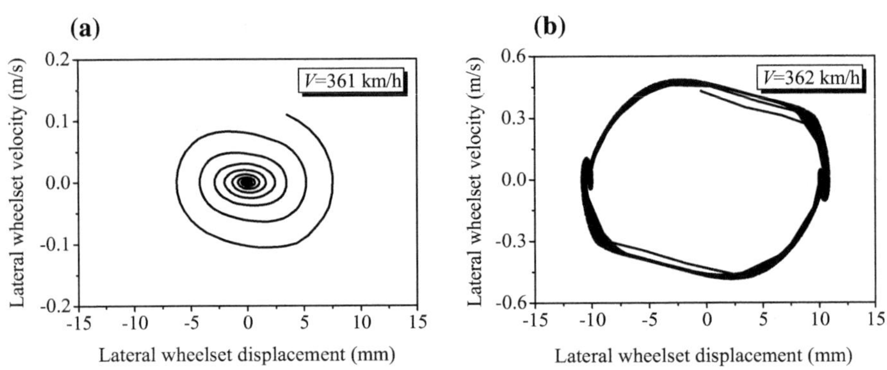

Fig. 7.3 Phase plane graph of wheelset lateral vibration near the critical state of vehicle instability: **a** equilibrium solution, and **b** periodic solution

7.1.2 Comparison of Calculated Critical Speeds Between the Coupled Model and the Traditional Model

Based on the above-mentioned method for computing the nonlinear critical speed, the theories of vehicle–track coupled dynamics and the traditional vehicle dynamics are applied to determine the nonlinear critical speeds of four typical vehicles employed in China, a high-speed passenger car, a freight car equipped with traditional Z8A bogie, a Chinese speedup locomotive, and a metro power car. Table 7.1 compares the results.

It can be seen from Table 7.1 that:

Table 7.1 Comparison of computed critical speeds based on the coupled model and the traditional model (unit: km/h)

Vehicle type	Coupled model	Traditional model
High-speed passenger car	362	391
Loaded freight car	113	118
Speedup locomotive	162	176
Metro car on floating slab track	188	210

(1) For the high-speed passenger car, the critical speed calculated through the coupled model is 362 km/h, while that of the traditional model is 391 km/h which is 8.01% larger.
(2) For the loaded freight car, the calculated critical speed of the traditional model is 118 km/h, which is 4.42% larger than that of the coupled model, 113 km/h.
(3) For the speedup locomotive, the calculated critical speed of the coupled model is 162 km/h while that of the traditional model is 176 km/h, which shows an increase of 8.64%.
(4) For the metro car running on floating slab track, the calculated critical speed of the coupled model is 210 km/h while that of the traditional model is 188 km/h, which shows an increase of 11.70%.

It can be further concluded that lower critical speeds are obtained by the coupled model with comparison to the case of the traditional model. The reason for the reduction in critical speeds of vehicles calculated by the vehicle–track coupled model is due to the consideration of track elasticity and damping in the coupled dynamics model. Since the rail has a lateral degree of freedom in the coupled model, the lateral vibrations of the left and right rails relieve the lateral constraint on wheelsets. The wheelsets are, therefore, prone to hunting, which subsequently leads to a reduction in the critical speed with respect to the traditional model accounting for rigid track structure. Field experience concurs with this conclusion.

7.1.3 Summary

(1) The critical speed of vehicle calculated by the traditional model is higher than that of the coupled model. The difference between these results by the two models is around 10%, which is consistent with the comparison results reported in Refs. [3, 4].
(2) Although there is no significant difference between the two models' simulation results, the traditional model exaggerates the lateral constraint of rails on wheelsets due to the assumption of the rigid track. The critical speed of the vehicle system is thus overestimated. The calculation result of the traditional model, therefore, has less safety margin, which must be considered in the actual design.

7.2 Comparison of Calculation Results on Vehicle Ride Comfort

To evaluate the simulation results of vehicle ride comfort based on the two models, three types of commonly used Chinese vehicles are considered, i.e., the high-speed passenger car, the speedup locomotive, and the fast freight car. The operation speeds of these vehicles are 300 km/h, 160 km/h, and 120 km/h, respectively. The adopted PSDs of random track irregularities are the low disturbance spectrum of German high-speed railway for the high-speed passenger car and the track spectrum of Chinese speedup trunk line for the speedup locomotive and the freight car.

Tables 7.2 and 7.3 compare the simulation results between the coupled model and the traditional model, in terms of the maximum lateral and vertical vibration accelerations of the car body (at the floor where 1 m away from the center plate in the lateral direction) and the corresponding ride comfort indexes.

For a detailed analysis of the difference in car body vibration simulation between the two models, the time histories of car body acceleration of the fast freight car are illustrated, as shown in Figs. 7.4 and 7.5.

The above simulation results indicate that the obtained vehicle vibration characteristics based on the coupled model and the traditional model show good agreement both in the waveform and the amplitudes, and both for vertical and lateral vibrations. The difference between the two models in lateral vibration is slightly larger than that in vertical vibration, and the ride comfort index calculated by the traditional model is a little higher compared with the coupled model.

Table 7.2 Car body accelerations between the coupled model and the traditional model (unit: m/s^2)

Vehicle type		Coupled model	Traditional model
High-speed passenger car	Lateral	0.49	0.50
	Vertical	0.73	0.76
Speedup locomotive	Lateral	1.56	1.60
	Vertical	1.07	1.08
Fast freight car	Lateral	4.21	4.59
	Vertical	4.51	4.54

Table 7.3 Ride comfort indexes between the coupled model and the traditional model

Vehicle type		Coupled model	Traditional model
High-speed passenger car	Lateral	2.33	2.36
	Lateral	2.61	2.63
Speedup locomotive	Lateral	3.19	3.20
	Vertical	2.90	2.91
Fast freight car	Lateral	3.62	3.64
	Vertical	3.84	3.87

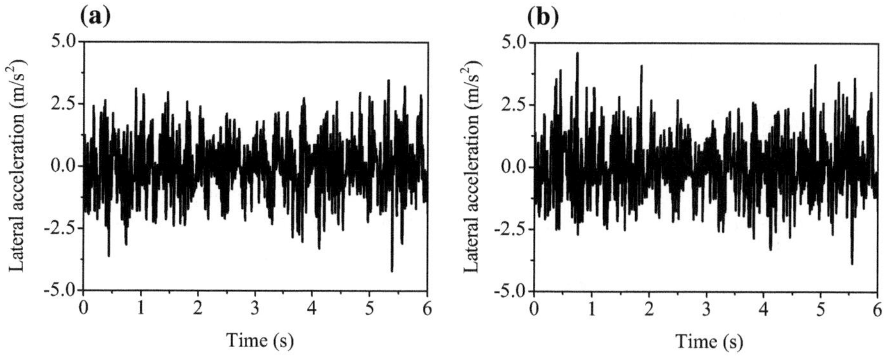

Fig. 7.4 Lateral car body accelerations of the fast freight car obtained from: **a** the coupled model, and **b** the traditional model

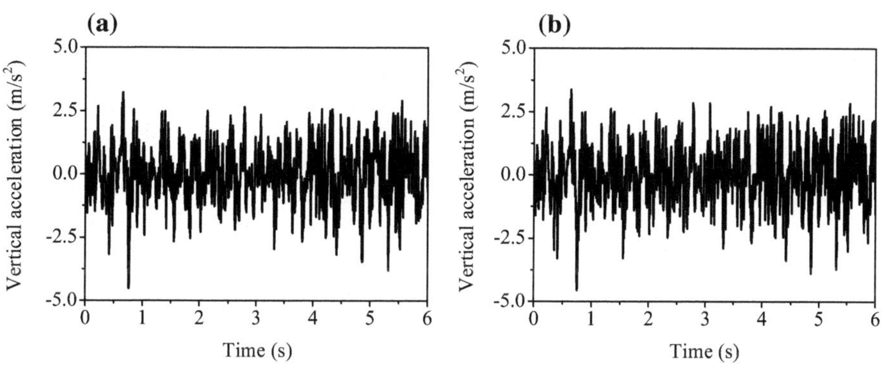

Fig. 7.5 Vertical car body accelerations of the fast freight car obtained from: **a** the coupled model, and **b** the traditional model

The good consistency of the results calculated by these two models is owing to the effective insulation of the high-frequency vibration via the primary and secondary suspensions. Influence of track vibration on car body vibration above the secondary suspension is significantly attenuated as a result.

On this account, both the vehicle–track coupled dynamics model and the traditional vehicle dynamics model can be adopted for analyzing and evaluating the ride comfort of vehicle system. Since the track vibration is omitted in the traditional model, it is more simple and efficient. However, to take account of the vibrations of the vehicle components below the secondary suspension (i.e. the bogie frames, wheelsets, traction motors, etc.), the results of the two models are still different [5].

7.3 Comparison of Calculation Results on Curving Performance

To dctcrminc thc differcnce in vehicle curving performance between the coupled model and the traditional model, the dynamic responses of two typical cases that a vehicle passes through a small radius (R = 250 m) curved track at low speed and a large radius (R = 6000 m) curved track at high speed are investigated. The comparison could provide a basis for dynamics analysis and safety assessment of railway vehicle curving performance.

7.3.1 Comparison of Vehicle Passing Through a Small Radius Curved Track at Low Speed

In the small radius curve case, 250 m is selected as the radius of the curve without loss of generality. The transition curve length is 80 m and the superelevation of the outer rail is 120 mm. A traditional Chinese passenger train with a running speed of 60 km/h is used as the train platform. The fifth-grade American track spectrum is set as the random track irregularity, which can simulate the track state with small radius curves in most cases.

The simulation results are compared in terms of the lateral wheel–rail force, the vertical wheel–rail force, the dynamic gauge widening and other curving safety indices.

1. *Lateral wheel–rail force*

Figure 7.6 shows the time histories of the lateral wheel–rail forces obtained by the two models. It is obvious that the result of the traditional model is larger than that of

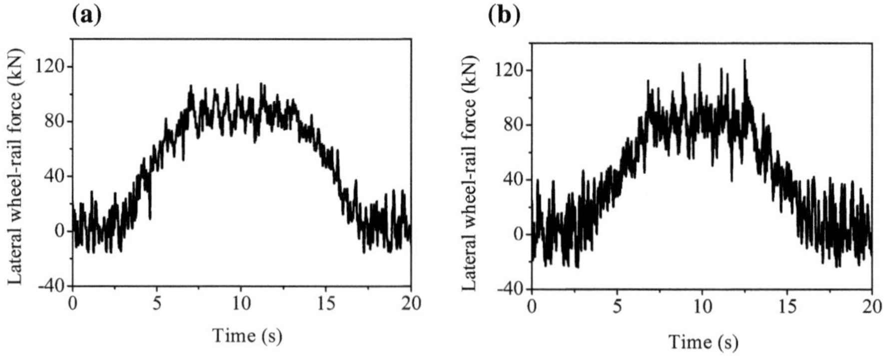

Fig. 7.6 Comparison of the calculated lateral wheel–rail forces (outer rail) between **a** the coupled model and **b** the traditional model

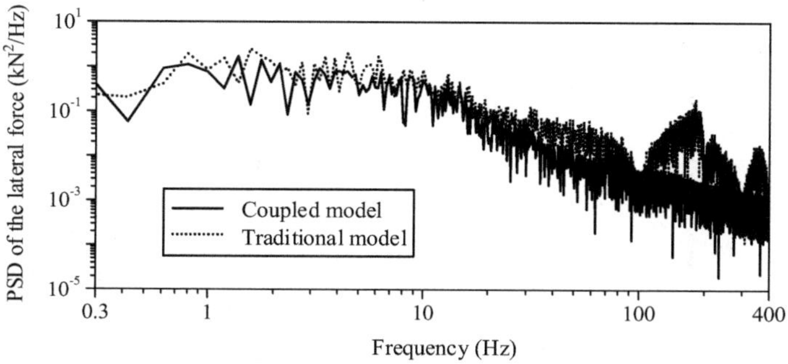

Fig. 7.7 Comparison of the PSDs of the lateral wheel–rail forces (outer rail) between two models

the coupled model, especially on the circular curve. The maximum value from the traditional model is 127.49 kN, which is 18.2% larger than the value 107.82 kN from the coupled model.

The PSDs of the lateral wheel–rail forces are shown in Fig. 7.7. The PSDs of the lateral wheel–rail force obtained from the two models show good agreement when the frequency is below 20 Hz. For frequency larger than 20 Hz, a significant difference occurs. Particularly at a frequency near 200 Hz, the result obtained from the traditional model is much larger. The reason lies in the consideration of track stiffness and damping effect in the coupled model, which can absorb the vibration with medium–high frequencies.

2. *Vertical wheel–rail force*

Figure 7.8 illustrates the time histories of the vertical wheel–rail forces. Compared with the coupled model, the result from the traditional model is much larger. The maximum force from the coupled model is 123.15 kN, while that from the traditional model increases by 10.1% to 135.54 kN.

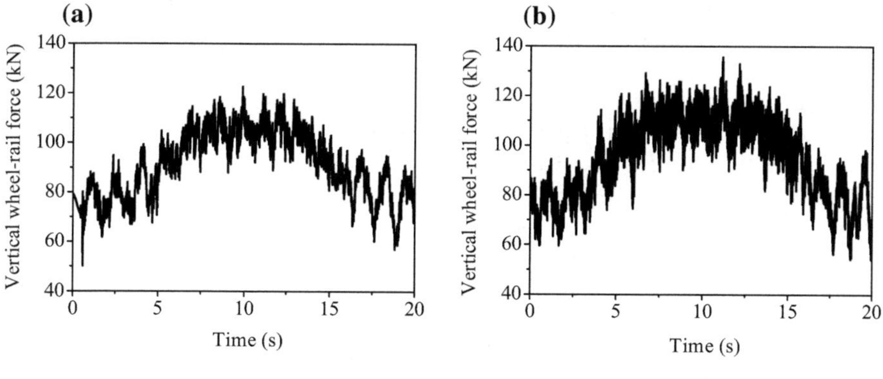

Fig. 7.8 Comparison of the calculated vertical wheel–rail force (outer rail) between **a** the coupled model and **b** the traditional model

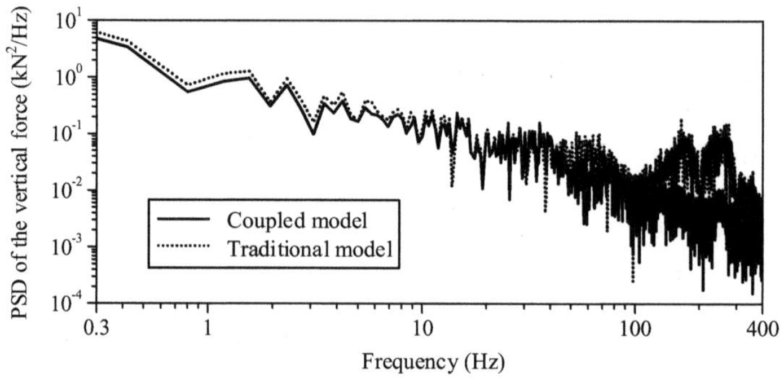

Fig. 7.9 Comparison of the PSDs of the vertical wheel–rail forces (outer rail) between two models

Figure 7.9 compares the corresponding PSDs of the vertical wheel–rail forces. There is no distinct difference between the two models in the frequency range below 40 Hz. With the increase of frequency, the result from the traditional model shows an enlarged tendency with respect to the coupled model, particularly in the frequency range higher than 100 Hz. These differences should be also caused by the track elasticity and damping effect considered in the coupled model.

3. *Dynamic gauge widening*

In practice, when railway vehicle passes through curved tracks, especially small radius curves, there will be elastic deformations and vibrations stimulated on track components due to the combined effect of vertical and lateral wheel–rail forces, including the torsion deformations of the rails (overturn) and the dynamic gauge widening (the difference of lateral vibration displacements between the left and right rails). These dynamic changes may pose threats to the running safety of vehicles. The traditional vehicle dynamics model is unavailable for dynamic deformations of track structures and their influence on the dynamic wheel–rail interaction due to the lack of consideration of track system vibration.

Taking the dynamic gauge widening as example, the calculated result of the vehicle–track coupled dynamic model is illustrated in Fig. 7.10. It shows that the gauge varies dynamically during the whole process of the vehicle passing through the curve. The maximum value of the dynamic gauge widening, in this case, is 1.32 mm. Relevant field test results can be found in Sect. 5.3.3 of Chap. 5.

Results in Ref. [6] indicated that the gauge widening had a large influence on the wheel–rail contact geometry. That's why the traditional model without consideration of rail dynamic deformation and gauge widening effect shows a distinct difference in simulating the wheel–rail contact geometry compared with the coupled model. The calculated wheel–rail forces and creep forces are therefore different. It provides another explanation of the difference in the calculated wheel–rail dynamic forces between the two models.

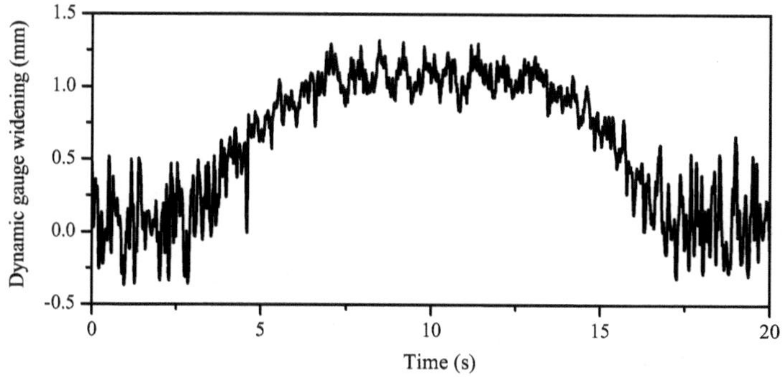

Fig. 7.10 Variation of dynamic gauge widening calculated by the coupled model

The above results are very similar to those reported in Ref. [7], where another small radius curve is analyzed ($R = 287$ m).

7.3.2 Comparison of Vehicle Passing Through a Large Radius Curved Track at High Speed

In calculation, the radius of the curve is 6000 m. The transit curve length is 180 m and the superelevation of the outer rail is 70 mm. A high-speed motor car named "China star" is simulated with a speed of 250 km/h. The low disturbance spectrum of German high-speed railway is used to describe the random track irregularity.

Several curving performance indices calculated by the coupled model and the traditional model are compared in Table 7.4. The lateral wheel–rail force, vertical wheel–rail force, and the lateral wheelset force from the traditional model are 28.03 kN, 180.79 kN, and 18.18 kN, respectively, which are 15%, 28.32%, and 15.28% larger than those from the coupled model. The calculated derailment

Table 7.4 Comparison of calculated dynamic indexes between the coupled model and the traditional model

Dynamic index	Coupled model	Traditional model
Lateral wheel–rail force (kN)	24.38	28.03
Vertical wheel–rail force (kN)	140.89	180.79
Wheelset lateral force (kN)	15.77	18.18
Derailment coefficient	0.29	0.36
Wheel unloading rate	0.31	0.46
Dynamic gauge widening (mm)	0.51	–[a]

[a]This index cannot be obtained from the traditional model

coefficient and wheel unloading rate from the traditional model are also larger than those from the coupled model. The dynamic gauge widening calculated by the coupled model is 0.51 mm, which is not available from the traditional model.

It can be concluded that for the case of a vehicle passing a large radius curve with a high speed, the calculated dynamics indices from the traditional model are much larger than those from the coupled model. The difference between the two models attains 15–20% and even 30%.

7.4 Conclusions

This chapter systematically demonstrates the differences between the computational results obtained with the vehicle–track coupled dynamics model and the traditional vehicle system dynamics model. From the perspectives of vehicle hunting stability, ride comfort, and curving performances, some conclusions can be obtained:

(1) The nonlinear hunting critical speed calculated by the traditional vehicle dynamics model is higher than that by the vehicle–track coupled dynamics model. The difference between the results of the two types of models is around 10%. Although the difference is not huge, the traditional vehicle dynamics overestimates the lateral stability of a vehicle running on an elastic track and therefore lead to less safety margin. This conclusion is very significant to the safety design of railway vehicles and should be noticed in the design stage.

(2) Little difference is observed in the ride comfort indices obtained by the traditional model and the coupled model due to the high insulation performance of the bogie suspensions. Therefore, both the vehicle–track coupled dynamics model and the traditional vehicle dynamics model can be used for analyzing and evaluating the ride comfort of railway vehicles under normal operational conditions. But if the focus is on the ride comfort of a vehicle passing imperfect track sections including bridge–embankment transitions, infrastructure settlement, and so on, the vehicle–track coupled dynamics model should be applied.

(3) Significant differences between the curving performances obtained from the traditional model and from the coupled model have been found for the cases considered in this chapter. The deviation of the wheel–rail dynamic forces is located within the range of 10–30%. The traditional vehicle dynamics model usually overestimates the wheel–rail dynamic interaction.

(4) The traditional model is only applicable to the analysis of the wheel–rail dynamic force in the low-frequency range. Due to the omitting of track vibration, elasticity and damping, the traditional model is not suitable for the medium–high frequency analysis of the wheel–rail interaction.

(5) The different results between the two models came from the applications of the different modeling option of track structure. The effects of the vibration of track components, and the elasticity and damping of the track system are considered in the vehicle–track coupled dynamics model. This modeling can reflect the

following two facts: (a) the track system can absorb part of the vibration energy induced by the wheel–rail interaction; (b) the dynamic gauge widening actually exists due to the lateral motion of rails, which can result in the change of the wheel–rail contact geometry and eventually influence the wheel–rail contact forces. However, these dynamic effects are not considered in the traditional vehicle dynamics model.

References

1. Hans T, Kaas PC. A bifurcation analysis of nonlinear oscillation in railway vehicles. In: Proceedings of the 8th IAVSD symposium. 1983. p. 320–9.
2. Wu P, Zeng J. A new method to determine linear and non-linear critical speed of the vehicle system. Rail Veh. 2000;38(5):1–4 (in Chinese).
3. Zhai WM, Wang KY. Lateral hunting stability of railway vehicles running on elastic track structures. J Comput Nonlinear Dyn ASME. 2010;5(4):041009-1 ~ 9.
4. Gialleonardo ED, Braghin F, Bruni S. The influence of track modelling options on the simulation of rail vehicle dynamics. J Sound Vib. 2012;331:4246–58.
5. Zhai WM. Vehicle-track coupled dynamics. 2nd ed. Beijing: China Railway Press; 2002 (in Chinese).
6. Wang KY, Zhai WM, Cai CB. Effect of the wheel–rail profile and system parameters on the wheel-rail space contact geometry relation. Rail Veh. 2002;40(2):14–8 (in Chinese).
7. Zhai WM, Wang KY, Cai CB. Fundamentals of vehicle–track coupled dynamics. Veh Syst Dyn. 2009;47(11):1349–76.

Chapter 8
Vibration Characteristics of Vehicle–Track Coupled System

Abstract This chapter shows the basic vibration characteristics of vehicle–track coupled systems under different wheel–rail system excitations obtained by using the vehicle–track coupled dynamics theory and the corresponding simulation software VICT and TTISIM. Seven typical wheel–rail excitations including smooth rail (no defect), local impact defects, local harmonic geometry defects, cyclic harmonic geometry defects, failure of system components, random track irregularities, and railway infrastructure settlement are considered. The vehicle and track dynamic responses and the wheel–rail interactions under these excitations will provide a general understanding of the dynamic characteristics of the vehicle–track coupled systems. The geometry and parameters of the railway vehicles and tracks adopted in the simulations presented in this chapter are given in the appendix, where the high-speed railway vehicle (Appendix A), the freight wagons (Appendix B), the high-speed ballasted track (Fig. C.1 and Table C.1 in Appendix C), the heavy-haul ballasted track (Table C.2 in Appendix C) and the high-speed ballastless slab track (Appendix D) are illustrated.

8.1 Steady-State Response of Vehicle–Track Interaction

The steady-state response analysis of vehicle–track interaction focuses on the vehicle and track responses in the case of a vehicle traveling on a smooth track with no irregularity or geometry defect. In such an ideal condition, the track geometry parameters along the railway line such as the sleeper spacing, the track curve radius, and the slope of the ramps are the excitation sources for the wheel–rail interaction. This section discusses the vibration characteristics of vehicle–track coupled systems under these steady-state excitations including the wheel–rail steady-state responses caused by sleeper span, the track steady-state responses under moving vehicle, and the wheel–rail interaction during steady-state curving.

W. Zhai, *Vehicle–Track Coupled Dynamics*,
https://doi.org/10.1007/978-981-32-9283-3_8

8.1.1 Steady-State Response Due to Sleeper Span

The periodic discrete sleepers along railway lines cause uneven support stiffness, which inevitably results in periodic fluctuations of wheel–rail interaction when a rail vehicle travels on the track. Figures 8.1, 8.2 and 8.3 show the periodic steady-state responses of wheel–rail forces and rail deformation due to the sleeper span in three typical vehicle–track systems.

Figures 8.1 and 8.2 illustrate the steady-state responses of the wheel–rail force and rail deformation when a passenger vehicle (axle load 14 t) runs on ballasted and ballastless tracks (rail pad stiffness 60 MN/m and 25 MN/m, respectively) at an operating speed of 300 km/h. For the two types of high-speed railway tracks, the periodic fluctuation amplitudes of the vertical wheel–rail force caused by sleeper span are approximately 2 kN and 1.5 kN, respectively. Figure 8.2b also depicts that the periodic discrete geometry characteristics of the track slabs in the ballastless track are reflected in the steady-state response of rail deformation. Figure 8.3 shows the steady-state responses of wheel–rail force and rail deformation when a freight

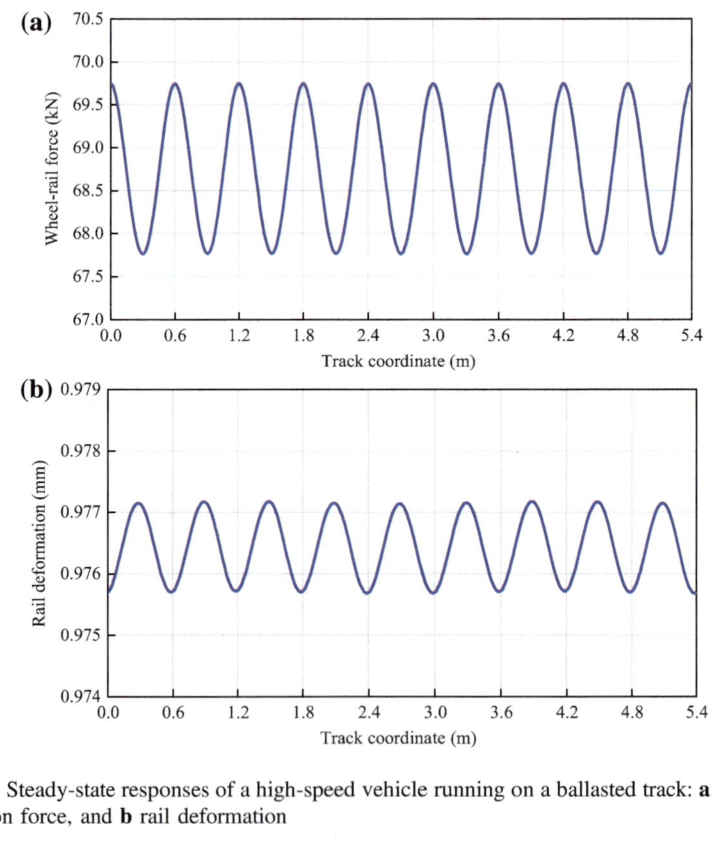

Fig. 8.1 Steady-state responses of a high-speed vehicle running on a ballasted track: **a** wheel–rail interaction force, and **b** rail deformation

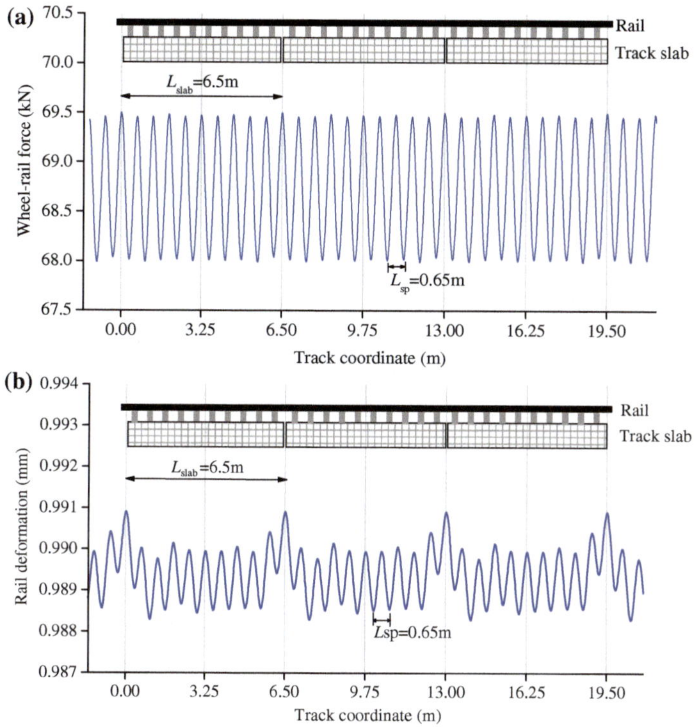

Fig. 8.2 Steady-state responses of a high-speed vehicle running on a ballastless track: **a** wheel–rail interaction force, and **b** rail deformation

wagon (axle load is 25 t) travels on a heavy-haul railway ballasted track (rail pad stiffness 160 MN/m) at a speed of 80 km/h. For the freight vehicle–track system, the periodic fluctuation amplitude of the wheel–rail force is approximately 2.5 kN. The above results indicate that the steady-state response of the rail deformation decreases as the rail pad stiffness increases and increases as the axle load increases. It is also observed that the steady-state periodic fluctuation amplitude of the wheel–rail force increases with the increase in rail pad stiffness and axle load. This is consistent with the phenomenon that the wheel–rail interaction caused by the sleeper span is conspicuous for the Harbin–Dalian high-speed railway system in winter.

8.1.2 Track Steady-State Response Under Moving Vehicle

Figures 8.4, 8.5 and 8.6 illustrate the steady-state responses of the rail deformation and the rail–sleeper interaction force of high-speed and freight vehicle–track

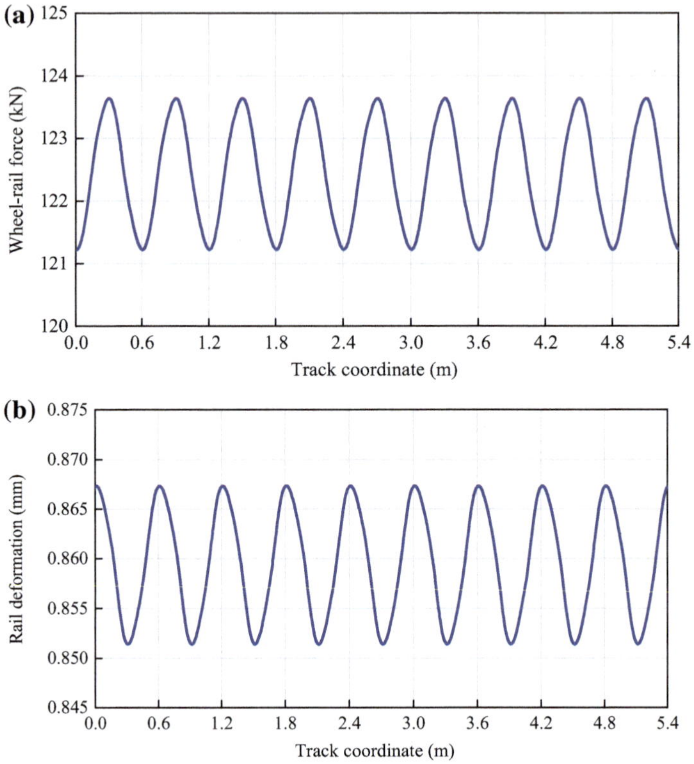

Fig. 8.3 Steady-state responses of a freight wagon running on a ballasted track: **a** wheel–rail interaction force, and **b** rail deformation

systems when the vehicle runs over a sleeper. Figures 8.4 and 8.5 show the track steady-state responses due to a passenger vehicle (axle load 14 t) traveling on ballasted and ballastless tracks (rail pad stiffness 60 MN/m and 25 MN/m, respectively) at an operating speed of 300 km/h. Steady-state responses of the ballasted track components under a freight wagon (axle load 25 t) running on a heavy-haul railway (rail pad stiffness 160 MN/m) at speed of 80 km/h are shown in Fig. 8.6.

From these figures, it can be seen that the peaks of the rail deformation and the rail–sleeper interaction force exhibit correspondence to the four wheelsets of the respective vehicles. For the high-speed vehicle–track systems, the vertical rail deformation and the rail–sleeper interaction force are approximately 1 mm and 25 kN, respectively; while for the freight wagon–track system, the two indices are 0.85 mm and 53 kN, respectively. This means a larger rail support stiffness results in a smaller rail deformation, and a larger axle load leads to a larger rail–sleeper interaction force.

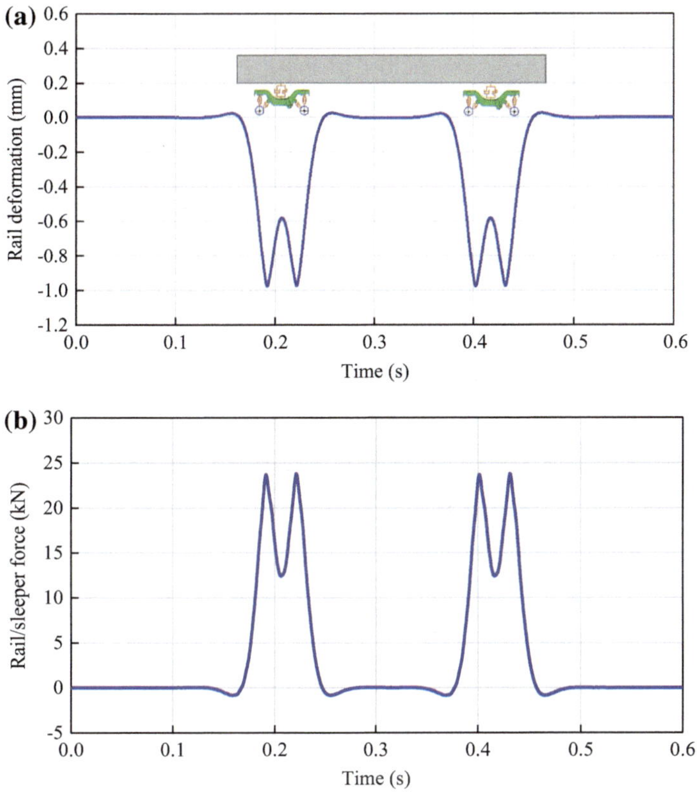

Fig. 8.4 Steady-state responses of a high-speed vehicle passing a ballasted track section: **a** rail deformation, and **b** rail–sleeper interaction force

8.1.3 Steady-State Curving Response

When a rail vehicle negotiates a curved track, large lateral forces are generated between the wheels and rails due to the centrifugal force and the change of track geometry. These lateral forces, in combination with small vertical forces, may cause wheel climbing and reduce ride comfort as the vehicle negotiates the curve. In this section, the steady-state curving performances of a high-speed passenger vehicle and a freight wagon are presented, as shown in Figs. 8.7 and 8.8, respectively.

Figure 8.7 illustrates the steady-state responses of the lateral and vertical wheel–rail forces, wheelset displacements, and lateral car body displacement of a high-speed vehicle negotiating a large radius curved track at a speed of 300 km/h. The high-speed curved track had a circular curve radius of 7000 m, a transition curve length of 670 m, an arc length of 400 m, and a super-elevation of 100 mm. The results show that the wheel–rail forces and the lateral vehicle displacements increase when the vehicle crosses the transition curve, and their peak responses

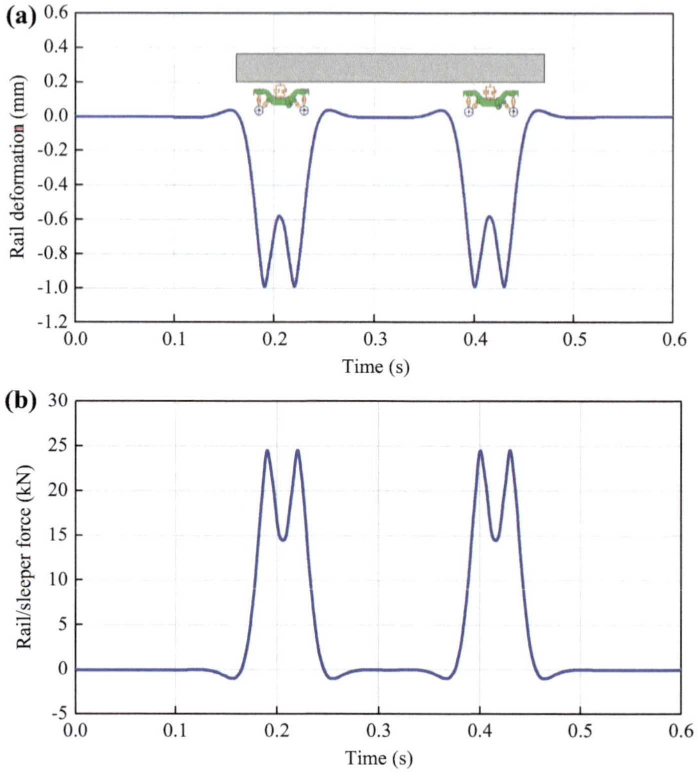

Fig. 8.5 Steady-state responses of a high-speed vehicle passing a ballastless track section: **a** rail deformation, and **b** rail–sleeper interaction force

appear at the circular curve. The lead wheelset has the largest lateral displacement of about 4 mm, and the lateral car body displacement increases to about 20 mm.

Figure 8.8 depicts the steady-state responses of the wheel–rail forces and lateral vehicle displacements of a freight wagon negotiating a small radius curved track at a speed of 80 km/h. The curved track had a curve radius of 250 m, a transition curve length of 80 m, an arc length of 100 m, and a super-elevation of 120 mm. It shows that the steady-state responses of the wheel–rail forces and lateral vehicle displacements of the freight wagon are similar to that of the high-speed vehicle, while the former has worse curving performance. The maximum lateral and vertical wheel–rail forces of the lead wheelset attain 25 kN and 160 kN, respectively, while the lateral wheelset and car body displacements have peak values of 10 mm and 28 mm, respectively. It can be seen that the curve with a smaller radius makes the vehicle curving performance poor.

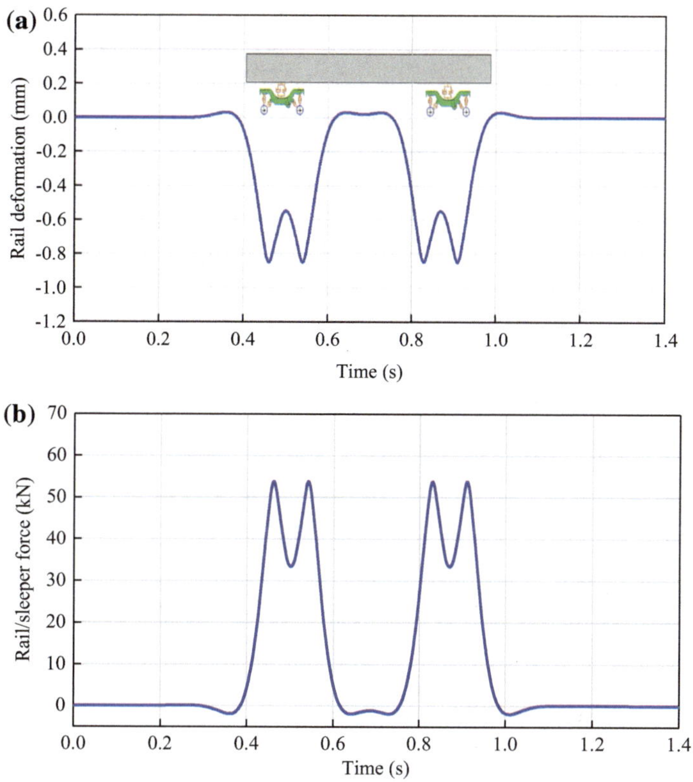

Fig. 8.6 Steady-state responses of a freight wagon traveling passing a ballasted track section: **a** rail deformation, and **b** rail–sleeper interaction force

8.2 Dynamic Response of Vehicle–Track Interaction Due to Local Geometry Defects

8.2.1 Dynamic Response to Vertical Impulsive Defects

There are multiform vertical impulsive excitation sources in the wheel–rail system, such as the rail weld joint, the wheel flat, and so on. These defects cause sudden transient impact loads to the wheel–rail system, resulting in intense vibration to the vehicle–track coupled system. In this section, the dynamic responses of railway vehicles passing rail dipped joints and weld joint irregularities on a ballasted track are provided as examples to illustrate the vibration characteristics of the vehicle–track coupled system under the excitation of vertical impulsive defects.

Figure 8.9 depicts the dynamic vertical wheel–rail force of a freight wagon negotiating a dipped rail joint (see Fig. 3.10 in Chap. 3) at a speed of 80 km/h. A total dip angle of $2a = 0.02$ rad was adopted in the simulation. Figure 8.9 shows

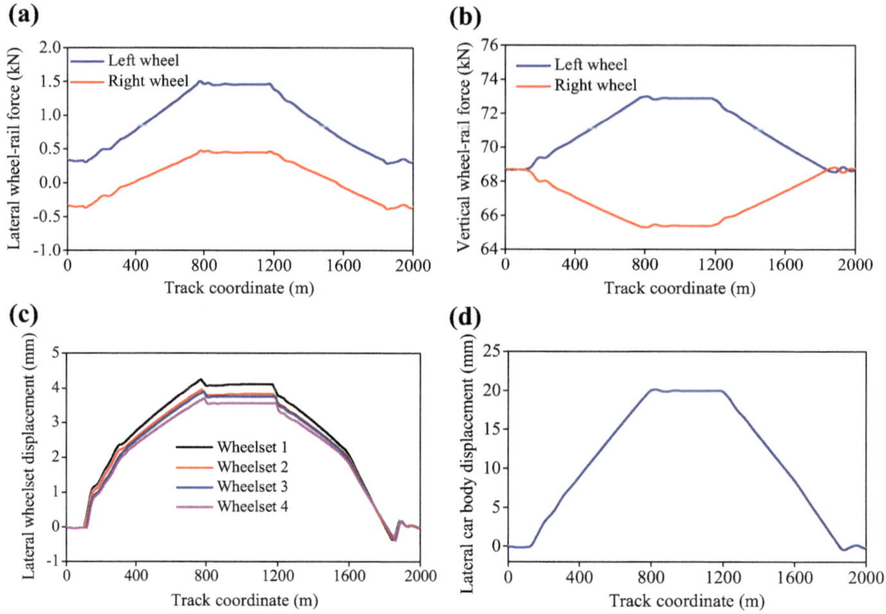

Fig. 8.7 Steady-state curving performance of a high-speed vehicle negotiating a large radius curved track: **a** lateral wheel–rail force, **b** vertical wheel–rail force, **c** lateral wheelset displacement, and **d** lateral car body displacement

that the time history of the wheel–rail impact force caused by the dipped rail joint exhibits two dominating peaks, which correspond to the P_1 and P_2 forces defined in the BR standard [1]. The P_1 force (short peak force) is a high-frequency force in the range of 200–1000 Hz, which is related to the natural frequency of the Hertzian contact between the unsprung mass of the vehicle and the rail. The P_1 force usually appears immediately (0.5 ms) after the wheel–rail impact and exists only for a very short duration. Therefore, the P_1 force will not be transmitted to the sprung mass of the vehicle system and the rail infrastructures. However, it can potentially damage the local contact region of the wheel tread and the railhead. On the contrary, the P_2 force (delayed peak force) is a low-frequency force in the range of 30–200 Hz, which is relevant to the vibration characteristics of the vehicle–track coupled system. The P_2 force usually acts for a relatively longer duration and hence affects the dynamic behavior of the vehicle sprung mass and the rail infrastructure. Therefore, the P_2 force has the potential to damage most of the vehicle and track components located in the vicinity of wheel–rail contact zone. As shown in Fig. 8.9, the P_1 and P_2 forces caused by a dipped rail joint with $2a$ = 0.02 rad to a freight wagon are approximately 265 kN and 190 kN, respectively.

Figure 8.10 shows the time histories of the vertical displacements of the wheelset and track system induced by the dipped rail joint. It can be seen that the wheel–rail displacements decay slowly and exhibit a low-frequency vibration,

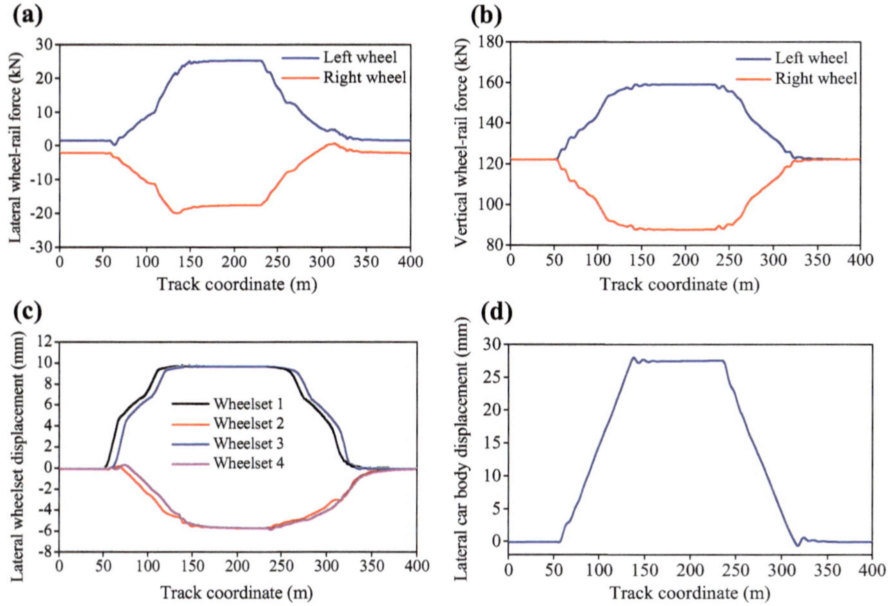

Fig. 8.8 Steady-state curving performance of a freight wagon negotiating a small radius curved track: **a** lateral wheel–rail force, **b** vertical wheel–rail force, **c** lateral wheelset displacement, and **d** lateral car body displacement

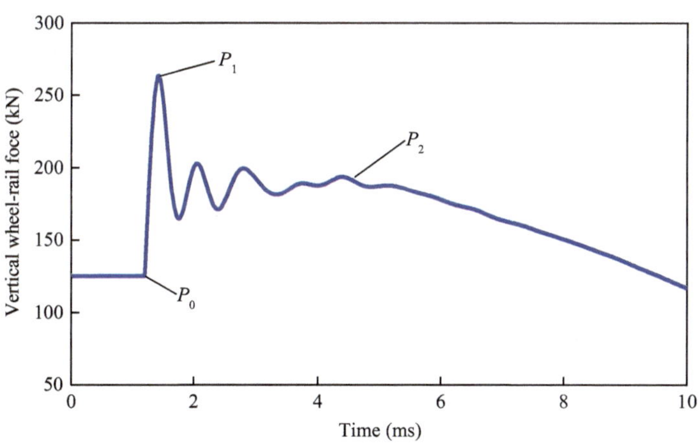

Fig. 8.9 Vertical wheel–rail force caused by dipped rail joint in heavy-haul railway system

which has a similar characteristic to the P_2 force. Besides, the maximum displacements of the wheelset and the track components appear at 4–6 ms after the wheel–rail impact. This also indicates that the P_2 force has a very important effect

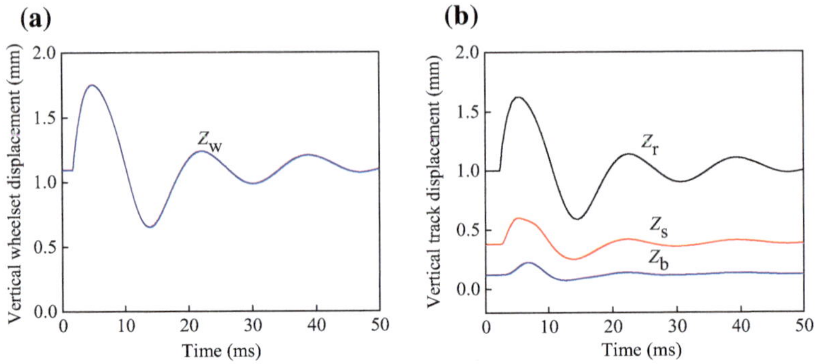

Fig. 8.10 Vertical displacements of **a** freight wagon wheelset and **b** track under the impact of dipped rail joint

on the dynamic response of the vertical displacements of the wheelset and track system. It is also evident from Fig. 8.10 that the impact vibration is transmitted from wheel–rail interface to rail infrastructure, which leads to that the vertical displacements of the wheelset, the rail, the sleeper, and the ballast (see Z_w, Z_r, Z_s, Z_b) decrease gradually.

The dynamic responses of the vertical accelerations of the wheelset and track system induced by the dipped rail joint are presented in Fig. 8.11. It shows that the time history of the vertical wheelset acceleration is basically consistent with that of the wheel–rail force, while the rail acceleration peak exists for a very short duration, which is similar to the P_1 force. The maximum vertical accelerations of the wheelset and the rail are approximately 30 *g* and 300 *g*, respectively, the latter is about 10 times larger than the former. It is also observed that the vertical acceleration amplitudes of the rail, the sleeper, and the ballast decrease gradually, their ratio remains approximately 90:7:1. The predominant frequencies of the vertical accelerations of the rail, the sleeper, and the ballast also decrease gradually; their ratio remains approximately 6:2:1. It is obvious that the vibration amplitude and frequency of track parts are highly related to the track structure parameters (such as mass, property of fastenings and rail pad, ballast damping, etc.). However, for different types of track structures, their vibration characteristics remain similar.

Figure 8.12 shows the dynamic vertical wheel–rail force of a high-speed vehicle negotiating a dipped rail joint at a speed of 300 km/h. It can be seen that the dynamic characteristics of the high-speed wheel–rail force are similar to that shown in Fig. 8.9, but it decays slower than that of the freight wagon. Figure 8.12 shows the P_1 and P_2 forces caused by a dipped rail joint with $2a = 0.006$ rad to the high-speed vehicle are approximately 210 kN and 160 kN, respectively.

The time histories of the vertical displacements of the high-speed wheelset and the track affected by the dipped rail joint are shown in Fig. 8.13, respectively. The results also show that the dynamic vertical displacements of the wheelset and track system are closely related to the P_2 force. It is also evident from Fig. 8.13 that the

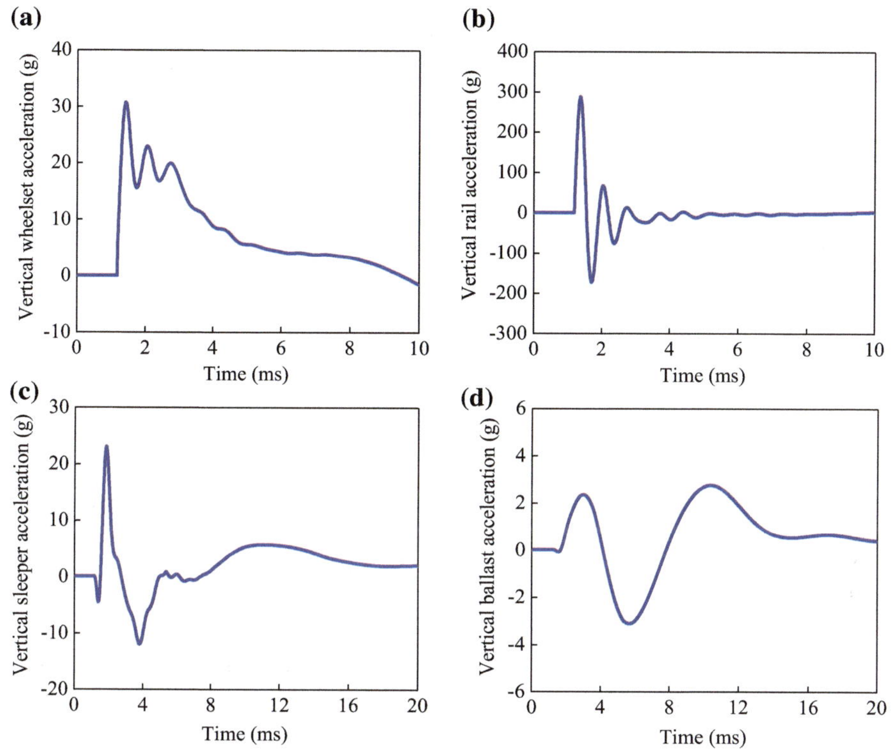

Fig. 8.11 Vertical accelerations of freight wagon–track system under the impact of dipped rail joint: **a** wheelset, **b** rail, **c** sleeper, and **d** ballast

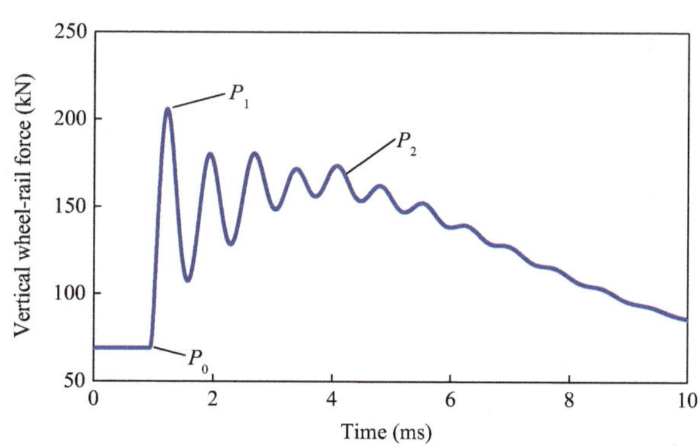

Fig. 8.12 Vertical wheel–rail force caused by dipped rail joint in high-speed railway system

impact vibration is transmitted from the wheel–rail interface to the rail infrastructure, which leads to that the vertical displacements of the wheelset, the rail, the sleeper, and the ballast (see Z_w, Z_r, Z_s, Z_b) decrease gradually.

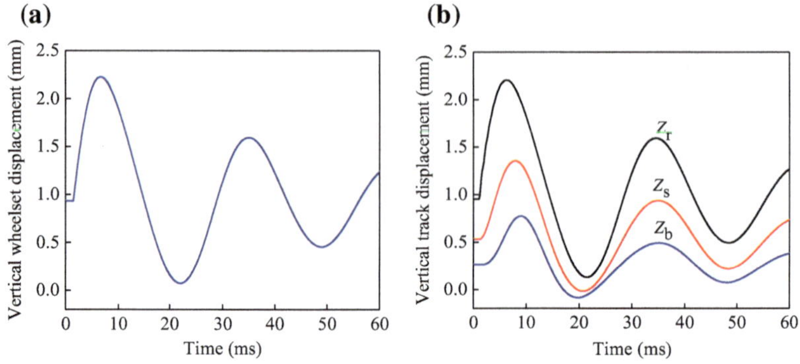

Fig. 8.13 Vertical displacements of **a** wheelset and **b** track under the impact of dipped rail joint

Figure 8.14 depicts the dynamic responses of vertical accelerations of the high-speed wheelset and track system induced by the dipped rail joint. It also shows that the dynamic response of the vertical wheelset acceleration is basically consistent with that of the wheel–rail force, while the rail acceleration response is highly related to the P_1 force. The maximum vertical accelerations of the wheelset and the rail are approximately 15 *g* and 300 *g*, respectively, the latter is about 20 times larger than that of the former. It is also observed that the vertical acceleration amplitudes of the rail, sleeper and ballast decrease gradually, their ratio remains approximately 60:3:1. The predominant frequencies of the vertical accelerations of the rail, sleeper, and ballast also decrease gradually. The results indicate that the vibration characteristics of the high-speed and the heavy freight railway systems induced by the dipped rail joint are similar although the vehicle and track structure parameters are different.

With the application of continuous welded rail in railway lines, the track geometry of the rail weld zone is greatly improved. Accordingly, the wheel–rail dynamic performance in the rail weld zone is also significantly improved. However, due to the limitation from welding materials, welding technology, track maintenance level, and other factors in practice, there are still surface damages such as fissures and squats generated in the rail weld zone under repeated rolling load from train wheels. Figure 8.15 illustrates a typical local defect in the rail weld zone of Chinese high-speed railways [2]. The basic feature of the weld rail joint irregularity is that the main cosine wave (wavelength L = 1 m, amplitude δ_1) is superimposed with a secondary short wave (wavelength λ = 0.1–0.4 m, amplitude δ_2), which can be called as complex-wave irregularity. The complex wave is substantially symmetric about the weld point, which is the reference position for the field measurement.

Figure 8.16 shows the wheel–rail dynamic responses of a high-speed vehicle negotiating a rail weld complex-wave irregularity in the rail weld zone of a ballastless track at a speed of 300 km/h. In the simulation, the weld complex-wave irregularity had the main cosine wave with a wavelength of L = 1 m and amplitude

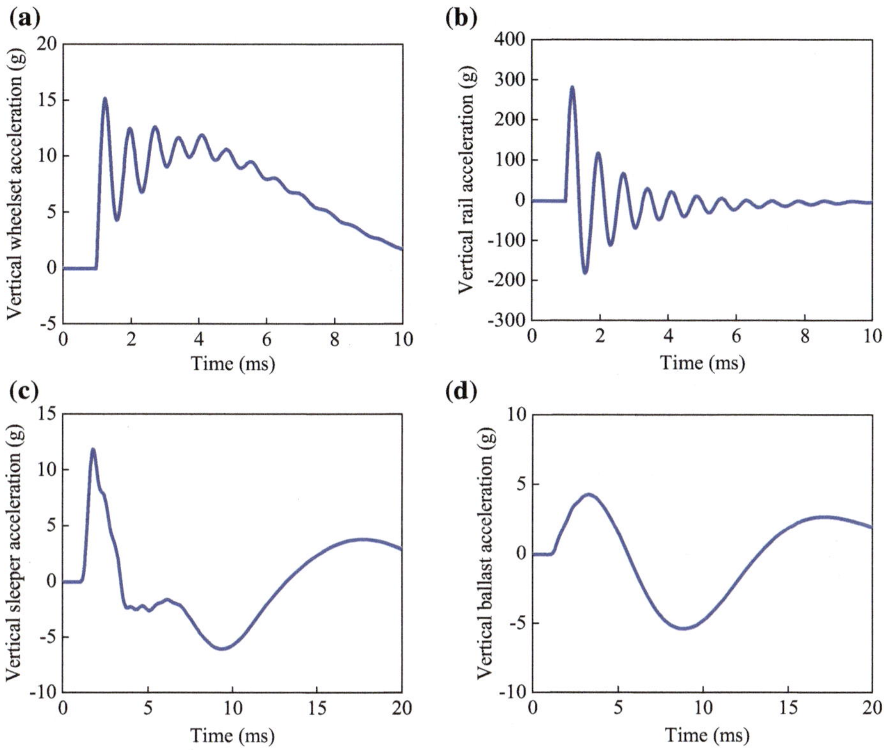

Fig. 8.14 Vertical accelerations of high-speed vehicle and track under the impact of dipped rail joint: **a** wheelset, **b** rail, **c** sleeper, and **d** ballast

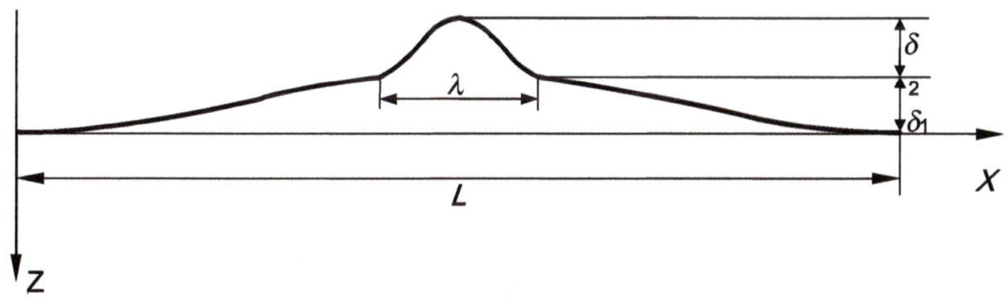

Fig. 8.15 Rail weld irregularity on Chinese high-speed railway lines

of $\delta_1 = 0.2$ mm, and a secondary short wave with a wavelength of $\lambda = 0.1$ m and amplitude of $\delta_2 = 0.1$ mm. It can be seen that the rail weld geometry irregularity can induce a large impact to the high-speed wheel–rail system. The peak dynamic responses of the wheel–rail force and the accelerations appear after the wheel passing over the superimposed short-wavelength irregularity. It means that the maximum dynamic responses of the wheel–rail system due to the rail weld

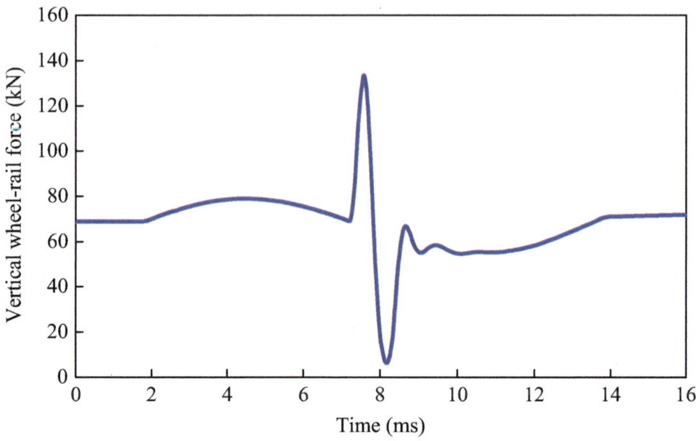

Fig. 8.16 Vertical wheel–rail force caused by rail weld irregularity in high-speed railway system

irregularity are mainly affected by the irregularity with a shorter length. Figure 8.16 shows the maximum force caused by a rail weld irregularity with a short wave (wavelength $\lambda = 0.1$ m and amplitude $\delta_2 = 0.1$ mm) to the high-speed wheel–rail system is approximately 135 kN (about 1 time of the static axle load), while the minimum wheel–rail force is only 5 kN. The results indicate that the short-wavelength irregularity existed in the high-speed railway weld zone can largely exacerbate the wheel–rail contact force and result in heavy unloading to the wheels, which can not only damage the local contact region of the wheel–rail system, but also adversely affect the running safety of high-speed trains.

The dynamic responses of the vertical accelerations of the high-speed wheel–rail system induced by the rail weld irregularity are shown in Fig. 8.17. It shows that the dynamic response of the vertical wheel–rail acceleration is basically consistent with that of the wheel–rail force. The maximum vertical accelerations of the

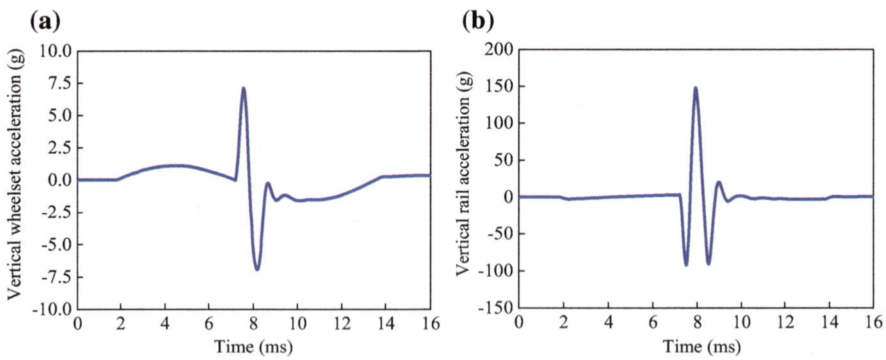

Fig. 8.17 Vertical accelerations of **a** wheelset and **b** rail under impact of rail weld irregularity

wheelset and the rail are approximately 7.5 *g* and 150 *g*, respectively, the latter is about 20 times of the former. This means that the wheel–rail impacts caused by the rail weld joint irregularities mainly damage the track system.

8.2.2 Dynamic Response to Lateral Impulsive Defects

Turnout is a very important component in railway track structures, which turns the running direction of trains from one railway line to another. When a high-speed train crosses the rail discontinuity in a turnout, strong lateral and vertical impacts between the wheels and rails occur [3, 4]. These impacts can potentially damage the local contact region of wheels and turnout components, and then shorten their service life. In this section, the dynamic response of a high-speed railway vehicle passing the rail discontinuity in a turnout is taken as an example to illustrate the vibration characteristics of vehicle–track coupled system under the excitation of lateral impulsive defects at railway turnout.

Figures 8.18 and 8.19 depict the dynamic responses of a high-speed wheel–rail system under the lateral impact of a high-speed wheel crossing a switch point rail (see Fig. 3.13 in Chap. 3). In the simulation, an angle of attack $\beta = 0.006$ rad was considered, which generated a lateral impact velocity of 0.5 m/s to the wheel. From Figs. 8.18 and 8.19, it is observed that the wheel–rail lateral impact at switch zone has a very important effect on the dynamic behavior of the high-speed train–track system, especially the lateral wheelset motion. Figure 8.18 shows that the maximum lateral wheel–rail force caused by the switch point impact is larger than 30 kN. This means such a strong lateral impact can induce wheel flange climbing the railhead. At the same time, the lateral wheelset displacement exhibits a low-frequency harmonic response and decays slowly, as shown in Fig. 8.19. The maximum lateral wheelset displacement attains approximately 8 mm. The above

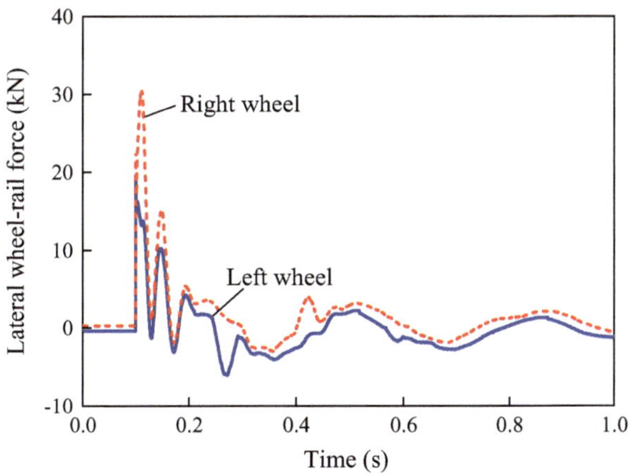

Fig. 8.18 Lateral wheel–rail force under lateral impact at rail switch in high-speed railway system

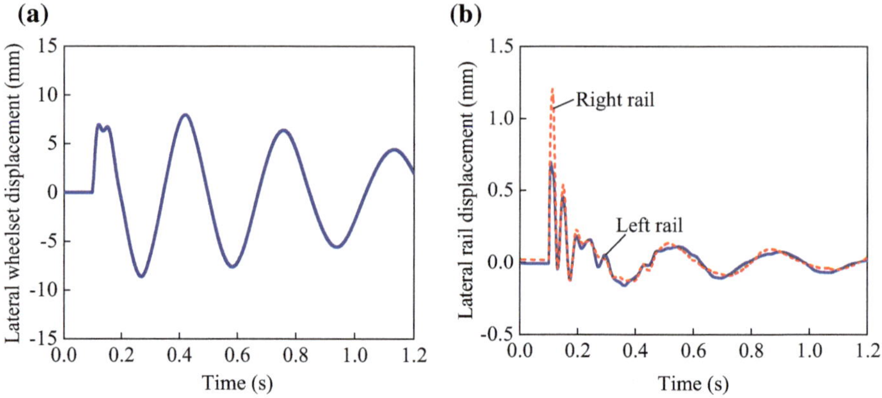

Fig. 8.19 Lateral displacements of high-speed **a** wheelset and **b** rail under lateral impact at rail switch

results show that the lateral impact at a railway switch can potentially trigger the hunting motion and wheel flange climbing behavior, which threatens the running safety of high-speed trains. Figure 8.19 shows that the lateral impact can also cause large deformation of rails; the lateral rail deformation attains a maximum value of approximately 1.1 mm. The dynamic response of lateral rail deformation is basically consistent with that of the lateral wheel–rail force is shown in Fig. 8.18.

8.2.3 Dynamic Response to Vertical Local Harmonic Geometry Defects

The vertical local harmonic geometry defects usually exist at rail joint zones, bridge–subgrade transition zones, rail infrastructure local settlement zones, and so on. Figures 8.20, 8.21 and 8.22 present the dynamic responses of a high-speed vehicle–track coupled system under the impact of a vertical local harmonic geometry defect (see Fig. 3.19 in Chap. 3) with a short wavelength of $L = 2$ m and an amplitude of $a = 2$ mm at a speed of 300 km/h.

It shows that the vibration characteristics of the vehicle–track coupled system induced by the vertical local harmonic geometry defect are quite different from that caused by the vertical impulsive defect. The dynamic vertical wheel–rail force caused by the local harmonic defect (see Fig. 8.20) no longer contains the P_1, P_2 forces as shown in Figs. 8.9 and 8.12, but exhibits two peaks of P_{max}, P_{min} forces. Under the excitation of the local harmonic defect, the wheel–rail force also shows a harmonic fluctuation, as shown in Fig. 8.20. It should be noted that the P_{min} force is very important for the evaluation of train running safety. If the P_{min} force reduces to zero, the wheel will lose contact with the rail. Such a situation will be a threat to the operating safety of high-speed trains. The P_{max} force shown in Fig. 8.20 is similar

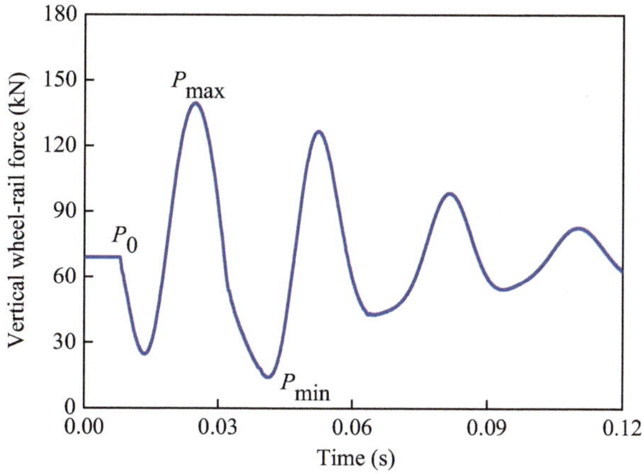

Fig. 8.20 Vertical wheel–rail force induced by a short local harmonic defect in high-speed railway system

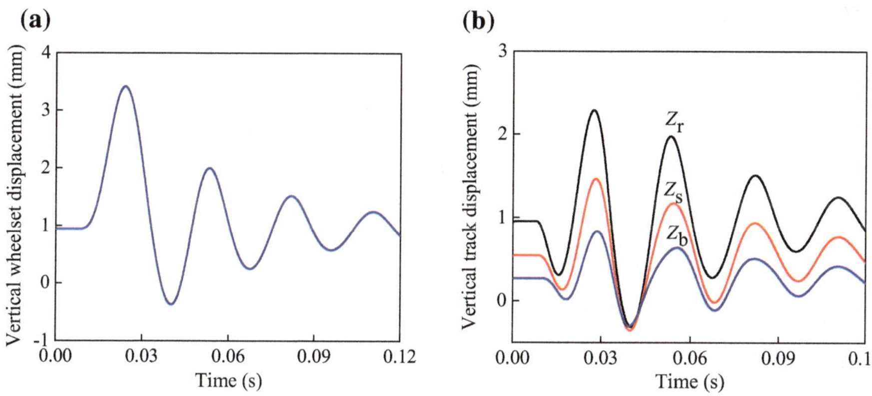

Fig. 8.21 Vertical displacements of **a** wheelset and **b** track induced by a short local harmonic geometry defect

to the P_2 force presented in Figs. 8.9 and 8.12, which is a low-frequency force and usually acts for a relatively long duration. The dynamic wheel–rail force induced by local harmonic geometry defects can affect the dynamic behavior of the vehicle sprung mass and the rail infrastructure significantly. Therefore, the dynamic responses of the vehicle and track systems also exhibit harmonic fluctuations as the wheel–rail force does as shown in Figs. 8.21 and 8.22.

Figure 8.21 shows the time histories of the vertical displacements of the wheelset and track system induced by the local harmonic geometry defect. It can be seen that the wheel–rail displacements decay slowly and exhibit a low-frequency vibration, which have similar characteristics to that shown in Fig. 8.13. It is also

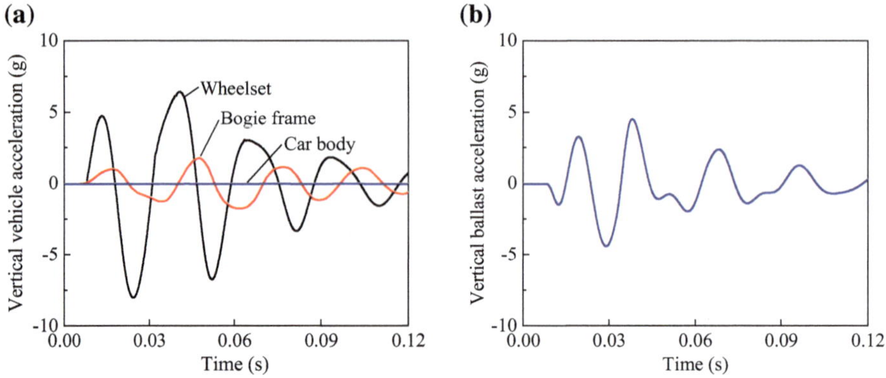

Fig. 8.22 Vertical accelerations of **a** vehicle and **b** rail induced by a short local harmonic geometry defect

evident that the impact vibration is transmitted from the wheel–rail interface to the rail infrastructure, which leads to that the vertical displacements of the wheelset, rail, sleeper, and ballast (see Z_w, Z_r, Z_s, Z_b) decrease gradually.

Figure 8.22 depicts the dynamic responses of the vertical accelerations of the high-speed vehicle system and ballast layer induced by the local harmonic geometry defect. It also shows that the dynamic response of the vertical wheelset acceleration is consistent with that of the wheel–rail force, while the car body acceleration is very small. It is also observed that the vertical acceleration amplitudes of the rail, sleeper and ballast decrease gradually, their ratio remains approximately at 2.1:1.6:1. The dynamic response of the ballast vertical acceleration is also similar to that of the vertical wheel–rail force, as shown in Fig. 8.22b. This means that the impact caused by the local harmonic geometry defect has a great influence on the vibration of the rail infrastructure.

Figures 8.23, 8.24, and 8.25 illustrate the high-speed wheel–rail dynamic responses under the impact of a vertical local harmonic geometry defect (see Fig. 3.19 in Chap. 3) with a long wavelength of $L = 20$ m and an amplitude of $a = 5$ mm at a speed of 300 km/h. Figure 8.23 shows that the wheel–rail force induced by a long-wavelength local harmonic defect no longer exhibits harmonic fluctuation as shown in Fig. 8.20. The reason should be that a local geometry defect with a wavelength larger than the distance between two axles of a bogie can trigger a significant pitch motion of the bogies, which reduce the wheel–rail impact force and then decrease the vibration amplitude of high-speed vehicle–track coupled system. From Fig. 8.23, it can also be seen that the wheel–rail force induced by a long vertical local harmonic geometry defect is small. However, this type of track defect has a significant effect on the car body vibration and ride quality, as shown in Fig. 8.24. Figure 8.25 demonstrates that the dynamic vertical displacement of the wheelset is very close to the shape of the local geometry defect, but the dynamic

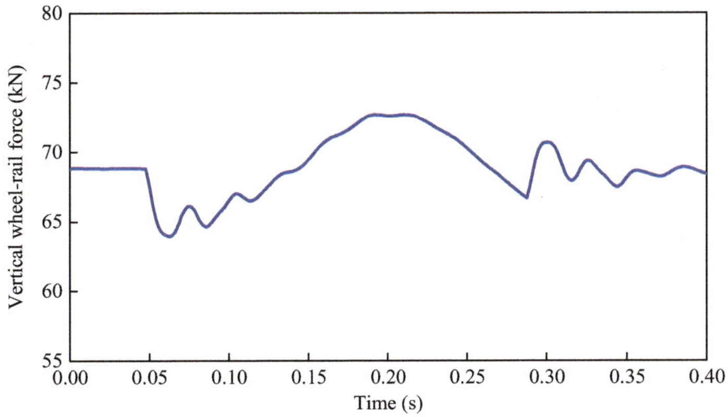

Fig. 8.23 Vertical wheel–rail force induced by a long local harmonic geometry defect in high-speed railway system

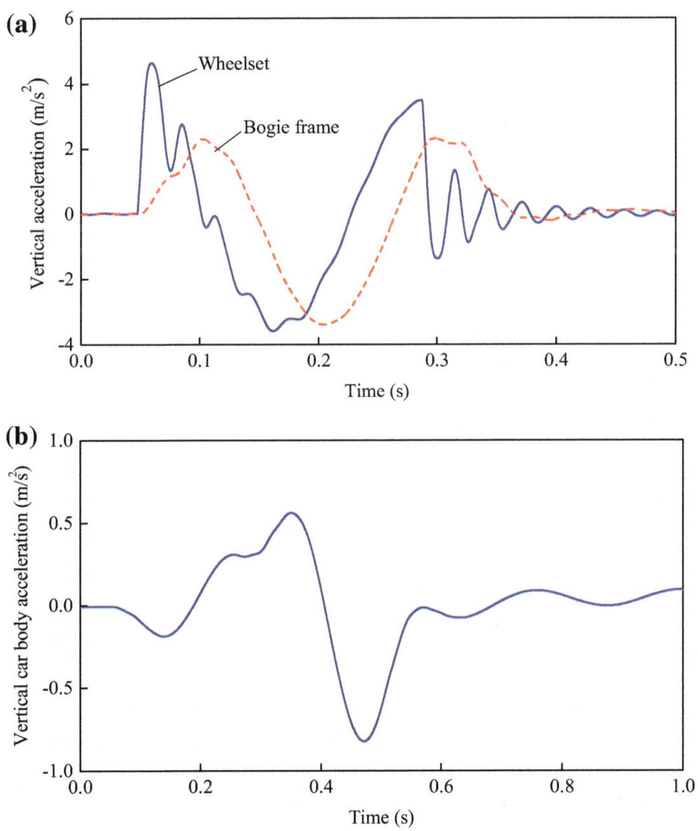

Fig. 8.24 Vertical accelerations of **a** bogie and **b** car body induced by a long local harmonic geometry defect

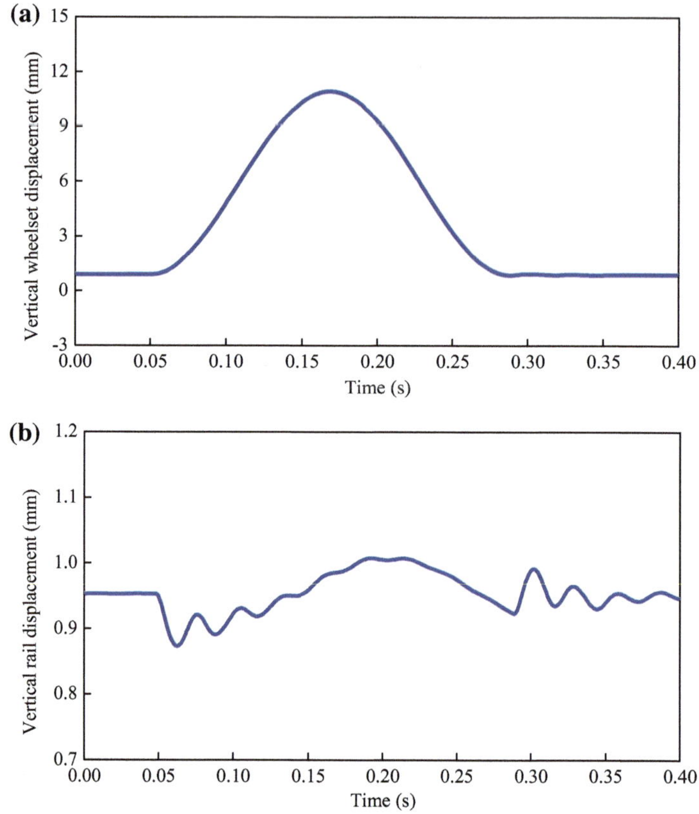

Fig. 8.25 Vertical displacements of **a** wheelset and **b** rail induced by a long local harmonic geometry defect

response of the vertical rail deformation is basically consistent with that of the vertical wheel–rail force is shown in Fig. 8.23.

8.2.4 Dynamic Response to Lateral Local Harmonic Geometry Defects

The dynamic characteristics of a vehicle–track coupled system subjected to the impact of lateral local harmonic geometry defects (such as rail buckle) overall exhibit a harmonic fluctuation form. The nonlinearities of the lateral wheel–rail geometry contact and wheel–rail creep force, however, generate more complexity to the dynamic responses of lateral wheel–rail interactions.

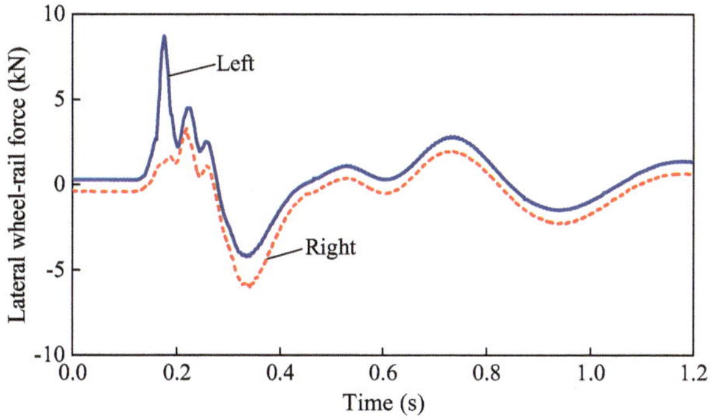

Fig. 8.26 Lateral wheel–rail force induced by a lateral local harmonic geometry defect in high-speed railway system

Figures 8.26, 8.27 and 8.28 present the dynamic responses of a high-speed vehicle–track coupled system under the impact of a lateral local harmonic geometry defect (see Fig. 3.19 in Chap. 3) with a wavelength of L = 10 m and an amplitude

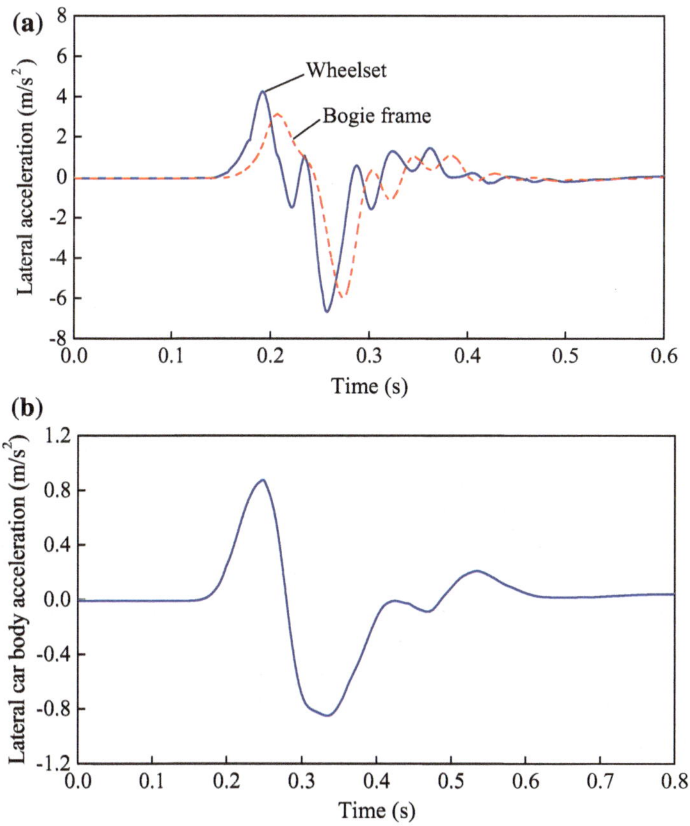

Fig. 8.27 Lateral accelerations of **a** bogie and **b** car body induced by a lateral local harmonic geometry defect

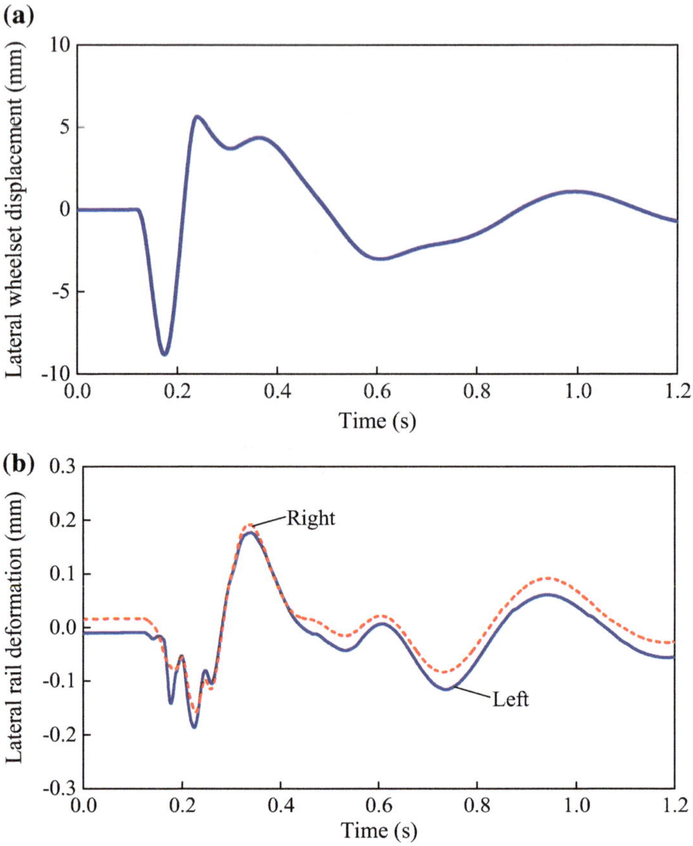

Fig. 8.28 Lateral displacements of **a** wheelset and **b** rail induced by a lateral local harmonic geometry defect

of $a = 5$ mm at a speed of 300 km/h. The time history of the lateral wheelset displacement in Fig. 8.28 shows that the lateral wheelset motion is highly related to the shape of the local harmonic geometry defect because the wheel–rail force (Fig. 8.26) and the rail deformation are small. The lateral wheelset displacement experiences a harmonic fluctuation after crossing the local defect. After a period of harmonic motion, the lateral wheelset displacement begins to decay to the equilibrium position. The lateral wheelset motion affects the wheel–rail contact geometry, which further influences the wheel–rail creep force, and finally, they make the dynamic response of the lateral wheel–rail force become complex (see Fig. 8.26). Figures 8.27 and 8.28 show that the lateral impact induced by the lateral local harmonic geometry defect can also cause important effects to the car body vibration and ride quality, which is similar to the results shown in Figs. 8.23, 8.24 and 8.25.

8.3 Dynamic Response of Vehicle–Track Interaction to Cyclic Geometry Defects

The cyclic geometry defects exist widely in the wheel–rail system, such as the rail corrugation, the wheel polygonization (wheel harmonic wear or wheel periodic out of roundness), etc. These cyclic geometry defects cause periodically forced vibration to the vehicle–track coupled system. In this section, the dynamic responses of railway vehicles running on ballasted and ballastless tracks with wheel polygonization or rail corrugation are taken as examples to illustrate the vibration characteristics of the vehicle–track coupled system under the excitation of cyclic geometry defects.

Figures 8.29, 8.30, and 8.31 illustrate the dynamic responses of high-speed vehicle–track coupled systems subjected to the impact of cyclic geometry defects at a speed of 300 km/h. In the simulation, a ballasted track (rail pad stiffness 60 MN/m) and a ballastless track (rail pad stiffness 25 MN/m) were considered. Specifically, Fig. 8.29 exhibits the time history of the dynamic wheel–rail force of a high-speed passenger vehicle (axle load is 14 t) containing polygonized wheels (see Fig. 3.24 in Chap. 3). The rolling circle of the polygonal wheels has 22 waves, which means the wavelength of the polygonal wear is 122.73 mm. Figures 8.30 and 8.31 show the

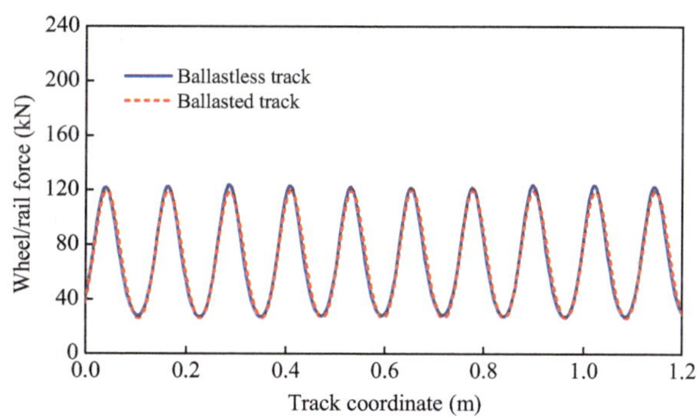

Fig. 8.29 Wheel–rail force induced by wheel polygonal wear defect

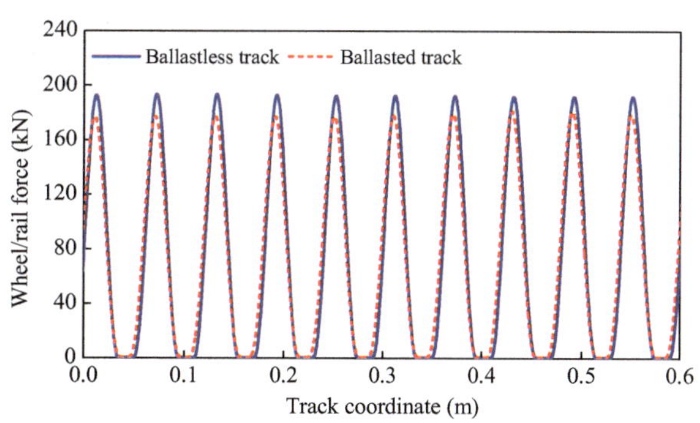

Fig. 8.30 Wheel–rail force induced by short-pitch rail corrugation

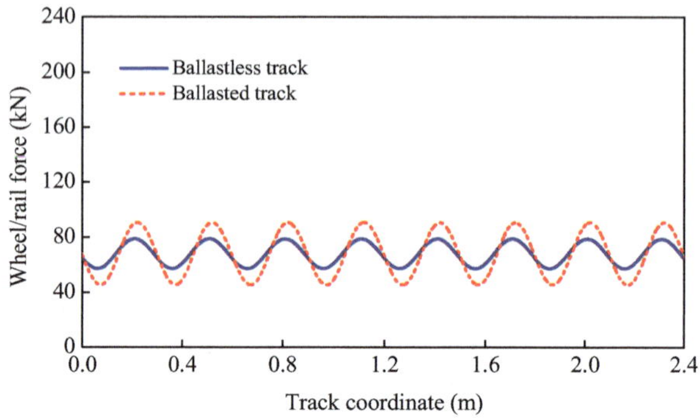

Fig. 8.31 Wheel–rail force induced by long-wavelength rail corrugation

time histories of the dynamic wheel–rail force of a high-speed passenger vehicle traveling on the ballasted and ballastless tracks with rail corrugations (see Fig. 3.18 in Chap. 3). The wavelengths of the rail corrugations considered in the two cases were 60 mm (Fig. 8.30) and 300 mm (Fig. 8.31), respectively. The amplitude of the wheel polygonal wear and the rail corrugations was set as 0.1 mm.

It can be seen that the wavelength of the cyclic geometry defects (rail corrugation and wheel polygonization) significantly affect the wheel–rail impact force and vibration behavior of the vehicle–track coupled system. The cyclic geometry defects with shorter wavelength will result in higher wheel–rail impact force and more drastic system vibration. Under the impact of short-pitch rail corrugation, the large wheel–rail force makes the wheel frequently loses contact with the rail, which exacerbates the wheel–rail impact in return, as shown in Fig. 8.30. Therefore, the cyclic geometry defects with short wavelength can seriously damage the vehicle and track components, especially for the components close to wheel–rail interface.

Figure 8.29 shows that the wheel–rail impact forces induced by the 22-order polygonal wheel defect are similar to the cases considering ballasted and ballastless tracks; however, the wheel–rail impact force on the ballasted track is smaller than that on the ballastless track for the cases of short-pitch rail corrugation, as shown in Fig. 8.30. On the other hand, Fig. 8.31 depicts that the wheel–rail force on the ballasted track is larger than that on the ballastless track for the cases of long-wavelength rail corrugation. The above results demonstrate that the ballasted track can better absorb the high-frequency vibration energy caused by the periodic geometry defects with short wavelength, while the ballastless track containing low-stiffness rail pad can reduce the low- and middle-frequency wheel–rail impacts induced by the cyclic geometry defects.

8.4 Dynamic Response of Vehicle–Track Interaction Due to Failure of System Component

Component failures of railway vehicle–track coupled system occur occasionally, especially due to the failures of bogie dampers, rail fastenings, etc. When a key component of a vehicle–track coupled system is disabled, the wheel–rail interaction would be intensified, which can enlarge the system vibration and reduce the dynamics performance of rail vehicles. In this section, the comparisons of the dynamic responses of the vehicle–track system with and without component failures are conducted to illustrate the vibration characteristics of the vehicle–track coupled system under the excitation of the failure of a system component. Three typical failure cases including the disabled lateral dampers on a high-speed bogie, the fatigue fracture of fastener clips, and the unsupported sleepers had been considered in the simulations.

8.4.1 Dynamic Response to Disabled Lateral Dampers on a High-Speed Bogie

The high-performance dampers (such as yaw dampers, secondary lateral dampers, vertical dampers) installed on high-speed bogies can not only reduce the vibration and improve the ride comfort, but also enhance the running safety of trains. Here, the dynamic response of a high-speed vehicle subjected to the failure of bogie secondary lateral dampers is presented as an example to illustrate the vibration characteristics of vehicle–track coupled system under the impact of disabled vehicle dampers. Figures 8.32 and 8.33 compare the dynamic responses of a high-speed vehicle containing normal and disabled lateral dampers under the impact of a local lateral geometry defect (same as considered in Sect. 8.2.4) at a speed of 300 km/h. The high-speed vehicle considered in the simulations had four secondary lateral dampers (each bogie installed with two lateral dampers).

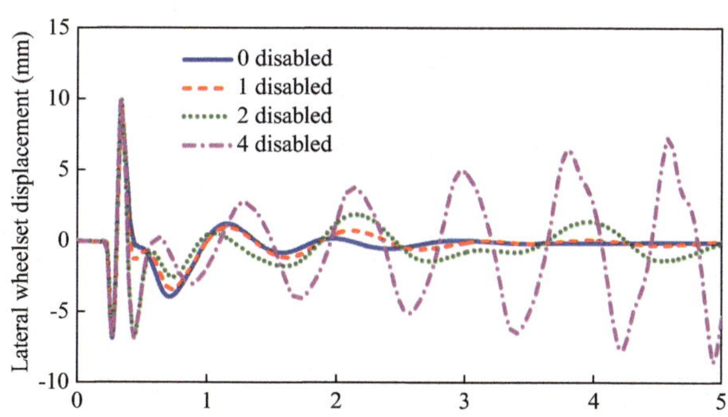

Fig. 8.32 Lateral wheelset displacement to disabled lateral dampers

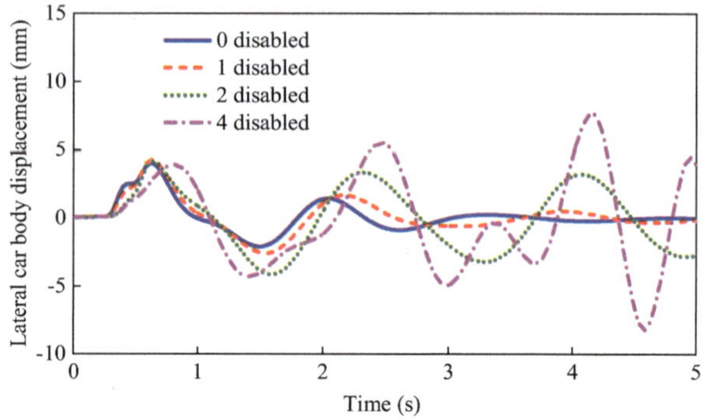

Fig. 8.33 Lateral car body displacement to disabled lateral dampers

It is evident from Figs. 8.32 and 8.33 that the failure of bogie secondary lateral dampers can largely increase the lateral displacements of the vehicle system and reduce the lateral stability of the high-speed vehicle. If only one lateral damper is disabled, the lateral dynamics and stability of the high-speed vehicle are only deteriorated a little, which does not affect the running safety. However, if two lateral dampers in a bogie are disabled, the lateral wheelset displacement largely increases and the hunting motion of the bogie is likely to occur. If all the 4 lateral dampers are disabled at the same time, the lateral displacements of both the wheelset and the car body are continually amplified. This means that the vehicle has already lost the lateral stability and it operates at great risk to derail. The above results show that the failure of vehicle dampers (such as lateral and yaw dampers) would greatly affect the dynamics performance and running safety of high-speed trains. Therefore, real-time monitoring of the service status of these key dampers and components is very important for ensuring the safe operation of high-speed trains.

8.4.2 Dynamic Response to Fracture of Fastener Clips

Fatigue fracture of fastener clips (see Fig. 3.37 in Chap. 3) has been observed frequently in track section with short-pitch rail corrugations, serious rail weld defects, etc., which greatly affect the running safety and ride comfort of rail vehicles. When the clip of a rail fastener fractures, the fastening system can no longer provide the lateral constraint to the rail, although the vertical support will prevail. In such a situation, the rail lateral deformation and vibration largely increase (see Fig. 8.34). The drastic fluctuation of wheel–rail force and the wheel load reduction commence (see Fig. 8.35) when a vehicle passes over the track section with the disabled fastener clips.

Figures 8.34 and 8.35 show the rail deformation and the dynamic vertical wheel–rail force of a high-speed vehicle negotiating a curved track section (same as considered in Sect. 8.1.3) with three adjacent disabled fastener clips at a speed of 300 km/h. It is observed that the lateral rail deformation increased approximately

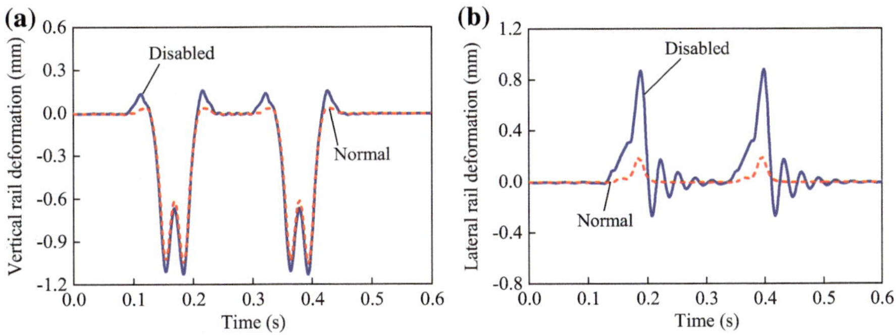

Fig. 8.34 Rail deformation due to fracture of fastener clips: **a** vertical and **b** lateral

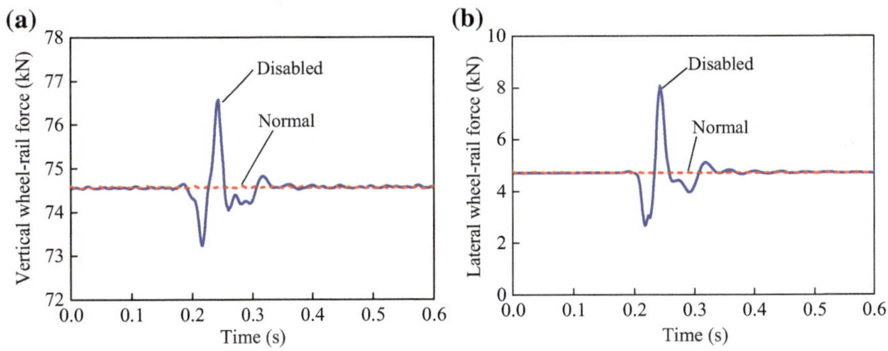

Fig. 8.35 Wheel–rail forces due to fracture of fastener clips: **a** vertical and **b** lateral

fivefold, with the associated enlargement of the vertical deformation. The vertical and lateral wheel–rail forces on the track with three adjacent disabled fastener clips increased by 5% and 1 time compared to the normal track. Due to the disabled fastener clips, the vertical and lateral accelerations of the rail increased by 30% and 5 times, respectively. The above results show that the failure of fastener clips can greatly aggravate the wheel–rail interaction, enlarge the rail deformation and vibration, and cause heavy damages to the railway tracks, which affect the ride comfort and running safety of rail vehicles.

8.4.3 Dynamic Response to Unsupported Sleepers

Unsupported sleepers or hanging sleepers (see Fig. 3.38 in Chap. 3) are very common in the ballasted track due to the nonuniform settlement of the ballast bed.

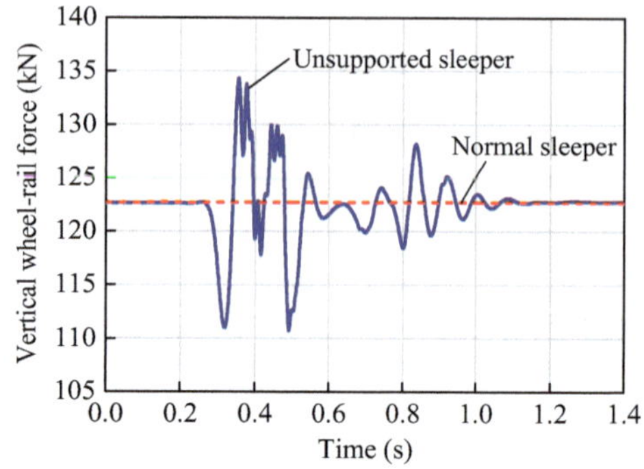

Fig. 8.36 Vertical wheel–rail force due to unsupported sleepers

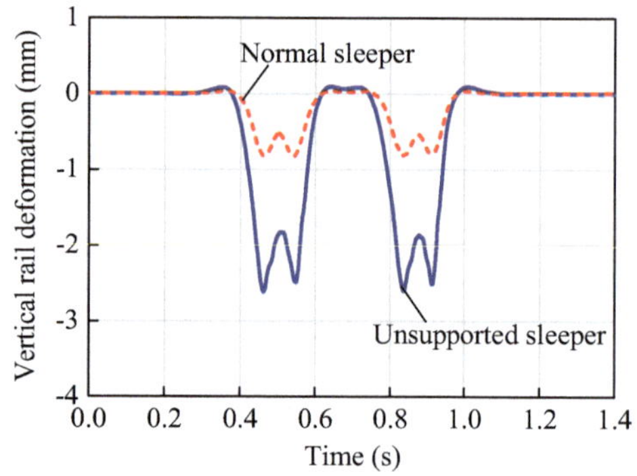

Fig. 8.37 Vertical rail deformation due to unsupported sleepers

The unsupported sleepers lower the rail support stiffness, increase the discontinuity to track system. It can also cause wheel load reduction and increase the rail deformation substantially under the vehicle operation, as shown in Figs. 8.36 and 8.37.

Figures 8.36 and 8.37 depict the dynamic vertical wheel–rail force and the rail deformation of a freight wagon negotiating a straight track section with two adjacent unsupported sleepers at a speed of 80 km/h. It can be seen that the vertical wheel–rail force on the track with two adjacent unsupported sleepers increase by 10% compared to the normal track, and the rail deformation increase about two times. This means that the hanging sleepers can aggravate the wheel–rail interaction and damage the track structures. It also reduces the ride comfort and influences the safe operation of rail vehicles.

8.5 Dynamic Response of Vehicle–Track Interaction to Random Irregularities

The geometry irregularities at wheel–rail interface are random and diversiform in the real railways. In this section, the dynamic response of a high-speed vehicle traveling on a ballastless track with random irregularities is presented as an example to illustrate the vibration characteristics of the vehicle–track coupled system under the excitation of random track irregularities. In the simulation, the Chinese high-speed train "CRH380A" and the Chinese high-speed ballastless track "CRTS-II" were considered. The train running speed was set as 300 km/h. The Chinese high-speed ballastless track spectrum (see Fig. 3.42 in Chap. 3) was applied to generate the random track irregularities. The dynamic responses of the high-speed vehicle–track system, including the frequency spectrum (PSD-power spectrum density) of the vertical and lateral accelerations of the car body, bogie frame, wheelsets, wheel–rail force, rail, and slab, are shown in Figs. 8.38, 8.39, 8.40, 8.41, 8.42, 8.43, 8.44, 8.45, 8.46, 8.47, 8.48 and 8.49.

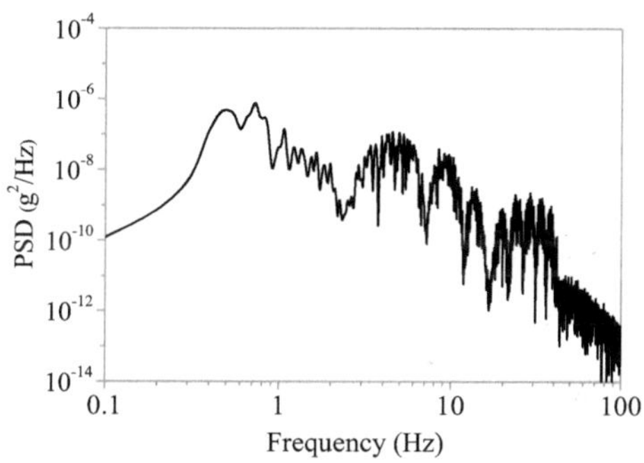

Fig. 8.38 Frequency spectrum of the vertical car body acceleration

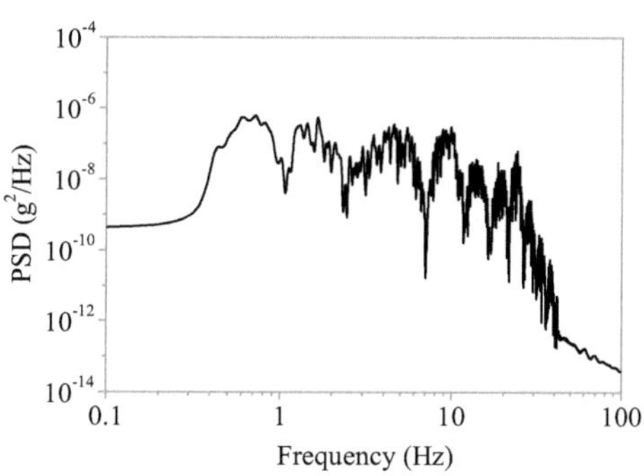

Fig. 8.39 Frequency spectrum of the lateral car body acceleration

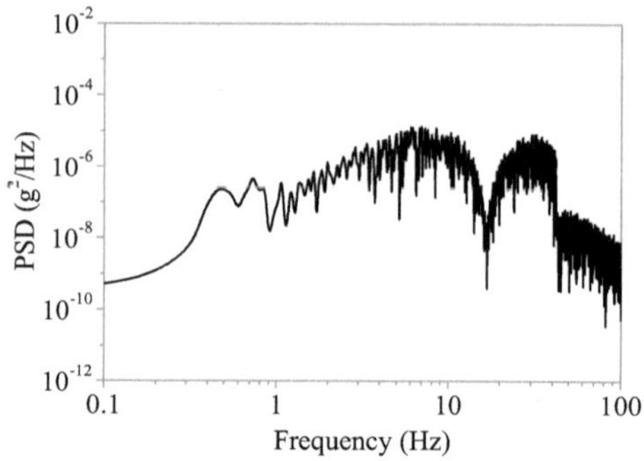

Fig. 8.40 Frequency spectrum of the vertical bogie frame acceleration

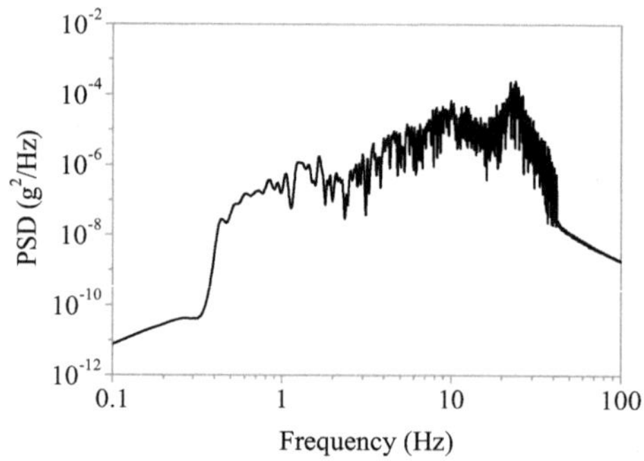

Fig. 8.41 Frequency spectrum of the lateral bogie frame acceleration

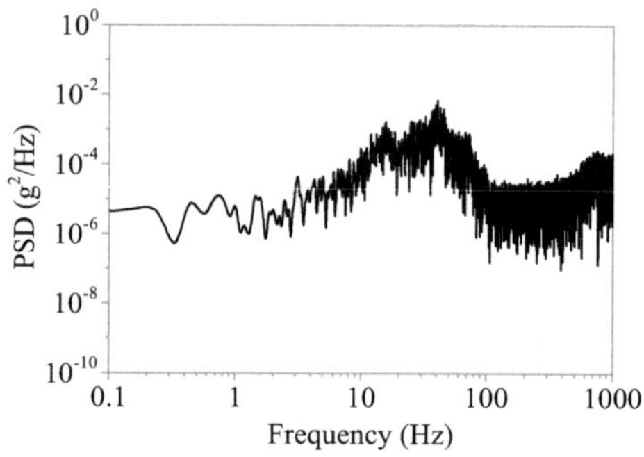

Fig. 8.42 Frequency spectrum of the vertical wheelset acceleration

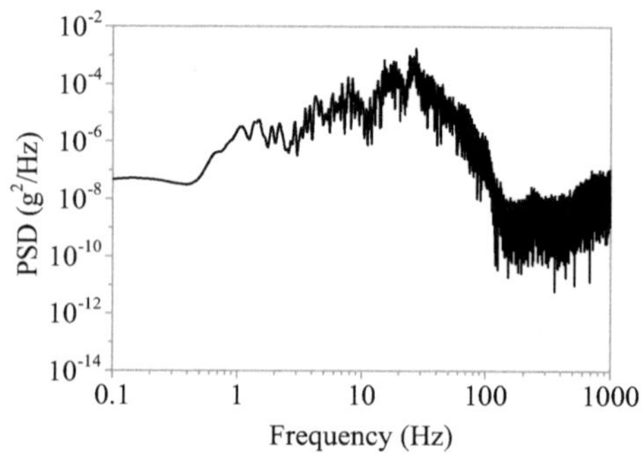

Fig. 8.43 Frequency spectrum of the lateral wheelset acceleration

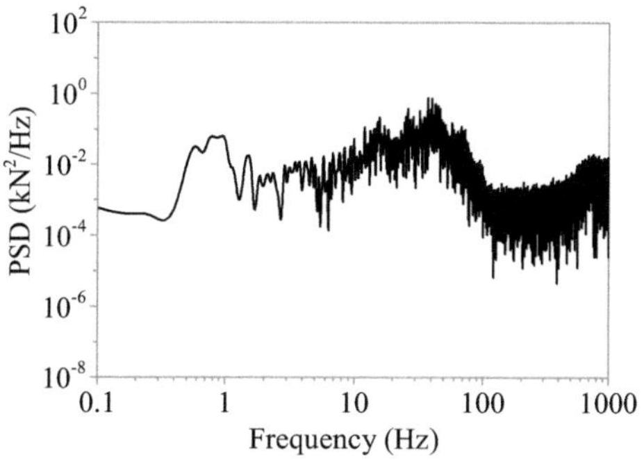

Fig. 8.44 Frequency spectrum of the vertical wheel–rail force

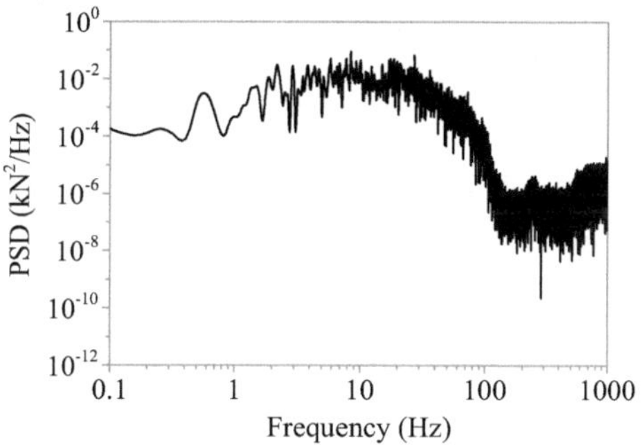

Fig. 8.45 Frequency spectrum of the lateral wheel–rail force

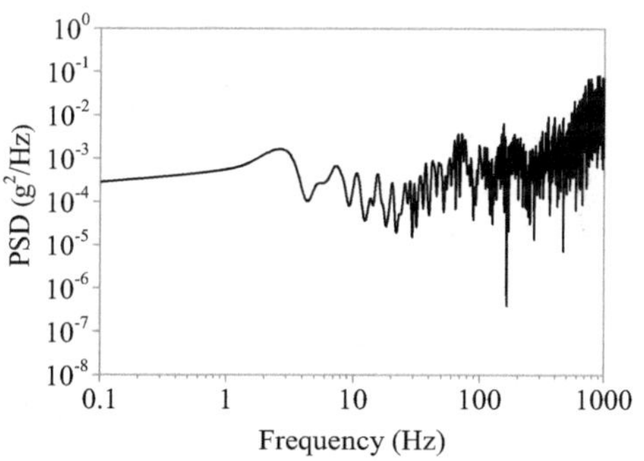

Fig. 8.46 Frequency spectrum of the vertical rail acceleration

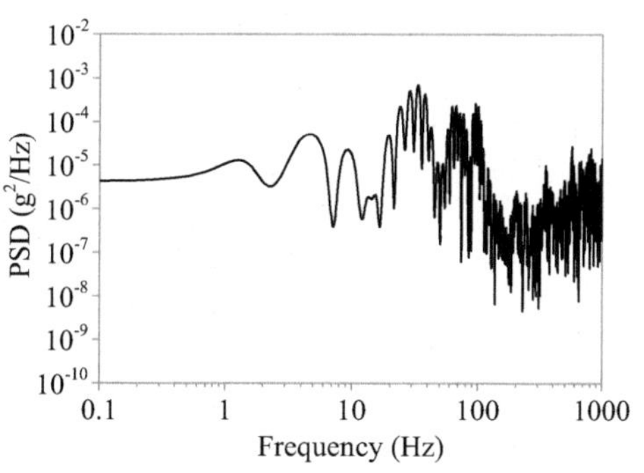

Fig. 8.47 Frequency spectrum of the lateral rail acceleration

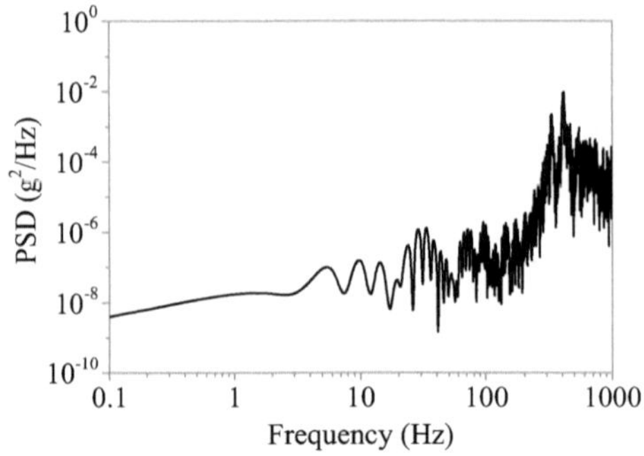

Fig. 8.48 Frequency spectrum of the vertical track slab acceleration

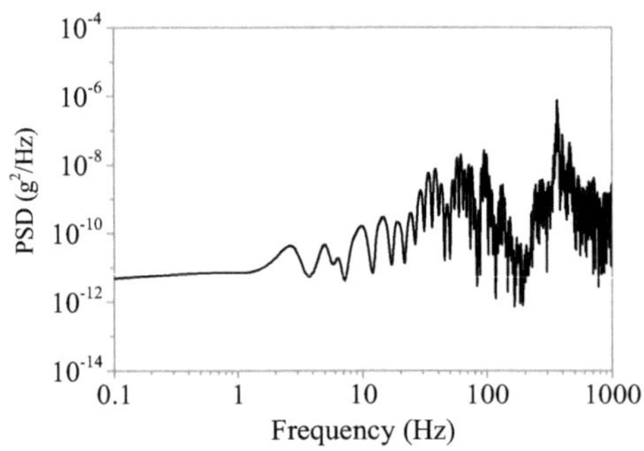

Fig. 8.49 Frequency spectrum of the lateral track slab acceleration

8.5.1 Vibration Characteristics of the Car Body

Figures 8.38 and 8.39 show the dynamic responses of the vertical and lateral accelerations of the car body in the frequency domain. As shown in Fig. 8.38, the frequency peaks of the vertical acceleration of the car body are in the range of 0.5–40 Hz. The first frequency peak is near 1 Hz, which represents the natural vibration frequency of the car body vertical and pitching motions [5, 6]. The other frequency peaks around 4, 7 and 10 Hz are mainly related to vibration modes of the bogie system. The frequency peaks around 30 Hz mainly reflect the forced vibration induced by the excitations from the vertical wheel–rail interaction. Figure 8.39 indicates that the distinct frequency peaks of the lateral acceleration of the car body are mainly in the range of 0.5–30 Hz. The first two frequency peaks are around 0.7 and 2.0 Hz, which are determined by the natural vibration frequency of the car body lateral and yawing motions [7]. The other frequency peaks near 4, 10, 13, 19 and 25 Hz reflect the vibration modes of the bogie system and the forced vibration induced by the excitations from the lateral wheel–rail interaction.

8.5.2 Vibration Characteristics of the Bogie Frame

Figures 8.40 and 8.41 depict the frequency spectrums of the vertical and lateral accelerations of the bogie frame. As shown in Fig. 8.40, the frequency peaks of the vertical vibration of the bogie frame are in the range of 0.5–40 Hz. The frequency peaks around 1, 7 and 30 Hz mainly reflect the vibration modes of the bogie system and the forced vibration induced by the excitations from vertical wheel–rail interaction. Figure 8.41 shows that the distinct frequency peaks of the lateral acceleration of the bogie frame are also in the range of 0.5–40 Hz. The frequency peaks around 2, 10, and 25 Hz are affected by the natural vibration frequency of the bogie system and the forced vibration induced by the excitations from the lateral wheel–rail interaction.

8.5.3 Vibration Characteristics of the Wheelset

Figures 8.42 and 8.43 show the dynamic responses of the vertical and lateral accelerations of the wheelset in the frequency domain. As presented in Fig. 8.42, the vibration frequency of the vertical acceleration of the wheelset is mainly in the range of 10–800 Hz. Due to the large contact stiffness between the wheel and the rail, the high-frequency contact vibration in the wheel–rail interface excited by the track irregularity is transmitted to the wheelset. Therefore, the frequency peaks of the vertical wheelset acceleration appeared at 30–50 Hz and 400–800 Hz mainly reflect the forced vibration induced by the excitations from the vertical wheel–rail interaction. Figure 8.43 indicates that the distinct frequencies of the lateral wheelset acceleration are mainly in the range of 1–100 Hz. The frequency peaks near 1.5, 8, 15, 25, and 230 Hz reflect the vibration modes of the bogie system and the forced vibration induced by the excitations from the lateral wheel–rail interaction. It is evident from Figs. 8.42 and 8.43 that the frequency spectrums of the vertical and lateral wheelset accelerations are quite different, which is determined by the characteristics of the vertical and lateral wheel–rail forces applied to the wheelset. The vertical force is the Hertzian contact force generated by the elastic deformation at the wheel–rail interface, while the lateral force is the creep force due to the relative slip between the wheel tread and the railhead. Obviously, the nonlinear characteristics of the Hertzian elastic contact force and the creep force are different.

8.5.4 Characteristics of the Wheel–Rail Forces

Figures 8.44 and 8.45 show the frequency spectrums of the vertical and lateral wheel–rail forces. Due to the high running speed and the large contact stiffness between the wheel and the rail, the high-frequency contact vibration in the wheel–

rail interface excited by the track irregularity is significant. The dynamic response of the vertical wheel–rail force is in a wide frequency range below 1000 Hz, which covers three distinct dominant frequency ranges, as shown in Fig. 8.44. The first main frequency is near 1 Hz, which represents the natural vibration frequency of the vertical car body suspension. The secondary frequency peak appears around 34 Hz, which is mainly related to the coupled vibration of the wheel–rail system. The third one is the high-frequency peak near 700 Hz, which reflects the high-frequency Hertzian contact vibration occurring at the wheel–rail interface. It can be seen that the frequency peak around 34 Hz has the maximum value. This means the wheel–rail coupled (or resonant) vibration has the biggest contribution to the vertical wheel–rail interaction force.

Figure 8.45 indicates that the lateral wheel–rail force is much smaller than the vertical one. The dynamic response of the lateral wheel–rail force is mainly distributed in the frequency range below 100 Hz. The frequency peaks near 0.6, 2 and 25 Hz reflect the vibration modes of the vehicle system and the forced vibration induced by the excitations from the lateral track irregularity. The above results show that the frequency spectrums of the vertical and lateral wheel–rail forces are quite different. Therefore, a clear understanding of the dynamic characteristics of the wheel–rail force is very important for developing design methods to minimize the wheel–rail interaction.

8.5.5 Vibration Characteristics of the Rail

Figures 8.46 and 8.47 illustrate the dynamic responses of the vertical and lateral accelerations of the rail in the frequency domain. From these figures, it can be observed that the vertical and lateral rail vibration frequencies are distributed in a wide range of frequency up to thousand Hz due to the high-frequency contact vibration at the wheel–rail interfaces. The lateral rail vibration is weaker than its vertical vibration. The lateral rail acceleration has several distinct frequency peaks in the range of 30–1000 Hz, while the frequency peaks of the vertical acceleration are in higher frequency range due to the vertical rail support stiffness being higher than the lateral stiffness. This is also because the vertical wheel–rail force is larger than the lateral force, as shown in Figs. 8.44 and 8.45. The above results are consistent with the experimental results reported in Ref. [8].

8.5.6 Vibration Characteristics of the Track Slab

Figures 8.48 and 8.49 show the frequency spectrums of the vertical and lateral accelerations of the track slab. It can be seen that the frequency peaks of the vertical acceleration of the track slab are mainly in the range of 300–700 Hz, while the vibration acceleration below 300 Hz is much lower. The lateral acceleration frequency is distributed in a wide frequency range of 30–500 Hz, which covers two distinct dominant frequency ranges of 30–140 Hz and 300–500 Hz. It is obvious that the vertical and lateral vibration characteristics of the track slab are quite different due to the differences in the wheel–rail interface and the rail vibrations upon the track slabs.

The aforementioned simulation results demonstrate that the primary and secondary suspension systems can effectively attenuate the vehicle system vibrations. Both the vertical and the lateral vibration amplitudes of the wheelset, the bogie frame, and the car body decrease successively when the wheel–rail vibration is transmitted upwards from the wheelset to the car body. The high-frequency vibrations of the bogie frames, as well as the car body high-frequency vibration above 40 Hz, are effectively suppressed. The wheel–rail vibration is also transmitted downwards from the rail to the track slab. Both the vertical and the lateral vibration amplitudes of the wheel–rail force, the rail, and the track slab decrease successively. However, for the high-speed railway tracks, the high-frequency vibrations at the wheel–rail interface can also have important effects on the dynamic behavior of rail infrastructure.

It is obvious that the vibration characteristics and frequency spectrums of the vehicle–track coupled system are highly related to the train and track structure parameters. But for different types of trains and track structures, their vibration characteristics are similar [9, 10]. Therefore, the simulated vibration characteristics of other train and track structures are not shown here.

8.6 Dynamic Response Due to Railway Infrastructure Settlement

8.6.1 Dynamic Response Due to Differential Subgrade Settlement

The superior integrity and stability have made the ballastless track being a preferred track form in high-speed railways. The use of the ballastless track on soil subgrade is still a major concern in view of the post-construction settlement, especially the differential subgrade settlement. Since soft soil ground with large compressibility and low permeability is widely distributed in China [11], the differential settlement on soft soil subgrade along the high-speed railway lines is a common occurrence due to the nonuniform soil properties, variation of groundwater-level and other defects of subgrade [12, 13]. Relevant monitoring has shown that in some regions, subgrade settlements of high-speed railways can be quite serious, sharply changing the original gradient of the railway line such that speed restrictions are warranted. Due to the limited adjustability of the ballastless track, the control of differential settlement becomes a key parameter of significance in the high-speed railways.

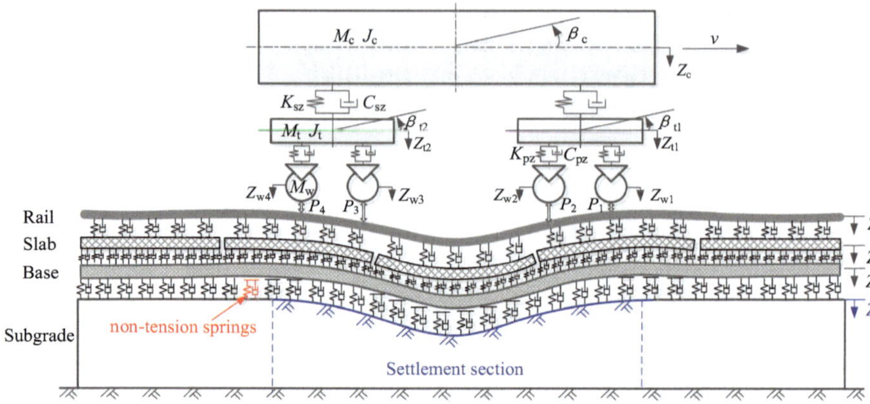

Fig. 8.50 Vehicle–slab track coupled dynamics model with differential subgrade settlement

1. *Vehicle–track coupled dynamics model with differential subgrade settlement*

To simulate the dynamic impact of the differential subgrade settlement on high-speed vehicle–ballastless track coupled system, the preceding vehicle–slab track coupled dynamics model is elaborated by taking into account the track weight and the contact between the concrete track and the soil subgrade as shown in Fig. 8.50. The differential subgrade settlement is introduced into the model as a given boundary condition, and the no-tension springs are adopted to account for the compression-only subgrade support. This model can be applied to the dynamic evaluation of the system affected by differential subgrade settlement, which is detailed in Ref. [14].

Taking the track weight into account, the second-order differential equations of the rail, the slab, and the base vibrations can be written as

$$\begin{cases} \ddot{q}_{\mathrm{r}k}(t) + \frac{E_\mathrm{r} I_\mathrm{r}}{m_\mathrm{r}}\left(\frac{k\pi}{l}\right)^4 q_{\mathrm{r}k}(t) = -\sum\limits_{i=1}^{N} F_{\mathrm{p}i}(t) Y_{\mathrm{r}k}(x_i) + \sum\limits_{j=1}^{4} p_j(t) Y_{\mathrm{r}k}(x_{\mathrm{w}j}) + C_{\mathrm{r}k}, & k = 1\text{–}NM_\mathrm{r} \\ m_\mathrm{s} l_\mathrm{s} \ddot{T}_k^{(m)}(t) + E_\mathrm{s} I_\mathrm{s} l_\mathrm{s} \beta_k^4 T_k^{(m)}(t) = \sum\limits_{i=1}^{N_1} F_{\mathrm{p}i}(t) X_k^{(m)}(x_i) - \sum\limits_{j=1}^{N_2} F_{\mathrm{s}j}(t) X_k^{(m)}(x_j) + D_k, & k = 1\text{–}NM_\mathrm{s} \\ \ddot{q}_{\mathrm{b}k}(t) + \frac{E_\mathrm{b} I_\mathrm{b}}{m_\mathrm{b}}\left(\frac{k\pi}{l}\right)^4 q_{\mathrm{b}k}(t) = \sum\limits_{i=1}^{N_2 \times N_\mathrm{s}} F_{\mathrm{s}i}(t) Y_{\mathrm{b}k}(x_i) - \sum\limits_{j=1}^{M} F_{\mathrm{b}j}(t) Y_{\mathrm{b}k}(x_j) + C_{\mathrm{b}k}, & k = 1\text{–}NM_\mathrm{b} \end{cases} \tag{8.1}$$

where $q_{\mathrm{r}k}(t)$, $T_k^{(m)}(t)$ and $q_{\mathrm{b}k}(t)$ are the generalized coordinate of the rail, slab, and base at time t; $Y_{\mathrm{r}k}(x)$, $Y_{\mathrm{b}k}(x)$, and $X_k^{(m)}(x)$ are the modal functions of the simply supported Euler beam (rail/base) and the mth free–free Euler beam (slab); $F_{\mathrm{p}i}(t)$, $F_{\mathrm{s}i}(t)$, and $F_{\mathrm{b}i}(t)$ are the forces of each fastener spring, mortar spring and subgrade spring; NM_r, NM_s, and NM_b are the orders of the vibration modes of the rail, slab and the base; C_{rk}, C_{sk} and D_{bk} are the addition items derived from track weight, which can be calculated using

$$\begin{cases} C_k = \frac{\sqrt{2ml}\cdot g}{k\pi}(1 - \cos k\pi) \\ D_k = m_s g \times \begin{cases} l_s, & \text{for } k = 1 \\ 0, & \text{for } k = 2 \\ \frac{\sinh(\beta_k l_s)}{\beta_k} + \frac{\sin(\beta_k l_s)}{\beta_k} - G_k\left[\frac{\cosh(\beta_k l_s)}{\beta_k} - \frac{\cos(\beta_k l_s)}{\beta_k}\right], & \text{for } k > 2 \end{cases} \end{cases} \tag{8.2}$$

where β_k and G_k are constant coefficients of the free–free beam.

Note that the subgrade support is modeled by no-tension springs, so the Heaviside function is employed here to describe the contact state between the track and the subgrade as follows:

$$H_{di}(t) = \begin{cases} 1, & \text{for } Z_b(x_i, t) > Z_0(x_i) \text{ (supported)} \\ 0, & \text{for } Z_b(x_i, t) \le Z_0(x_i) \text{ (unsupported)} \end{cases} \tag{8.3}$$

where $Z_b(x, t)$ is the vertical displacement of the concrete base; $Z_0(x)$ is the differential subgrade settlement. Thus, the force of the subgrade spring at each node can be written as

$$F_{bi}(t) = H_{di}\{k_b[Z_b(x_i, t) - Z_0(x_i)] + c_{bi}\dot{Z}_b(x_i, t)\} \tag{8.4}$$

Commonly, the cosine curve is utilized as the typical settlement pattern as shown in Fig. 8.51. In the settlement section, the subgrade displacement $Z_0(x_i)$ is defined as

$$Z_0(x_i) = \begin{cases} \frac{A}{2}\left[1 + \cos\left(\frac{2\pi}{s}\cdot\left(x_i - \frac{l}{2}\right)\right)\right], & \text{for } \frac{l}{2} - \frac{s}{2} < x_i < \frac{l}{2} + \frac{s}{2} \\ 0, & \text{others} \end{cases} \tag{8.5}$$

2. *Dynamic responses of vehicle–track coupled system induced by differential subgrade settlement*

The above model was applied to investigate the effect of differential subgrade settlement on the dynamic performance of the coupled system. A typical Chinese high-speed vehicle (Appendix A) and the CRTS-II ballastless track (Appendix D) were employed in the simulation. Since the distribution of the differential subgrade settlement is multifarious in practice, and the concrete track structure has a higher stiffness than the soil subgrade, which can resist the subgrade settlement to some

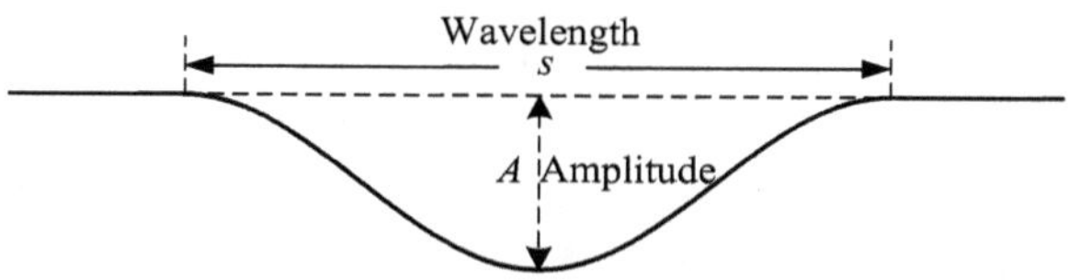

Fig. 8.51 Sketch of the differential subgrade settlement

extent. In this analysis, different combinations of settlement wavelength (10–30 m) and amplitude (10–30 mm) are simulated with a train speed of 300 km/h. To highlight the influence of the differential settlement, the random track irregularity is omitted in the simulation.

(1) Typical characteristics of the settlement-induced dynamic responses

The typical characteristics of the dynamic responses of the system due to differential subgrade settlement are illustrated where the differential settlement is 10 mm/15 m.

Figure 8.52 shows the variations of the vertical wheel–rail force of the 1st wheelset and the car body acceleration as the vehicle passing over the differential settlement section. The wheel–rail force in Fig. 8.52a shows an obvious drop when the wheelset moves into the settlement section, corresponding to a process of wheel unloading, and the maximum wheel unloading rate is approximately 0.10. Then a sharp rise of the wheel–rail force occurs when the wheelset climbs out of the settlement section, with a peak value of 76.17 kN around the settlement center. Due to the influence of the subsequent wheelsets, there is still a slight oscillation after the wheelset moves out of the settlement section. As shown in Fig. 8.52b, the vertical car body acceleration is quite sensitive to the differential settlement. Drastic vibration shows up as the two bogies of the vehicle successively pass through the settlement section, resulting in two complete sinusoidal cycles, with a maximum value of about 0.26 m/s^2. Due to the damping effect, the vibration of car body attenuates after the vehicle leaving the settlement section and will finally level off. Analysis reveals that there is a positive correlation between the decay rate of the car body vibration and the train speed.

Dynamic responses of the track structure induced by the combined effect of the subgrade settlement and moving vehicle are illustrated in Fig. 8.53, including the variations of the rail displacement and the fastening force at the settlement center, as well as the contact force between the track and subgrade.

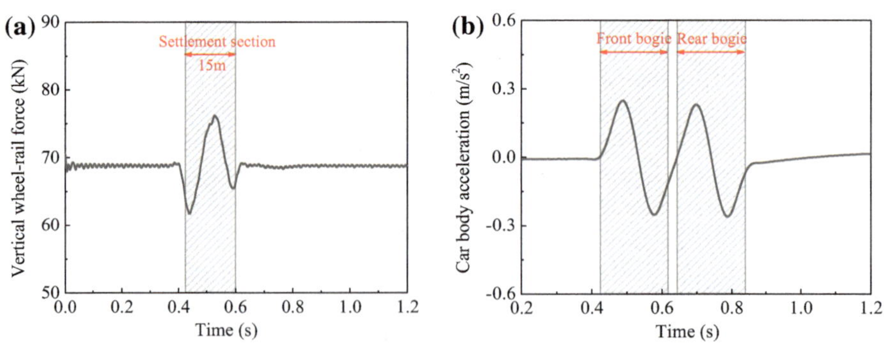

Fig. 8.52 Vehicle dynamic responses induced by differential subgrade settlement: **a** vertical wheel–rail force, and **b** car body acceleration

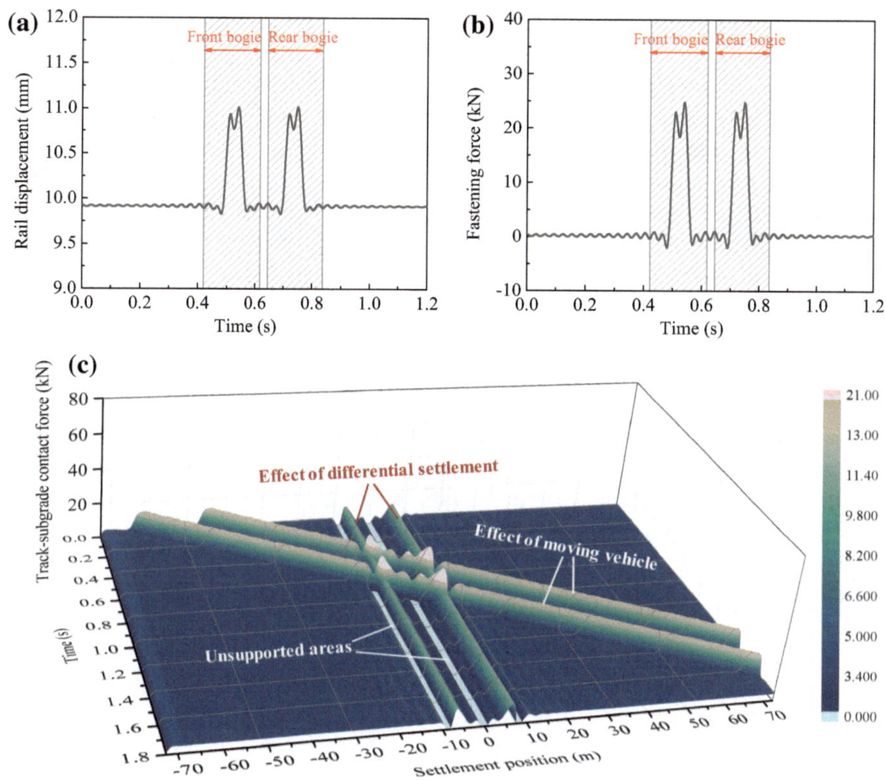

Fig. 8.53 Track dynamic responses induced by differential subgrade settlement: **a** rail displacement, **b** fastening force, and **c** track–subgrade contact force

From Fig. 8.53a, it can be seen that the initial rail displacement is not zero because the track settles along with the subgrade under the action of self-weight before the vehicle's arrival. However, due to the high bending stiffness of the concrete track, the track deflections vary as the ballastless track generally loses contact with the subgrade at some positions. The magnitude of the rail deflection above the center point of the settlement section, in this case, is a little lower than 10 mm. During the vehicle's passage, the wheelset-induced rail displacement is about 1 mm, so the amplitude of rail vibration is close to 11 mm.

Figure 8.53b shows the dynamic rail–slab contact force (i.e., fastening force). There are four peaks corresponding to the four wheelsets of the vehicle on the curve. The resistance force of the fastener at the settlement center also has an initial value caused by the differential settlement, but it is quite small compared to the vehicle-induced fastening force. The initial fastening force is only 0.6 kN, while that caused by the four wheelsets increases to about 22 kN.

Figure 8.53c shows the variation of the track–subgrade contact force during the entire process of the passage of the vehicle. Different from the fastening force connecting the rail and the slab, it shows that the contact force between the track and the subgrade is more affected by the differential settlement. In Fig. 8.53c, two apparent waves corresponding to the two bogies of the vehicle move over time, and two stationary wave crests located between −7.5 and 7.5 m are ascribed to the differential settlement section. Three unsupported areas along the track corresponding to the hanging sleepers around the settlement center are evident; in these areas, the track is separated from the subgrade and near the settlement boundaries the track is slightly arched. The track–subgrade contact forces in these unsupported areas are 0 as shown in the figure. Before the vehicle moves into the settlement zone, the wheelset-induced track–subgrade contact force is about 14 kN, and the maximum force caused by the settlement is 11 kN. When the wheelset moves close to the settlement area, the track–subgrade contact force shows significant variation. For those unsupported areas, the contact between track and subgrade is reestablished and lost again periodically during the passage of the vehicle. The track–subgrade contact forces at two settlement boundaries at ±7.5 m display stress concentrations significantly amplified by the moving vehicle. The peak force due to the superposition of the dynamic vehicle impact and the stationary settlement effect is approximately 20 kN.

(2) Influence of settlement wavelength

Figure 8.54 shows the effect of the settlement wavelength on the dynamic responses of the coupled system. The settlement wavelength has been increased from 10 to 30 m, and the settlement amplitude is set as 10 mm.

The variations of the dynamic indexes are nonlinear with an increase in the settlement wavelength. From the vertical wheel–rail force, wheel unloading rate and car body acceleration shown in Fig. 8.54a, b, it is clear that the settlement wavelength of 10 m exacerbates the wheel–rail interaction and the vehicle vibration the most. The maximum vertical wheel–rail force at 10 m is about 116.56 kN (increasing by about 70%) and the wheel unloading rate is 0.29. The maximum car body acceleration is 0.39 m/s^2. With the settlement wavelength expands to 15 m and larger, the settlement-induced maximum vertical wheel–rail force drops below 80 kN, and both the wheel–rail interaction and the car body acceleration gradually decline with the increase in settlement wavelength.

For the rail displacement at the settlement center in Fig. 8.54c, the settlement wavelength of 10 m also leads to an abnormal response. The initial rail deflection is only 7.06 mm, resulting from the significant contact failure between the track and subgrade around the settlement center. When the vehicle moves on, the rail vibration is intensified, and the maximum rail displacement increases to 11.63 mm. When the settlement wavelength is 15 m, the initial rail deflection is almost close to the subgrade deformation with a slight difference, so the vehicle-induced maximum rail displacement becomes normal, which is only 1.15 mm larger than the initial displacement. With the progressive increase in wavelength, the rail displacements remain stable, due to the track being unsupported.

The track–subgrade contact force at the settlement center in Fig. 8.54d shows that the initial force for the settlement wavelength of 10 m is 0, which means that

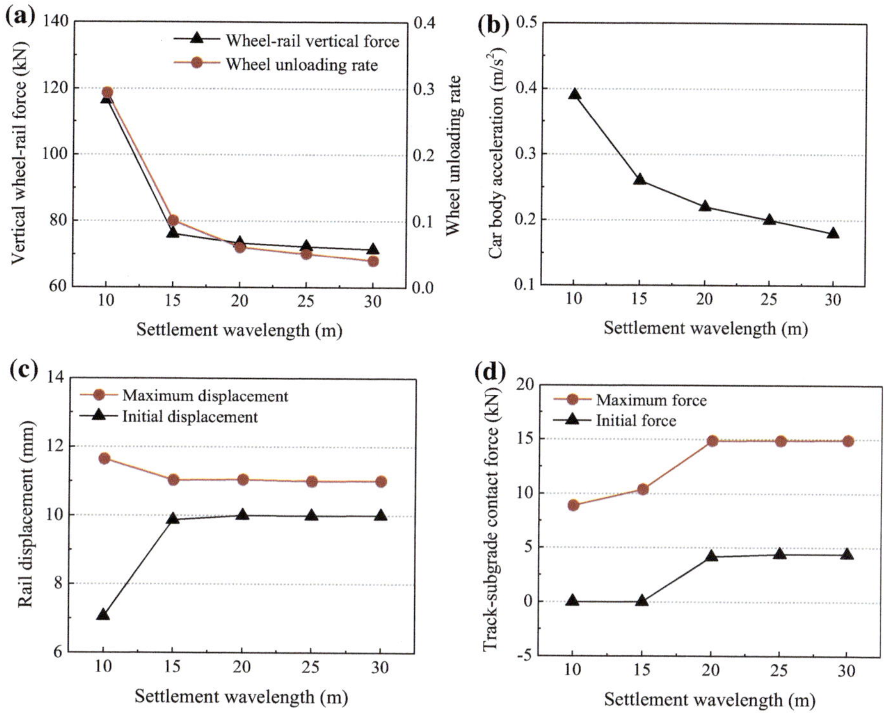

Fig. 8.54 Influence of settlement wavelength on system dynamic responses: **a** vertical wheel–rail interaction, **b** car body acceleration, **c** rail displacement, and **d** track–subgrade contact force

the initial contact between the track and the subgrade here is lost. Due to the hanging track structure, the vehicle-induced dynamic track–subgrade contact force is smaller, which is only 8.86 kN. With the increase of the settlement wavelength, the unsupported areas are gradually eliminated, and the difference between the vehicle-induced maximum track–subgrade contact force and the settlement-induced initial track–subgrade contact force stabilize around 10 kN.

The above results indicate that the dynamic performance of the vehicle–slab track system suffers slightly from the wide-range settlement. Besides, there exists a particular range of settlement wavelength which may excite the resonance of the vehicle structure to an extent and will significantly exacerbate the wheel–rail interaction. For the specific conditions considered in this analysis, it is around 10 m.

(3) Influence of settlement amplitude

Similarly, the influence of the settlement amplitude is illustrated in Fig. 8.55, with a constant settlement wavelength of 15 m.

Figures 8.55a, b show almost linear growth of the vertical wheel–rail force, wheel unloading rate and car body acceleration as the settlement amplitude

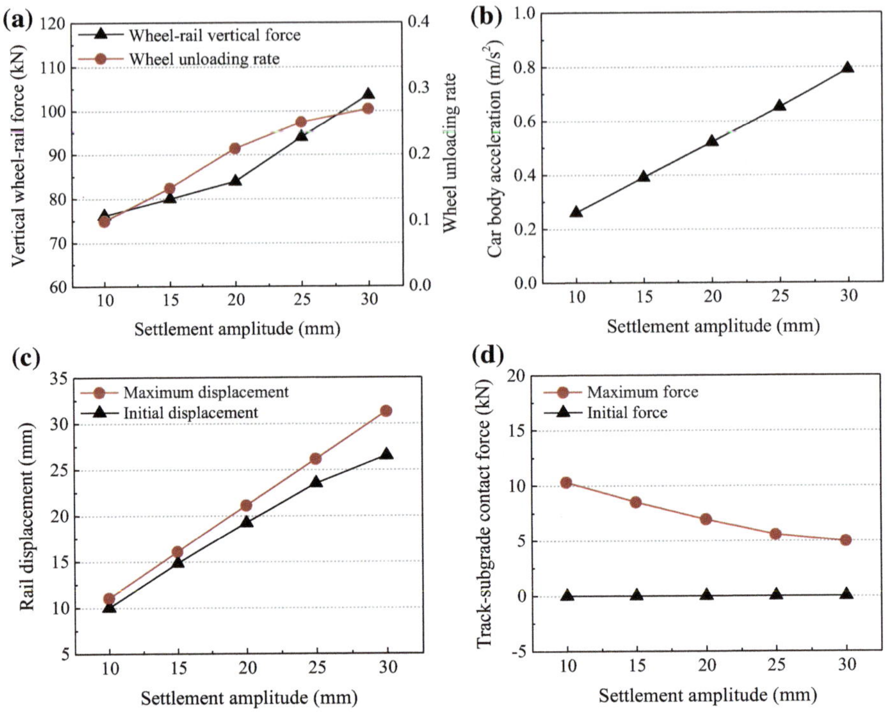

Fig. 8.55 Influence of settlement amplitude on system dynamic responses: **a** vertical wheel–rail interaction, **b** car body acceleration, **c** rail displacement, and **d** track–subgrade contact force

amplifies from 10 to 30 mm. It implies that the wheel–rail interaction and the ride comfort will be likely to exceed the critical limit if the settlement amplitude continues to worsen.

In Fig. 8.55c, it is visible that the initial rail displacement induced by the differential settlement without vehicle load also increases with the settlement amplitude but gradually slows down. For the settlement case of 30 mm/15 m, the initial rail deflection at the settlement center is 26.44 m. It indicates that the unsupported area between the track and the subgrade is increasing significantly. Additionally, the wheel–rail force is increasing, so the corresponding dynamic rail displacement shows a corresponding increase, and the difference between the maximum rail displacement and the initial displacement gradually increases from 1.09 to 4.82 mm.

It can be seen from Fig. 8.55d that at the settlement center, the initial track–subgrade contact force is always nil, it indicates that there is always a separation between the track and the subgrade. Due to the high rigidity of track structure, the gap between the track and the subgrade around the settlement center widens with the increasing settlement amplitude. Consequently, the dynamic impact from the vehicle load here is significantly attenuated as depicted by the red dotted line.

8.6.2 Dynamic Response Due to Differential Ballast Settlement

The settlement of the ballasted railway track mainly depends on the plastic deformation of the subgrade and the degradation of the ballast accounting for train dynamic loading as well as the combination of geologic conditions and ambient conditions [15]. When the external factors are different along the track, the differential settlement of the subgrade and ballast occurs [16]. The rail and the sleepers settle along with the ballast due to gravity. The rail surface deflection finally turns into a track irregularity [17]. With an increase in the ballast settlement, the separation of the sleepers from the ballast bed may appear in local areas, which will lead to the formation of unsupported sleepers [18]. Track irregularity and unsupported sleepers resulting from the settlement have a significant influence on the dynamic response of the vehicle–track coupled system.

1. *Vehicle–track coupled dynamics model with differential ballast settlement*

Since ballast is composed of granular material, the relationship between the subgrade settlement and the ballast settlement is affected by many complex factors. An improved vehicle–track coupled dynamics model [19] that concerns the effect of the differential ballast settlement is shown in Fig. 8.56.

Similar to the differential subgrade settlement model proposed in Sect. 8.6.1, three key points are included in this ballast settlement model: (1) the weight of rail and sleeper; (2) no-tension springs accounting for the compression-only ballast support; (3) the differential ballast settlement as a given boundary condition and additional forces to keep the ballast blocks in the given settlement positions when

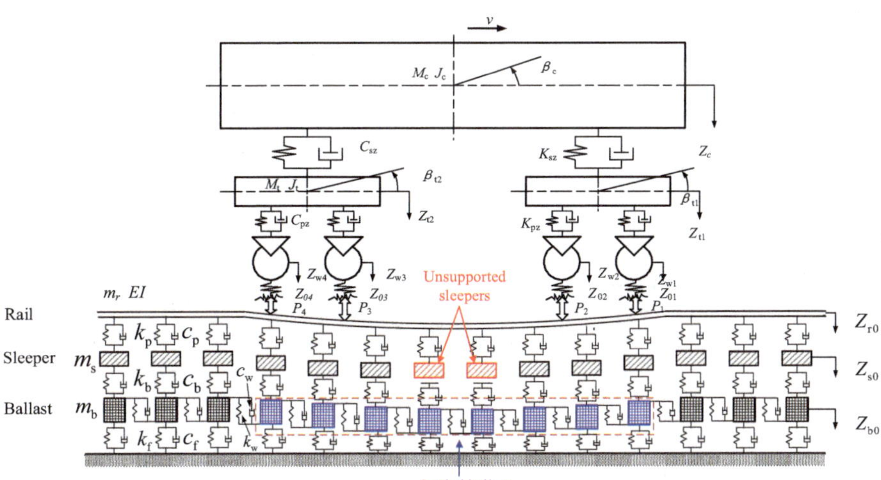

Fig. 8.56 Vehicle–track coupled dynamics model with differential ballast settlement

the track system is in equilibrium. Taking these three key points into consideration, the equations of motion of the track subsystem can be modified as follows:

(1) Vertical rail motion

Taking the weight of rail into account, the equation of the vertical rail motion can be written as

$$\ddot{q}_k(t) + \frac{EI}{m_\mathrm{r}}\left(\frac{k\pi}{l}\right)^4 q_k(t) = -\sum_{i=1}^{N} F_{\mathrm{rs}i}(t)Y_k(x_i) + \sum_{j=1}^{4} p_j(t)Y_k\left(x_{\mathrm{w}j}\right) + C_k, \quad (k = 1\text{–}NM) \tag{8.6}$$

with

$$C_k = \frac{\sqrt{2m_\mathrm{r}l}\cdot g}{k\pi}(1-\cos k\pi) \tag{8.7}$$

(2) Vertical sleeper motion

Taking the weight of sleeper into account, the equation of the vertical sleeper motion can be written as

$$m_\mathrm{s}\ddot{Z}_{\mathrm{s}i}(t) = m_\mathrm{s}g + F_{\mathrm{bs}i}(t) - F_{\mathrm{rs}i}(t) \quad (i = 1\text{–}N) \tag{8.8}$$

where $F_{\mathrm{bs}i}(t)$ is the contact force between the ith sleeper and the ballast. As the ballast support is modelled by the non-tension springs and only the fully supported or unsupported sleepers are considered, $F_{\mathrm{bs}i}(t)$ can be described as

$$F_{\mathrm{bs}i}(t) = H_{di}(t)\left\{k_\mathrm{b}[Z_{\mathrm{s}i}(t) - Z_{\mathrm{b}i}(t)] + c_\mathrm{b}\left[\dot{Z}_{\mathrm{s}i}(t) - \dot{Z}_{\mathrm{b}i}(t)\right]\right\} \tag{8.9}$$

where H_{di} is Heaviside function, which can be written as follows:

$$H_{di}(t) = \begin{cases} 1, & \text{for } Z_{\mathrm{s}i}(t) > Z_{\mathrm{b}i}(t) \text{ (supported)} \\ 0, & \text{for } Z_{\mathrm{s}i}(t) \le Z_{\mathrm{b}i}(t) \text{ (unsupported)} \end{cases} \tag{8.10}$$

(3) Vertical ballast motion

The vibration differential equation of the ballast block can be described as

$$\begin{aligned} &M_\mathrm{b}\ddot{Z}_{\mathrm{b}i}(t) + (2c_\mathrm{w} + c_\mathrm{f})\dot{Z}_{\mathrm{b}i}(t) + (2k_\mathrm{w} + k_\mathrm{f})Z_{\mathrm{b}i}(t) - c_\mathrm{w}\dot{Z}_{\mathrm{b}(i-1)}(t) - c_\mathrm{w}\dot{Z}_{\mathrm{b}(i+1)}(t) \\ &- k_\mathrm{w}Z_{\mathrm{b}(i-1)}(t) - k_\mathrm{w}Z_{\mathrm{b}(i+1)}(t) - F_{\mathrm{bs}i}(t) - F_{\mathrm{b}i0} = 0 \qquad (i = 1\text{–}N), \end{aligned} \tag{8.11}$$

where $F_{\mathrm{b}i0}$ is the additional force applied to the ith ballast block to keep the ballast blocks in the given settlement position, which can be calculated by

$$F_{\mathrm{b}i0} = k_{\mathrm{b}} H_{di0}(Z_{\mathrm{b}i0} - Z_{\mathrm{s}i0}) + k_{\mathrm{f}} \cdot Z_{\mathrm{b}i0} + k_{\mathrm{w}}\left(2Z_{\mathrm{b}i0} - Z_{\mathrm{b}i-1,0} - Z_{\mathrm{b}i+1,0}\right) \quad (i = 1\text{–}N) \tag{8.12}$$

where $Z_{\mathrm{b}i0}$ is the given ballast settlement, $Z_{\mathrm{s}i0}$ is the vertical displacement of the sleeper induced by the ballast settlement when the whole track system is in equilibrium state without vehicle load. Detailed description of the solution method to calculate $Z_{\mathrm{b}i0}$ and $Z_{\mathrm{s}i0}$ can be seen in Ref. [9].

The cosine curve shown in Fig. 8.51 is used to describe the ballast settlement. $Z_{\mathrm{b}i0}$ is defined as

$$Z_{\mathrm{b}i0} = \begin{cases} \frac{A}{2}\left[1 + \cos\left(\frac{2\pi}{s} \cdot \left(x_i - \frac{l}{2}\right)\right)\right], & \text{for } \frac{l}{2} - \frac{s}{2} < x_i < \frac{l}{2} + \frac{s}{2} \\ 0, & \text{others} \end{cases} \tag{8.13}$$

2. *Dynamic responses of vehicle–track coupled system induced by differential ballast settlement*

The dynamic performance of the coupled system when a Chinese high-speed vehicle (Appendix A) passes through the differential ballast settlement section is investigated for a train speed of 250 km/h. The track is a high-speed ballasted track (Fig. C.1 and Table C.1 in Appendix C). The rail length is set as 501 times the sleeper spacing. The total mode number of the rail mode functions (*NM*) is set as 500, equals to the number of the sleeper. The random track irregularity is ignored in the simulation. The typical characteristics of the dynamic responses of the system due to the differential ballast settlement are illustrated where the differential settlement is 10 mm/10 m.

The responses of the vertical wheel–rail force of the first wheelset and the car body acceleration are shown in Fig. 8.57. The variations of both the vertical wheel–rail force and the car body acceleration are similar to those shown in Fig. 8.52. The vertical wheel–rail force decreases first when the wheelset just moves into the settlement section, and then increases sharply until the vertical force reaches the peak value around the settlement center. The maximum wheel unloading rate and the vertical wheel–rail force are approximately 0.15 and 79.67 kN, respectively. The variation of the car body acceleration is similar to the two complete sinusoidal cycles which correspond to two bogies passing through the settlement zone. The car body acceleration attenuates slowly after the vehicle leaves the settlement zone.

Figure 8.58 shows the effect of the ballast settlement and moving vehicle on the response of the track system, including the dynamic gap between sleeper–ballast and the rail displacement at the settlement center, as well as the contact force between the sleeper and ballast. Figure 8.58a shows the dynamic gap between the sleeper and the ballast at the settlement center. The sleeper is at first unsupported when the vehicle is far away from the settlement zone, and the hanging gap size is

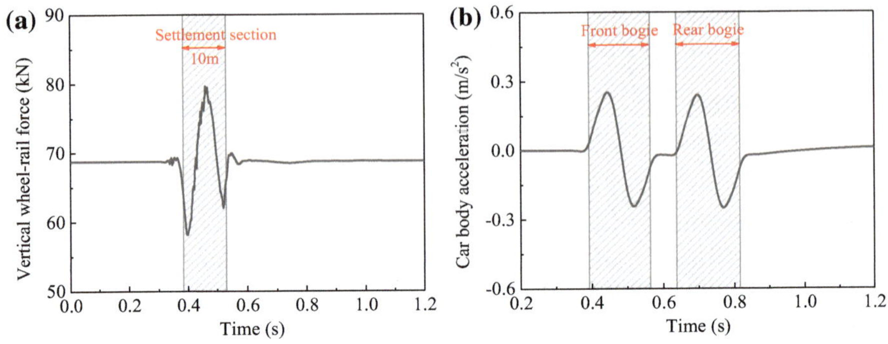

Fig. 8.57 Vehicle dynamic responses induced by differential ballast settlement: **a** vertical wheel–rail force and **b** car body acceleration

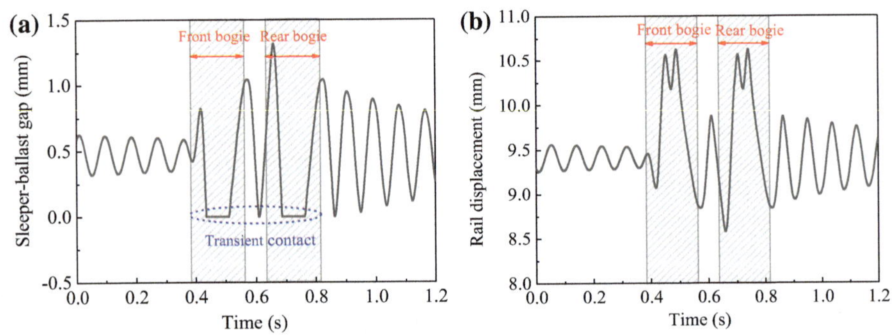

Fig. 8.58 Track dynamic responses induced by differential ballast settlement: **a** sleeper–ballast gap and **b** rail displacement

about 0.5 mm. The unsupported sleeper is transiently in contact with the ballast during the successive passages of two adjacent bogies through the settlement zone. It can be seen from Fig. 8.58a that the ballast settlement and moving vehicle are two main factors influencing the contact state of the unsupported sleeper during the vehicle's passage. Figure 8.58b shows the response of the rail displacement. The initial rail displacement is about 9.5 mm due to the track deflection under the action of self-weight. During the two bogies passing through the settlement zone, the amplitude of the rail vibration increases to 10.5 mm under the combined influence of the ballast settlement and the wheel–rail contact force. Due to lack of effective vibration reduction measures for the unsupported sleepers, the vibration of rail attenuates very slowly relative to that shown in Fig. 8.53a.

When the vehicle passes through the settlement zone, the contact states between the sleepers and the ballast progressively vary due to the settlement effect. Figure 8.59 shows the sleeper–ballast contact state induced by the differential ballast settlement and moving vehicle. In this figure, the horizontal axis shows the

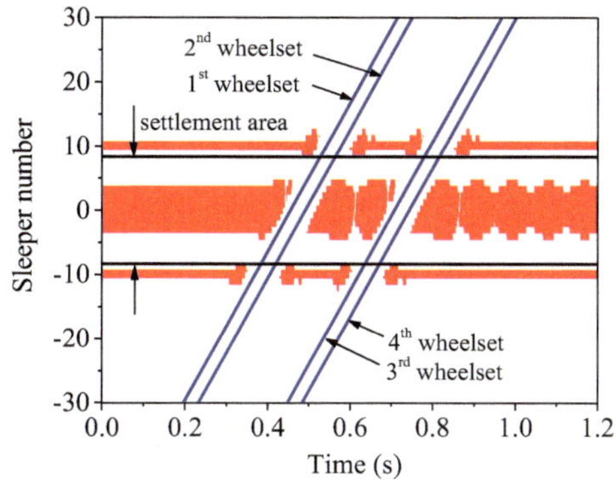

Fig. 8.59 Dynamic contact state between sleepers and ballast induced by differential ballast settlement

time range, and the vertical axis shows the sleeper number. The two black lines represent the boundary of the ballast settlement, and the middle area is the settlement zone. The four blue lines depict the locations of the four wheelsets at different times. The running direction of the vehicle is from sleeper −30 to sleeper +30. The full contact between sleeper and ballast is shown in white color, while the unsupported state is red-colored. It can be seen from Fig. 8.59 that there are three unsupported areas corresponding to the hanging sleepers at the settlement center zone and two arched areas near the settlement boundaries where the track is slightly arched. The initial unsupported sleepers instantaneously maintain contact with the ballast when the wheelset moves over, and immediately recover to the unsupported state after the wheelset moves off.

The sleeper–ballast contact force induced by both the differential ballast settlement and the moving vehicle is shown in Fig. 8.60. The variations of the contact force between the sleepers and the ballast of the settlement area are similar to the variations of the track–subgrade contact force that depicted in Fig. 8.53c. For each sleeper, the contact force reaches a peak value when one of the four wheelsets pass

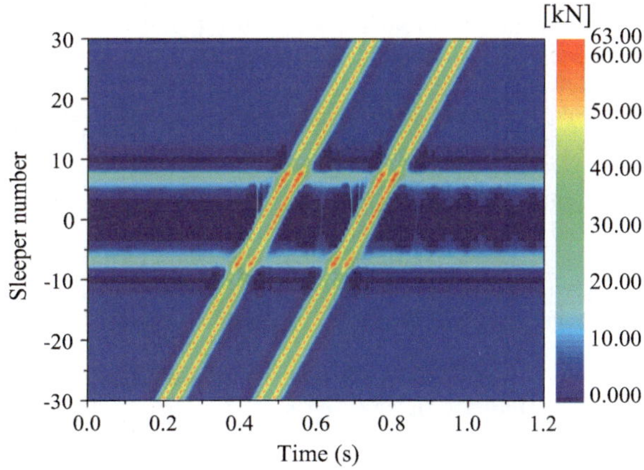

Fig. 8.60 Sleeper–ballast contact force induced by differential ballast settlement

through, as shown by the four tilted strips in the figure. The sleepers with larger contact forces are at the boundary of the settlement zone. This is because the sleepers beside the settlement boundary are all unsupported, and those well-supported sleepers at the boundary need to carry the gravity and the inertial forces of the rail and the nearby sleepers. At the boundaries of the settlement zone, the maximum sleeper–ballast contact force is 63.0 kN, which is about 54.7 kN at the non-settlement area.

References

1. Jenkins HH, et al. The effect of track and vehicle parameters on wheel–rail vertical dynamic forces. Railw Eng J. 1974;3(1):2–16.
2. Gao JM, Zhai WM, Guo Y. Wheel–rail dynamic interaction due to rail weld irregularity in high-speed railways. Proc Inst Mech Eng Part F: J Rail Rapid Transit. 2018;232(1):249–61.
3. Wang P. Study on wheel–rail system dynamics in switch area. Ph.D. thesis. Chengdu: Southwest Jiaotong University; 1997 (in Chinese).
4. Ren ZS. Study on vehicle–turnout coupled system dynamics. Ph.D. thesis. Chengdu: Southwest Jiaotong University; 2000 (in Chinese).
5. Wang TF. Vehicle system dynamics. Beijing: China Railway Press; 1994 (in Chinese).
6. Garg VK, Dukkipati RV. Dynamics of railway vehicle systems. Ontario: Academic Press Canada; 1984.
7. Zhai WM, Liu PF, Lin JH, Wang KY. Experimental investigation on vibration behaviour of a CRH train at speed of 350 km/h. Int J Rail Transp. 2015;3(1):1–16.
8. Shenton MJ. The vibrational environment of the rail, sleeper and ballast. Technical Memorandum. Derby: British Rail Research; 1974.
9. Zhai WM. Simulation and experiment of railway wheel/rail impact vibrations. Chin J Comput Mech. 1999;16(1):93–9 (in Chinese).
10. Chen G. Analysis of the random vibration of the vehicle–track coupling system. Ph.D. thesis. Chengdu: Southwest Jiaotong University; 2000 (in Chinese).
11. Wang BL. Subgrade and track engineering of high-speed railway. Shanghai: Tongji University Press; 2015 (in Chinese).
12. Chen RP, Chen JM, Zhao X, et al. Cumulative settlement of track subgrade in high-speed railway under varying water levels. Int J Rail Transp. 2014;2(4):205–20.
13. Olivier B, Connolly DP, Costa PA, et al. The effect of embankment on high speed rail ground vibrations. Int J Rail Transp. 2016;4(4):229–46.
14. Guo Y, Zhai WM, Sun Y. A mechanical model of vehicle–slab track coupled system with differential subgrade settlement. Struct Eng Mech. 2018;66(1):15–25.
15. Ngo NT, Indraratna B, Rujikiatkamjorn C. Stabilization of track substructure with geo-inclusions—experimental evidence and DEM simulation. Int J Rail Transp. 2017;5(2):63–86.
16. Dahlberg T. Some railroad settlement models—a critical review. Proc Inst Mech Eng Part F: J Rail Rapid Transit. 2001;215(4):289–300.
17. Li X, Nielsen JC, Pålsson BA. Simulation of track settlement in railway turnouts. Veh Syst Dyn. 2014;52(sup1):421–39.
18. Lundqvist A, Dahlberg T. Load impact on railway track due to unsupported sleepers. Proc Inst Mech Eng Part F: J Rail Rapid Transit. 2005;219(2):67–77.
19. Sun Y, Guo Y, Chen ZG, et al. Effect of differential ballast settlement on dynamic response of vehicle track coupled systems. Int J Struct Stab Dyn. 2018;18(7):1850091-1–29.

Chapter 9
Principle and Method of Optimal Integrated Design for Dynamic Performances of Vehicle and Track Systems

Abstract In this chapter, concept, principle, and method of optimal integrated design for dynamic performances of vehicle and track systems are proposed based upon the vehicle–track coupled dynamics theory and its simulation software. Two case studies are provided for demonstrating the implementation of the proposed method, i.e., optimal design of suspension parameters of a heavy-haul locomotive, HX_D2C, for minimizing lateral wheel–rail dynamic interaction on small radius curves, and design of a steep gradient section on Guangzhou–Shenzhen–Hong Kong high-speed railway to ensure vehicle ride comfort and running safety.

9.1 Principle of Optimal Integrated Design for Dynamic Performances of Vehicle and Track Systems

It is well known that the operation of railway transportation networks relies on the interactions between wheel and rail. Dynamic wheel–rail forces play a key role in this function, since they are the main causes of vibrations, impacts, fatigue, and damage in the vehicle and track systems. They also result in degradation and failure of the wheel–rail system. Therefore, mitigating dynamic wheel–rail interactions is crucial to ensuring long-term efficient operations of railway transportation.

In order to alleviate the dynamic interactions between vehicles and tracks, the author has proposed the optimal integrated design concept for the dynamic performance of vehicles and tracks [1]. The "optimal integrated design" refers to the concept that the designs for the railway vehicle system and the track structure system should be adapted and adjusted with full consideration of each other. In this way, optimal results for the overall dynamic performance of the whole system can be achieved.

In order to achieve optimal integration of the dynamic performances of vehicle and track systems, the systematic design concept must be adopted. The vehicle system and the track system should be regarded as an integrated system with their interactive and coupled behaviors included. A comprehensive optimized design should be aimed at the satisfaction of overall vehicle–track system dynamic

W. Zhai, *Vehicle–Track Coupled Dynamics*,
https://doi.org/10.1007/978-981-32-9283-3_9

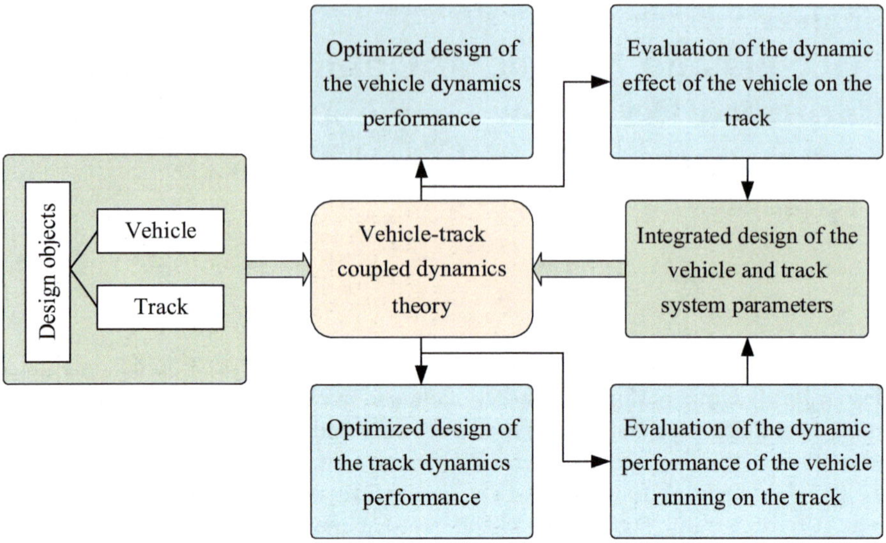

Fig. 9.1 Flowchart of the optimal integrated design principle for dynamic performances of vehicle and track systems

performance indexes. This is the general principle of the vehicle–track integrated design. Nevertheless, in a specific design process, the designed subject is still a subsystem (vehicle or track) rather than an interactive design for both subsystems (which is limited by the current professional divisions and other factors). In contrast to the traditional design process of subsystems, it is necessary for the integrated design process to consider the dynamic effect of each subsystem on the other, i.e., to determine whether each designed system is suited to its related system or not. This vehicle–track integrated design concept can be realized with the aid of the vehicle–track coupled dynamics theory and its simulation software.

Figure 9.1 illustrates the fundamental principle of the vehicle–track integrated design. The design subject can be either the vehicle or the track. Once the design subject is determined, the other one is regarded as the dynamic environment of the design subject. Using the vehicle–track coupled dynamics theory, the dynamic influence of the other subsystem can be taken into consideration in the optimized design process for the dynamic performance of the subject. Meanwhile, the dynamic effect of the design subject on the other system can be investigated simultaneously. After this, the parameters of the design subject should be optimized according to the evaluation result of its dynamic interaction with the other system. Following the same procedure, the dynamic performance of the design subject should be reassessed and the dynamic influence of the subject on the other system should be reevaluated repeatedly until the optimal dynamic performance of the overall system is achieved.

It is necessary to note that, if the design subject is the vehicle system, its dynamic impact indexes on the track system can be set as the dynamic wheel–rail force, the dynamic deformation of the track structure due to dynamic wheel loads, etc. In contrast, if the design subject is the track system, its dynamic influence indexes on the vehicle system can be specified as the vehicle's running performance indexes, such as the ride comfort, running safety, hunting stability and so on.

9.2 Method of Optimal Integrated Design for Dynamic Performances of Vehicle and Track Systems

Based on the integrated design principle described in Sect. 9.1, the dynamic design methods for the optimal integrated designs of vehicle and track systems can be realized using the simulation platform of vehicle–track coupled dynamics.

9.2.1 Dynamic Design Method for Vehicle System Based on the Optimal Integrated Design Principle

As shown in Fig. 9.2, the initial design scheme of the railway vehicle system (the parameters of the vehicle dynamics model), together with its operational conditions from the track (the parameters of the track structure and those of track horizontal alignment and vertical profile), should be the input of the simulation system for vehicle–track coupled dynamics analysis. The dynamic responses of the vehicle can be analyzed and predicted by using the simulations, including vehicle hunting stability, ride comfort and curve negotiation performance. Meanwhile, the dynamic loading indexes of the vehicle on the track (dynamic wheel–rail force, dynamic track deformation, and dynamic track stress) can be obtained. According to the criteria for vehicle dynamic performance and dynamic wheel–rail interaction, the dynamic response indexes can be evaluated comprehensively and the feasibility of the vehicle design can be determined correspondingly. If the design cannot be accepted, the inferior indexes and their related sensitive parameters of the vehicle (such as suspension stiffness and damping, unsprung mass, etc.) can be identified and revised. In this way, optimized parameters can eventually be determined and used in the simulation design platform for dynamic performance analysis and evaluation. This design process can be implemented repeatedly until an optimized design is achieved. The final vehicle design should satisfy the vehicle's dynamic performance requirements and the track loading requirements simultaneously.

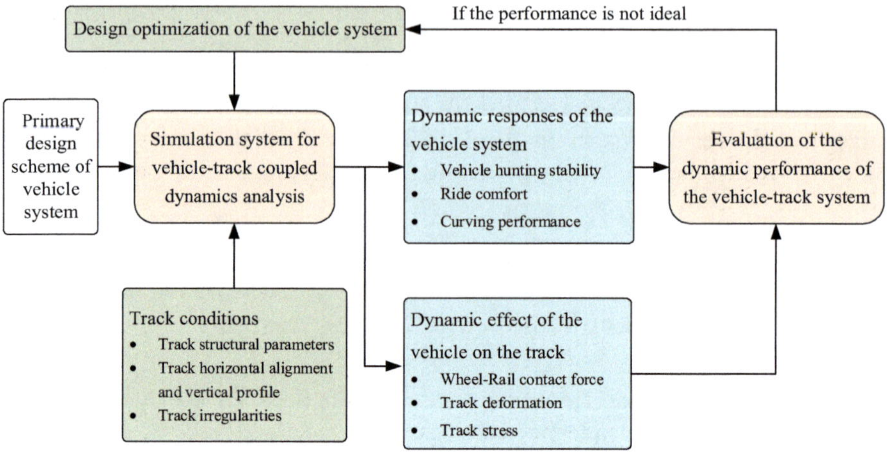

Fig. 9.2 Dynamic design method for railway vehicle system based on the optimal integrated design principle

9.2.2 Dynamic Design Method for Track System Based on the Optimal Integrated Design Principle

The dynamic design procedure of the track system based on the principle of the optimal integrated design concept is shown in Fig. 9.3. The initial design of the track (horizontal alignment and vertical profile parameters and the track structural parameters), together with the conditions of the vehicles to be operated (vehicle dynamics parameters and operating speed) should be regarded as the inputs of the simulation system for vehicle–track coupled dynamics analysis. The dynamic responses of the track structure, including track vibration, loading and deformation characteristics, can be predicted and analyzed. Meanwhile, the vehicle's running performance indexes on the designed track (such as ride comfort index and dynamic wheel–rail safety index) can be obtained. According to relevant evaluation criteria (or standards) for the performances of the vehicle and track, the dynamic response indexes can be assessed in a comprehensive way and the feasibility of the track design can be investigated. If the design is unacceptable, the unsatisfactory performance indexes and relevant sensitive design parameters (such as minimum curve radius, cant, transition curve length, track stiffness parameters, etc.) should be identified and revised. The structural parameters optimized can be utilized in the simulation platform for dynamic performance analysis and evaluation. The design process should be carried out repeatedly until the optimized design is satisfactorily obtained. An optimized design means both track dynamic performance and running behavior of the vehicles on the designed track can be guaranteed.

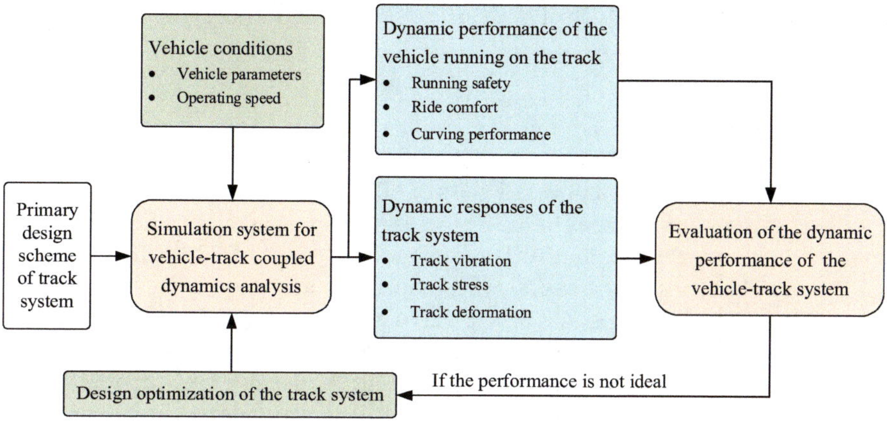

Fig. 9.3 Dynamic design method for track system based on the optimal integrated design principle

9.3 Case Study I: Optimal Design of Suspension Parameters of a Heavy-Haul Locomotive

The heavy-haul locomotive, HX_D2C, is a new type of mainstream AC drive freight electric locomotive, which has been widely used in China at the speed level of 120 km/h. It is produced by CRRC Corporation Limited with the axle form of C_0–C_0 and axle load of 25 t. In 2010, the first dynamics performance test for the first prototype locomotive was performed, including an operational test on some small radius curves (R = 300 m) with a maximum running speed of 70 km/h. The test was carried out on the Taiyuan–Jiaozuo railway line between Moon Mountain and Jincheng North. The results of the first test indicated that the designed locomotive has a rather large lateral wheel–rail force and the derailment coefficient sometimes exceeds the safety threshold in these small radius curves. This could not guarantee the locomotive's long-term safety running in regions with plenty of small radius curves. Consequently, the curve negotiation performance of the locomotive needed to be improved before mass production.

In order to solve this problem, the dynamic design method for vehicle system based on the optimal integrated design principle, described in Sect. 9.2.1, could be adopted. In this regard, the vehicle–track coupled dynamics theory and the corresponding dynamics simulation software TTISIM were employed to investigate the effect of the locomotive suspension parameters on curving performance of the HX_D2C locomotive passing through small radius curves. The results indicated a problem in the matching between the longitudinal and the lateral stiffness of the prototype locomotive primary suspension. Then, a series of reasonable suspension stiffness values were proposed through parameter optimization to improve the curve negotiation dynamics performance of the HX_D2C locomotive. This practical engineering problem was eventually solved successfully.

9.3.1 Operation Safety Analysis of HX_D2C Prototype Locomotive Through Small Radius Curves

Thc simulation conditions are the same as that of the field test: the curve radius is 300 m; the rail height difference is 120 mm; the track structure is of the existing common ballasted track with the rail mass of 60 kg/m, and the sleeper is made from concrete; the rail irregularities of the Chinese speedup main railway line are applied as the excitations; the locomotive is running at the speed of 70 km/h.

The time histories of the lateral wheel–rail force and the derailment coefficient when the locomotive is negotiating the curve are presented in Fig. 9.4. The lateral wheel–rail dynamic force increases gradually when the locomotive is moving into the transition curve. When the locomotive runs at the positions connecting the transition curve and the circular curve, the dynamics indexes representing the locomotive curve negotiation performance, reach their maximum values. Specifically, the mean value of the lateral wheel–rail force is up to 90 kN with a maximum wheel–rail force of 123 kN, and the corresponding maximum value of the derailment coefficient has reached 0.91 which has exceeded the threshold value of 0.9 given in the Chinese standard TB/T 2360-93 [2].

In practice, it has been also found that, in the line service test of the HX_D2C prototype locomotive, the maximum outside (on the high rail) lateral wheel–rail force of the guiding wheelset exceeded 110 kN. The derailment coefficient has instances where the value is beyond the threshold, although most values of this index are below 0.9. Results from both theoretical calculation and field test indicate that intensified lateral wheel–rail dynamic interactions are observed when the HX_D2C prototype locomotive is negotiating the small radius curves, which brings great challenges to the safe operation of the locomotive and this issue must be alleviated.

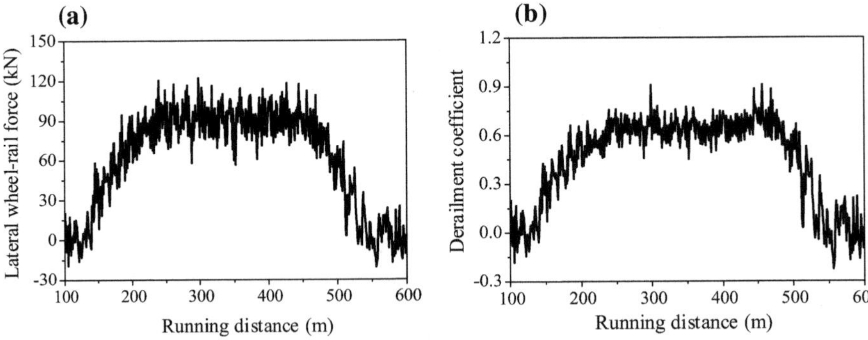

Fig. 9.4 Calculated wheel–rail dynamic indexes for the HX_D2C locomotive running safety: **a** lateral wheel–rail force, and **b** derailment coefficient

9.3.2 Optimization Scheme to Improve Curve Negotiation Performance of HX_D2C Heavy-Haul Locomotive

There exist many potential ways to improve the locomotive curve negotiation performance. However, based on the practical locomotive conditions and theoretical analysis, our research group had found that the previously designed axle-box positioning stiffness of the HX_D2C locomotive is high, and reducing the longitudinal and the lateral axle-box positioning stiffness has an evident effect on the alleviation of the lateral wheel–rail dynamic interaction. Consequently, improvement of the axle-box positioning stiffness is regarded as the key link of the following optimization process.

The designed and the measured axle-box positioning stiffness of the HX_D2C prototype locomotive are shown in Table 9.1. It can be seen that the designed longitudinal and lateral axle-box positioning stiffness values are 199 kN/mm and 6.89 kN/mm, respectively. However, the measured practical values for these are 236 kN/mm and 10.01 kN/mm, respectively. This indicates that the practical axle-box positioning stiffness values are higher than the designed values, namely, with the ratio of 1.19 and 1.45 for longitudinal and lateral positioning stiffness, respectively.

Variations of the simulated lateral wheel–rail force versus the locomotive primary suspension stiffness when the locomotive is negotiating the curve are displayed in Fig. 9.5. The results indicate that reduction of longitudinal and lateral stiffness of the primary suspension can effectively reduce the lateral wheel–rail dynamic interaction in small radius curves. Further, the variation of the lateral wheel–rail force is more sensitive to the change of lateral stiffness. Consequently, optimizing the primary suspension lateral stiffness will be the major measure to reduce the lateral wheel–rail forces.

It can be seen from Fig. 9.5a that a smaller primary suspension lateral stiffness will benefit the reduction of lateral wheel–rail force. However, smaller stiffness could also deteriorate locomotive lateral stability. Thus, the principle for selecting primary suspension lateral stiffness is that a smaller stiffness value is preferred in the premise of guaranteeing good motion stability. Besides, it is worthy of notice in Fig. 9.5a that the lateral wheel–rail force varies gently with the lateral stiffness in the region of soft stiffness below 2.65 kN/mm, while a prompt increase in the lateral force can be observed when the lateral stiffness is greater than 2.65 kN/mm. Hence, the reasonable primary suspension lateral stiffness of the HX_D2C locomotive should be 2.65 kN/mm in theory.

Table 9.1 Designed and measured values of HX_D2C locomotive axle-box positioning stiffness

Positioning stiffness	Vertical (kN/mm)	Lateral (kN/mm)	Longitudinal (kN/mm)
Designed value	1.71	6.89	199
Measured value	1.41	10.01	236

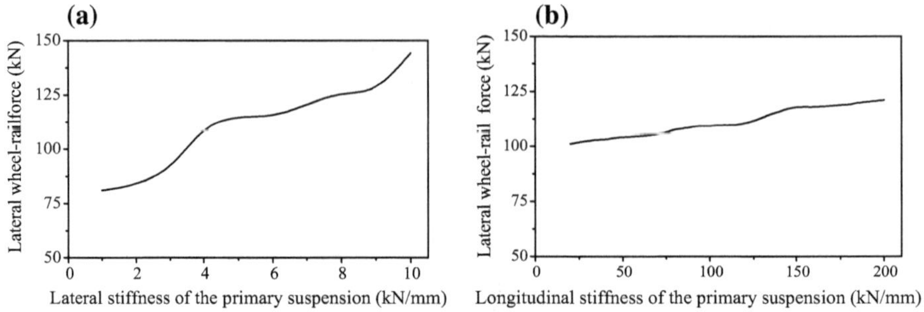

Fig. 9.5 Lateral wheel–rail force variation versus **a** lateral, and **b** longitudinal stiffness of primary suspension

For the determination of primary longitudinal stiffness, the results in Fig. 9.5b show that the lateral wheel–rail force decreases with the reduction of lateral stiffness. However, the sensitivity of the lateral force to lateral stiffness is low. Consequently, the primary suspension longitudinal stiffness should not be too small due to the consideration of its high traction force delivery function and so as to also guarantee the locomotive motion stability. In addition, the structural shape of the rubber joint will affect the longitudinal and lateral stiffness of the primary suspension simultaneously as it is an independent entirety in the structural design of the primary suspension. In other words, there exists a certain relation of interdependence between the lateral and longitudinal stiffness of the primary suspension. Based on the aforementioned analysis, the theoretical longitudinal stiffness of the locomotive primary suspension is selected as 40 kN/mm after the determination of lateral stiffness value. Further, due to the complexity of the rubber joint manufacturing technique and limitations of rubber spring manufacturing conditions, the final primary suspension stiffness values for practical usage are determined as longitudinal stiffness is 52 kN/mm while the lateral stiffness is 2.6 kN/mm.

In order to verify the reasonability of the stiffness parameter optimization, comparisons of the locomotive dynamic performance are made using the primary suspension stiffness parameters before and after optimization. The time histories of the lateral wheel–rail forces and derailment coefficient before and after parameter optimization are illustrated in Figs. 9.6 and 9.7. In these simulations, the locomotive is also negotiating the small curve with a radius of 300 m at the speed of 70 km/h. A conclusion can be drawn from the results that the lateral wheel–rail force and the derailment coefficient have significantly decreased after parameter optimization. The greatest decrement ratios are 20% and 18.7% for the lateral wheel–rail force and the derailment coefficient, respectively.

The smaller primary suspension stiffness after parameter optimization may affect locomotive motion stability. Therefore, the nonlinear critical speed of the HX_D2C locomotive running on elastic track has been further verified. The lateral displacement responses of the first wheelset are shown in Fig. 9.8 where the locomotive is running on a straight line at the speed of 237 km/h and 238 km/h,

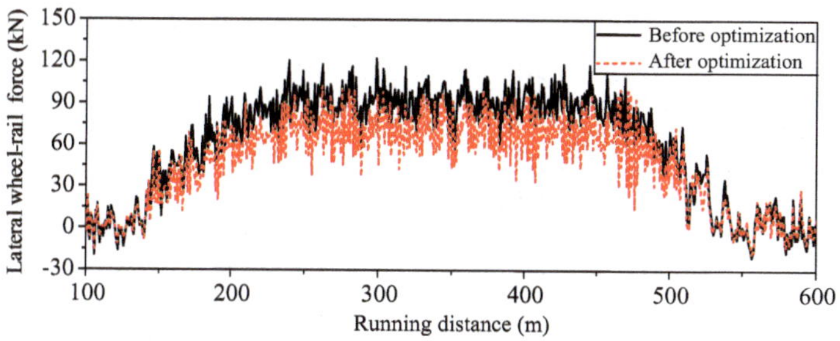

Fig. 9.6 Comparison of HX_D2C locomotive lateral wheel–rail force during curve negotiation before and after parameter optimization

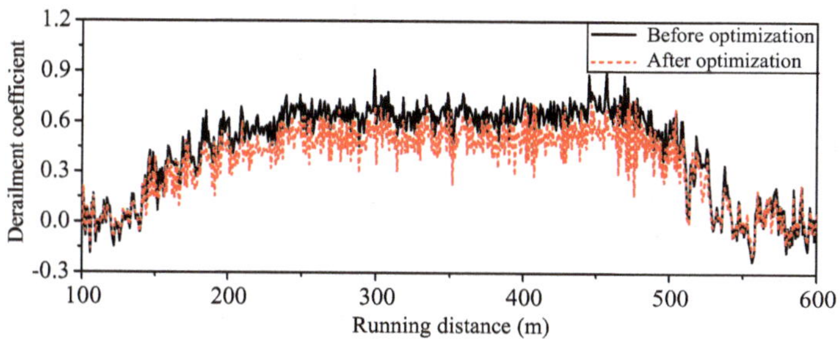

Fig. 9.7 Comparison of HX_D2C locomotive derailment coefficient during curve negotiation before and after parameter optimization

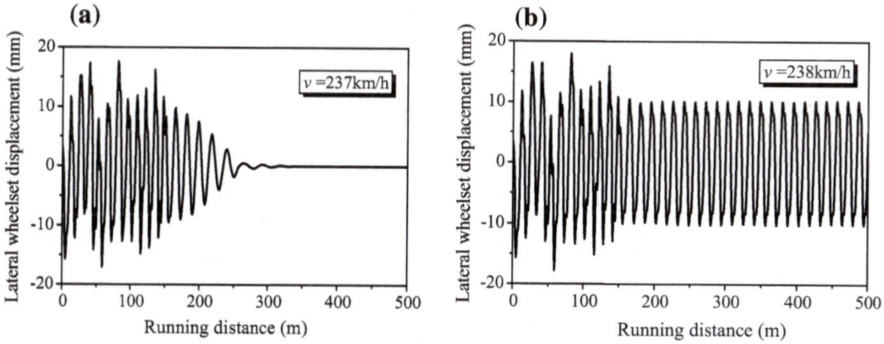

Fig. 9.8 Determination of nonlinear critical speed of HX_D2C locomotive after optimization design: **a** stable solution, and **b** periodic solution

respectively. The results indicate that the nonlinear critical speed of the locomotive after parameter optimization is 238 km/h which is much higher than its designed speed (120 km/h). The optimized parameters can meet the requirement of running stability.

9.3.3 *Application of HX_D2C Heavy-Haul Locomotive After Design Optimization*

In order to examine the practical operation performance of the locomotive after parameter design optimization, a field test of the full-scale train was again conducted on the same railway line section as the previous test for the locomotive before the design parameter optimization. This test was carried out by the locomotive manufacturing company together with China Academy of Railway Sciences in December 2010. The detailed information of the testing railway line is reported as the curve operation performance test was carried out in the railway section of the Taiyuan–Jiaozuo line between Yueshan and Jincheng North, and the operation performance on a straight line was performed in the railway section of Beijing–Guangzhou line between Xinxiang and Anyang; the maximum running speed for the locomotive test was 132 km/h.

1. *Comparison between the tested dynamics performance of HX_D2C locomotive before and after design improvement*

The lateral wheel–rail force was especially focused on in this test so as to reveal whether the problem that large lateral wheel–rail forces were generated by the HX_D2C locomotive during negotiating a small radius curve in the previous test had been solved. The tested lateral wheel–rail force variations of the first wheelset versus the change of curvature are compared in Fig. 9.9 for the locomotive before and after improvement of the design. The results show that the lateral wheel–rail force is greater when the locomotive negotiates a smaller radius curve (i.e., larger curvature), indicating a more intensified lateral wheel–rail interaction as radius decreases. The maximum wheel–rail force approaches 100 kN before the improved design of the locomotive, while this value decreases to 88 kN after the improved design. Further, the decrement in the lateral wheel–rail force was larger in the

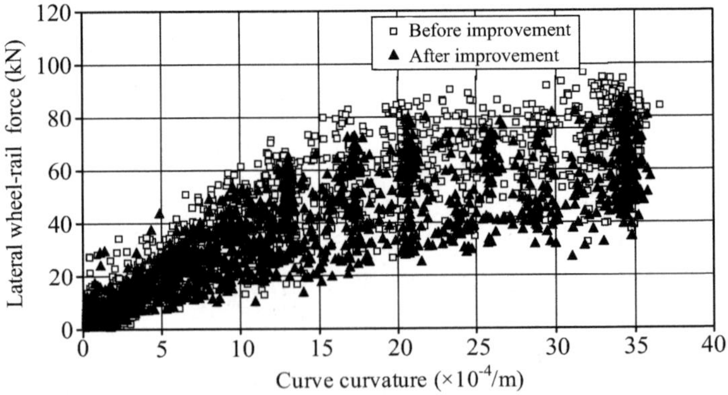

Fig. 9.9 Variation of lateral wheel–rail force versus curve curvature before and after the improvement of the primary suspension parameters

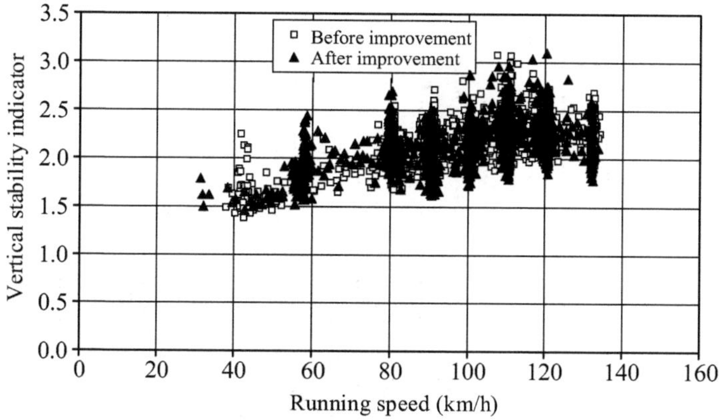

Fig. 9.10 Vertical stability of HX_D2C locomotive before and after improvement of locomotive design

smaller radius curve. Thus, the curve negotiation performance, especially for small radius curves, has been improved greatly after the improved design of the HX_D2C locomotive.

At the same time, the locomotive running stability was again examined in this test so as to check whether the change of primary suspension stiffness will bring any adverse effect to the locomotive operation quality. Variations of the vertical and lateral stability indexes of the locomotive on a straight line versus the running speed are presented in Figs. 9.10 and 9.11, respectively. It can be seen that the locomotive running stability is changed slightly after improvement. The running stability of the locomotive is in a good state for both unimproved and improved designs. The maximum vertical stability index shown in Fig. 9.10 is about 3.1 which is of the

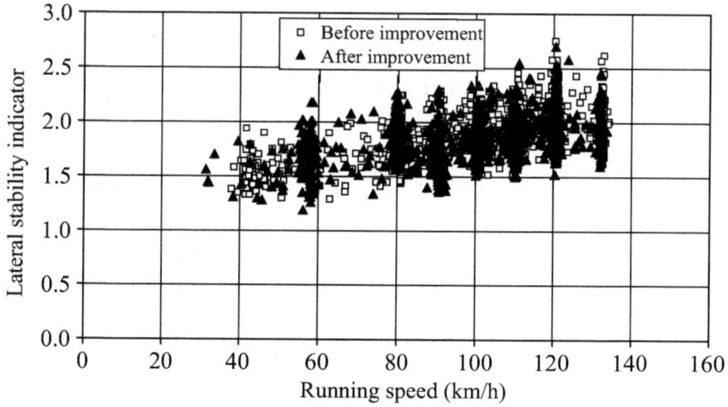

Fig. 9.11 Lateral stability of HX_D2C locomotive before and after improvement of locomotive design

Fig. 9.12 HX_D2C heavy-haul freight locomotive in operation

"fine" level according to testing standards. The lateral stability index versus running speed is shown in Fig. 9.11 and it reaches its maximum value of 2.75 when the locomotive runs at the speed of 120 km/h. For this case, the lateral stability index is also in the range of the "fine" level, while, in the other running speed range, the lateral stability index is in the "excellent" level. Thus, a conclusion can be drawn that the change in the primary suspension stiffness of the improved design will not cause apparent variations to the HX_D2C locomotive running stability.

The aforementioned results of the full-scale locomotive running test indicate that the curve negotiation performance of the HX_D2C heavy-haul locomotive has been improved effectively through the design optimization process. Specifically, the lateral wheel–rail force decreases significantly, and the running stability of the locomotive has also been guaranteed. This improvement can benefit the operation of this locomotive in small radius curves.

2. *Practical application of the HX_D2C heavy-haul locomotive*

The HX_D2C locomotive after the improved design has been put into mass production (see Fig. 9.12). Currently, there have been 250 locomotives produced and deployed to Xinxiang locomotive depot. This type of locomotive has been the prime freight locomotive for the railway lines, such as the Houyue line and the Xinhe line, and the locomotives are operating with good performance. The development of this locomotive has met the requirements during capacity expansion and speedup development of conventional Chinese railway lines.

9.4 Case Study II: Design of a Steep Gradient Section of a High-Speed Railway

9.4.1 Engineering and Research Background

The Guangzhou–Shenzhen–Hong Kong high-speed railway is the extension line of the Beijing–Guangzhou high-speed railway in the major network of the Chinese “four longitudinal and four lateral” high-speed railways and is also the backbone of the intercity express rail transportation network in the Pearl River Delta.

The Guangzhou–Shenzhen–Hong Kong high-speed railway has to cross the Shiziyang area of the Pearl River as shown in Fig. 9.13. The Shiziyang section is located in the wide and flat plain area of the Pearl River Delta. The riverbed has a width of around 4 km with undulating topography, and includes Xiaohu Island, Shazai Island, Haiou Island, etc. For the special geographical conditions in the Shiziyang area, the design institution proposed two line-selection schemes, crossing both Shazai Island and Haiou Island. Two options were proposed for each scheme, namely the long tunnel and the bridge–tunnel as shown in Figs. 9.14 and 9.15. These schemes involved some vertical profile sections with large slopes of 20, 30, and 34‰, which were the first time that Chinese high-speed railway faced such a complex and large gradient line. Undoubtedly, the Shiziyang area was the most difficult section of the whole railway line from an engineering perspective.

The feasibility research for the Guangzhou–Shenzhen section of the Guangzhou–Shenzhen–Hong Kong high-speed railway started in 2004 when the

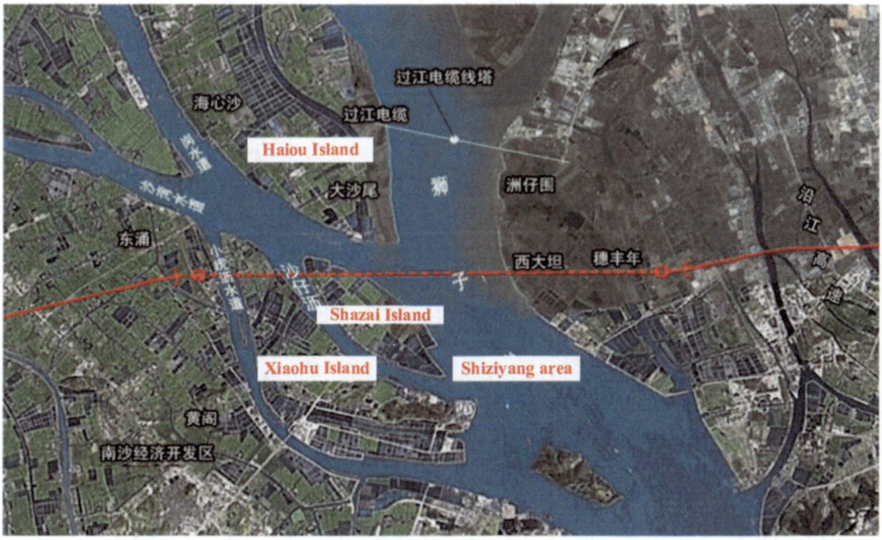

Fig. 9.13 Sketch map of Guangzhou–Shenzhen–Hong Kong high-speed railway crossing Shiziyang area of Pearl River

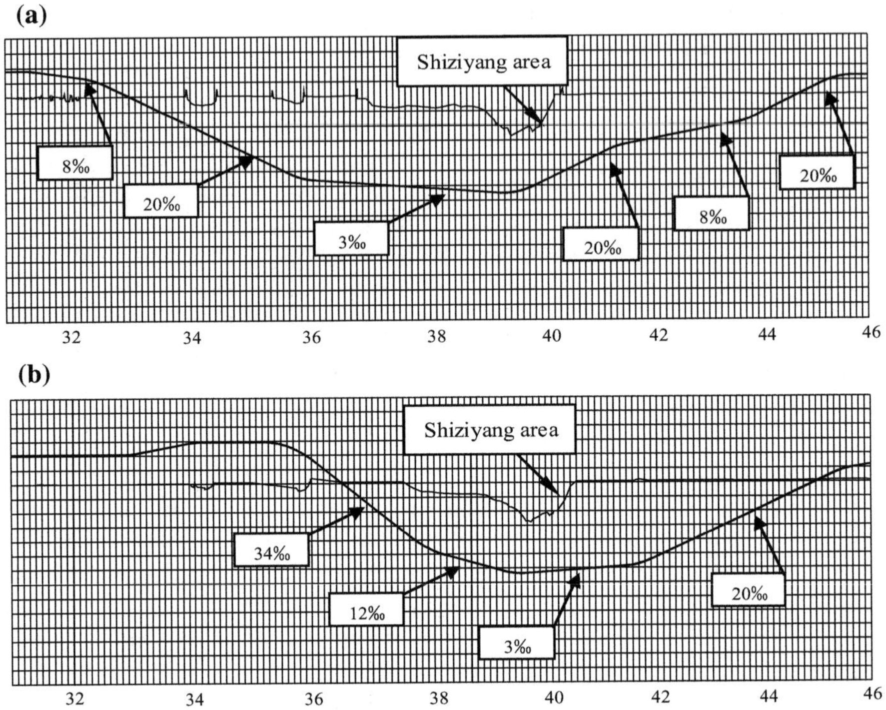

Fig. 9.14 Vertical profiles of two design schemes in the section of Shazai Island: **a** long tunnel scheme, and **b** bridge–tunnel scheme

design speed was set at 300 km/h. However, could a high-speed train safely and smoothly pass through such steep gradients at a speed of 300 km/h? This was the first challenging problem relating to the compatible design of train and track subsystems in construction engineering of this Chinese high-speed line, and was a major technical problem that involved the economy of line-selection technology and the running safety and ride comfort of high-speed trains which had to be solved during the design stage.

Designated by the design institution of the Guangzhou–Shenzhen–Hong Kong high-speed railway, we used the dynamic design method for track system based on the optimal integrated design principle presented in Sect. 9.2.2 to solve the abovementioned compatible design problem. Complete dynamic simulations were carried out for the running safety and ride comfort analysis of high-speed trains passing at 300 km/h through the horizontal alignment and vertical profile sections of the proposed four line-selection schemes in the Shiziyang area. Through safety evaluation and scheme comparison according to the evaluation code of vehicle dynamic performance, the best scheme suggestion was finally put forward [4].

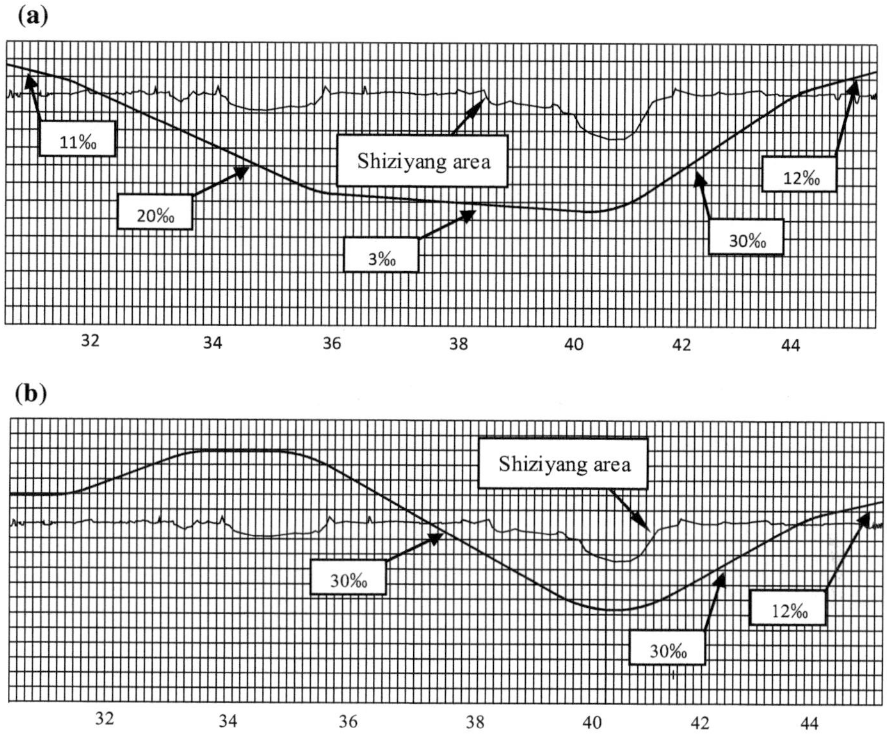

Fig. 9.15 Vertical profiles of two design schemes in the section of Haiou Island: **a** long tunnel scheme, and **b** bridge–tunnel scheme

9.4.2 Comparison of High-Speed Running Performance Between Long Tunnel Scheme and Bridge–Tunnel Scheme for Shazai Island

In accordance with the design speed of 300 km/h for the Guangzhou–Shenzhen–Hong Kong high-speed railway and the practice of high-speed trains internationally, the Germany ICE350 high-speed EMU[1] was adopted in the simulation analysis, and the low interference track spectrum of the German high-speed railway was employed as the excitation input.

When the high-speed train passes Shazai Island through the long tunnel scheme and bridge–tunnel scheme, the indexes of running safety and ride comfort show different characteristics. Figures 9.16 and 9.17 present the variation of derailment coefficient and lateral acceleration of the car body, respectively, for both schemes.

[1]ICE350 high-speed EMU is the improved vehicle of ICE3. The top speed is 350 km/h. The train consists of 8 passenger cars and 431 passenger places in total.

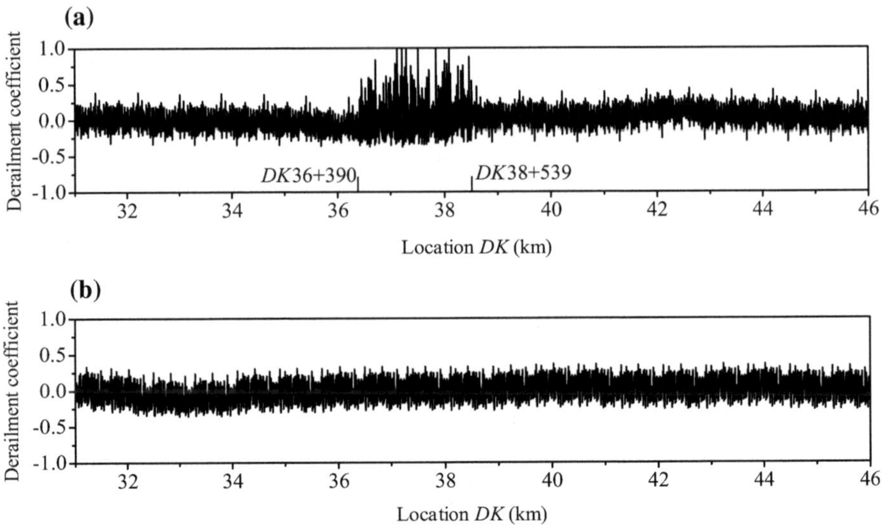

Fig. 9.16 Variation of derailment coefficient of high-speed train for **a** bridge–tunnel and **b** long tunnel design schemes of the Shazai Island

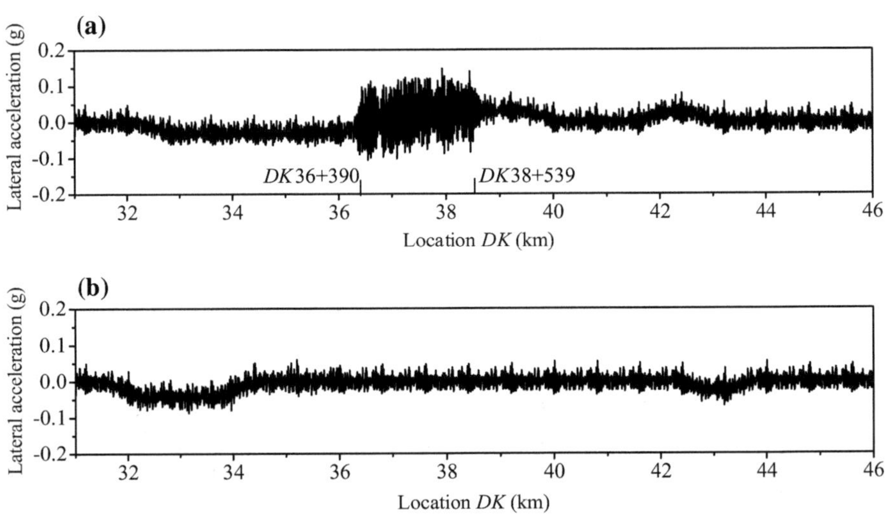

Fig. 9.17 Lateral car body acceleration of high-speed train for **a** bridge–tunnel and **b** long tunnel design schemes of the Shazai Island

As can be seen in Fig. 9.16, for the bridge–tunnel scheme, with the high-speed train running at location DK36+390 (grade change point from a vertical curve to a 34‰ downhill slope), the derailment coefficient increases abruptly from the normal state under the combined influence of the vertical profile with a slope of 34‰ and

7000 m radius horizontal transition curve. This process continues to the location of DK38+539 (grade change point between a 12‰ downhill slope and a 3‰ uphill slope). Clearly, the maximum value of the derailment coefficient reaches 0.98, which is larger than the dynamic safety limit of the high-speed train (0.8), while the derailment coefficient for the long tunnel scheme remains smaller than the safety limit. As can be seen from Fig. 9.17, the lateral car body acceleration in the bridge–tunnel scheme has abnormal fluctuations at the location of DK36+390–DK38+539. The peak acceleration is 0.15 *g*, exceeding the ride comfort limit of the high-speed train (0.1 *g*). This performance index of the bridge–tunnel scheme is larger than that of the long tunnel scheme by about 66.7%. The lateral running stability of the bridge–tunnel scheme also increases to 3.0 which just reaches the edge of the "qualified" level, while this value is found to be 2.46 for the long tunnel scheme which belongs to the "excellent" level. Additionally, the wheel unloading rates for the long tunnel scheme and bridge–tunnel scheme are 0.47 and 0.64, respectively. Obviously, the former one is smaller than the qualified limit of 0.65, and the latter one is close to the qualified limit [4].

As can be seen, the requirements for indexes of running safety and ride comfort of the high-speed train can be satisfied for the long tunnel scheme of Shazai Island. However, the running safety and ride comfort of the high-speed train cannot be ensured for the bridge–tunnel scheme.

9.4.3 Comparison of High-Speed Running Performance Between Long Tunnel Scheme and Bridge–Tunnel Scheme for Haiou Island

When the high-speed train passes Haiou Island through the long tunnel scheme and bridge–tunnel scheme, the dynamic behavior of the high-speed train shows similar characteristics with that for Shazai Island. The calculated maximum value of dynamic responses, such as indexes of running safety and ride comfort, are summarized in Table 9.2.

By comparing the various safety indexes, the maximum derailment coefficient is 0.53 for the long tunnel scheme, which is smaller than the safety limit of 0.8. The maximum derailment coefficient for the bridge–tunnel scheme is 0.92, which exceeds the safety limit. The maximum wheel unloading rate for the long tunnel scheme is 0.41, which is smaller than the limit value of 0.65, and this value is found to be 0.98 for the bridge–tunnel scheme which is, therefore, not a qualified scheme. The maximum overturning coefficient is calculated as 0.61 for the long tunnel scheme, which is lower than the limit value of 0.8, whereas this maximum value is 1.0 for the bridge–tunnel scheme, which is greater than the safety limit.

For the comparison of stability indexes, it can be seen that the lateral stability index of the car body for the long tunnel scheme is 2.44 which is indicating an excellent level, while the same index is found to be 2.94 (just a qualified level) for

Table 9.2 Peak dynamic responses of two design schemes for Haiou Island

Dynamic performance indexes	Long tunnel scheme	Bridge–tunnel scheme
Lateral wheel–rail force (kN)	27.89	30.97
Derailment coefficient	0.53	0.92
Wheel unloading rate	0.41	0.98
Overturning coefficient	0.61	1.00
Lateral car body acceleration (*g*)	0.08	0.16
Vertical car body acceleration (*g*)	0.05	0.07
Lateral stability index	2.44	2.94
Vertical stability index	2.25	2.33

the bridge–tunnel scheme. The vertical stability indexes are all excellent for the long tunnel scheme and the bridge–tunnel scheme.

As can be seen from above analyses, when the high-speed train passes through Haiou Island at a speed of 300 km/h, all safety indexes and comfort indexes for the long tunnel scheme can meet the requirements of high-speed operation and have sufficient safety margins. However, for the bridge–tunnel scheme, under the combined effect of the horizontal curve with a 7000 m radius and the vertical profile with a slope of 30‰, the running safety index of the train exceeds the allowable limit for part of the operation, and the ride comfort index is greatly reduced and does not meet the requirements.

9.4.4 Comparison and Selection Between Shazai Island Scheme with Long Tunnel and Haiou Island with Long Tunnel

It can be seen from the above analysis that the running safety and ride stability indexes of the bridge–tunnel scheme for Shazai Island and Haiou Island are all worse than those of the long tunnel scheme for these two islands. The bridge–tunnel scheme cannot meet the requirements for running safety and ride comfort, while the long tunnel scheme for the two islands can satisfy these requirements. Therefore, the long tunnel scheme is recommended as the solution.

To determine the best design scheme, the dynamic performance indexes for the Shazai Island scheme and the Haiou Island scheme that can meet the requirements of high-speed operation are listed in Table 9.3.

As can be seen from Table 9.3, when the high-speed train passes through the Shazai Island when adopting the long tunnel scheme, the key safety indexes (derailment coefficient and lateral wheel–rail force) are all smaller than those of Haiou Island when adopting the long tunnel scheme. The wheel unloading rates and overturning coefficients of the two schemes are quite close to each other, and there is no obvious difference in the ride comfort index. Therefore, the best design

Table 9.3 Comparison of high-speed running performance indexes of the long tunnel scheme for Shazai Island and Haiou Island

Dynamic performance indexes	Shazai Island	Haiou Island
Lateral wheel–rail force/kN	18.85	27.89
Vertical wheel–rail force/kN	97.67	105.21
Derailment coefficient	0.37	0.53
Wheel unloading rate	0.47	0.41
Overturning coefficient	0.69	0.61
Lateral car body acceleration (*g*)	0.09	0.08
Vertical car body acceleration (*g*)	0.06	0.05
Lateral stability index	2.46	2.44
Vertical stability index	2.24	2.25

scheme we finally recommend was the long tunnel scheme crossing Shazai Island, which provided scientific guidance for the design department's decision-making.

9.4.5 Project Implementation and Operation Practice

The above-recommended scheme was adopted for the practical project of the Shiziyang section of Guangzhou–Shenzhen–Hong Kong high-speed railway (Fig. 9.18), which solved the complicated problem for line selection with large slopes that was first encountered in Chinese high-speed railway design at that time.

On December 18, 2005, the Guangzhou–Shenzhen section of the high-speed railway line officially started construction, and the construction of Shiziyang tunnel began on November 9, 2007, and was completed on March 12, 2011. The total length of this tunnel is 10.8 km and it carries the fastest trains of any underwater railway tunnel in the world.

On December 26, 2011, the Guangzhou–Shenzhen–Hong Kong high-speed railway was officially opened. From the experiment results before opening and the practical operation experience, the high-speed trains have a good running performance which fully satisfies the requirements of high-speed running safety and ride

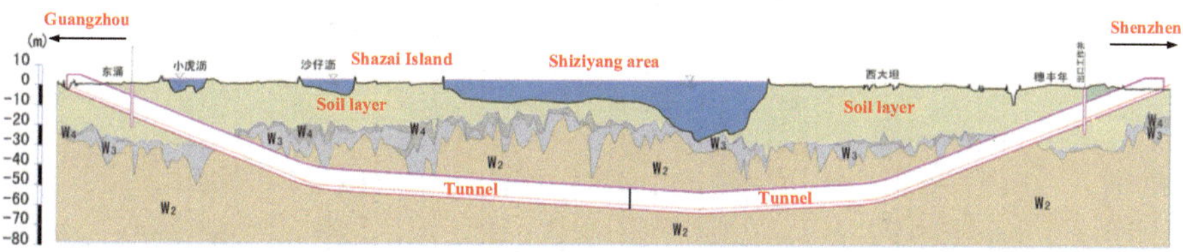

Fig. 9.18 Vertical profile of practical project of Shiziyang tunnel

comfort. From the comparisons and recommendations of design schemes in 2005 and the practical operation in 2011, the validity and reliability of the recommended scheme (long tunnel scheme crossing the Shazai Island) is proved by the 6-year engineering practice.

References

1. Zhai WM. Principle, method and engineering practice of optimal integrated designs of railway vehicle and track. China Railw Sci. 2006;27(2):60–5 (in Chinese).
2. TB/T2360-93. Evaluation method and standard of dynamic performance test of railway locomotive. Beijing: China Railway Publishing House; 1993 (in Chinese).
3. Zhai WM, Wang KY, Yang YL, et al. Applications of the theory of vehicle–track coupling dynamics to the design of modern locomotives and rolling stocks (in Chinese). J China Railw Soc. 2004;26(4):24–30.
4. Zhai WM, Wang KY. Evaluation of running safety and ride comfort of trains passing through the Pearl river section of Guangzhou-Shenzhen-Hong Kong high-speed railway line. Report No. TTRI-2005-09. Chengdu: Train and Track Research Institute, Southwest Jiaotong University; 2005 (in Chinese).

Chapter 10
Practical Applications of the Theory of Vehicle–Track Coupled Dynamics in Engineering

Abstract The vehicle–track coupled dynamics theory has been successfully applied to solve tremendous practical problems in railway engineering, involving engineering projects in high-speed railways, heavy-haul railways, and speedup railways. This chapter introduces some representative practical application cases in the rapid development process of Chinese railways, including redesign of dynamic performance of a speedup locomotive, reduction of rail side wear on heavy-haul railway curves, safety control of coupler swing angle of a heavy-haul long train, and design of a shared high-speed passenger, and freight railway.

10.1 Redesign of Dynamic Performance of a Speedup Locomotive

10.1.1 Engineering Background

The type SS_{7E} locomotive is a C_0–C_0 passenger electric locomotive developed by China North Locomotive and Rolling Stock Corporation in order to meet the requirement of the fifth speedup of Chinese railways. The designed speed of this type of locomotive is 170 km/h and the axle load is 21 t. Its bogie frame has two end beams and two middle beams. The wheelset is equipped with two entire rolled steel wheels at the ends and a hollow wheelset axle so as to reduce the wheel load and unsprung mass. The traction motor is suspended from the bogie frame, and the traction rod assemblies include bilateral pull rods together with connecting lever and rods. A double hollow shaft with six connecting rods is applied for the mechanical transmission equipment and the unit braking technique is used in the primary brake rigging. The primary suspension consists of an axle-box pull rod and a coil spring on each side of the bogie, and the secondary suspension has the structure of a highly flexible coil spring connecting in series with dual rubber pads. In addition, anti-yaw dampers are installed between the bogie and the car body in order to resist the hunting instability of the bogie at high speed.

W. Zhai, *Vehicle–Track Coupled Dynamics*,
https://doi.org/10.1007/978-981-32-9283-3_10

At the end of 2002, the prototype of the locomotive went through an operational assessment of 200,000 km. However, it was found in the trial operation that the locomotive lateral swaying vibration was much intensified in the speed range of 100–150 km/h. In this speed range, the lateral acceleration index exceeded the limit value specified in the Chinese national standard. However, the lateral vibration of the locomotive running at other speeds was relatively moderate, and the vertical running stability was kept in a good state for the complete running speed range. Thus, the abnormal lateral vibration problem of the locomotive had to be solved before commencing mass production.

Under this situation, commissioned by the manufacturer (Datong Electric Locomotive Company), the author's team applied the theory of vehicle–track coupled dynamics to perform in-depth analysis and studies on this practical engineering problem. The locomotive suspension parameters were optimized with the consideration of lateral dynamic interactions between the locomotive and the track. Using this method, an effective solution was proposed without changing the locomotive structure.

10.1.2 Simulation on Abnormal Lateral Vibration of SS_{7E} Locomotive Prototype

The simulation software TTISIM was used to simulate and analyze the lateral vibration performance of the SS_{7E} locomotive prototype running on tangent track. The range of locomotive running speed for the simulation was 70–180 km/h with an interval of 10 km/h. The operational conditions of the Chinese mainline railway were applied and the common ballast track structure was used. The rail size was 60 kg/m, and the rail was discretely supported by concrete sleepers.

Variations of the maximum lateral displacements of the car body and the bogie versus locomotive running speed are presented in Figs. 10.1 and 10.2, respectively. It can be seen that the lateral vibrations of the car body near the speed of 100 km/h are quite drastic, and similar for those of the bogie near the speed of 120 km/h.

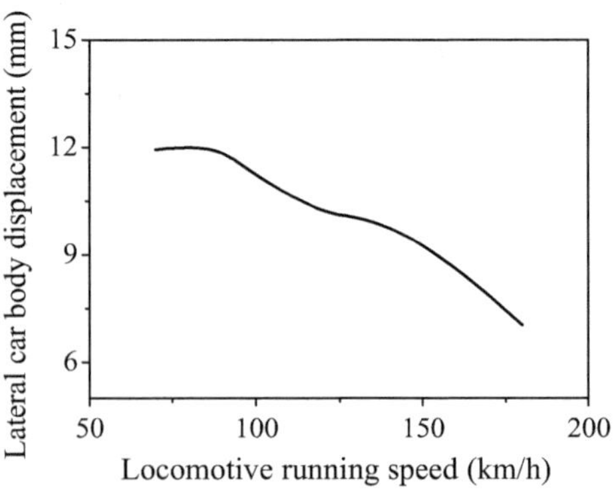

Fig. 10.1 Variation of lateral car body displacement versus locomotive running speed

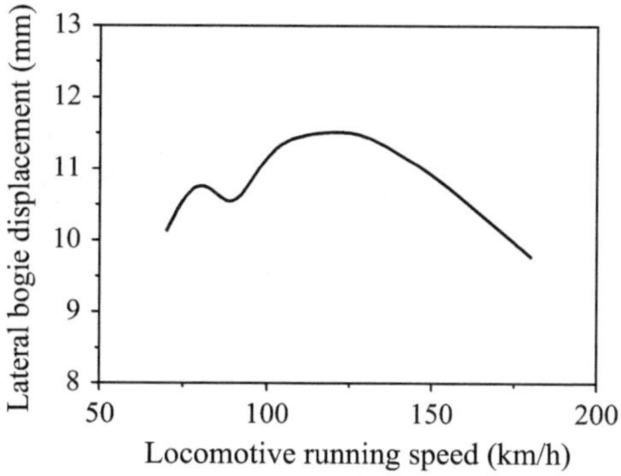

Fig. 10.2 Variation of lateral bogie displacement versus locomotive running speed

In order to illustrate the lateral vibration behaviors of this locomotive running on tangent track, the lateral car body displacement at the speed of 80 km/h is compared with that at 180 km/h (see Fig. 10.3). In addition, comparisons of lateral bogie displacements are made between the locomotive running speeds of 120 and 180 km/h (see Fig. 10.4). It can be seen that both lateral car body vibrations at the speed of 80 km/h and lateral bogie vibrations at the speed of 120 km/h are more intensified than their vibrations at the speed of 180 km/h.

It can be seen that the results of the aforementioned theoretical analyses agree well with the abnormal lateral vibration phenomenon observed in the trial running of the SS_{7E} prototype locomotive on the Longhai railway line. It also indicates that using the vehicle–track coupled dynamics theory is capable of reproducing the lateral nonlinear vibration behavior of this locomotive running on the track.

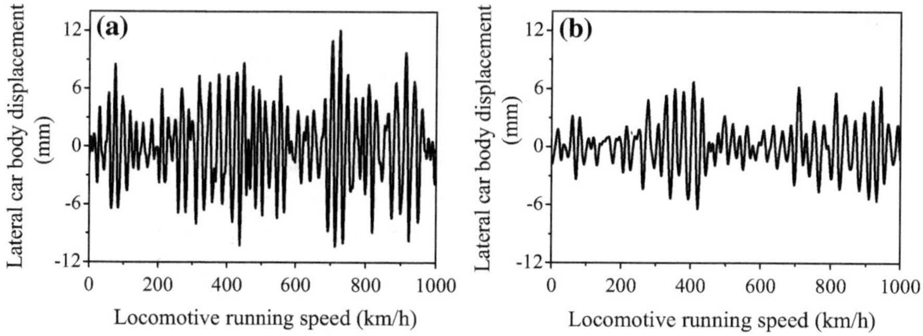

Fig. 10.3 Time histories of lateral car body displacement for the locomotive running at the speed of: **a** v = 80 km/h, and **b** v = 180 km/h

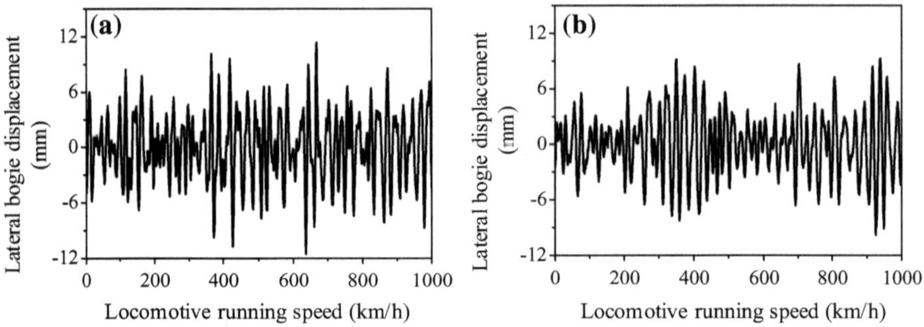

Fig. 10.4 Time histories of lateral bogie displacement for the locomotive running at the speed of: **a** v = 120 km/h, and **b** v = 180 km/h

10.1.3 Technical Proposal for Improving the Lateral Vibration Performance of SS_{7E} Locomotive

The main reason to cause the aforementioned phenomenon is believed to be the deficient design of the locomotive lateral suspension parameters. There are many approaches to improve rolling stock lateral motion stability, such as matching and optimization of the structure and suspension parameters of the rolling stock. The author and team have proposed six sets of optimized suspension parameters for this locomotive. By comparing the effectiveness of the six proposals and considering the realization complexity of the proposal as well as the time requirement, the damping parameter of the yaw absorber and the lateral stiffness of the secondary suspension are mainly selected for detailed optimization design.

The effect of the damping coefficient of the yaw absorber on lateral car body vibration is shown in Fig. 10.5. It can be seen that the maximum lateral displacement of the car body will decrease with the increase of yaw absorber damping coefficient. The lateral car body displacement will progressively decrease as the

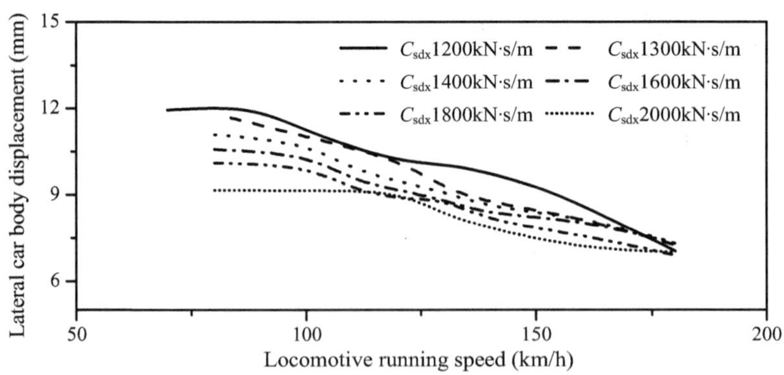

Fig. 10.5 Effect of yaw damping on car body lateral vibration

damping increases from 1200 to 2000 kN s/m. In particular, it provides the best performance to resist the lateral car body vibration in the low-speed range when the damping coefficient reaches 2000 kN s/m.

Figure 10.6 displays the effect of the damping coefficient of the yaw absorber on the lateral vibration of the bogie frame. When the locomotive runs at a speed lower than 140 km/h, the lateral displacement of the bogie frame could be effectively reduced by increasing the yaw absorber damping, especially at the speed of 120 km/h when the damping increases to be 1400 kN s/m or greater. Additionally, the maximum value of the lateral bogie frame displacement will generally reduce with the increase of the damping when the locomotive speed lies in the range of 140–180 km/h, because the lateral bogie frame displacement actually increases in some cases (1400 kN s/m, for example) in this speed range. However, the displacement value will increase with the increase of the damping for minor conditions. In general, the lateral bogie frame vibrations of the locomotive could be effectively suppressed when the damping is 1800 kN s/m or greater.

Furthermore, the effects of the lateral stiffness of the secondary suspension on the lateral displacements of the car body and bogie frame are presented in Figs. 10.7 and 10.8, respectively. It can be seen that the decrease of the secondary suspension lateral stiffness can not only reduce the lateral car body displacement in the entire speed range (see Fig. 10.7), but also can effectively resist the lateral bogie frame vibration in the speed range of 100–140 km/h (see Fig. 10.8).

Based on the aforementioned results, two improved designs are proposed. Comparisons of the lateral car body and bogie frame vibrations are made between the proposed designs and the original design, and the results are shown in Fig. 10.9. It can be seen that the abnormal lateral vibration of the SS_{7E} prototype locomotive has been suppressed effectively through the application of the improved designs. Comparisons between the two proposed designs reveal that both of them have similar performance with regard to mitigating the lateral vibration of the bogie frame, while the second design is better for resisting the lateral car body vibration in the low-speed range.

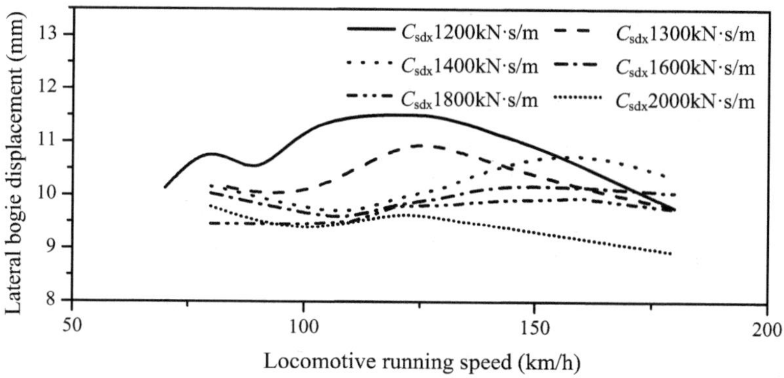

Fig. 10.6 Effect of the yaw damping on bogie frame lateral vibration

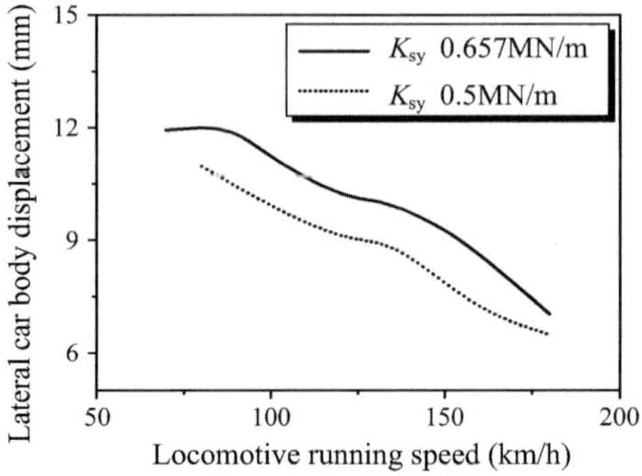

Fig. 10.7 Effect of secondary suspension lateral stiffness on lateral car body displacement

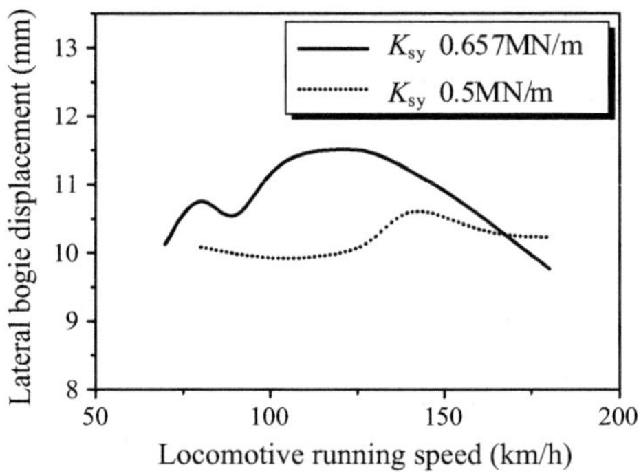

Fig. 10.8 Effect of secondary suspension lateral stiffness on lateral bogie frame displacement

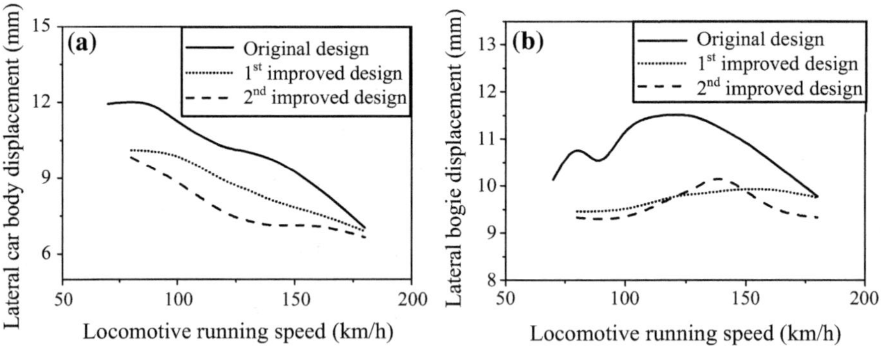

Fig. 10.9 Locomotive lateral vibration performances for different designs: **a** car body, and **b** bogie frame

10.1.4 Practical Performance and Application Status of the Improved SS_{7E} Speedup Locomotive

Due to the limitation from time requirements of the fifth Chinese railway speedup plan, the first improved design was finally applied into practice by the locomotive manufacturer although the second design had a better performance than the first one. In order to assess the practical performance of the locomotive after the initial improvement, the line running tests on the riding comfort of the SS_{7E} locomotive were performed on the Longhai and Jingguang lines by Southwest Jiaotong University together with Datong Electric Locomotive Company. The tested results were then compared with that of the prototype locomotive before improvement as tested by China Academy of Railway Sciences.

1. *Comparison of locomotive lateral vibrations before and after improvements*

The statistical test results for the lateral car body vibration acceleration and the lateral stability index before and after improvement are presented in Figs. 10.10 and 10.11, respectively. In these figures, the locomotive running speed ranges from 0 to 160 km/h.

It can be seen from Fig. 10.10 that the lateral car body vibration accelerations of the locomotive before improvement have seriously exceeded the limit value of 0.25 *g* in the speed range of 80–160 km/h, and most of them are even up to 0.25–0.35 *g* with the maximum value of 0.37 *g*. However, for the locomotive after improvement, the lateral car body vibrations are reduced greatly with most of the values below 0.2 *g* which meets the eligibility requirements. It should be noted that

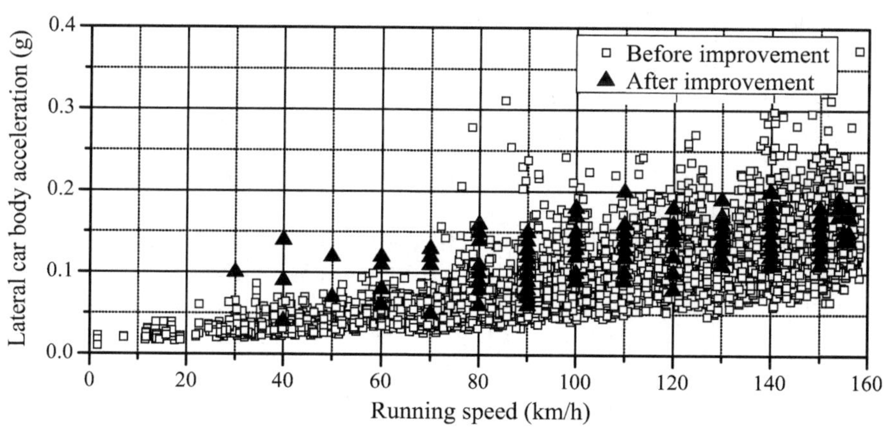

Fig. 10.10 Variation of lateral car body vibration acceleration versus locomotive running speed

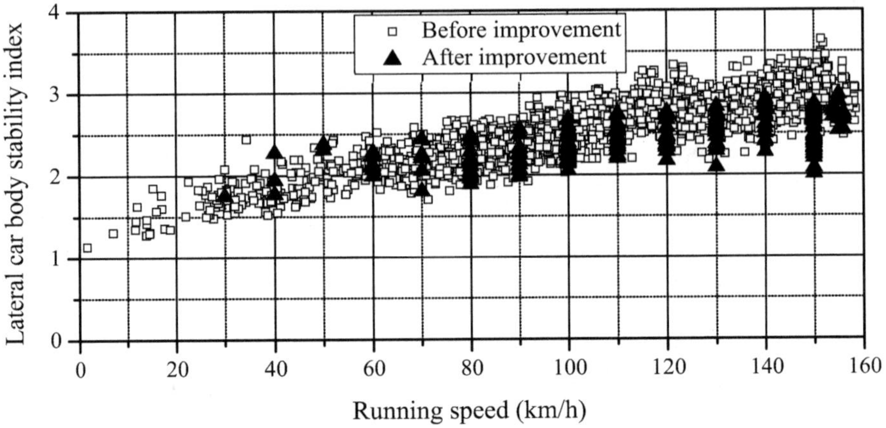

Fig. 10.11 Variation of lateral car body stability index versus locomotive running speed

the lateral car body vibrations after improvement have been suppressed for the speed range of 80–160 km/h, while the amplitudes of the lateral car body vibrations are increased for the speed range of 30–70 km/h. However, it has to be pointed out that the increases were very small, and the vibrations were still below 0.15 *g* which belongs to an "excellent" level. Consequently, the locomotive after the improvement has a good lateral operational quality in the entire designed speed range.

Figure 10.11 shows that the car body stability index of the improved locomotive was reduced when compared with the results before improvement for the entire speed range. This is especially true for the higher speed range (100–160 km/h) with a maximum value of 3.0, which is ranked in the "good" level. While, the maximum car body stability index before improvement is up to 3.6, which belongs to the disqualification grade.

2. *Practical application status of the* SS_{7E} *speedup locomotive*

The improved SS_{7E} locomotive, as shown in Fig. 10.12, had been put into mass production ahead of schedule in 2003 to meet the time requirement from the fifth Chinese railway speedup plan that began on April 18, 2004. Up to now, about 140 SS_{7E} locomotives have been manufactured and put into service. They have become the main passenger locomotives for the speedup of existing railway lines such as the Longhai and Jingguang railways. It has demonstrated that these locomotives have run on the two lines safely for a long period, which brings considerable social and economic benefits.

Fig. 10.12 SS_{7E} speedup passenger electric locomotive after improvement

10.2 Reducing Rail Side Wear on Heavy-Haul Railway Curves

In this section, an extended application of the theory of vehicle–track coupled dynamics will be introduced. In this application, the theory was used as an optimization strategy to design rail-grinding profiles on heavy-haul railway curves. The purpose of the optimization is to reduce the rail side wear on curves. A design methodology of rail asymmetric-grinding profiles is proposed based on detailed analysis of wheel–rail dynamic interaction by using the theory of vehicle–track coupled dynamics. As a practical application example, the rail asymmetric-grinding profiles were designed for a curve with 600 m radius on the Chinese Shuohuang heavy-haul railway. The rails on a test curve were ground according to the redesigned profiles. Wheel–rail dynamics indexes and rail side wear were measured before and after the rail grinding. The experimental results show that the wheel–rail dynamic interaction is clearly improved and the rail side wear is alleviated by 30–40% after rail grinding.

10.2.1 The Problem of Rail Wear on Curves of Heavy-Haul Railways

The wear of wheels and rails, especially the wheel flange wear and rail side wear on curves, is a long-standing problem of heavy-haul railways. With the increasing train axle load and transport capacity, the wheel–rail interaction is inevitably aggravated.

Severe wheel–rail dynamic interaction will induce severe wear of wheels and rails, especially in curved track sections.

This section takes the Shuohuang railway (one of the Chinese major heavy-haul railway lines) as an example. The rail wear state and characteristics of the railway are discussed in this section. The Shuohuang heavy-haul railway is a dedicated line for coal transportation from west to east. At present, the annual traffic volume of this railway has exceeded 300 million gross tonnes (MGt). Severe rail wear occurs on curves of this railway. Onsite wear measurements were carried out for rails on curves with three different radii, namely 500, 600, and 1000 m [1]. The measurement locations of rail wear are illustrated in Fig. 10.13a while Fig. 10.13b shows the statistics of the measurements of rail side wear and vertical wear.

It is shown from Fig. 10.13b that the predominant wear is the side wear of outer rails on small radius curves. The outer rail side wear increases rapidly with the decrease of curve radius, and reaches a maximum of 22.14 mm on the 500 m radius curve. As the curve radius increases to 1000 m, the rail side wear drops down greatly; the maximum value is 7.05 mm. The values of the rail wear were measured at a traffic volume of about 355 MGt. Compared with the outer rail, the inner rail has much less side wear (only 1–2 mm). For the vertical wear, there are small differences between the inner rail and outer rail. The vertical wear is much smaller than the side wear on the outer rail of sharp curves.

On sharp curves, a common defect that occurs on inner rails is the spalling defect. This defect is mainly found distributed along the center region of the rail crown, as shown in Fig. 10.14a. On the outer rail, severe side wear is found and the profile nearly conforms to the wheel flange shape (see Fig. 10.14b, c). The wheel–rail contact under this condition usually leads to an increase in load on the gauge side of the outer rail. With a continuous expansion in railway traffic volume, the wear rate of wheels and rails presents an increasing tendency. Therefore, it is of significance to seek technical measures to alleviate wear at the wheel–rail contact interface in heavy-haul railways.

Numerous studies have proved that rail profile grinding is an effective measure to slow down the wear and to prolong rail service life. In the 1970s, a rail grinding

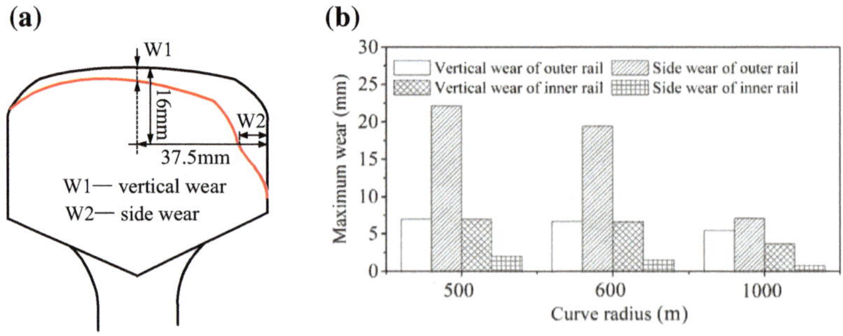

Fig. 10.13 Rail wear on Shuohuang heavy-haul railway curves: **a** measurements, and **b** statistics (Reprinted from Ref. [1], Copyright 2014, with permission from Taylor & Francis.)

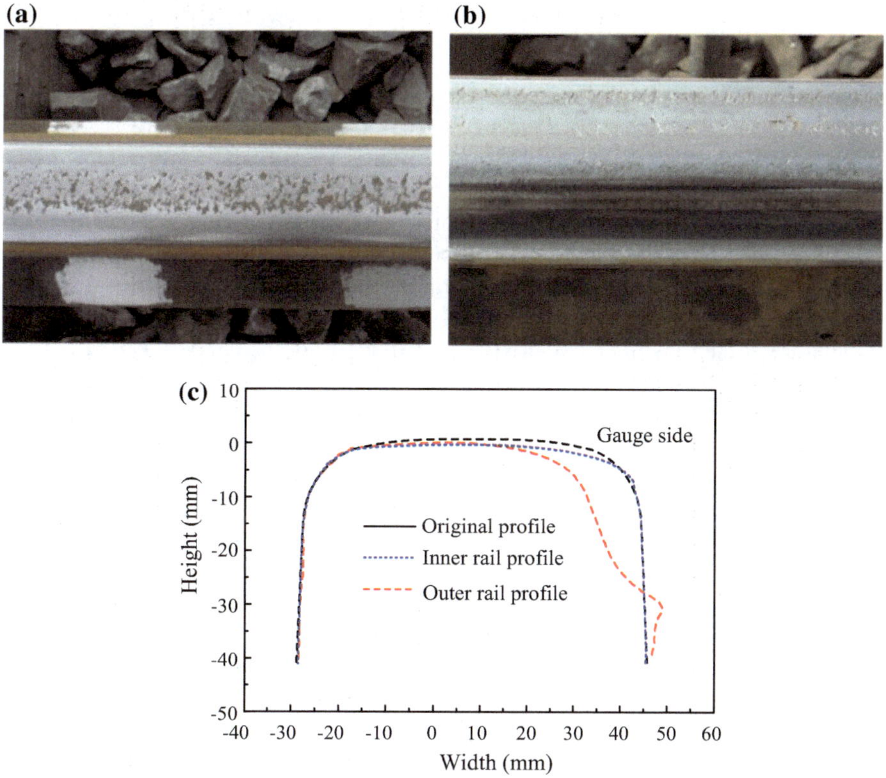

Fig. 10.14 Rails on sharp curve of Shuohuang heavy-haul railway: **a** rail defect of inner rail, **b** rail defect of outer rail, and **c** measured rail profiles (Reprinted from Ref. [1], Copyright 2014, with permission from Taylor & Francis.)

company operating in Western Australia successfully alleviated rail side wear through rail profile grinding [2]. In the early 1980s, the rail asymmetric-grinding technique on curves was proposed by Australian heavy-haul railways, and was then adopted by railways of USA, Canada, etc. [3–5]. In China, some experimental and theoretical studies on rail wear of curved tracks have also been carried out [5], but studies and applications of the rail asymmetric-grinding technique are still at the initial stage. In particular, there are few studies on the design method of rail profiles as well as the relationship between the rail profile and the wheel–rail interaction dynamics.

Concerning on the severe rail side wear found on many Chinese heavy-haul railway curves, the author has presented a design methodology for rail asymmetric-grinding profiles based on detailed analysis of wheel–rail dynamic interaction by using the theory of vehicle–track coupled dynamics [1].

10.2.2 Design Methodology of Rail Asymmetric-Grinding Profiles for Curves

Field surveys show that the wear of inner and outer rails presents an obvious asymmetric feature which is caused by the asymmetries of wheel–rail contact points and wheel–rail interactive forces on the inner and outer rails. In order to alleviate the rail side wear, rail asymmetric-grinding for the outer and inner rails may be an effective way to change the wheel–rail contact status and eventually improve the dynamic performance of wheel–rail interaction. For this purpose, it becomes very important to accurately understand the wheel–rail contact relationships and the vehicle–track interaction characteristics on curved sections. On curved sections, the vehicle–track dynamic interaction is more complicated than that on tangent tracks due to the wheel–rail contact points on curves generally being distributed over a wider area on both wheel and rail profiles. In order to make a detailed study on vehicle–track interactive performance, it is necessary to synthetically consider the vehicle system, the track system, and the wheel–rail contact interface. To this end, the vehicle–track coupled dynamics theory and its simulation techniques, described in Chaps. 2 and 4, provide an effective tool in the design process of rail asymmetric-grinding profiles. Figure 10.15 gives an overview of the design process of rail asymmetric-grinding profiles for curved tracks [1].

The implementing procedures illustrated in Fig. 10.15 are explained as follows. First, the actual profiles of wheels and rails are measured in the field. Second, the wheel–rail contact geometry is analyzed and assessed using the measured wheel and rail profiles. Third, dynamic simulation of the whole vehicle–track system is carried out using the theory of vehicle–track coupled dynamics. In the simulations, the measured wheel and rail profiles, as well as the real operational conditions, are used. Next, the dynamic performance of the wheel–rail interaction is evaluated

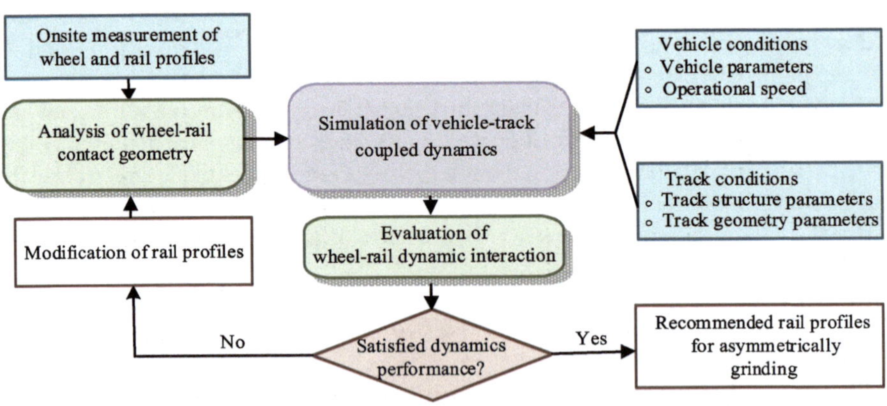

Fig. 10.15 Design process of rail asymmetric-grinding profiles based on analysis of wheel–rail interaction (Reprinted from Ref. [1], Copyright 2014, with permission from Taylor & Francis.)

based on the simulation results of vehicle–track coupled dynamics. The evaluations are mainly focused on lateral dynamics indexes such as the wheel–rail lateral forces and the wheelset angles of attack on curves. If the rail profiles are not able to meet the requirements of achieving low wheel–rail dynamic interaction, in particular, low lateral interaction, a modification of the rail profiles and a re-evaluation of the wheel–rail contact geometry and the corresponding wheel–rail dynamics performances should be done. The rail profile modification criteria are discussed later in this section. This process is repeated until a satisfactory wheel–rail dynamics performance is achieved. Finally, the recommended asymmetric-grinding profiles of the inner and outer rails are obtained. The recommended profiles should ensure not only good wheel–rail contact geometry but also low wheel–rail dynamic interaction.

The design process of rail asymmetric-grinding profiles involves many key elements. Among these, the design criteria for curved rail profiles and the wheel–rail dynamic interaction are the fundamental issues.

In general, two basic criteria should be followed in the design process of rail asymmetric-grinding profiles on curves. First, increasing the rolling radius difference (RRD) between the outer and inner wheels as much as possible can improve wheelset steering ability, so that the creep forces could help the bogie turn smoothly along the curve. As a result, both the wheelset angle of attack and the rail side wear will be reduced. Secondly, the rail profiles need to match the wheel profiles to reduce wheel–rail contact stress so that the rolling contact fatigue of wheels and rails can be decreased.

It is necessary to understand the wheel–rail contact characteristics for various segments of the wheel and rail profiles before progressing the design or modification of rail profiles. Generally, the contact point distribution can be divided into five zones [6] as shown in Fig. 10.16. It is important to highlight these wheel–rail contact zones: Zone *A* is the contact region between the rail gauge corner and the wheel flange; Zone *B* is the contact region between the rail gauge corner and the wheel flange root; Zone *C* is the contact region between the rail crown and wheel tread; Zone *D* is the contact region between the near field side of both the rail and

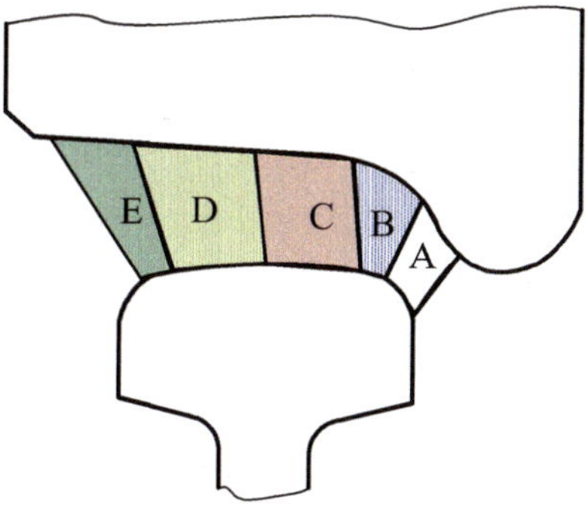

Fig. 10.16 Wheel–rail contact zones (Reprinted from Ref. [1], Copyright 2014, with permission from Taylor & Francis.)

the wheel and Zone *E* is the contact region between the far-field side of both the rail and the wheel.

On curves, the side wear of the outer rail mainly occurs in Zones *A* and *B*. There are several principles that can be used to reduce the outer rail side wear through the grinding of rail profiles:

- Grind Zone *A* to avoid contact between the wheel flange and the gauge corner.
- Modify Zone *B* to reduce the discontinuous distribution of the contact points and to provide conformal contact.
- Modify Zone *C* to make the rail crown radius of curvature slightly smaller than that of the wheel tread so that the wheel–rail contact stress can be reduced.
- Grind Zone *D* to avoid contact in this zone and move contact points to Zone *C*. In this way, the rolling radius of the outer wheel is increased and the self-steering ability can be improved on curves.

For the inner rail, the contact points are mainly centered along the top of the rail [3]. In order to reduce the side wear of the outer rail and restrict the vertical wear on the inner rail simultaneously, Zones *B*, *C* and *D* on the inner rail also need to be modified:

- Modify Zones *B* and *C* to avoid contact in the gauge corner, and to make contact points in Zone *C* move toward Zone *D*, resulting in the decrease of the rolling radius of the inner wheel.
- Modify Zone *D* to ensure contact points to be distributed in Zone *C* or *D* uniformly.

After the modification of rail profiles, the wheel–rail contact geometry should be analyzed and evaluated. Through the analysis of static wheel–rail contact geometry including both contact point distributions and contact geometry parameters of rail, the appropriate asymmetric-grinding profiles can be obtained. Among them, the RRD between the outer and inner wheels is the most important parameter affecting the curve negotiation performance of railway vehicles.

By a compatible design of wheel–rail contact geometry, the rail-grinding profiles with a good contact geometry state may be obtained. However, another important issue must be paid attention to, namely the wheel–rail dynamic interaction performance of a heavy-haul freight vehicle passing through a curved track with the redesigned rail-grinding profiles. Here, the lateral wheel–rail force is the key factor which is mainly responsible for rail side wear. On the curved track, the wheelset is steered by the lateral forces acting on the wheels. The steering mainly depends on the longitudinal creep forces ($F_{x\mathrm{L}i}$, $F_{x\mathrm{R}i}$) and the lateral creep forces ($F_{y\mathrm{L}i}$, $F_{y\mathrm{R}i}$), as shown in Fig. 10.17 [7]. If the wheel and rail profiles are redesigned to be able to provide enough RRD, the creep forces will push the wheelset to adjust itself to the radial position without wheel flange contact. Conversely, small RRD results in large angles of attack, and the wheel flange usually contacts with the rail gauge side. Once the wheelset is steered by the wheel flange, severe rail side wear may happen.

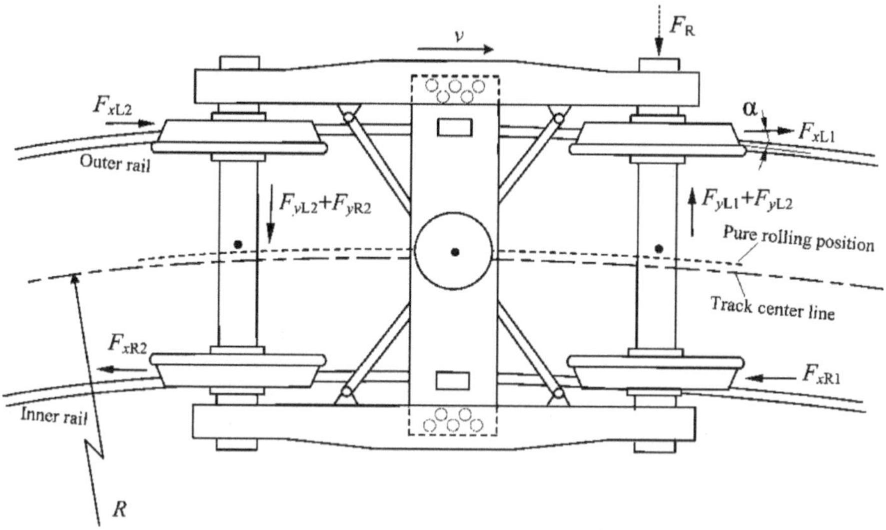

Fig. 10.17 Basic wheel–rail forces acting on bogie in curve (Reprinted from Ref. [1], Copyright 2014, with permission from Taylor & Francis.)

In the actual track structure, rails are connected with sleepers by fasteners while sleepers are supported by the ballast. All the components form an elastic-damping vibration system. Subjected to a large lateral and vertical wheel–rail forces on curved track, the lateral, vertical, and torsional vibration displacements of rails cannot be ignored, especially the lateral and torsional displacements of the outer rail. Undoubtedly, rail movements will affect wheel–rail contact geometry as well as the dynamic forces. Therefore, the track vibrations should be taken into account in solving the wheel–rail dynamic interaction on curves. Here, the wheel–rail coupled model established under the framework of vehicle–track coupled dynamics in Chap. 2 is used. The wheel–rail coupled model is capable of considering rail vibrations in lateral, vertical and torsional directions.

10.2.3 Numerical Implementation for Design of Rail Asymmetric-Grinding Profiles on a Practical Railway Curves

Based on the techniques of the above methodology and analysis, the rail asymmetric-grinding profiles can be redesigned for curves. As an example, the rail asymmetric-grinding profiles were redesigned for a curve with 600 m radius on the Chinese Shuohuang heavy-haul railway. On this curve, the superelevation of the outer rail is 75 mm, and the length of the transition curve is 140 m. The rail cant is

1:20. The track structure includes 75 kg/m size rail, type III sleepers, type II elastic fasteners and a ballast bed with the thickness of 0.3 m. The main vehicle running on the line is a freight wagon of type C_{70} that is using the Chinese LM wheel profile. All of the vehicle and track components comply with the Chinese railway industry standards.

Based on the standard profile of 75 kg/m size rail, a number of asymmetric-grinding profiles were redesigned. Among them, a pair of rail asymmetric profiles with lower wheel–rail interaction and lesser grinding requirement was selected to be the recommended profiles for rail grinding. Figure 10.18 gives the redesigned asymmetric-grinding profiles for the inner and outer rails. Comparing with the inner rail profile, the grinding range of the outer rail profile is larger and its grinding zones are more widely spread over the rail gauge face, gauge corner, and rail crown. The inner rail profile is mainly ground in the local region of the rail crown.

Using the profile data, the RRD functions of the wheelset before and after rail grinding are calculated, as shown in Fig. 10.19. Here, the wheelset lateral displacement ranges from −15 to 15 mm. The lateral wheelset displacement is defined as negative if it moves laterally toward the outer rail and positive if toward the inner rail. In the case of the LM wheel profile matching with the redesigned rail profiles, there exists obvious asymmetric characteristic in the RRD curve (Fig. 10.19). This is due to the asymmetries of the left and right rail profiles. When the lateral

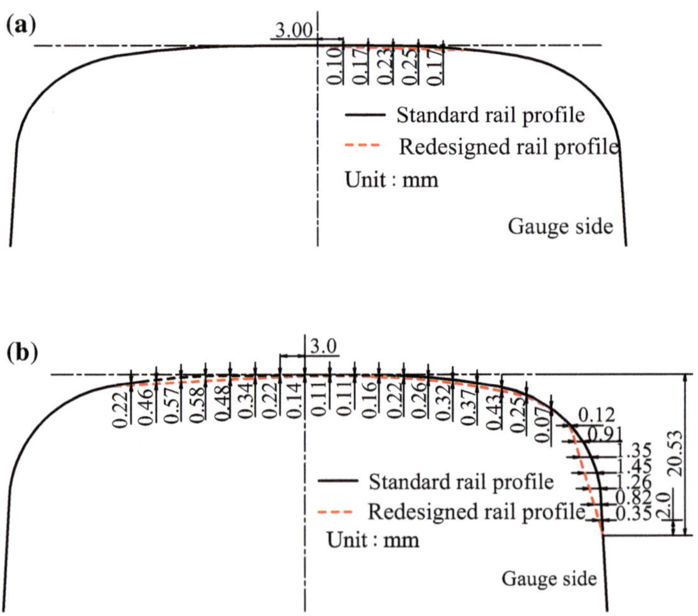

Fig. 10.18 Redesigned rail-grinding profiles for **a** inner and **b** outer rails of the 600 m radius curve of Shuohuang heavy-haul railway (Reprinted from Ref. [1], Copyright 2014, with permission from Taylor & Francis.)

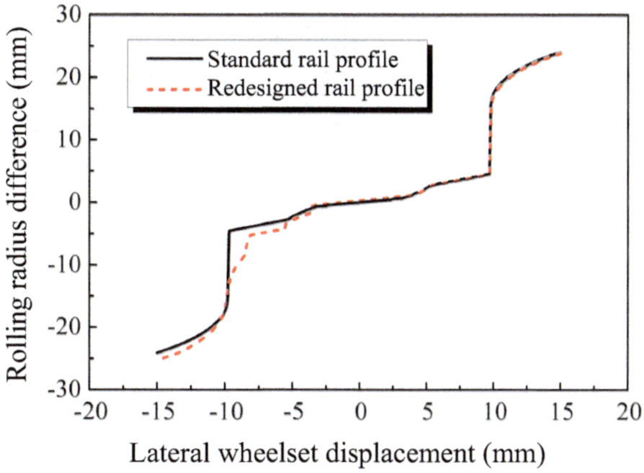

Fig. 10.19 Rolling radius differences of wheelset with the LM wheel profile (Reprinted from Ref. [1], Copyright 2014, with permission from Taylor & Francis.)

displacement of the wheelset varies within −6 to −10 mm, the RRD of the redesigned rail profiles is larger than that of the standard rail profile. In this case, the steering ability of the wheelset can be improved. The modification of rail profiles will influence the contact angle difference of the wheelset simultaneously. The analysis results in Fig. 10.20 indicate that, if the wheelset moves laterally from −6 to −10 mm, the contact angle difference value of the redesigned rail profile is larger than that of the standard rail profile. For the redesigned rail profile with a larger contact angle difference, the lateral component of the wheel–rail normal force can provide appropriate steering ability, which prevents the wheelset moving excessively towards the outer rail. Consequently, the probability of contact between the wheel flange and the rail gauge side will be reduced.

The distributions of wheel–rail contact points reflect the regions where the wheel and rail frequently contact with each other. In order to reduce the rail side wear, the number of contact points on the rail gauge face should be limited to as few as possible. Figure 10.21 compares the wheel–rail contact point distributions on the

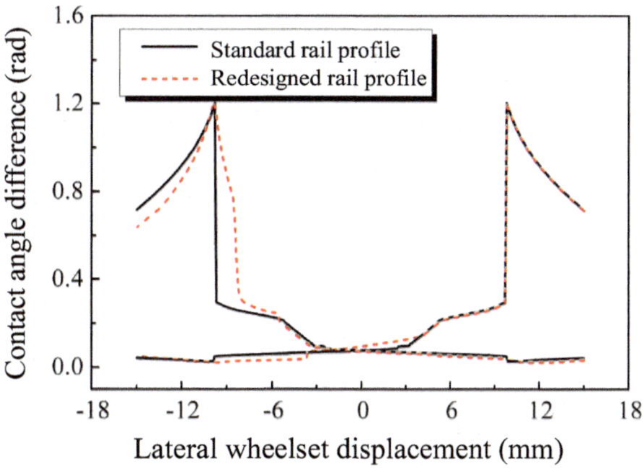

Fig. 10.20 Contact angle differences of wheelset with the LM wheel profile (Reprinted from Ref. [1], Copyright 2014, with permission from Taylor & Francis.)

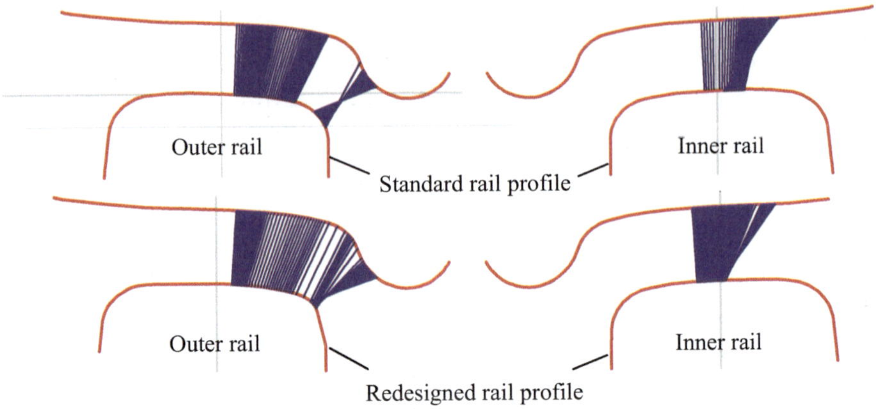

Fig. 10.21 Wheel–rail contact point distributions of LM wheel profile with standard rail profile and redesigned rail-grinding profiles (Reprinted from Ref. [1], Copyright 2014, with permission from Taylor & Francis.)

surfaces of the standard rail profile and the designed rail-grinding profiles. In comparison, the lateral displacement of the wheelset is from 0 to −15 mm. It can be seen from Fig. 10.21 that the wheel–rail contact geometry property is greatly improved after rail grinding. For the outer rail, the contact points spread over a wider zone after the rail grinding. According to Fig. 10.16, the wheel and outer rail rarely contact in Zone *A* and Zone *D*. Even if a large lateral displacement of the wheelset occurs, the contact is still at the gauge corner, not the rail gauge face. The contact points in Zone *B* distribute continuously if the rails are ground. For the inner rail, the contact points distribute more uniformly in Zone *C* after the grinding. It is proven theoretically that the rail profiles after asymmetric grinding could meet the demand of the wheel–rail contact geometry required in the design criteria.

To evaluate the wheel–rail dynamic interaction after rail grinding, the vehicle–track coupled dynamics simulation system described in Chap. 4 is used. The curve negotiation performances of a freight wagon on a section of track with the standard and ground rail profiles are analyzed, respectively. In the simulation, the measured track irregularities are used as the input excitation of the system. Figure 10.22 compares the wheel–rail dynamics responses at a normal curve negotiation speed of 70 km/h before and after rail grinding. The results in Fig. 10.22a, b show that, compared with the standard rail profile, the use of the redesigned rail profile reduces the maximum lateral displacement of the wheelset by 14.6% and the maximum wheelset angle of attack by 24.5%. Figure 10.22a indicates that the redesigned rail profiles are helpful to prevent the wheelset moving laterally with large displacement, so that the probability of wheel flange contact can be reduced. Figure 10.22b reflects that the wheelset with the redesigned rail profiles tends to be at the radial position of the curve. Figure 10.22c and d show that, after rail grinding, the maximum values of the lateral wheel–rail force and the frictional power of the outer rail are reduced by 26.7% and 27%, respectively.

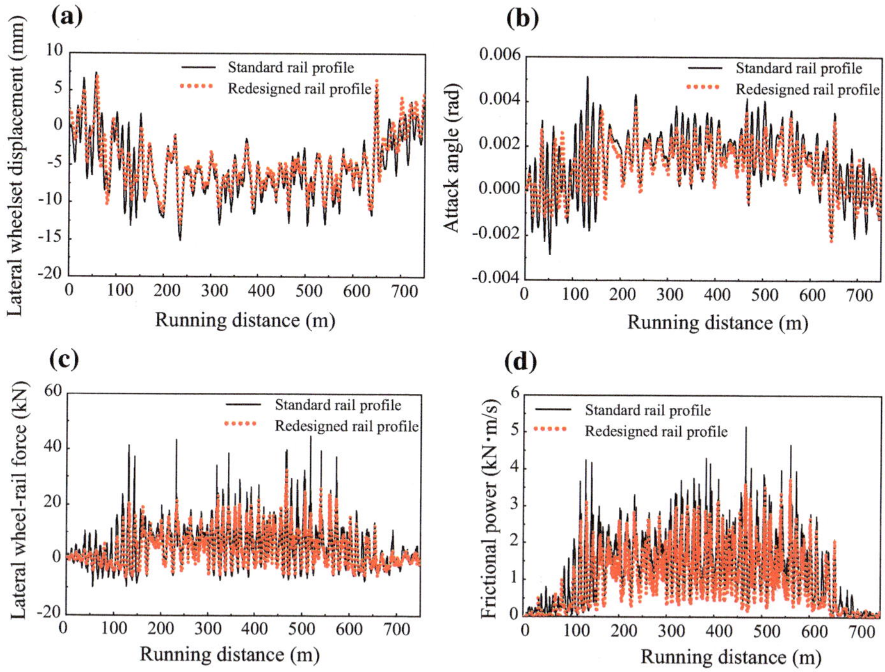

Fig. 10.22 Comparison of dynamics indexes before and after rail grinding: **a** lateral wheelset displacement, **b** angle of attack, **c** lateral wheel–rail force, and **d** frictional power (Reprinted from Ref. [1], Copyright 2014, with permission from Taylor & Francis.)

It is concluded from the above theoretical analysis that, if the redesigned rail asymmetric-grinding profiles are adopted, the wheel–rail contact geometry and the wheel–rail dynamics performance will be effectively improved, and the rail side wear on heavy-haul railway curves can be significantly reduced.

10.2.4 Engineering Practice and Implementation Effect

In order to validate the actual effect of the method proposed in this chapter, the design example of the rail asymmetric-grinding profiles illustrated in Sect. 10.2.3 was put into practice on the Shuohuang heavy-haul railway.

1. *Asymmetric grinding of rail profiles on curve*

A test section was chosen at a curve with a radius of 600 m on the Shuohuang railway. Using the redesigned rail profiles in Fig. 10.18, the inner and outer rails on the test section were ground by the RR48-HP4 grinding train in August 2008.

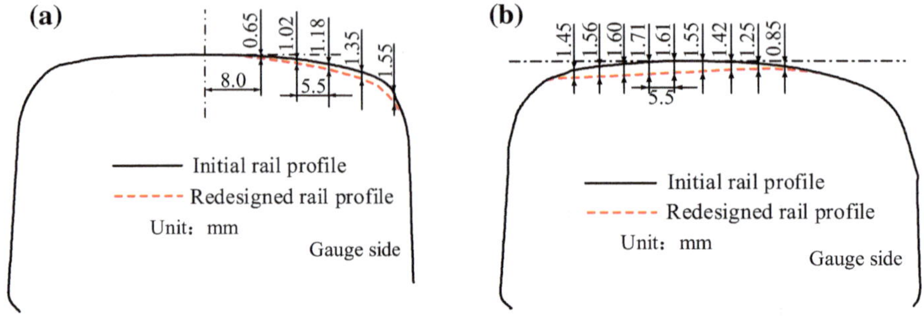

Fig. 10.23 Redesigned rail profiles for corrective grinding in the test curve: **a** inner rail, and **b** outer rail (Reprinted from Ref. [1], Copyright 2014, with permission from Taylor & Francis.)

Based on the results of 8 months of track observation and measurement of the rail profiles, some changes from the originally redesigned profiles were found. The rail profiles were therefore redesigned with corrections based on these observations and ground once again in April 2009 to the profiles as shown in Fig. 10.23. The grinding position is mainly centered in the region of the rail top corresponding to the Zones *C* and *D* as shown in Fig. 10.16. For the inner rail, the grinding sections scatter within the range from the gauge corner to the centerline of the rail top, corresponding to the contact Zones *A*, *B*, and *C* that are shown in Fig. 10.16.

2. *Field test of wheel–rail dynamics performance*

In order to examine the effectiveness of the grinding application, field tests were carried out on the tested curve before and after the corrective grinding for rail profiles. The field tests are focused on the wheel–rail dynamic performance. Figure 10.24 illustrates the measurement arrangements for track structure vibration and wheel–rail forces.

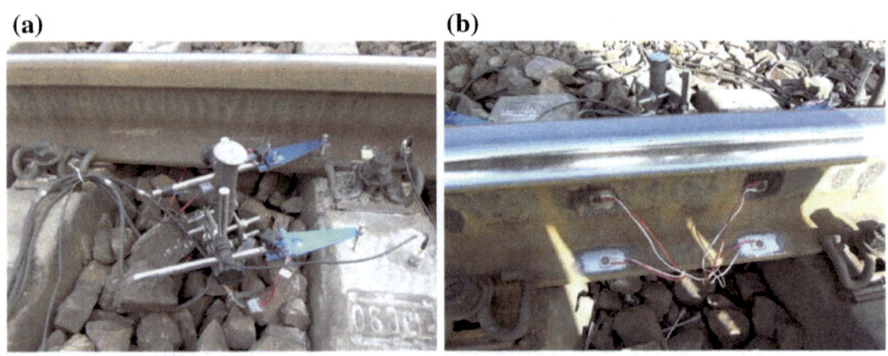

Fig. 10.24 Dynamic performance measurements: **a** track structure vibration, and **b** wheel–rail forces (Reprinted from Ref. [1], Copyright 2014, with permission from Taylor & Francis.)

In the test, some key dynamic performance indexes such as the wheel–rail vertical force, wheel–rail lateral force, track gauge dynamic widening and rail displacements were obtained before and after rail grinding. As an example, Fig. 10.25 compares the wheel–rail forces and the outer rail displacements measured before and after the corrective grinding. With the same conditions of the vehicle and the track, the average lateral wheel–rail force decreased obviously after grinding. The average lateral and vertical rail displacements also decreased by 26% and 24%, respectively, after grinding. This clearly demonstrates that, when the rails in the test section were ground according to the designed schemes, the wheel–rail contact conditions were improved so that the lateral wheel–rail dynamic interaction was alleviated.

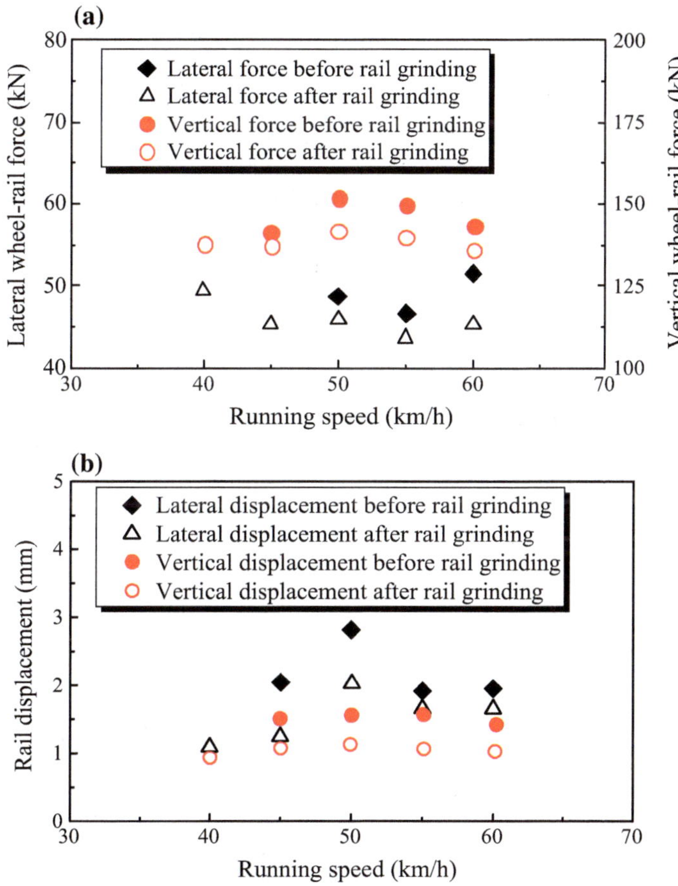

Fig. 10.25 Comparisons of wheel–rail dynamics indexes of outer rail before and after rail grinding in test section: **a** lateral and vertical wheel–rail forces, and **b** lateral and vertical rail displacements (Reprinted from ref. [1], Copyright 2014, with permission from Taylor & Francis.)

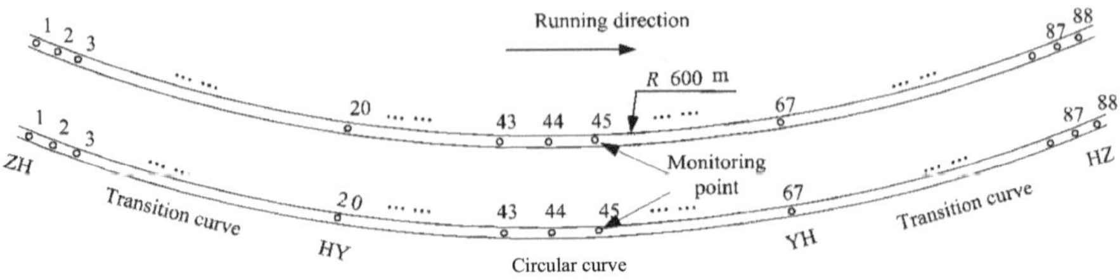

Fig. 10.26 The measuring point arrangement of rail profiles in track test section (Reprinted from Ref. [1], Copyright 2014, with permission from Taylor & Francis.)

3. *Measurement of rail wear*

For the wear measurement process, the wear of the inner and outer rails at 176 measuring points were periodically measured in the test section (see Fig. 10.26). The longitudinal spacing between every two adjacent points was 10 m. The measurement time interval was 1 month in the early stage and 2 months in the later stage. 16 months in total were spent to complete this measurement process.

From the statistical analysis for rail wear at every measuring point, the overall wear information on the rails in the test section was obtained; therefore, the implementation effect of rail asymmetric grinding could be evaluated. As an example, the measured evolution of the outer rail side wear in the circular curve is plotted in Fig. 10.27. For a comparative purpose, the historically recorded data of the rail side wear on the same curve without rail grinding is also presented in the figure. The results showed that the average rail side wear of the outer rail decreased by 30–40% after the rail asymmetric grinding.

It can be concluded that the wheel–rail dynamic interaction performance was improved and the rail side wear was reduced using the design method for grinding asymmetric rail profiles on heavy-haul railway curves.

Fig. 10.27 Comparison of outer rail side wear in test track (Reprinted from Ref. [1], Copyright 2014, with permission from Taylor & Francis.)

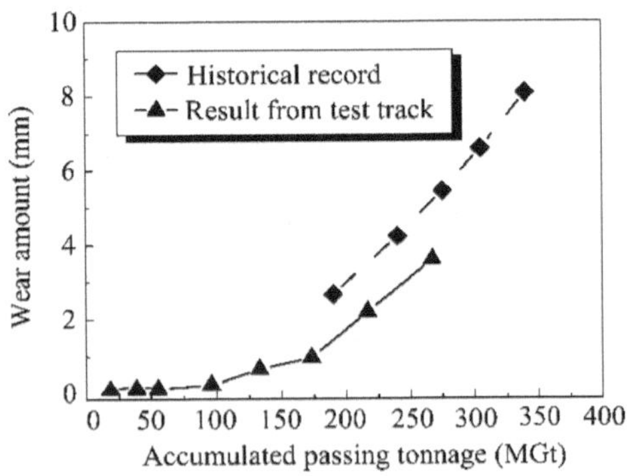

10.3 Safety Control of the Coupler Swing Angle of a Heavy-Haul Long Train

10.3.1 Engineering Background

A railway coupler is used to connect a vehicle to a locomotive or to another vehicle so as to transmit traction forces or impact forces in a controlled fashion. Couplers are also used to maintain a certain distance between adjacent vehicles. Consequently, couplers are usually subjected to complicated forces during train operations, including alternating tensile and compressive forces, impact forces, and bending moments. For heavy-haul trains heavier than about 10,000 t, the traction forces and braking forces will substantially increase due to the significant growth of the train mass and length. Heavier trains will result in more serious interactions between vehicles through the couplers. This is likely to cause a series of significant problems that endanger train operational safety without a scientific design or a reasonable control of the operation of the train. Actually, some train separations, coupler breakage, and/or even derailments due to intensified longitudinal impacts have been observed. All these issues have seriously disturbed the normal transportation order and have directly affected the train operational safety. They also adversely affect the economic benefits of the railway transportation. For example, serious train derailments on tangent track happened many times during the traction and braking tests of Chinese long heavy-haul trains (see Fig. 10.28). In these derailments, rails were completely overturned and damaged.

Fig. 10.28 Heavy-haul locomotive derailment

In order to solve this practical problem and improve operational safety of the heavy-haul trains, the author and team were commissioned by the manufacturer of the heavy-haul locomotives to conduct comprehensive investigations and studies on this problem. The analyzed results indicate that the cause for the rail overturning is the occurrence of a large lateral wheel–rail forces which may come from the lateral component of longitudinal coupler force. Under this situation, the established locomotive–track spatially coupled dynamics model is employed for detailed simulation and longitudinal coupler forces are also considered in the model. Based on this model, the effect of longitudinal coupler forces on lateral dynamic interactions of the locomotive–track coupled system is analyzed. The influencing mechanism of the coupler free swing angle on the locomotive operational safety demonstrates that there is a causal relationship between them. On this basis, a limit value for the coupler free swing angle design is then proposed and applied in the final coupler system design to successfully solve this significant practical problem. The main study results are going to be introduced as follows.

10.3.2 Analysis of Wheel–Rail Dynamic Interaction with Large Coupler Free Swing Angle

The coupler may swing in the horizontal plane when it is under compressive actions. The maximum possible swing angle is defined as the coupler free swing angle φ which is illustrated in Fig. 10.29. This coupler free swing angle is limited physically due to the action of various kinds of auxiliary elements, such as the secondary lateral stop.

The FT prototype coupler imported to China was applied in the locomotive shown in Fig. 10.28 which depicts how the locomotive was derailed in a heavy-haul train test. This type of coupler has a large free swing angle, and it is likely to generate relatively large coupler angles with a maximum value exceeding 10° under longitudinal compressive forces. First, the wheel–rail dynamic interaction performance of the locomotive with the prototype coupler is analyzed when the coupler swing angle is 8°. Note that this swing angle is still smaller than the

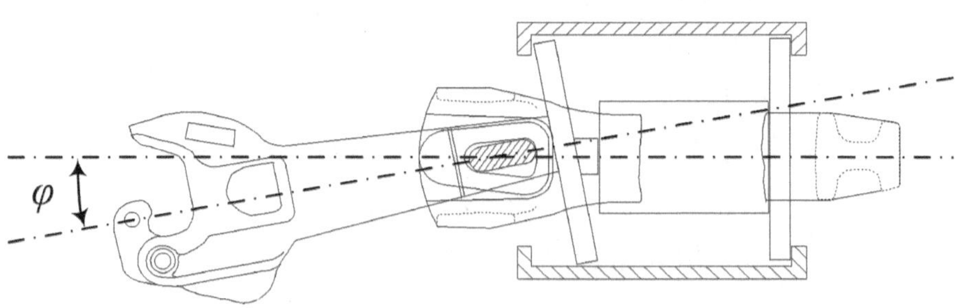

Fig. 10.29 Coupler free swing angle

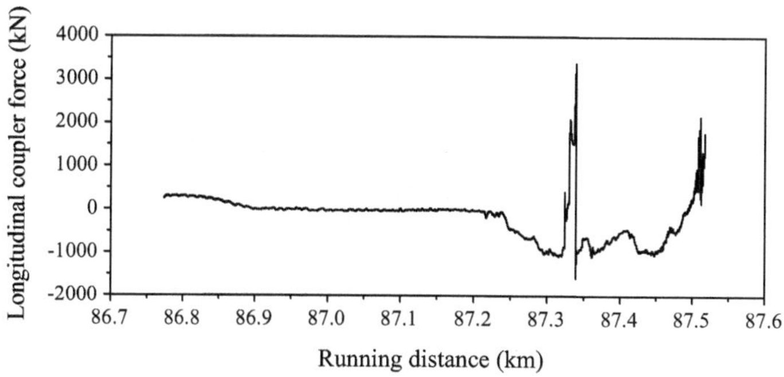

Fig. 10.30 Tested longitudinal coupler force of heavy-haul locomotive

maximum coupler angle that the coupler can reach. For the longitudinal coupler force, its value is assigned with 1500 kN (far smaller than its maximum value) according to the tested coupler longitudinal force results of the heavy-haul locomotive (see Fig. 10.30).

The calculated lateral wheel–rail force results are presented in Fig. 10.31. During the simulation, the heavy-haul locomotive is running on tangent track at its maximum operational speed of 80 km/h. Here, the coupler swing angle is set to be 8°. It can be seen that most of the lateral wheelset forces are greater than 120 kN; and the maximum value is even up to 174.7 kN which is far higher than the safety threshold of 79.3 kN [8].

In addition, the corresponding calculated derailment coefficient is given in Fig. 10.32. The results indicate that there are so many occurrences with amplitudes larger than 0.9, and the maximum value of 1.2 was observed during the operational process. This does not meet the required limit value of 0.9 specified in the Chinese railway standard entitled "Identification methods and evaluation standard for the test of railway locomotive dynamics performance" (TB/T2360-93) [9].

The analyzed results indicate that the lateral wheel–rail interactions are much intensified, and the safety indexes, such as the lateral wheelset force and the derailment coefficient, significantly exceed their safety thresholds. This locomotive with this type of coupler with a large swing angle cannot meet the requirements for safe

Fig. 10.31 Calculated lateral wheelset force of heavy-haul locomotive

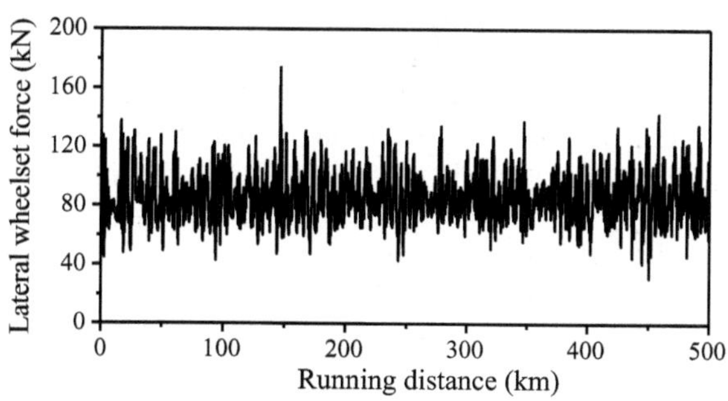

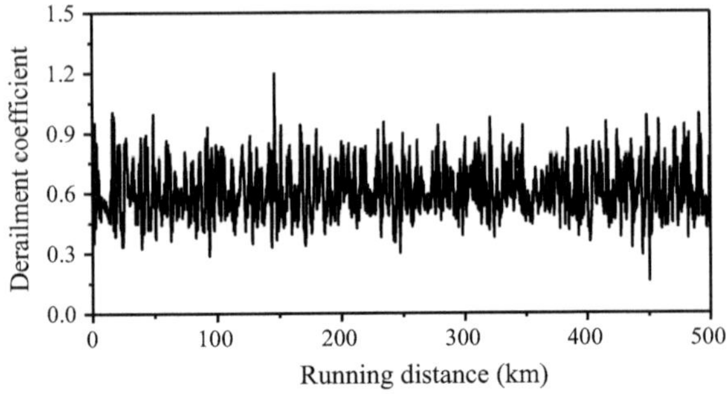

Fig. 10.32 Calculated results of heavy-haul locomotive derailment coefficient

operation. Consequently, application of this locomotive is likely to cause the serious accidents or derailments and endanger the operational safety of heavy-haul trains.

10.3.3 Effect of Coupler Free Swing Angle on Heavy-Haul Locomotive Running Safety and Its Safety Design

The analyzed results in the last subsection demonstrate that the possible reason for the heavy-haul locomotive derailments is the large coupler free swing angle which generates a large lateral coupler force component. The large lateral force component can cause abnormal lateral wheel–rail dynamic interaction forces and rail overturning. Thus, the author and team further investigate the relationship between the coupler free swing angle and locomotive running safety. The study was conducted to determine the safety threshold of the coupler free swing angle and supply some scientific guidance for improving the coupler design.

The maximum values of the calculated lateral wheelset force under various coupler free swing angles are shown in Fig. 10.33. In these simulations, the simulated locomotive is braking on tangent track at the speed of 80 km/h that is the practical maximum operational speed when the locomotive is used for 20,000 t heavy-haul trains. The lateral wheelset force increases nonlinearly with the increase of the coupler free swing angle. When the coupler free swing angle is less than 3°, the lateral wheelset force has a slight increase with the increase of the coupler free angle. In the angle range of 3–6°, the lateral wheelset force increases promptly with the coupler free swing angle, especially the lateral wheelset force is beyond its safety threshold when the coupler free swing angle is larger than 3.7°.

The corresponding maximum values of the locomotive derailment coefficient are displayed in Fig. 10.34. Similarly, the derailment coefficient also increases nonlinearly with the increase of coupler free swing angle. When the coupler free swing angle is less than 4°, the derailment coefficient has a gradual increase with the growth of the coupler free swing angle. When the coupler free swing angle exceeds 4°, the derailment coefficient increases rapidly with the increase of the coupler free

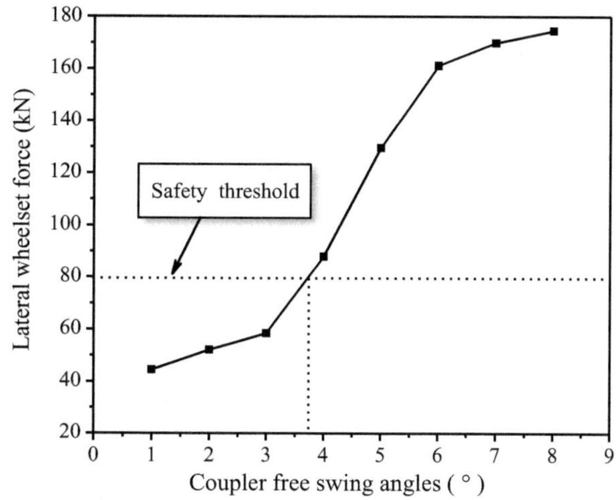

Fig. 10.33 Variation of lateral wheelset force versus coupler free swing angle

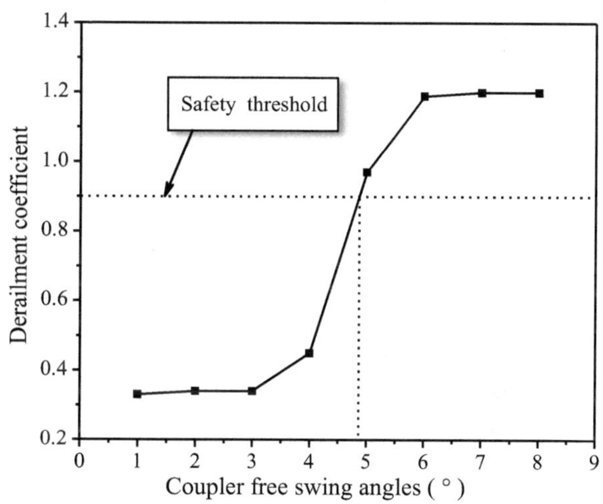

Fig. 10.34 Variation of derailment coefficient versus coupler free swing angle

swing angle, and it reaches the safety threshold when the coupler free swing angle is increased to 4.9°.

It can also be seen from the analysis that both lateral wheelset force and derailment coefficient indicate that the locomotive running safety could meet the safety requirements when the coupler free swing angle is smaller than 3.7°. Considering a reasonable safety margin, it is suggested that the maximum free swing angle of the heavy-haul locomotive coupler should be controlled to no more than 3°.

This proposal had subsequently been applied to the improved design of the prototype coupler by the locomotive manufacturer. This newly designed coupler has a maximum coupler free swing angle of 3°. It had been put into practice in July

2008. The practical application demonstrates that the operational safety of the heavy-haul locomotive can be guaranteed using the improved coupler.

10.4 Application and Practice for Design of Fuzhou–Xiamen Shared High-Speed Passenger and Freight Railway

10.4.1 Engineering and Research Background

The Fuzhou–Xiamen high-speed railway (Fig. 10.35) in China is an important component of the Hangzhou–Fuzhou–Shenzhen passenger dedicated line. It is also one of the Chinese "four longitudinal lines and four lateral lines" of the main railway network, and is an important part of transport channels in this Chinese coastal area.

The feasibility study of the Fuzhou–Xiamen high-speed railway began in 2002 and it was approved by the Chinese State Council on July 2004. At that time, the design orientation is as follows: passenger transportation is the priority, but freight

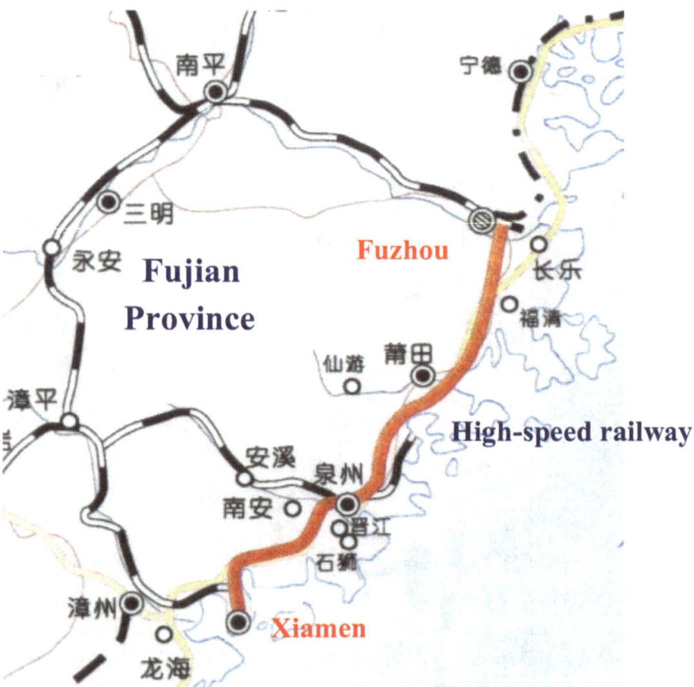

Fig. 10.35 Fuzhou–Xiamen high-speed railway

transportation must be taken into consideration. Specifically, this line should not only satisfy the operational requirement of high-speed passenger train running at a maximum speed of 200–250 km/h, but also meet the operational requirement of freight trains with a hauling mass of 3500 t. This railway line was the first shared high-speed passenger and freight railway in China.

Shared high-speed passenger and freight railways are only employed in a few countries such as Germany, Italy, etc.; the design of this kind of railways was still a challenging task for railway engineers. On the one hand, it lacked simple design criteria for the determination of various parameters (minimum curve radius, superelevation, transition curve length, longitudinal slope) of horizontal and vertical profiles for the high and low speed shared line because there was an insufficient theoretical basis for the influence of these parameters on the overall running safety and ride comfort at both high and low speeds. On the other hand, the dynamic effects of high-speed passenger trains and low-speed freight trains on track structures are quite different. When a freight train with a larger axle load runs at a low speed, and a high-speed passenger train with a smaller axle load runs at high speed, how can the dynamic effects on the track structures for the two cases be reconciled? Is it possible to realize a compatible design? Quite a few studies with regard to these issues have been published.

For this practical engineering problem, the author and team collaborated with the design institution of Fuzhou–Xiamen Railway (China Railway Eryuan Engineering Group Co. Ltd.) to comprehensively investigate the running safety and stability [10] of the track–vehicle system, as well as the dynamic interaction between vehicle and track [11] by combining the line design parameters with vehicle dynamic performance using the vehicle–track coupled dynamics theory and the principle and method for optimal integrated design of vehicle and track systems in Chap. 9. The design work was completed at the beginning of 2003, thus realizing the optimization of the line design and safety pre-evaluation of the Fuzhou–Xiamen shared passenger and freight railway. The design work also provided the necessary theoretical basis for the development of design standards for shared high-speed passenger and freight railways in China.

As examples, the determination of parameters for horizontal curves, the parameter matching of horizontal and vertical sections, and the evaluation of dynamic interaction between vehicle and track will be briefly discussed for the speeds of 200/120 km/h. More detailed research on dynamic performance under other speeds can be found in Refs. [10, 11].

Based on the actual conditions at that time, the "China Star" high-speed test train and the 120 km/h speedup freight train were selected as the high- and low-speed operational conditions, respectively. For a clearer expression, the motor car and trailer car of the "China Star" high-speed train is referred to as ZH-D and ZH-T, respectively. HJ-120 and ZK_4 are referred to as speedup freight locomotive and freight vehicle are referred to as HJ-120 and KZ_4, respectively.

10.4.2 Effect of Key Parameters of Horizontal Curve on Dynamic Performance of High- and Low-Speed Trains

Existing studies show that the important horizontal curve parameters are mainly the minimum curve radius, superelevation and transition curve length, which have important influences on the dynamic performance of high- and low-speed trains. According to the design requirements, the minimum curve radius is divided into four different levels under each speed matching scheme, and the superelevation is considered as a constant value that complies with existing standards. For the speed match of 200/120 km/h, the abovementioned parameters are listed in Table 10.1.

Using the track irregularity that was measured from the high-speed test section in the Zhengzhou–Wuhan railway, the dynamic performance indexes of train running safety and stability, and the wheel–rail wear index are calculated and summarized in Table 10.2. This table shows the results when the "China Star" train with a speed of

Table 10.1 Parameters of horizontal curves under the speed match of 200/120 km/h

Curve radius (m)	Level 1	Level 2	Level 3	Level 4
	4,000	2,800	2,200	1,800
Transition curve length (m)	120	180	240	240
Superelevation (mm)	70	110	145	150

Table 10.2 Dynamic performance of passenger and freight train passing through track curves under speed matching scheme of 200/120 km/h

Speed (km/h)	Passenger car 200		Freight car 120	
Curve radius range (m)	4,000 (Level 1), 2,800 (Level 2), 2,200 (Level 3), 1,800 (Level 4)			
Vehicle type	ZH-D	ZH-T	HJ-120	ZK_4
Transition curve length (m)	130	200	260	270
Lateral wheel–rail force (kN)	34.58–47.70	26.91–38.59	31.42–43.36	30.73–31.92
Vertical wheel–rail force (kN)	134.0–154.4	99.98–113.8	152.2–169.9	138.4–143.8
Lateral wheelset force (kN)	35.39–42.49	28.52–37.19	41.01–54.83	29.74–34.92
Rollover coefficient	0.44–0.53	0.46–0.73	0.27–0.43	0.29–0.36
Derailment coefficient	0.27–0.34	0.27–0.44	0.25–0.27	0.26–0.29
Wheel unloading rate	0.45–0.53	0.57–0.79	0.44–0.58	0.30–0.38
Wheel–rail wear index (N m/m)	108.8–158.4	89.18–144.9	99.21–227.9	106.2–128.7
Lateral car body acceleration (g)	0.11–0.15	0.08–0.13	0.10–0.24	0.23–0.24
Vertical car body acceleration (g)	0.07–0.09	0.06–0.06	0.06–0.09	0.20–0.20
Lateral stability index	2.36–2.45	1.88–2.08	2.09–2.48	2.47–2.65
Vertical stability index	1.98–2.11	1.76–1.85	1.66–1.92	2.32–2.43

200 km/h and the speedup freight train with a speed of 120 km/h passes through curved tracks with different radius levels. More detailed data can be seen in reference [10]. Conclusions that can be drawn from Table 10.2 are as follows:

(1) The safety indexes of all cases are within the range of qualified and allowable values. Some indexes, such as the lateral and vertical wheel–rail forces and the derailment coefficient have large safety margins, and most of the stability indexes are of excellent levels.
(2) Dynamic performance indexes all increase to a certain extent with the decrease of curve radius. For example, when the curve radius reduces from 4,000 to 1,800 m, the lateral wheel–rail force increases by about 10 kN. The increase of vertical wheel–rail force is relatively small, with a maximum increase of about 11%. For the safety indexes, the increments are significant. Specifically, the rollover coefficient and wheel unloading rate increase by 20–35%, the wheel–rail wear index grows by 30–220%, and the lateral and vertical car body accelerations rise by 20–40%.
(3) For the "China Star" high-speed test train, the lateral wheel–rail force, wheel–rail wear index, lateral and vertical car body accelerations of the ZH-D motor car are all larger than those of the ZH-T trailer car. However, the rollover coefficient, derailment coefficient, and wheel unloading rate of the ZH-D are all smaller than those of the ZH-T.
(4) When the speedup freight train passes through the track curves at the speed of 120 km/h, the safety indexes and wheel–rail wear index of the HJ-120 (except for the derailment coefficient) are all larger than those of the ZK_4, whereas the lateral and vertical stability index of the HJ-120 are smaller compared with those of the ZK_4.
(5) The lateral wheel–rail interaction force and wear index of the HJ-120 are greater than those of the ZH-D. However, the rollover coefficient and derailment coefficient of the former are smaller. In several curved tracks, the lateral car body acceleration of the former are about 60% larger than that of the latter. The vertical car body accelerations of the two cases are almost the same.
(6) The vertical wheel–rail interaction force of the speedup train is 30% greater than that of the ZH-T. The rollover coefficient, derailment coefficient, wheel unloading rate, and the wear index of the former are smaller. The lateral and vertical car body accelerations of the former are 1–2 times larger than those of the latter.

In summary, the safety indexes of the freight train are smaller than those of the passenger train, but they all meet the safety limit requirement. The ride comfort indexes of the motor car are larger than those of the trailer car for the passenger train. The result for the freight train is these indexes are higher for the wagons than for the locomotives, but they are all at an excellent level. The ZH-D has the largest wheel–rail wear index, followed by the HJ-120, the ZH-D, and the ZK_4.

In conclusion, from the viewpoints of train running safety and ride comfort, the minimum radius of the horizontal curve can be selected as 1,800 m. Accordingly, the minimum transition curve length can be set as 240 m for a shared high-speed passenger and freight railway under the speed matching scheme of 200/120 km/h.

10.4.3 Optimal Integrated Design of Horizontal and Vertical Profiles for the Shared Passenger and Freight Railway

Whether the vertical curves are compatible with the horizontal curves or not is an important research question in the matching design of horizontal and vertical profiles. The setting rules described in relevant standards issued by the Ministry of Railways limit the overlapping of horizontal and vertical profile curves. However, these rules mainly consider the workload of the railway line and the associated maintenance difficulty; the rules do not specifically account for train running safety and ride comfort. Obviously, in the case of overlapped horizontal and vertical curves, if the dynamic performance indexes meet the running safety and comfort requirements on the shared high-speed passenger and freight railway line, this design will greatly reduce the project cost and generate a good economic benefit.

This section takes a length of the Fuzhou–Xiamen high-speed railway as an example, and the effects of key parameters of vertical and horizontal curves on the dynamic performance of a high-speed passenger train and low-speed freight train are investigated. The purpose of the investigation is to guide a reasonable matching design of the railway horizontal and vertical profile curves. Only the locomotive that has a larger dynamic interaction on the track structure is adopted in the calculation [10]. The 200 km/h high-speed train is represented by the ZH-D, and the 120 km/h speedup freight train is represented by the HJ-120.

A schematic diagram of the railway horizontal and vertical profiles is shown in Fig. 10.36, and the corresponding curve parameters are listed in Table 10.3 where i is the vertical curve slope; L_1 and L_2 are the slope lengths; ΔL_1 and ΔL_2 are the shortest lengths of transition curves that meet the requirements of train running safety and ride comfort; R_s is the radius of the vertical curve; l_s is the length of the vertical curve; R is the radius of the horizontal curve; l is the transition curve length

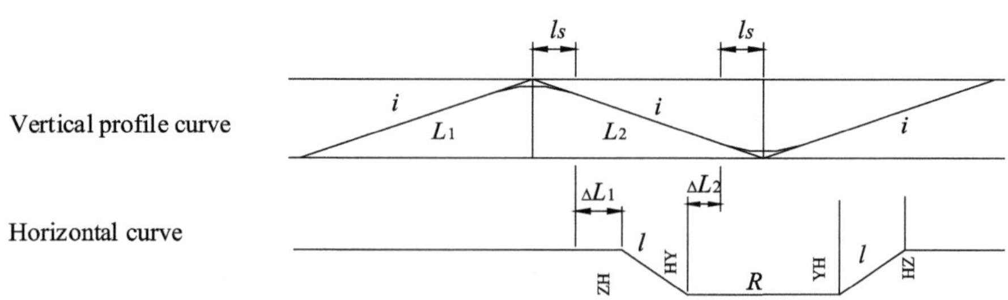

Fig. 10.36 Schematic diagram of horizontal and vertical profile curves

Table 10.3 Parameters of horizontal and vertical profile curves for 200/120 km/h speed match design

Vertical curve				Horizontal curve		
i (‰)	L_1 (m)	L_2(m)	R_s (m)	R (m)	l (m)	h (mm)
6	400	420	15,000	1800	240	150

of the horizontal curve; h is the superelevation of the horizontal curve; ZH is the transition point from tangent track to transition curve; HY is the transition point from transition curve to circular curve; YH is the transition point from circular curve to transition curve; HZ is the transition point from transition curve to tangent track.

For the case with zero distance between the start/endpoint of a vertical profile curve and the HY/YH or ZH/HZ point of a horizontal curve, namely when $\Delta L_1 = \Delta L_2 = 0$, the dynamic performance indexes of the high-speed passenger train and low-speed freight train are listed in Table 10.4 when the trains are passing through the combined profile. In the simulation, track irregularities that were measured from a section of high-speed test railway were used.

It can be seen from Table 10.4 that all the safety indexes are smaller than the corresponding limit values. For example, the maximum lateral wheel–rail forces of the ZH-D and the HJ-120 are 44.89 kN and 49.03 kN, respectively, and their limit values are 78 kN and 92 kN, respectively. The derailment coefficient and wheel unloading rate are 0.41 and 0.59, respectively, which are also within the safety range. In addition, the vertical and lateral accelerations satisfy the ride comfort requirement, and all stability indexes show good ride comfort for the trains.

It is indicated that on the shared high-speed passenger and freight railway line that has the selected horizontal and vertical profiles, the running safety and ride comfort indexes of the trains can satisfy all the requirements. The design scheme has effectively reduced the project cost for the Fuzhou–Xiamen high-speed railway line.

Table 10.4 Train dynamic performance for the case with zero distance connection between vertical profile curve and horizontal curve

Speed (km/h)		200	120
Locomotive type		ZH-D	HJ-120
Lateral wheel–rail force (kN)		44.89	49.03
Vertical wheel–rail force (kN)		166.07	162.36
Wheelset force (kN)		43.08	49.92
Rollover coefficient		0.55	0.53
Derailment coefficient		0.40	0.41
Wheel unloading rate		0.59	0.47
Wear index (N m/m)		193.04	226.70
Car body acceleration (g)	Lateral	0.17	0.19
	Vertical	0.09	0.09
Stability index	Lateral	2.46	2.50
	Vertical	2.14	2.12

10.4.4 Dynamic Effects of High- and Low-Speed Trains on Track Structures

In this section, dynamic effects of high- and low-speed trains on track structures of the Fuzhou–Xiamen shared passenger and freight railway are compared using vehicle–track coupled dynamics simulations. Safety assessments are further conducted according to the corresponding evaluation criteria.

1. *Dynamic interaction analysis under the excitation of random track irregularities*

Under the excitation of random track irregularities, the vertical dynamic effects on track structures induced by passenger and freight train loads are summarized in Table 10.5.

As it can be seen from the table, the vertical wheel–rail contact force, rail supporting force, and ballast bed surface stress induced by the HJ-120 running at the speed of 120 km/h are greater than those induced by the ZH-D running at the speed of 200 km/h. For example, the vertical wheel–rail contact force increases by about 15%, the ballast bed surface stress increases by around 10%. Other dynamics indexes of the train vehicles in the table are very close to each other. The vertical wheel–rail force of the ZK_4 is smaller than that of the ZH-T; the ZK_4 and ZH-T show almost the same level in the rail supporting force, ballast bed surface stress, and rail displacement. The ballast bed acceleration and subgrade surface stress are found to be smaller under low-speed freight train operations.

Compared with the evaluation criteria for track dynamic effect, it can be known that all the maximum values of the vertical wheel–rail dynamic responses do not exceed the safety control limits.

2. *Dynamic interaction analysis under the excitation of turnout*

Studies show that the vertical dynamic effect is more severe when locomotives pass through turnouts compared with that of other vehicles. Figure 10.37 compares the

Table 10.5 Dynamic effect indexes of track structures induced by passenger and freight train loads

Speed (km/h)	Speed of passenger train 200		Speed of freight train 120	
Locomotive vehicle type	ZH-D	ZH-T	HJ-120	ZK_4
Vertical wheel–rail force (kN)	164.9	130.6	191.5	125.3
Rail supporting force (kN)	55.9	44.0	64.5	47.0
Wheel–rail contact stress (MPa)	1041.0	1014.4	1024.9	1033.0
Ballast bed surface stress (MPa)	0.275	0.216	0.317	0.231
Subgrade surface stress (MPa)	0.089	0.073	0.090	0.069
Ballast bed acceleration (m/s^2)	16.51	17.74	7.63	6.39
Vertical rail displacement (mm)	1.44	1.12	1.67	1.23

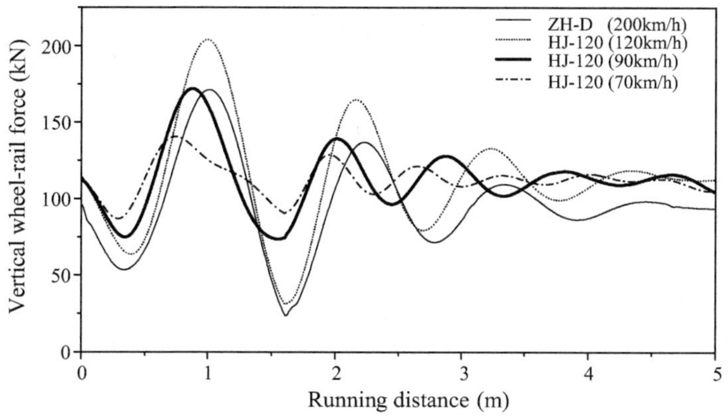

Fig. 10.37 Comparison of vertical wheel–rail force induced by ZH-D and HJ-120 passing through turnout

wheel–rail dynamic response when the ZH-D and HJ-120 pass through a movable-point turnout with different running speeds. As can be seen from the figure, the dynamic force induced by the HJ-120 passing the turnout at 120 km/h is significantly greater than that induced by the ZH-D passing the turnout at 200 km/h; an increase of 20% can be identified for the former case, but it still meets the safety operational standard.

3. *Dynamic interaction analysis under the excitation of rail joint*

As shown in Fig. 10.38, the wheel–rail force induced by the ZH-D passing through the rail joint at the speed of 200 km/h is much larger than that induced by the HJ-120 with the speed of 120 km/h. Correspondingly, the rail supporting force, wheel–rail contact stress and ballast–bed surface stress have similar characteristics [11]. This is due to the fact that wheel–rail interaction is quite sensitive to train running speeds, and an increase of the running speed would result in a dramatic increase of wheel–rail impact force. The dynamic effects of passenger and freight vehicles basically have effects to those of the locomotives (see Fig. 10.39).

Generally, the rail displacement and subgrade surface stress induced by freight locomotives are greater than those induced by a motor car of the high-speed passenger train. This is mainly due to the fact that the axle load of the HJ-120 (23 t) is larger than that of the ZH-D (19.5 t). However, simulation results show that the amount of increase is not significant at around 10–20% [11].

All dynamic behavior indexes under the speed match design can meet the safety operational standard.

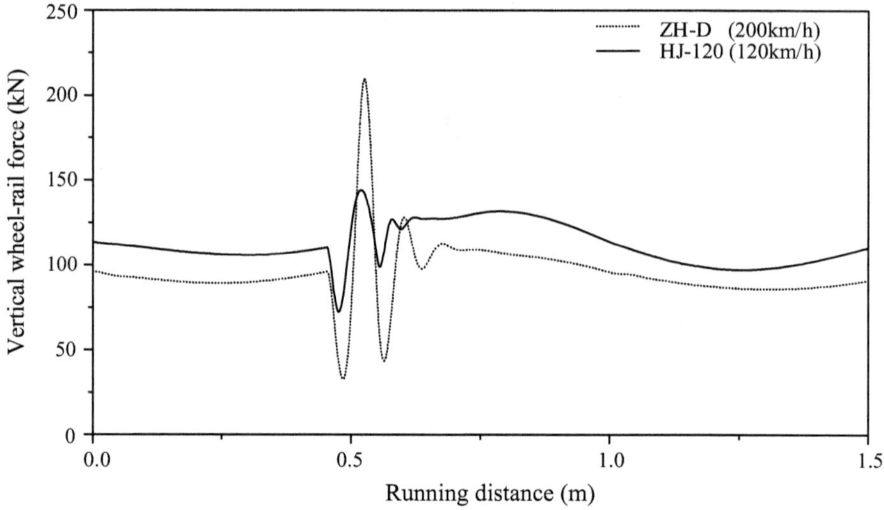

Fig. 10.38 Comparison of vertical wheel–rail contact force induced by ZH-D and HJ-120 passing through rail joint

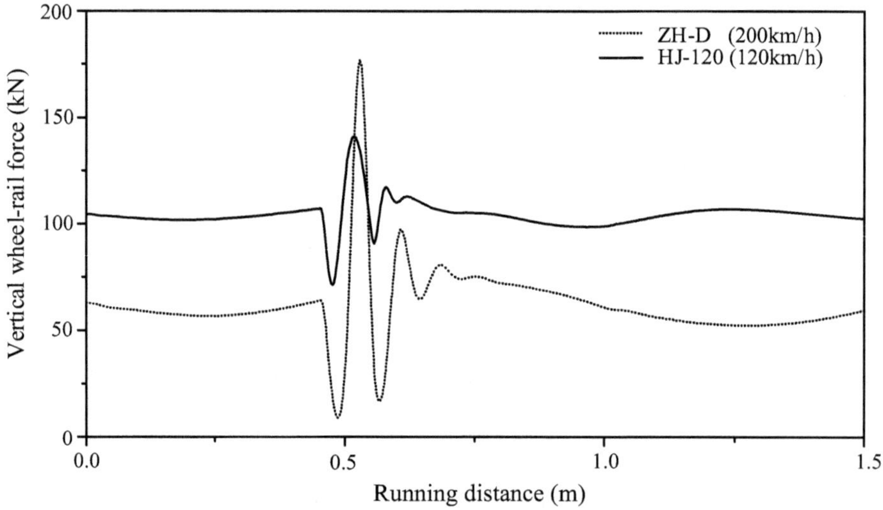

Fig. 10.39 Comparison of vertical wheel–rail contact force induced by ZH-T and ZK4 passing through rail joint

10.4.5 *Technical Measures for Mitigating Dynamic Effects of Freight Train on Shared Passenger and Freight Railway Track*

The above results show that the dynamic effects of the HJ-120 at the speed of 120 km/h on track structures are generally greater than that of the ZH-D at the speed of 200 km/h. Therefore, it is necessary to study how to reduce the dynamic interactions between speedup freight trains and track. This section gives an example of the parameter optimization scheme of railway line design based on dynamic analysis and economic feasibility studies.

1. *Improved design scheme for horizontal curve parameters*

To illustrate the mitigation effect of the improved design scheme on lateral wheel–rail forces, Table 10.6 lists the comparison of lateral wheel–rail forces of the HJ-120 between the original scheme (minimum curve radius $R = 1800$ m, transition curve length $l = 240$ m, superelevation $h = 150$ mm) and the improved scheme ($R = 2800$ m, $l = 180$ m, $h = 110$ mm) under the mixed operational mode of 200/120 km/h. It can be seen that when the minimum horizontal curve radius changes from 1800 to 2800 m, the lateral wheel–rail force drops from 44.39 to 37.23 kN, a reduction of about 16%; the lateral wheelset force decreases by around 17%; the derailment coefficient decreases from 0.28 to 0.26; the wheel–rail wear index declines by 26%; and the lateral rail displacement reduces from 1.42 to 1.28 mm. Clearly, the improved design scheme is proven to be an effective measure for dynamic interaction mitigation.

2. *Improved design scheme for track parameters*

To illustrate the reduction effect of the improved design scheme on vertical wheel–rail contact forces, Table 10.7 lists the comparison of vertical wheel–rail contact forces between the original scheme (rail pad stiffness of 60–80 MN/m) and the

Table 10.6 Comparison of lateral dynamic interaction between speedup freight locomotive and track after improvement of horizontal curve parameters

Dynamic performance index	Lateral wheel–rail force (kN)	Lateral wheelset force (kN)	Derailment coefficient	Wear index (N · m/m)	Lateral rail displacement (mm)
Original scheme of horizontal curve parameters ($R = 1800$ m, $l = 240$ m, $h = 150$ mm)	44.39	54.92	0.28	230.05	1.42
Improved scheme of horizontal curve parameters ($R = 2800$ m, $l = 180$ m, $h = 110$ mm)	37.23	45.72	0.26	170.31	1.28
Decline in level of index (%)	16	17	7	26	10

Table 10.7 Comparison of vertical dynamic interaction between speedup freight locomotive and track after the improvement of rail pad stiffness

Dynamic performance index	Vertical wheel–rail force (kN)	Rail supporting force (kN)	Wheel–rail contact stress (MPa)	Surface stress of ballast bed (MPa)	Surface stress of subgrade (MPa)	Acceleration of ballast bed (g)	Vertical rail displacement (mm)
Original rail pad stiffness (K_p = 60–80 MN/m)	203.8	69.3	1046.5	0.341	0.096	6.95	1.81
New rail pad stiffness (K_p = 50–60 MN/m)	198.1	63.6	1036.1	0.313	0.088	6.35	2.16
Decline level of index (%)	2.8	8.2	1.0	8.2	8.3	8.6	−19

improved scheme (rail pad stiffness of 50–60 MN/m) when the HJ-120 locomotive passes through a No. 18 turnout at the speed of 120 km/h. As can be seen from the table, all wheel–rail dynamic indexes decrease to a certain extent with the decrease of rail pad stiffness, especially for the infrastructure. For example, the rail supporting force drops by 8.2%, the acceleration and stress of ballast bed and subgrade stress reduce by 8–10%. This is of great importance to relieve the dynamic effect of freight trains on the track structure on a shared passenger and freight railway line.

10.4.6 Project Implementation and Practical Operation Effect

The above research results were applied directly in the design of the 200–250 km/h Fuzhou–Xiamen shared high-speed passenger and freight railway. The proposed design schemes and technical standards were adopted based on practical engineering concepts which strongly supported the design and construction of China's first shared high-speed passenger and freight railway in 2003 at a time when the corresponding design standard was limited. The research results provided the theoretical basis for the development of the design standards of shared high-speed passenger and freight railways in China.

The construction of Fuzhou–Xiamen high-speed railway began on September 30, 2005, and was completed on July 20, 2009. It officially opened to traffic on April 26, 2010 (Fig. 10.40). The upgraded operation shortened the running time on that route from 13 h to around 1 h and 40 min. The operational practice has proven

Fig. 10.40 Practical operation of Fuzhou–Xiamen shared railway for high-speed passenger and freight trains

that the high and low speed (passenger and freight) trains can achieve the desired effect and run safely and smoothly on the same corridor.

The four engineering application examples presented in this chapter has demonstrated the effectiveness of the vehicle–track coupled dynamics theory. With the development of the modern railway, especially for the higher speed passenger transportation and heaver freight transportation, the operational environment of trains appears to be more and more complicated, and the dynamic interaction between the vehicles and the tracks becomes more intensified. The vehicle–track coupled dynamics theory will play more active roles in the modern railway train and track design process.

References

1. Zhai WM, Gao JM, Liu PF, et al. Reducing rail side wear on heavy-haul railway curves based on wheel–rail dynamic interaction. Veh Syst Dyn. 2014;52(Suppl):440–54.
2. Zarembski AM. The evolution and application of rail profile grinding. Rep. AREA Bull. 1988;718(89):149–68.
3. Sato Y. Design of rail head profiles with full use of grinding. Wear. 1991;144:363–72.
4. Frick A. Rail grinding operations in Sweden. Track Signal. 2007;11(4):16–9.
5. Jin XS, Du X, Guo J, et al. State of arts of research on rail grinding. J Southwest Jiaotong Univ. 2010;45(1):1–11 (in Chinese).
6. Longson BH, Lamson ST. Development of rail profile grinding at Hamersley Iron. In: Proceedings of 2nd international heavy-haul railway conference, Colorado Springs, CO, USA. 1982.
7. International Heavy Haul Association. Guidelines to best practices for heavy haul railway operations: wheel and rail interface issues. Virginia Beach, VA, USA: International Heavy Haul Association; 2001.
8. GB5599-85. Specifications for dynamic performance evaluation and test identification of railway vehicles. Beijing: China Railway Publishing House; 1985 (in Chinese).
9. GB5599-85. Evaluation method and standard of railway locomotive dynamic performance test. Beijing: China Railway Publishing House; 1993 (in Chinese).
10. Wang KY. Study on dynamic performance of mixed transportation of high and medium speed passenger cars and low speed freight wagons on Fuzhou-Xiamen railway. TTRI-2003-07. Chengdu: Train and Track Research Institute, Southwest Jiaotong University; 2005 (in Chinese).
11. Zhai WM. Study on the influence of mixed transportation of high-speed passenger trains and low-speed freight trains on track structure dynamics of Fuzhou-Xiamen railway. TTRI-2003-05. Chengdu: Train and Track Research Institute, Southwest Jiaotong University; 2003 (in Chinese).

Appendices

This appendix provides the geometry and parameters of the railway vehicles and tracks adopted in the simulations in this book, which may be helpful for researchers who want to analyze the vehicle–track coupled dynamics.

Appendix A: Geometry and Parameters of the High-Speed Train

The configuration and main parameters of a typical Chinese high-speed railway vehicle, namely CRH380A, are presented in this appendix since they have been adopted in the simulation examples in Chap. 8. The maximum operating speed of the CRH380A high-speed railway vehicle can be up to 380 km/h and its regular operating speed is 350 km/h. The configuration of the CRH380A high-speed railway vehicle is shown in Fig. A.1, while its main parameters are listed in Table A.1.

W. Zhai, *Vehicle–Track Coupled Dynamics*,
https://doi.org/10.1007/978-981-32-9283-3

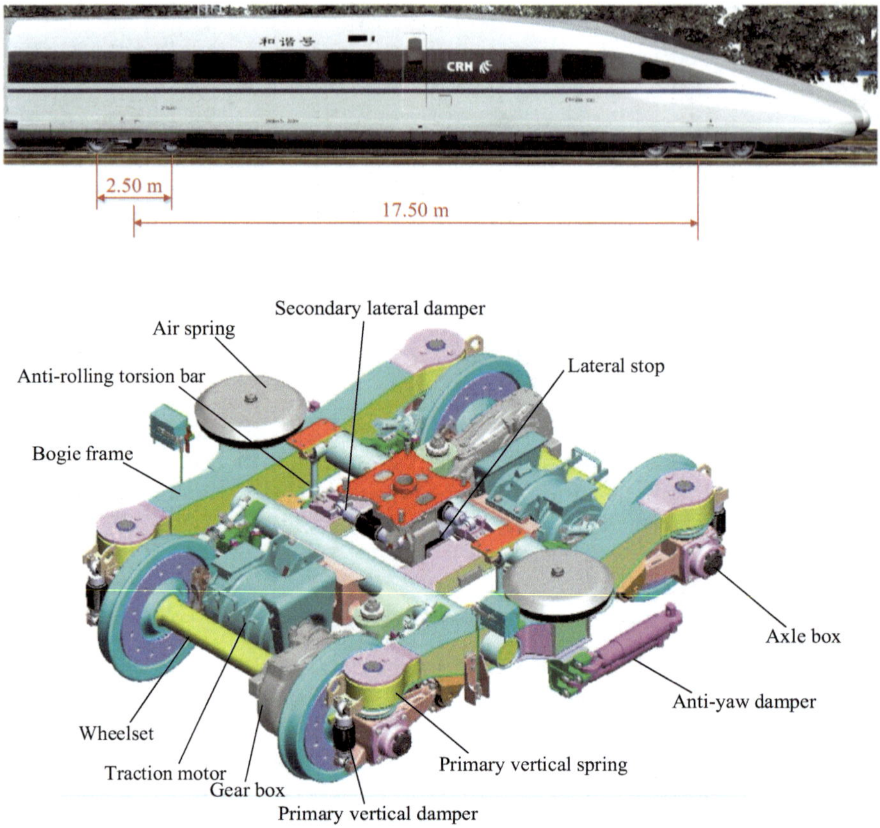

Fig. A.1 Configuration of the CRH380A high-speed railway vehicle

Table A.1 Main parameters of the high-speed railway vehicle

Notation	Parameter	Value	Unit
Mass/inertia			
M_c	Car body mass	43,862.5	kg
M_t	Bogie mass	2,400.0	kg
M_w	Wheelset mass	1,850.0	kg
I_{cx}	Mass moment of inertia of car body about *X*-axis	1.094×10^5	$kg \cdot m^2$
I_{cy}	Mass moment of inertia of car body about *Y*-axis	1.654×10^6	$kg \cdot m^2$
I_{cz}	Mass moment of inertia of car body about *Z*-axis	1.561×10^6	$kg \cdot m^2$
I_{tx}	Mass moment of inertia of bogie about *X*-axis	1,944	$kg \cdot m^2$

(continued)

Table A.1 (continued)

Notation	Parameter	Value	Unit
I_{ty}	Mass moment of inertia of bogie about *Y*-axis	1,314	$kg \cdot m^2$
I_{tz}	Mass moment of inertia of bogie about *Z*-axis	2,400	$kg \cdot m^2$
I_{wx}	Mass moment of inertia of wheelset about *X*-axis	967	$kg \cdot m^2$
I_{wy}	Mass moment of inertia of wheelset about *Y*-axis	123	$kg \cdot m^2$
I_{wz}	Mass moment of inertia of wheelset about *Z*-axis	967	$kg \cdot m^2$
Primary suspension			
K_{px}	Stiffness coefficient along *X*-axis	14.680	MN/m
K_{py}	Stiffness coefficient along *Y*-axis	6.470	MN/m
K_{pz}	Stiffness coefficient along *Z*-axis	1.176	MN/m
C_{pz}	Damping coefficient along *Z*-axis	13 ($\|\Delta v_{sx}\| < 0.01$ m/s)	$kN \cdot s/m$
		6.5 ($\|\Delta v_{sz}\| \geq 0.15$ m/s)	
Secondary suspension			
K_{sx}	Stiffness coefficient along *X*-axis	0.160	MN/m
K_{sy}	Stiffness coefficient along *Y*-axis	0.160	MN/m
K_{sz}	Stiffness coefficient along *Z*-axis	0.190	MN/m
C_{sz}	Damping coefficient along *X*-axis	4.9 ($\|\Delta v_{sx}\| < 0.015$ m/s)	$MN \cdot s/m$
		0.01 ($\|\Delta v_{sx}\| \geq 0.015$ m/s)	
C_{sy}	Damping coefficient along *Y*-axis	58.8 ($\|\Delta v_{sz}\| < 0.15$ m/s)	$kN \cdot s/m$
		6.10 ($\|\Delta v_{sz}\| \geq 0.15$ m/s)	
C_{sz}	Damping coefficient along *Z*-axis	40	$kN \cdot s/m$
Dimension			
l_c	Semi-distance between bogies	8.75	m
l_t	Semi-distance between wheelsets in bogie	1.25	m
R_0	Wheel radius	0.43	m
d_w	Lateral semi-span of primary suspensions	1.0	m
d_s	Lateral semi-span of secondary suspensions	1.23	m
H_{tw}	Height of bogie center from wheelset center	0.08	m
H_{Bt}	Height of body center from secondary suspension	0.29	m
H_{cB}	Height of secondary suspension from bogie center	0.54	m

Appendix B: Geometry and Parameters of the Freight Wagons

Considering the usage of the parameters in the simulation examples in Chaps. 8 and 10, the configuration and main parameters of two typical Chinese three-piece freight wagons, namely C80 and C62, are presented in this appendix. The maximum operating speed of the wagons can attain 80 km/h. The configuration of the C80 freight wagon is shown in Fig. B.1, while its main parameters are listed in Table B.1. The main parameters of the C62 freight wagon are listed in Table B.2.

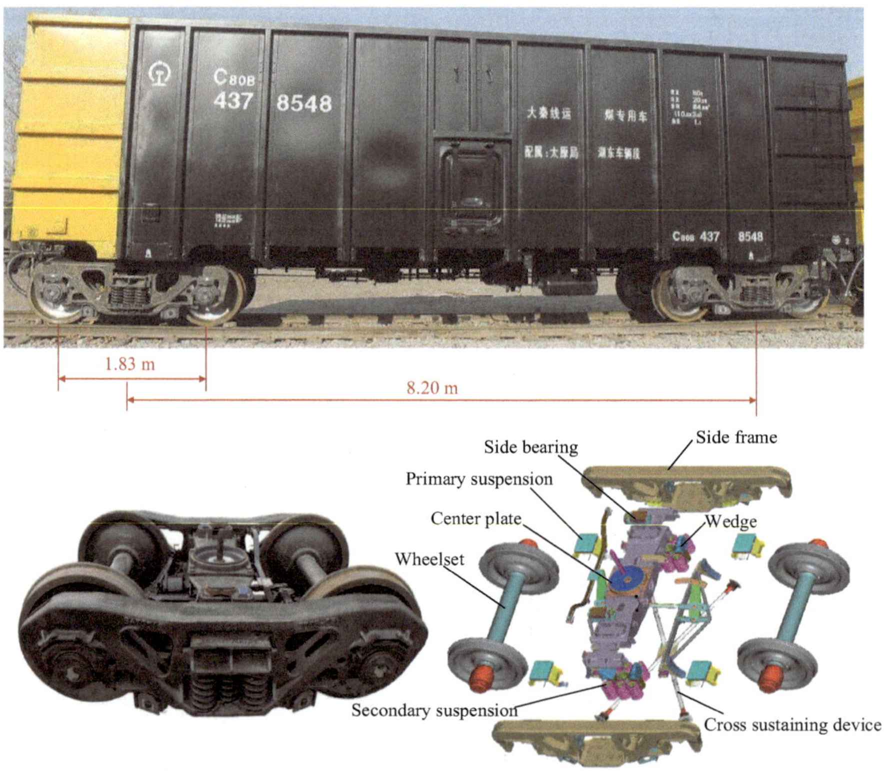

Fig. B.1 Configuration of the freight wagon C80

Table B.1 Main parameters of the freight wagon C80 with three-piece bogies

Notation	Parameter	Value[a]	Unit
Mass/inertia			
M_c	Car body mass	91,838	kg
M_t	Side frame mass	745	kg
M_B	Bolster mass	497	kg
M_w	Wheelset mass	1,171	kg
I_{cx}	Mass moment of inertia of car body about *X*-axis	2.163×10^5	$kg \cdot m^2$
I_{cy}	Mass moment of inertia of car body about *Y*-axis	0.996×10^6	$kg \cdot m^2$
I_{cz}	Mass moment of inertia of car body about *Z*-axis	0.984×10^6	$kg \cdot m^2$
I_{Bz}	Mass moment of inertia of bolster about *Z*-axis	258	$kg \cdot m^2$
I_{ty}	Mass moment of inertia of side frame about *Y*-axis	188	$kg \cdot m^2$
I_{tz}	Mass moment of inertia of side frame about *Z*-axis	173	$kg \cdot m^2$
I_{wx}	Mass moment of inertia of wheelset about *X*-axis	700	$kg \cdot m^2$
I_{wy}	Mass moment of inertia of wheelset about *Y*-axis	140	$kg \cdot m^2$
I_{wz}	Mass moment of inertia of wheelset about *Z*-axis	700	$kg \cdot m^2$
Primary suspension			
K_{px}	Stiffness coefficient along *X*-axis	13.0	MN/m
K_{py}	Stiffness coefficient along *Y*-axis	11.0	MN/m
K_{pz}	Stiffness coefficient along *Z*-axis	160.0	MN/m
Secondary suspension			
K_{sx}	Stiffness coefficient of secondary suspension along *X*-axis	3.127	MN/m
K_{sy}	Stiffness coefficient of secondary suspension along *Y*-axis	3.127	MN/m
K_{sz}	Stiffness coefficient of secondary suspension along *Z*-axis	4.235	MN/m
K_{px}	Stiffness coefficient of a cross joint bar	14.8	MN/m
Dimension			
l_c	Semi-longitudinal distance between bogies	4.10	m
l_t	Semi-longitudinal distance between wheelsets in bogie	0.915	m
d_w	Lateral semi-span of primary suspensions	0.9905	m
d_s	Lateral semi-span of secondary suspensions	0.9905	m
R_0	Wheel radius	0.42	m

[a]Loaded case

Table B.2 Main parameters of the freight wagon C62 with three-piece bogies

Notation	Parameter	Value[a]	Unit
Mass/inertia			
M_c	Car body mass	77,000	kg
M_t	Side frame mass	330	kg
M_B	Bolster mass	470	kg
M_w	Wheelset mass	1,200	kg
I_{cx}	Mass moment of inertia of car body about *X*-axis	1.0×10^5	$kg \cdot m^2$
I_{cy}	Mass moment of inertia of car body about *Y*-axis	1.2×10^6	$kg \cdot m^2$
I_{cz}	Mass moment of inertia of car body about *Z*-axis	1.07×10^6	$kg \cdot m^2$
I_{Bz}	Mass moment of inertia of bolster about *Z*-axis	190	$kg \cdot m^2$
I_{ty}	Mass moment of inertia of side frame about *Y*-axis	100	$kg \cdot m^2$
I_{tz}	Mass moment of inertia of side frame about *Z*-axis	80	$kg \cdot m^2$
I_{wx}	Mass moment of inertia of wheelset about *X*-axis	740	$kg \cdot m^2$
I_{wy}	Mass moment of inertia of wheelset about *Y*-axis	100	$kg \cdot m^2$
I_{wz}	Mass moment of inertia of wheelset about *Z*-axis	740	$kg \cdot m^2$
Secondary suspension			
K_{sx}	Stiffness coefficient of secondary suspension along *X*-axis	4.14	MN/m
K_{sy}	Stiffness coefficient of secondary suspension along *Y*-axis	4.14	MN/m
K_{sz}	Stiffness coefficient of secondary suspension along *Z*-axis	5.32	MN/m
Dimension			
l_c	Semi-longitudinal distance between bogies	4.25	m
l_t	Semi-longitudinal distance between wheelsets in bogie	0.875	m
R_0	Wheel radius	0.42	m

[a]Loaded case

Appendix C: Geometry and Parameters of the Ballasted Tracks

The typical Chinese ballasted tracks are adopted in the simulation examples in Chaps. 8, 9 and 10. Thus, their configuration and main parameters are presented in this appendix. The configuration of the Chinese high-speed ballasted track is shown in Fig. C.1, while its main parameters are listed in Table C.1. The main parameters of the Chinese heavy-haul railway ballasted track are listed in Table C.2.

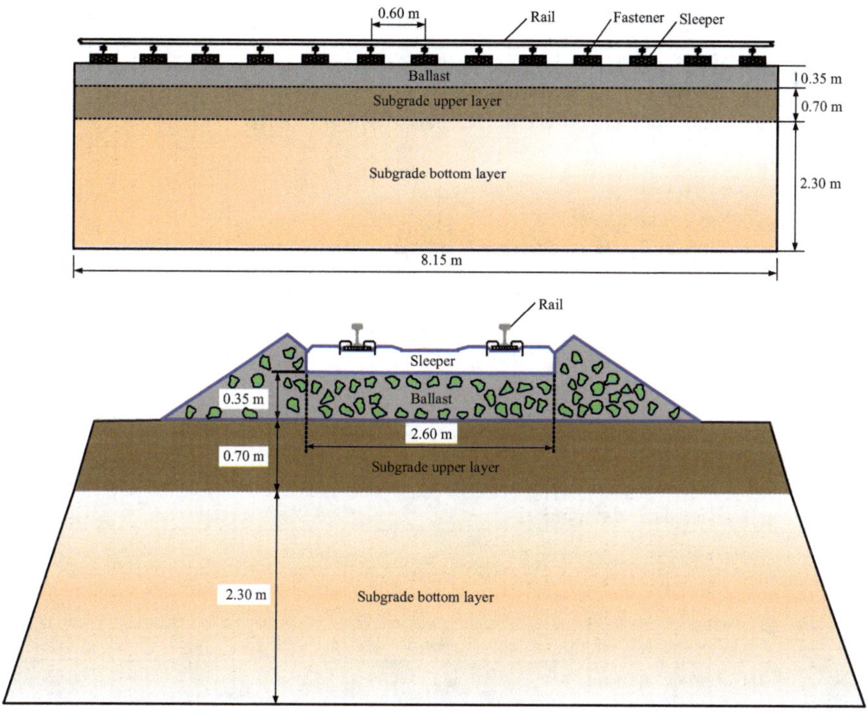

Fig. C.1 Configuration of the high-speed ballasted track

Table C.1 Main parameters of the high-speed railway ballasted track

Notation	Parameter	Value (per rail seat)	Unit
Rail			
E	Elastic modulus of rail	2.059×10^{11}	N/m^2
ρ	Density of rail	7.86×10^3	kg/m^3
I_0	Torsional inertia of rail	3.741×10^{-5}	m^4
I_y	Rail second moment of area about *Y*-axis	3.217×10^{-5}	m^4
I_z	Rail second moment of area about *Z*-axis	5.24×10^{-6}	m^4
GK	Rail torsional stiffness	1.9587×10^5	$N \cdot m/rad$
m_r	Rail mass per unit length	60.64	kg/m
Fastener			
K_{pv}	Fastener stiffness in vertical direction	6.0×10^7	N/m
K_{ph}	Fastener stiffness in lateral direction	4.0×10^7	N/m
C_{pv}	Fastener damping in vertical direction	5.0×10^4	$N \cdot s/m$
C_{ph}	Fastener damping in lateral direction	4.0×10^4	$N \cdot s/m$
Sleeper			
l_s	Sleeper spacing	0.60	m
M_s	Sleeper mass (half)	170	kg
l_e	Effective support length of half sleeper	1.175	m
l_b	Sleeper width	0.290	m
Ballast			
M_b	Ballast mass	340	kg
ρ_b	Ballast density	1.75×10^3	kg/m^3
E_b	Elastic modulus of ballast	1.2×10^8	Pa
C_{bv}	Ballast damping	6.0×10^4	$N \cdot s/m$
K_w	Ballast shear stiffness	7.84×10^7	N/m
C_w	Ballast shear damping	8.0×10^4	$N \cdot s/m$
α	Ballast stress distribution angle	35	°
h_b	Ballast thickness	0.35	m
Subgrade			
E_f	Subgrade K_{30} modulus	1.9×10^8	Pa/m
C_{fv}	Subgrade damping	1.0×10^5	$N \cdot s/m$

Table C.2 Main parameters of the heavy-haul railway ballasted track

Notation	Parameter	Value (per rail seat)	Unit
Rail			
E	Elastic modulus of rail	2.059×10^{11}	N/m^2
ρ	Density of rail	7.86×10^3	kg/m^3
I_0	Torsional inertia of rail	3.741×10^{-5}	m^4
I_y	Rail second moment of area about Y-axis	3.217×10^{-5}	m^4
I_z	Rail second moment of area about Z-axis	5.24×10^{-6}	m^4
GK	Rail torsional stiffness	1.9587×10^5	$N \cdot m/rad$
m_r	Rail mass per unit length	60.64	kg/m
Fastener			
K_{pv}	Fastener stiffness in vertical direction	7.8×10^7	N/m
K_{ph}	Fastener stiffness in lateral direction	5.0×10^7	N/m
C_{pv}	Fastener damping in vertical direction	5.0×10^4	$N \cdot s/m$
C_{ph}	Fastener damping in lateral direction	4.0×10^4	$N \cdot s/m$
Sleeper			
l_s	Sleeper spacing	0.545	m
M_s	Sleeper mass (half)	125.5	kg
l_e	Effective support length of half half-sleeper	1.250	m
l_b	Sleeper width	0.273	m
Ballast			
M_b	Ballast mass	660.0	kg
ρ_b	Ballast density	1.75×10^3	kg/m^3
E_b	Elastic modulus of ballast	1.2×10^8	Pa
K_{bv}	Ballast stiffness	1.5×10^8	N/m
C_b	Ballast damping	5.88×10^4	$N \cdot s/m$
K_w	Ballast shear stiffness	7.80×10^7	N/m
C_w	Ballast shear damping	8.0×10^4	$N \cdot s/m$
α	Ballast stress distribution angle	35	°
h_b	Ballast thickness	0.45	m
Subgrade			
K_{fv}	Subgrade stiffness	1.7×10^8	N/m
C_{fv}	Subgrade damping	3.1×10^4	$N \cdot s/m$

Appendix D: Geometry and Parameters of the High-Speed Ballastless Slab Track

The configuration and main parameters of a typical Chinese high-speed ballastless track, namely CRTS-II, are presented in this appendix. The configuration of the CRTS-II high-speed ballastless track is shown in Fig. D.1, while its main parameters are listed in Table D.1. These parameters have been adopted in the simulation examples shown in Chap. 8.

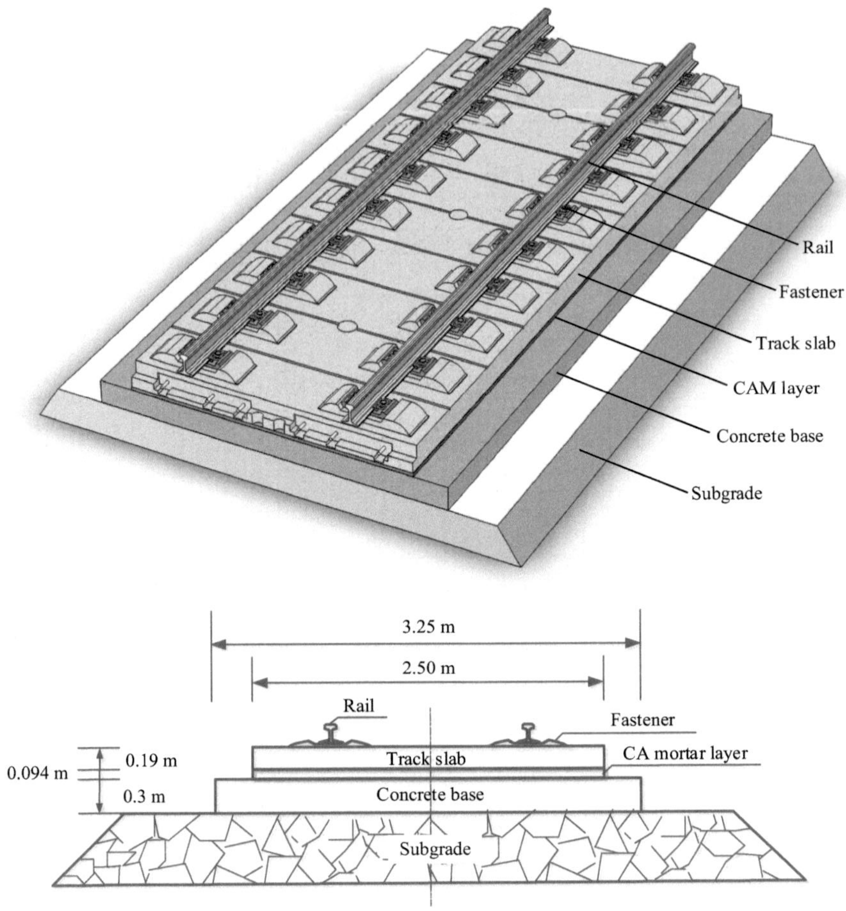

Fig. D.1 Configuration of the CRTS II high-speed ballastless slab track

Table D.1 Main parameters of the high-speed ballastless slab track

Notation	Parameter	Value (per rail seat)	Unit
Rail			
E	Elastic modulus of rail	2.059×10^{11}	N/m^2
ρ	Density of rail	7.86×10^3	kg/m^3
I_0	Torsional inertia of rail	3.741×10^{-5}	m^4
I_y	Rail second moment of area about *Y*-axis	3.217×10^{-5}	m^4
I_z	Rail second moment of area about *Z*-axis	5.24×10^{-6}	m^4
GK	Rail torsional stiffness	1.9587×10^5	$N \cdot m/rad$
m_r	Rail mass per unit length	60.64	kg/m
Fastener			
K_{pv}	Stiffness coefficient along *Z*-axis	2.5×10^7	N/m
K_{ph}	Stiffness coefficient along *Y*-axis	2.0×10^7	N/m
C_{pv}	Damping coefficient along *Z*-axis	5.0×10^4	$N \cdot s/m$
C_{ph}	Damping coefficient along *Y*-axis	4.0×10^4	$N \cdot s/m$
l_s	Fastener spacing	0.65	m
Track slab			
l_e	Length along *X*-axis	6.45	m
D_s	Width along *Y*-axis	2.50	m
H_s	Thickness along *Z*-axis	0.19	m
ρ_s	Density	2.5×10^3	kg/m^3
E_s	Elastic modulus	3.6×10^{10}	N/m^2
CAM layer			
D_c	Width along *Y*-axis	2.50	m
H_c	Thickness along *Z*-axis	0.094	m
ρ_c	Density	2.4×10^3	kg/m^3
E_c	Elastic modulus	3.0×10^{10}	N/m^2
Concrete base			
D_b	Width along *Y*-axis	3.25	m
H_b	Thickness along *Z*-axis	0.30	m
ρ_b	Density	2.5×10^3	kg/m^3
E_b	Elastic modulus	2.55×10^8	N/m^2